PHINEAS REDUX

ANTHONY TROLLOPE'S PALLISER NOVELS

General Editor: W. J. Mc Cormack

CAN YOU FORGIVE HER? (1864–5)

Introduced by Kate Flint and edited by Andrew Swarbrick, with a Preface by Norman St. John-Stevas

PHINEAS FINN (1869)

Introduced and edited by Jacques Berthoud

THE EUSTACE DIAMONDS (1873)

Introduced and edited by W. J. Mc Cormack

PHINEAS REDUX (1874)

Introduced by F. S. L. Lyons and edited by John C. Whale

THE PRIME MINISTER (1876)

Introduced by John McCormick and edited by Jennifer Uglow

THE DUKE'S CHILDREN (1880)

Introduced and edited by Hermione Lee

ANTHONY TROLLOPE

Phineas Redux

INTRODUCED BY

F. S. L. LYONS

AND EDITED BY

JOHN C. WHALE

WITH ILLUSTRATIONS BY

T. L. B. HUSKINSON

OXFORD UNIVERSITY PRESS

Oxford New York

Oxford University Press

Oxford New York Toronto
Delhi Bombay Calcutta Madras Karachi
Petaling Jaya Singapore Hong Kong Tokyo
Nairobi Dar es Salaam Cape Town
Melbourne Auckland

and associated companies in
Berlin Ibadan

This edition first published 1983 as a World's Classics paperback
Reprinted 1985, 1986, 1988, 1990

This cloth edition issued in 1991 by Oxford University Press, Inc.
200 Madison Avenue, New York, New York 10016

British Library Cataloguing in Publication Data
Trollope, Anthony
Phineas redux.—(The World's classics)
I. Title. II. Whale, John C.
823'.8[F] PR5684.P1
ISBN 0-19-281589-X

Library of Congress Cataloging in Publication Data
Trollope, Anthony, 1815-1882.
Phineas redux.
(The Centenary of Anthony Trollope's Palliser Novels)
(The World's classics)
Bibliography: p.
I. Lyons, F. S. L.: (Francis Stewart Leland), 1923–
II. Whale, John C. III. Title.
IV. Series, Trollope, Anthony, 1815-1882. Palliser novels.
PR5684.P4 1983 823'.8 82-14094
ISBN 0-19-281589-X (pbk.)
ISBN 0-19-520898-6

2 4 6 8 10 9 7 5 3 1

Printed in the United States of America
on acid-free paper.

CONTENTS

INTRODUCTION

A SADDER AND A WISER MAN

TROLLOPE may have been right in supposing that few people would read his entire series of 'semi-political' (his expression) Palliser novels in the correct sequence, or indeed in any sequence. But whoever wants to get the best out of them will certainly require to read the two central stories that deal with Phineas Finn as though they were the one book which in his *Autobiography* he describes them as being. Although there was a long gap between them—*Phineas Finn* was first serialized in the *St. Paul's Magazine* in 1867 and *Phineas Redux* in the *Graphic* in 1873–4, and they were published as two-volume novels in 1869 and 1874 respectively—Trollope preserved their continuity so well that whenever he writes or speaks of 'Phineas Finn' it is generally the whole canvas he has in mind and not just this or that section of it.

He has himself explained why continuity was more important to him in this long and apparently rambling novel (though appearances can be deceptive) than in almost any other. 'In writing *Phineas Finn*', he recorded afterwards, 'I had constantly before me the necessity of progression in character—of marking the changes in men and women which would naturally be produced by the lapse of years'. And he continued: 'The happy motherly life of Violet Effingham which was due to the girl's honest but long-restrained love: the tragic misery of Lady Laura, which was equally due to the sale she made of herself in her wretched marriage; and the long suffering but final success of the hero, of which he had deserved the first by his vanity and the last by his constant honesty, had been foreshadowed to me from the first.'[1]

[1] Trollope, *An Autobiography*, The World's Classics (Oxford, 1980), pp. 318–20.

Nevertheless, while Trollope did carry out this grand design with a large measure of success, the gap which so often opens in his novels between his clear conception of character and his decidedly unclear development and resolution of plot and incident, is manifest from the outset. Part of the trouble which this caused him is his own fault, but part of it, the more interesting part, springs from the way in which his central character develops almost in spite of his creator. The difficulty that Trollope brought upon himself arose from what he himself called the 'blunder' of having as his hero a young Irish politician on the make. At the start of *Phineas Finn*, Phineas is a raw recruit indeed. Worse than that, he is already something of an oddity on the Irish scene, as he is assuredly a rarity on the English scene. He is an oddity in Ireland because he cuts across the lines of what was even then becoming a sharply demarcated society. He is the son of a religiously mixed marriage which meant, according to the Irish custom of those days, that while his sisters were brought up in their mother's Anglican creed, he followed his father's side and remained a Roman Catholic.

This, however, did not prevent him from attending Trinity College, Dublin, and cutting his debating teeth in that nursery of fledgeling politicians, the College Historical Society. Phineas preceded by a few years the prohibition by the Irish Roman Catholic bishops upon their flock entering that then overwhelmingly Protestant institution, though his parents, because of their mixed marriage, would probably have ignored the ban had it existed. But his political views, as we see them slowly evolving, would certainly not have been acceptable in his Alma Mater. It was bad enough that he should turn out a liberal rather than a conservative, but that he should have taken up towards the end of the first novel the cause of tenant-right, however loosely defined, showed dangerous leanings towards nationalism. Worse still, in *Phineas Redux* he dabbles in church disestablishment. True, Trollope presents this as a British rather than an Irish issue, but since the Irish Church was disestablished in 1869—the very year in which *Phineas Finn* was published in book form—for Phineas to be taking up such a sensitive issue would not have been regarded as respectable behaviour for a Trinity College man. More than that, his tendency to be found in the neighbourhood of such explosive questions at the wrong time was bound to raise serious doubts among his friends about his political reliability.

And this, as Trollope realized when it was too late, was precisely where his hero's Irishness made the novelist's task much harder than it need have been. At one level the two books together chronicle the rise of an obscure young man up the English political ladder. Lacking a landed background (his father is a country doctor), dependent for his livelihood either on the bounty of others or on his own uncertain earnings, he succeeds by a combination of good fortune, charm, genuine ability, and the friendship of some illustrious and influential ladies—Lady Laura Standish (Kennedy by marriage), Lady Glencora Palliser (Duchess of Omnium for most of *Phineas Redux*), and Madame Max Goesler. This could have been the life history of any political gigolo, but Trollope, by making him an Irishman, complicated the issue excruciatingly. 'There was nothing to be gained by the peculiarity', he admitted later, 'and there was an added difficulty in obtaining sympathy and affection for a politician belonging to a nationality whose politics are not respected in England.'[2]

Honest man that he was, Trollope, having saddled his hero with his Irishness, has to expose him to the full consequences of that deplorable condition. When, towards the end of *Phineas Finn*, Phineas is planning to throw up office and break with his party over tenant-right, two of the colleagues whom he most detests let him have it right in the face. 'I've always had a fear about you, Finn', says Ratler, the liberal Whip, 'that you would go over the traces some day. Of course, it's a very grand thing to be independent.' 'The fact

[2] Trollope, *An Autobiography*, The World's Classics (Oxford, 1980), p. 318.

is, Finn', says Mr. Bonteen, joining in the assault, 'you are made of clay too fine for office. I've always found it has been so with men from your country. You are the grandest horses in the world to look at on a prairie, but you don't like the slavery of harness.'[3] And when in the later book Phineas's name is scandalously linked with Lady Laura Kennedy, so that the maddened Kennedy tries to murder him, Finn apparently pays a compound price for this notoriety because of his Irishness. 'I never liked him from the first', says Mr. Bonteen, 'and always knew he would not run straight. No Irishman ever does.'[4]

Why did Trollope make his hero an Irishman, knowing that this was the sort of reaction he was likely to evoke? His own explanation—that the scheme of the book came to him on a visit to Ireland—is quite inadequate. We have to remember what a special place Ireland had in Trollope's affections and how untypical his attitude towards that country was among Englishmen of his day. It was Ireland that gave him his first breakthrough in his professional career in the postal service, Ireland that made him a hunting addict, Ireland that showed him friendship and good humour, and it was from Ireland that he chose the wife who transformed the rest of his life. Moreover, the Phineas novels are yet another version of that perennial nineteenth-century theme—the young man from the provinces breaching the ranks of the establishment and imposing himself by a combination of charm, ability, and fortune. Trollope could hardly have found a more complete outsider than an Irish politician and since Irishmen were even then deemed to be almost preternaturally lucky, there was a peculiar fitness in the fact that, as Lady Glencora says of him, 'Mr. Finn is one of those Irish gentlemen who always seem to be under some special protection'.[5]

[3] Trollope, *Phineas Finn*, The World's Classics, (Oxford, 1982), II, pp. 295–6.

[4] Trollope, *Phineas Redux*, The World's Classics (Oxford, 1983), I, p. 280.

[5] Ibid., I, p. 266.

But Phineas created problems for Trollope over and above his Irishness. Trollope's old schoolfellow and lifelong friend, Sir William Gregory, complained bitterly that Phineas Finn was a libel on the Irish gentleman, of which, incidentally, Sir William himself was the *beau idéal*.[6] A modern critic has said much the same, though even more harshly, when he describes Phineas as 'a case of a character outstripping the author's intention, since he emerges as a nasty, philandering young man whose true nature seems to have baffled people in the story as well as generations of readers'.[7]

This is not entirely fair. Or rather, it is fair about the earlier stages of Phineas's career, while leaving no scope for that progression of character by which Trollope set so much store. In *Phineas Finn*, it has to be admitted, Phineas passes with great rapidity from the love of Lady Laura, to that of Violet Effingham, and thence to that of Madame Max Goesler. Lady Laura does not marry him because she cannot afford to do so. Violet Effingham does not marry him because her heart is set on marrying Lord Chiltern. And while Madame Max in a very powerful scene offers her own hand and her large fortune to Phineas, he resists what, at that stage of his development, he would have regarded as a marriage of convenience without, on his part, a basis of true love. In the event, having already decided to resign office and his seat in parliament, he returns to Ireland and marries his childhood sweetheart, Mary Flood Jones, whom he has discarded and picked up again as the exigencies of the story have demanded. Phineas's amatory progress—less linear, perhaps, than circular in the manner of *La Ronde*—is not very edifying. Indeed, one has at times the disagreeable feeling that Trollope himself did not realize how un-edifying it really was, since his chief regret was that in his usual hit and miss fashion he had not worked out a more satisfactory ending for

[6] T. H. S. Escott, *Anthony Trollope: his work, associates and literary originals* (London, 1913), p. 266.

[7] R. C. Terry, *Anthony Trollope: the artist in hiding* (London, 1977), pp. 106–7.

the first novel. 'As I fully intended to bring my hero again into the world, I was wrong to marry him to a simple pretty Irish girl, who could only be felt as an encumbrance on such return. When he did return I had no alternative but to kill the pretty simple Irish girl,—which was an unpleasant and awkward necessity.'[8]

One has to say that in the sequel this hard necessity doesn't seem to have bothered either Phineas or any of his friends. One of his more objectionable traits in the earlier book had been to keep the existence of Mary Flood Jones hidden from the great ladies whom he was pursuing or who were pursuing him. With her conveniently out of the way, Phineas is welcomed back into his old circle as if he had never left it. So at least it seems at first, but we do not get very far into *Phineas Redux* before we realize that the process of resurrection has changed our hero in several profound ways. First and most obviously, his connection with Ireland, which had been crucial in the earlier novel, has virtually ceased. In *Phineas Finn*, Phineas had sat for part of the time for an Irish constituency, some of the novel's tension had sprung from the alternations between England and Ireland, and it was, after all, an Irish issue which had brought his career to a not dishonourable crisis. Now, however, there is an almost palpable irony. While his enemies, Bonteen and the ineffable newspaper editor, Quintus Slide, still refer to him in pointedly anti-Irish terms, these rebound from the armour of his indifference. His whole development throughout the book is within an English context and when the issues are finally resolved they are this time resolved with absolutely no reference to Ireland.

But there is a second and more important sense in which Phineas is transformed by his resurrection. He is quite simply both a sadder and a wiser man. Although drawn back to politics apparently irresistibly, he never approaches them with the zest of earlier days. He finds himself obliged to fight the difficult seat of Tankerville and he encounters all the

[8] Trollope, *An Autobiography*, The World's Classics (Oxford, 1980), p. 318.

bribery and skulduggery which Trollope himself had encountered at the Beverley election of 1868 between the writing of the two Phineas books. Phineas, like Trollope, loses the election, but unlike Trollope he wins the seat on petition. Yet, although getting in for Tankerville seems a flash of Phineas's old good fortune, the election leaves a bad taste and from this early incident we learn that for Phineas things are not what they were. Once again, as in the first book, he makes difficulties for himself. At the hustings he had come out strongly for church disestablishment and at once finds himself at odds with what Parnell would later call 'the liberal wire-pullers'. While disestablishment was of course a liberal cause (or at any rate a cause beloved by the nonconformist wing of the liberal party), much depended upon right timing and in this matter, as earlier with the tenant question, Phineas is hopelessly ahead of his party. And when for tactical reasons it becomes necessary to vote against a conservative proposal to tackle the problem, there is a critical moment when it seems as if Phineas will yet again give his conscience full rein on a matter of principle.

In fact, he does not do this, and for once opts for expediency, but the incident contributes to his disenchantment. In *Phineas Redux* Trollope was anxious to subordinate the political to the human interest more completely than in *Phineas Finn*, but the division was never absolute and it is Phineas's human predicament which does more than anything else to finish him with politics. When he returns to London society he finds that Lady Laura is still living apart from her husband, the fanatical Presbyterian, Robert Kennedy. Partly at Kennedy's request, Phineas visits Lady Laura in her exile at Dresden. But so far from urging her to return to Kennedy, as the latter has desired him to do, Phineas has to listen to her declaration of love for *him*, a declaration as hopeless as it is passionate. Back in London, the half-crazed Kennedy gives way to his jealousy of Phineas and narrowly fails to shoot him. Phineas does not call the police, but his attempts to hush things up are inevitably foiled by Quintus Slide and he finds himself in the deeply compromising position of having his name coupled with that of a married woman whom he no longer loves and who herself is a target for gossip through having left her husband.

Phineas's involvement in this ugly scandal is the hinge on which the book turns. He had taken it for granted that on his return to politics he would pick up where he had left off and would again hold office as he had so successfully done in *Phineas Finn*. But he finds avenue after avenue closed to him and unworthy persons being appointed over his head. All this happens without a word being said directly to him, and in a masterly chapter, 'The World Becomes Cold', Phineas experiences what it is to fall foul of the establishment. The worst cut of all comes when he learns that his particular enemy, Mr. Bonteen, is to be promoted to the Chancellorship of the Exchequer, while there is still nothing for Phineas—'and there was a general feeling, not expressed, but understood, that his affair with Mr. Kennedy stood in his way'.[9] In his disappointment, Phineas goes as formerly to his women friends. 'He had always gone to some woman', says Trollope in a revealing passage, 'in old days to Lady Laura, or to Violet Effingham, or to Madame Goesler. By them he could endure to be petted, praised, or upon occasion even pitied. But pity or praise from any man had been distasteful to him.'[10]

Once again, in short, he becomes 'the ladies' pet'. Lady Glencora, now Duchess of Omnium, takes up his case, but while she is able to block Bonteen's rise to the Chancellorship she cannot conjure an appointment for Phineas. The scene is thus set for the violent quarrel between Phineas and Bonteen which is followed immediately by Bonteen's murder. The circumstantial evidence against Phineas is so strong that he is arrested and brought to trial, a great set piece which occupies much of the second volume of the novel. Nothing is more characteristic of Trollope than the way in which, hav-

[9] Trollope, *Phineas Redux*, The World's Classics (Oxford, 1983), I, p. 285
[10] Ibid., I, p. 287.

ing led up to the murder, he then loses interest in the crime and reserves all his best effects for the trial. Where Wilkie Collins would have made a mystery and built his whole novel round it, Trollope has no compunction in revealing at the outset that the murderer is the Reverend Mr. Emilius, who is being pursued by Mr. Bonteen on suspicion of bigamy. The interest of the succeeding events centres upon how the case against Phineas is demolished by his lawyer, the marvellous Chaffanbrass, and upon the way in which Trollope prepares for the ultimate union of those two outsiders, Phineas and Marie Goesler, after the latter has intrepidly travelled to central Europe to find the evidence which shall be enough to save Phineas, even if not to hang the real culprit.

The changed character of Phineas is emphasized by the effect upon him of having been on trial for his life. Bearing up manfully while the ordeal in court is actually in progress, he breaks down afterwards and finds it hard even to meet his friends. Worse than that, he finds his zest for politics so completely gone that when he is at last asked to join the government he staggers the Prime Minister and nearly everyone else, including Marie Goesler, by an almost contemptuous refusal. Trollope's analysis of this critical moment shows how clear-sightedly he could regard his muddle-headed hero:

> Of his own feelings in regard to the offer which was about to be made to him he had hardly succeeded in making her understand anything. That a change had come upon himself was certain, but he did not at all believe that it had sprung from any weakness caused by his sufferings in regard to the murder. He rather believed that he had become stronger than weaker from all that he had endured. He had learned when he was younger,—some years back,—to regard the political service of his country as a profession in which a man possessed of certain gifts might earn his bread with more gratification to himself than in any other. The work would be hard, and the emolument only intermittent; but the service would in itself be pleasant; and the rewards of that service,—should he be so successful as to obtain reward,—would be dearer to him than anything which could accrue to him from other labours. To sit in the Cabinet

for one Session would, he then thought, be more to him than to preside over the Court of Queen's Bench as long as did Lord Mansfield. But during the last few months a change had crept across his dream,—which he recognized but could hardly analyse. He had seen a man whom he despised promoted, and the place to which the man had been exalted had at once become contemptible in his eyes. And there had been quarrels and jangling, and the speaking of evil words between men who should have been quiet and dignified. No doubt Madame Goesler was right in attributing the revulsion in his hopes to Mr. Bonteen and Mr. Bonteen's enmity; but Phineas Finn himself did not know that it was so.[11]

She remains loyal to him, of course, and, after a bitter and intense scene with Lady Laura, he comes to rest eventually in Marie Goesler's arms. We are to suppose that the two outsiders will live happily ever after. Perhaps they do, but Phineas never again re-enters the charmed inner circle of power. Later in the series we catch a glimpse of him as a middle-aged MP in comfortable circumstances (living on his wife's money, not to put too fine a point on it), but of the old Phineas no trace is left.

As the story of Phineas himself *Phineas Redux* has something autumnal about it. Trollope has subjected his hero to such a series of ordeals that nothing of the golden boy remains and what we are given is a sad, at times almost sombre, progress towards maturity and self-wisdom. But Trollope, here as always, paints on a wide canvas and if the fates of Phineas and of Lady Laura cast a shade, the development of the other characters along their predictable lines is sheer joy. The married life of the Chilterns at Harrington Hall, though it seems a loss of direction for Violet, is in fact idyllically happy and she is shown at her best as a wife and mother who has learnt to manage her 'savage lord' so that without realizing it he is as completely broken in as one of his own hunters. The sub-plot of Adelaide Palliser's marriage, which Violet and Glencora arrange between them, is beauti-

[11] Trollope, *Phineas Redux*, The World's Classics (Oxford 1983), II, pp. 332–3.

fully done, while Glencora's incorrigible itch to interfere in what does not concern her, brings her once more into collision—but in a strange way, loving collision—with her husband, Plantagenet. He is unhappy at having to leave the House of Commons when he becomes Duke of Omnium, but we are never in any doubt that Glencora will make a formidable Duchess.

Her friendship with Marie Goesler flowers in *Phineas Redux*. At the end of *Phineas Finn* Marie had resisted the old Duke's offer to make her his mistress or even his wife and this had won Glencora's heart. The two women are present at the old Duke's death-bed, where his infatuation all too obviously still persists. But Marie behaves impeccably and her absolute refusal to accept his legacy of jewels not only makes possible Adelaide Palliser's future happiness, but cements the alliance between Glencora and herself. Both women are studies in the vivacity which Trollope could convey so well, utterly different from each other and having in common only their creator's unstinting prodigality. Alas, that prodigality seems to have faltered in the case of Lady Laura whom he himself regarded as 'the best character' in the two books. She is indeed a genuinely tragic figure and the scene at Königstein where she declares her love for Phineas, and still more the scene near the end where Phineas reveals to her that he is going to marry Marie Goesler, are among the most powerful Trollope ever wrote. Nevertheless, she is monotonous, precisely because she is exactly what the word implies, all on one tone. Her self-pity, though itself piteous, beats upon poor Phineas like a dentist's drill. True, he had ill-used her from time to time, but we come to feel, as he evidently feels himself, that he had never quite deserved this.

There is a case for arguing that *Phineas Redux*, although perhaps less well-known and popular than *Phineas Finn*, is in some ways a greater achievement. In it Trollope displays, at times almost negligently, his two greatest strengths—his power of characterization and his capacity to dramatize the ordinary. Phineas Finn began life almost too good to be true and it was never easy to take his early romantic escapades quite as seriously as we were supposed to do. But in *Phineas Redux*, with its darker tone and more astringent style, it is recognizably a real world with which we are dealing and we pay it the ultimate tribute of feeling a genuine sense of loss when we leave it.

NOTE ON THE TEXT

Phineas Redux was serialized in the *Graphic; An Illustrated Weekly Newspaper* between 19 July 1873 and 10 January 1874. Of the twenty-six instalments, all but the last two contained three chapters each; these two contained four chapters each. Each instalment included an illustration by Frank Holl. The novel was published in two volumes by Chapman and Hall in December 1873, and an American edition (published by Harper) appeared in one volume in March 1874. As the proprietors of the *Graphic* had bought foreign and colonial rights, the novel did not appear in the Tauschnitz series (generally favoured by Trollope for continental issue) but in the rival Asher series.

The text of the present edition was produced by R. W. Chapman and revised in 1951 by Robert H. Taylor for The Oxford Trollope, under the general editorship of Michael Sadleir and Frederick Page.

SELECT BIBLIOGRAPHY

THERE is a daily increase in the critical material written about Trollope, and a note such as this can only be useful by being highly selective. Readers should consult George H. Ford, *Victorian Fiction; a Second Guide to Research* (Modern Language Association of America, New York, 1978), and the bibliographies published annually in *Victorian Studies*.

Michael Sadleir's *Trollope: A Commentary* (Oxford University Press, 3rd edn., 1961) is still the best biography, though it has been augmented by James Pope Hennessy, *Anthony Trollope* (Little, Brown, Boston, 1971) and C. P. Snow, *Trollope: His Life and Art* (Macmillan, 1975). In *Trollope: A Bibliography* (Dawson, 2nd edn., 1964) Sadleir also provides an account of Trollope's original publications. Donald Smalley, *Trollope: the Critical Heritage* (Routledge, 1969) and David Skilton, *Anthony Trollope and his Contemporaries* (Longman, 1972) very fully document his reception.

Of the many general studies recently published, the following should be noted—A. O. J. Cockshut, *Anthony Trollope: A Critical Study* (Collins, 1955); Ruth apRoberts, *Trollope: Artist and Moralist* (Chatto & Windus, 1971); James R. Kincaid, *The Novels of Anthony Trollope* (Clarendon Press, 1977). Robert Tracy's *Trollope's Later Novels* (University of California Press, Berkeley, 1978), though it does not deal with any of the Palliser novels at length, should not be overlooked.

The most important studies to deal specifically with the political fiction are Arthur Pollard, *Trollope's Political Novels* (University of Hull, 1968), which argues that the author's own political engagement is the root of his power in these novels; John Halperin's *Trollope and Politics: A Study of the Pallisers and Others* (Macmillan, 1977), and Juliet McMaster's *Trollope's Palliser Novels: Theme and Pattern* (Macmillan, 1978). Halperin takes a conventional view of politics, while McMaster is inclined to depoliticize the fiction even as to content as well as form.

One or two more marginal books deserve notice. *A Guide to*

Trollope by W. G. and J. T. Gerould (Princeton University Press, Princeton, 1948), is a dictionary which can aid those who have not memorized every character in the forty-seven novels. John W. Clark's *Language and Style of Anthony Trollope* (Deutsch, 1975) scandalously lacks an index but is otherwise excellent. In *Trollope and his Illustrators* (Macmillan, 1980), N. John Hall discusses two Palliser novels—*Can You Forgive Her?* and *Phineas Finn*—and reproduces illustrations by Hablot K. Brown and J. E. Millais.

Critical articles in journals and elsewhere are legion. A useful historical context is provided by J. A. Banks, 'The Way They Lived Then: Anthony Trollope and the 1870s', *Victorian Studies*, 12, 1968, pp. 177–200. A useful foreign perspective is present in Ludwig Borinski's 'Trollope's Palliser Novels', *Die Neueren Sprachen*, 12, 1963, pp. 389–407.

In connection with this edition of *Phineas Redux* two nineteenth-century books which influenced Trollope's thinking should be studied—Henry Taylor, *The Statesman* (1836), and Walter Bagehot, *The English Constitution* (1865, 1872 rev. ed.). John Vincent, *The Formation of the Liberal Party 1857–1868* (Constable, 1966) and W. L. Burn, *The Age of Equipoise* (Allen & Unwin, 1964) provide a scholarly and readable perspective upon the period. For the specifically Irish background history see F. S. L. Lyons, *Ireland Since the Famine* (Collins, 1971) and, with particular attention to the anomalies of the United Kingdom and their consequences, Oliver MacDonagh, *Ireland: the Union and its Aftermath* (rev. ed. Allen & Unwin, 1977).

J. R. Dinwiddy, 'Who's Who in Trollope's Political Novels', *Nineteenth Century Fiction* (1967) is a great more than an identification of originals, and the same author's 'Elections in Victorian Fiction', *Victorian Newsletter* (1974) relates Phineas's re-election to Trollope's sustained political ambitions.

A CHRONOLOGY OF ANTHONY TROLLOPE

Virtually all Trollope's fiction appeared first in serial form, with book production timed to coincide with the final instalment of the serial. In this chronology the titles are dated as on the title-page of the first book edition. On a very few occasions the book edition appeared in December of the year previous to that indicated on the title-page, so as to catch the Christmas sales.

1815 (24 Apr.) Born at 6 Keppel Street, Bloomsbury, the fourth son of Thomas and Frances Trollope.

1822 To Harrow as a day-boy.

1825 To a private school at Sunbury.

1827 To school at Winchester.

1830 Removed from Winchester and returned to Harrow.

1834 The family moves to Bruges.
(Autumn) He accepts a junior clerkship in the General Post Office.

1841 (Aug.) Deputy Postal Surveyor at Banagher, King's County, Ireland.

1843 (Autumn) Begins work on his first novel, *The Macdermots of Ballycloran*.

1844 (11 June) Marries Rose Heseltine.
Transferred to Clonmel, County Tipperary.

1845 Promoted to the office of Surveyor, and transferred to Mallow, County Cork.

1845–7 Famine and epidemic throughout Ireland, especially the south and west with which Trollope was familiar.

1847 *The Macdermots of Ballycloran*, published in 3 vols. (Newby).

1848 Rebellion in Ireland, concentrated in Cork and Tipperary.
The Kellys and the O'Kellys; or Landlords and Tenants 3 vols. (Colburn).

1850 Writes *The Noble Jilt* (published 1923).
La Vendée; an Historical Romance 3 vols. (Colburn).

1851 Transferred to England.

1853 Returns to Ireland; completes *The Warden* (the first of the Barsetshire novels) in Belfast.

1854 (Autumn) Leaves Belfast and settles outside Dublin at Donnybrook.

1855 *The Warden* 1 vol. (Longman).

1857 *Barchester Towers* 3 vols. (Longman).
(Sept.) Visits his mother in Florence.

1858 *Doctor Thorne* 3 vols. (Chapman & Hall).
The Three Clerks 3 vols. (Bentley).
(Feb.) Departs for Egypt on Post Office business.
(Mar.) Removes from Egypt to Palestine.
(Apr.–May) Returns via Malta, Gibraltar, and Spain.
(May–July) Visits Scotland and north of England on business.
(Aug.–Oct.) At home in Ireland.
(Nov.) On Post Office business in the West Indies.

1859 *The Bertrams* 3 vols. (Chapman & Hall).
The West Indies and the Spanish Main 1 vol. (Chapman & Hall).
(Sept.) Holiday in the Pyrenees.
(Dec.) Leaves Ireland; settles at Waltham Cross.

1860 *Castle Richmond* 3 vols. (Chapman & Hall).
(Oct.) With his wife he visits his mother and brother in Florence; makes the acquaintance of Kate Field, a beautiful twenty-year-old American with whom he falls in love.

1861 *Framley Parsonage* 3 vols. (Smith, Elder).
Tales of All Countries—first series, 1 vol. (Chapman & Hall).
(Spring) Elected a member of the Garrick Club; some two years later is elected to the committee to fill the vacancy caused by Thackeray's death.
(Aug.) To America on official business.

1862 *Orley Farm* 2 vols. (Chapman & Hall).
North America 2 vols. (Chapman & Hall).
The Struggles of Brown, Jones and Robinson; by One of the Firm 1 vol. (New York, Harper—an American piracy).
(Spring) Returns home from America.

1863 *Rachel Ray* 2 vols. (Chapman & Hall).
Tales of All Countries—second series 1 vol. (Chapman & Hall).
(6 Oct.) Death of his mother, Mrs Frances Trollope.
(Dec.) Death of W. M. Thackeray.

1864 *The Small House at Allington* 2 vols. (Smith, Elder).
Can You Forgive Her? 2 vols. (Chapman & Hall).
(Summer) Elected a member of the Athenaeum Club.

1865 *Miss Mackenzie* 1 vol. (Chapman & Hall).
Hunting Sketches 1 vol. (Chapman & Hall).

1866 *The Belton Estate* 3 vols. (Chapman & Hall).

Travelling Sketches 1 vol. (Chapman & Hall).
Clergymen of the Church of England 1 vol. (Chapman & Hall).

1867 *Nina Balatka* 2 vols. (Blackwood).
The Last Chronicle of Barset 2 vols. (Smith, Elder).
The Claverings 2 vols. (Smith, Elder).
Lotta Schmidt and Other Stories 1 vol. (Strahan).
(1 Sept.) Resigns from the Civil Service.

1868 *Linda Tressel* 2 vols. (Blackwood).
(Mar.) Leaves London for the United States on business involving copyright, in touch again with Kate Field.
(July) Returns from America.
(Nov.) Stands unsuccessfully as Liberal candidate for Beverley, Yorkshire, losing £2,000 in the enterprise.

1869 *Phineas Finn; the Irish Member* 2 vols. (Virtue & Co.).
He Knew He was Right 2 vols. (Strahan).
Did He Steal It? A Comedy in Three Acts—a version of *The Last Chronicle of Barset* 1 vol. (printed by Virtue & Co.).

1870 *The Vicar of Bullhampton* 1 vol. (Bradbury, Evans).
An Editor's Tales 1 vol. (Strahan).
The Commentaries of Caesar 1 vol. (Blackwood).

1871 *Sir Harry Hotspur of Humblethwaite* 1 vol. (Hurst & Blackett).
Ralph the Heir 3 vols. (Hurst & Blackett).
(Apr.) Gives up house at Waltham Cross.
(May) Sails to Australia to visit his son.
(20 July) Arrives at Melbourne.

1872 *The Golden Lion of Granpere* 1 vol. (Tinsley).
(Jan.–Oct.) Travelling in Australia and New Zealand.
(Apr.) A dramatized (and pirated) version of *Ralph the Heir* produced by Charles Reade.
(Dec.) Returns via the United States, and settles in Montagu Square, London.

1873 *The Eustace Diamonds* 3 vols. (Chapman & Hall).
Australia and New Zealand 2 vols. (Chapman & Hall).
(Winter) Hunting actively.

1874 *Phineas Redux* 2 vols. (Chapman & Hall).
Lady Anna 2 vols. (Chapman & Hall).
Harry Heathcote of Gangoil, a Tale of Australian Bush Life 1 vol. (Sampson Low).

1875 *The Way We Live Now* 2 vols. (Chapman & Hall).
(Feb.) Travels to Ceylon via Brindisi and the Suez Canal, once again on the way to Australia.
(Mar.–Apr.) In Ceylon.

(June) Arrives in Australia.
(Aug.–Oct.) Sailing homewards.
(Oct.) Begins work on his *Autobiography*.

1876 *The Prime Minister* 4 vols. (Chapman & Hall).

1877 *The American Senator* 3 vols. (Chapman & Hall).
Christmas at Thompson Hall 1 vol. (New York, Harper).
(June) Leaves London for South Africa.
(Dec.) Sails for home.

1878 *Is He Popenjoy?* 3 vols. (Chapman & Hall).
South Africa 2 vols. (Chapman & Hall).
How the 'Mastiffs' Went to Iceland 1 vol. (Virtue & Co.).
(June–July) Travels to Iceland in the yacht 'Mastiff'.

1879 *An Eye for an Eye* 2 vols. (Chapman & Hall).
John Caldigate 3 vols. (Chapman & Hall).
Cousin Henry 2 vols. (Chapman & Hall).
Thackeray 1 vol. (Macmillan).

1880 *The Duke's Children* 3 vols. (Chapman & Hall).
(July) Settles at Harting Grange, near Petersfield.

1881 *Dr. Wortle's School* 2 vols. (Chapman & Hall).
Ayala's Angel 3 vols. (Chapman & Hall).

1882 *Why Frau Frohmann Raised Her Prices; and Other Stories* 1 vol. (Isbister).
Kept in the Dark 2 vols. (Chatto & Windus).
Marion Fay 3 vols. (Chapman & Hall).
The Fixed Period 2 vols. (Blackwood).
Palmerston 1 vol. (Isbister).
(May) Visits Ireland to collect material for a new Irish novel.
(Aug.) Returns to Ireland a second time.
(Sept.) Moves to London for the winter.
(6 Dec.) Dies in London.

1883 *Mr. Scarborough's Family* 3 vols. (Chatto & Windus).
The Landleaguers (unfinished) 3 vols. (Chatto & Windus).
An Autobiography 2 vols. (Blackwood).

1884 *An Old Man's Love* 2 vols. (Blackwood).

1923 *The Noble Jilt* (a play) 1 vol. (Constable).

1927 *London Tradesmen* 1 vol. (Elkin Mathews).

1972 *The New Zealander* 1 vol. (Oxford University Press).

CONTENTS

Volume I

CONTENTS

Volume II

PHINEAS REDUX*

VOLUME I

CHAPTER I
Temptation

THE circumstances of the general election of 18—*will be well remembered by all those who take an interest in the political matters of the country. There had been a coming in and a going out of Ministers previous to that,—somewhat rapid, very exciting, and, upon the whole, useful as showing the real feeling of the country upon sundry questions of public interest. Mr. Gresham had been Prime Minister of England, as representative of the Liberal party in politics. There had come to be a split among those who should have been his followers on the terribly vexed question of the Ballot.* Then Mr. Daubeny for twelve months had sat upon the throne distributing the good things of the Crown amidst Conservative birdlings, with beaks wide open and craving maws, who certainly for some years previous had not received their share of State honours or State emoluments. And Mr. Daubeny was still so sitting, to the infinite dismay of the Liberals, every man of whom felt that his party was entitled by numerical strength to keep the management of the Government within its own hands.

Let a man be of what side he may in politics,—unless he be much more of a partisan than a patriot,—he will think it well that there should be some equity of division in the bestowal of crumbs of comfort. Can even any old Whig wish that every Lord Lieutenant of a county should be an old Whig? Can it be good for the administration of the law that none but Liberal lawyers should become Attorney-Generals, and from thence Chief Justices or Lords of Appeal? Should no Conservative Peer ever represent the majesty of England in India, in Canada, or at St. Petersburgh? So arguing, moderate Liberals had been glad to give Mr. Daubeny and his merry men a chance. Mr. Daubeny and his merry men had not neglected the chance given them. Fortune favoured them, and they made their hay while the sun shone with an energy that had never been surpassed, improving upon Fortune, till their natural enemies waxed impatient. There had been as yet but one year of it, and the natural enemies, who had at first expressed themselves as glad that the turn had come, might have endured the period of spoliation with more equanimity. For to them, the Liberals, this cutting up of the Whitehall cake by the Conservatives was spoliation when the privilege of cutting was found to have so much exceeded what had been expected. Were not they, the Liberals, the real representatives of the people, and, therefore, did not the cake in truth appertain to them? Had not they given up the cake for a while, partly, indeed, through idleness and mismanagement, and quarrelling among themselves; but mainly with a feeling that a moderate slicing on the other side would, upon the whole, be advantageous? But when the cake came to be mauled like that—oh, heavens! So the men who had quarrelled agreed to quarrel no more, and it was decided that there should be an end of mismanagement and idleness, and that this horrid sight of the weak pretending to be strong, or the weak receiving the reward of strength, should be brought to an end. Then came a great fight, in the last agonies of which the cake was sliced manfully. All the world knew how the fight would go; but in the meantime lord-lieutenancies were arranged; very ancient judges retired upon pensions; vice-royal Governors were sent out in the last gasp of the failing battle; great places were filled by tens, and little places by twenties; private secretaries were established here and there; and the hay was still made even after the sun had gone down.

In consequence of all this the circumstances of the election of 18— were peculiar. Mr. Daubeny had dissolved the House, not probably with any idea that he could thus retrieve his fortunes, but feeling that in doing so he was occupying the last normal position of a properly-fought Constitutional battle. His enemies were resolved, more firmly than they were resolved before, to knock him altogether on the head at the general election which he had himself called into existence. He had been disgracefully out-voted in the House of Commons on various subjects. On the last occasion he had gone into his lobby with a minority of 37, upon a motion brought forward by Mr. Palliser, the late Liberal Chancellor of the Exchequer, respecting decimal coinage.* No politician, not even Mr. Palliser himself, had expected that he would carry his Bill in the present session. It was brought forward as a trial of strength; and for such a purpose decimal coinage was as good a subject as any other. It was Mr. Palliser's hobby, and he was gratified at having this further opportunity of ventilating it. When in power, he had not succeeded in carrying his measure, awed, and at last absolutely beaten, by the infinite difficulty encountered in arranging its details. But his mind was still set upon it, and it was allowed by the whole party to be as good as anything else for the purpose then required. The Conservative Government was beaten for the third or fourth time, and Mr. Daubeny dissolved the House.

The whole world said that he might as well have resigned at once. It was already the end of July, and there must be an autumn Session with the new members. It was known to be impossible that he should find himself supported by a majority after a fresh election. He had been treated with manifest forbearance; the cake had been left in his hands for twelve months; the House was barely two years old; he had no 'cry' with which to meet the country; the dissolution was factious, dishonest, and unconstitutional. So said all the Liberals, and it was deduced also that the Conservatives were in their hearts as angry as were their opponents. What was to be gained but the poor interval of three months? There were clever men who suggested that Mr. Daubeny had a scheme in his head—some sharp trick of political conjuring, some 'hocus-pocus presto' sleight of hand, by which he might be able to retain power, let the elections go as they would. But, if so, he certainly did not make his scheme known to his own party.

He had no cry with which to meet the country, nor, indeed, had the leaders of the Opposition. Retrenchment, army reform, navy excellence, Mr. Palliser's decimal coinage, and general good government gave to all the old-Whig moderate Liberals plenty of matter for speeches to their future constituents. Those who were more advanced could promise the Ballot, and suggest the disestablishment of the Church. But the Government of the day was to be turned out on the score of general incompetence. They were to be made to go, because they could not command majorities. But there ought to have been no dissolution, and Mr. Daubeny was regarded by his opponents, and indeed by very many of his followers also, with an enmity that was almost ferocious. A seat in Parliament, if it be for five or six years, is a blessing; but the blessing becomes very questionable if it have to be sought afresh every other Session.

One thing was manifest to thoughtful, working, eager political Liberals. They must have not only a majority in the next Parliament, but a majority of good men—of men good and true. There must be no more mismanagement; no more quarrelling; no more idleness. Was it to be borne that an unprincipled so-called Conservative Prime Minister should go on slicing the cake after such a fashion as that lately adopted?

Old bishops had even talked of resigning, and Knights of the Garter had seemed to die on purpose. So there was a great stir at the Liberal political clubs, and every good and true man was summoned to the battle.

Now no Liberal soldier, as a young soldier, had been known to be more good and true than Mr. Finn, the Irishman, who had held office two years ago to the satisfaction of all his friends, and who had retired from office because he had found himself compelled to support a measure* which had since been carried by those very men from whom he had been obliged on this account to divide himself. It had always been felt by his old friends that he had been, if not ill-used, at least very unfortunate. He had been twelve months in advance of his party, and had consequently been driven out into the cold. So when the names of good men and true were mustered, and weighed, and discussed, and scrutinised by some active members of the Liberal party in a certain very private room not far removed from our great seat of parliamentary warfare; and when the capabilities, and expediencies, and possibilities were tossed to and fro among these active members, it came to pass that the name of Mr. Finn was mentioned more than once. Mr. Phineas Finn was the gentleman's name—which statement may be necessary to explain the term of endearment which was occasionally used in speaking of him.

'He has got some permanent place,' said Mr. Ratler, who was living on the well-founded hope of being a Treasury Secretary under the new dispensation; 'and of course he won't leave it.'

It must be acknowledged that Mr. Ratler, than whom no judge in such matters possessed more experience, had always been afraid of Phineas Finn.

'He'll lave it fast enough, if you'll make it worth his while,' said the Honourable Laurence Fitzgibbon, who also had his expectations.

'But he married when he went away, and he can't afford it,' said Mr. Bonteen, another keen expectant.

'Devil a bit,' said the Honourable Laurence; 'or, anyways,

the poor thing died of her first baby before it was born. Phinny hasn't an impidiment, no more than I have.'

'He's the best Irishman we ever got hold of,' said Barrington Erle—'present company always excepted, Laurence.'

'Bedad, you needn't except me, Barrington. I know what a man's made of, and what a man can do. And I know what he can't do. I'm not bad at the outside skirmishing. I'm worth me salt. I say that with a just reliance on me own powers. But Phinny is a different sort of man. Phinny can stick to a desk from twelve to seven, and wish to come back again after dinner. He's had money left him, too, and 'd like to spend some of it on an English borough.'

'You never can quite trust him,' said Bonteen. Now Mr. Bonteen had never loved Mr. Finn.

'At any rate we'll try him again,' said Barrington Erle, making a little note to that effect. And they did try him again.

Phineas Finn, when last seen by the public, was departing from parliamentary life in London to the enjoyment of a modest place under Government in his own country, with something of a shattered ambition. After various turmoils he had achieved a competency, and had married the girl of his heart. But now his wife was dead, and he was again alone in the world. One of his friends had declared that money had been left to him. That was true, but the money had not been much. Phineas Finn had lost his father as well as his wife, and had inherited about four thousand pounds. He was not at this time much over thirty; and it must be acknowledged in regard to him that, since the day on which he had accepted place and retired from London, his very soul had sighed for the lost glories of Westminster and Downing Street.

There are certain modes of life which, if once adopted, make contentment in any other circumstances almost an impossibility. In old age a man may retire without repining, though it is often beyond the power even of the old man to do so; but in youth, with all the faculties still perfect, with the body still strong, with the hopes still buoyant, such a change as that which had been made by Phineas Finn was more than he, or than most men, could bear with equanimity. He had revelled in the gas-light, and could not lie quiet on a sunny bank. To the palate accustomed to high cookery, bread and milk is almost painfully insipid. When Phineas Finn found himself discharging in Dublin the routine duties of his office, —as to which there was no public comment, no feeling that such duties were done in the face of the country,—he became sick at heart and discontented. Like the warhorse out at grass he remembered the sound of the battle and the noise of trumpets. After five years spent in the heat and full excitement of London society, life in Ireland was tame to him, and cold, and dull. He did not analyse the difference between metropolitan and quasi-metropolitan manners; but he found that men and women in Dublin were different from those to whom he had been accustomed in London. He had lived among lords, and the sons and daughters of lords; and though the official secretaries and assistant commissioners among whom his lot now threw him were for the most part clever fellows, fond of society, and perhaps more than his equals in the kind of conversation which he found to be prevalent, still they were not the same as the men he had left behind him,—men alive with the excitement of parliamentary life in London. When in London he had often told himself that he was sick of it, and that he would better love some country quiet life. Now Dublin was his Tibur, and the fickle one found that he could not be happy unless he were back again at Rome.* When, therefore, he received the following letter from his friend, Barrington Erle, he neighed like the old warhorse, and already found himself shouting 'Ha, ha,' among the trumpets.

'—— *Street, 9th July*, 18—.

'MY DEAR FINN,

'Although you are not now immediately concerned in such trifling matters you have no doubt heard that we are all to be sent back at once to our constituents, and that there will be a general election about the end of September. We are sure that we shall have such a majority as we never had before;

but we are determined to make it as strong as possible, and to get in all the good men that are to be had. Have you a mind to try again? After all, there is nothing like it.

'Perhaps you may have some Irish seat in your eye for which you would be safe. To tell the truth we know very little of the Irish seats—not so much as, I think, we ought to do. But if you are not so lucky I would suggest Tankerville in Durham. Of course there would be a contest, and a little money will be wanted; but the money would not be much. Browborough has sat for the place now for three Parliaments, and seems to think it all his own. I am told that nothing could be easier than to turn him out. You will remember the man—a great, hulking, heavy, speechless fellow, who always used to sit just over Lord Macaw's shoulder. I have made inquiry, and I am told that he must walk if anybody would go down who could talk to the colliers every night for a week or so. It would just be the work for you. Of course, you should have all the assistance we could give you, and Molescroft would put you into the hands of an agent who wouldn't spend money for you. £500 would do it all.

'I am very sorry to hear of your great loss, as also was Lady Laura, who, as you are aware, is still abroad with her father. We have all thought that the loneliness of your present life might perhaps make you willing to come back among us. I write instead of Ratler, because I am helping him in the Northern Counties. But you will understand all about that.

'Yours, ever faithfully,
'BARRINGTON ERLE.

'Of course Tankerville has been dirty. Browborough has spent a fortune there. But I do not think that that need dishearten you. You will go there with clean hands. It must be understood that there shall not be as much as a glass of beer. I am told that the fellows won't vote for Browborough unless he spends money, and I fancy he will be afraid to do it heavily after all that has come and gone. If he does you'll have him out on a petition. Let us have an answer as soon as possible.'

He at once resolved that he would go over and see; but, before he replied to Erle's letter, he walked half-a-dozen times the length of the pier at Kingston meditating on his answer. He had no one belonging to him. He had been deprived of his young bride, and left desolate. He could ruin no one but himself. Where could there be a man in all the world who had a more perfect right to play a trick with his own prospects? If he threw up his place and spent all his money, who could blame him? Nevertheless, he did tell himself that, when he should have thrown up his place and spent all his money, there would remain to him his own self to be disposed of in a manner that might be very awkward to him. A man owes it to his country, to his friends, even to his acquaintance, that he shall not be known to be going about wanting a dinner, with never a coin in his pocket. It is very well for a man to boast that he is lord of himself, and that having no ties he may do as he pleases with that possession. But it is a possession of which, unfortunately, he cannot rid himself when he finds that there is nothing advantageous to be done with it. Doubtless there is a way of riddance. There is the bare bodkin.* Or a man may fall overboard between Holyhead and Kingston in the dark, and may do it in such a cunning fashion that his friends shall think that it was an accident. But against these modes of riddance there is a canon set, which some men still fear to disobey.

The thing that he was asked to do was perilous. Standing in his present niche of vantage he was at least safe. And added to his safety there were material comforts. He had more than enough for his wants. His work was light: he lived among men and women with whom he was popular. The very fact of his past parliamentary life had caused him to be regarded as a man of some note among the notables of the Irish capital. Lord Lieutenants were gracious to him, and the wives of judges smiled upon him at their tables. He was encouraged to talk of those wars of the gods at which he had been present, and was so treated as to make him feel that he was somebody in the world of Dublin. Now he was invited to give all this up; and for what?

He answered that question to himself with enthusiastic eloquence. The reward offered to him was the thing which in all the world he liked best. It was suggested to him that he should again have within his reach that parliamentary renown which had once been the very breath of his nostrils. We all know those arguments and quotations, antagonistic to prudence, with which a man fortifies himself in rashness. 'None but the brave deserve the fair.' 'Where there's a will there's a way.' 'Nothing venture nothing have.' 'The sword is to him who can use it.' 'Fortune favours the bold.' But on the other side there is just as much to be said. 'A bird in the hand is worth two in the bush.' 'Look before you leap.' 'Thrust not out your hand further than you can draw it back again.' All which maxims of life Phineas Finn revolved within his own heart, if not carefully, at least frequently, as he walked up and down the long pier of Kingston Harbour.

But what matter such revolvings? A man placed as was our Phineas always does that which most pleases him at the moment, being but poor at argument if he cannot carry the weight to that side which best satisfies his own feelings. Had not his success been very great when he before made the attempt? Was he not well aware at every moment of his life that, after having so thoroughly learned his lesson in London, he was throwing away his hours amidst his present pursuits in Dublin? Did he not owe himself to his country? And then, again, what might not London do for him? Men who had begun as he begun had lived to rule over Cabinets, and to sway the Empire. He had been happy for a short twelvemonth with his young bride,—for a short twelvemonth,—and then she had been taken from him. Had she been spared to him he would never have longed for more than Fate had given him. He would never have sighed again for the glories of Westminster had his Mary not gone from him. Now he was alone in the world; and, though he could look forward to possible and not improbable events which would make that future disposition of himself a most difficult question for him, still he would dare to try.

As the first result of Erle's letter Phineas was over in London early in August. If he went on with this matter, he must, of course, resign the office for holding which he was now paid a thousand a year. He could retain that as long as he chose to earn the money, but the earning of it would not be compatible with a seat in Parliament. He had a few thousand pounds with which he could pay for the contest at Tankerville, for the consequent petition which had been so generously suggested to him, and maintain himself in London for a session or two should he be so fortunate as to carry his election. Then he would be penniless, with the world before him as a closed oyster to be again opened, and he knew,—no one better,—that this oyster becomes harder and harder in the opening as the man who has to open it becomes older. It is an oyster that will close to again with a snap, after you have got your knife well into it, if you withdraw your point but for a moment. He had had a rough tussle with the oyster already, and had reached the fish within the shell. Nevertheless, the oyster which he had got was not the oyster which he wanted. So he told himself now, and here had come to him the chance of trying again.

Early in August he went over to England, saw Mr. Molescroft, and made his first visit to Tankerville. He did not like the look of Tankerville; but nevertheless he resigned his place before the month was over. That was the one great step, or rather the leap in the dark,—and that he took. Things had been so arranged that the election at Tankerville was to take place on the 20th of October. When the dissolution had been notified to all the world by Mr. Daubeny an earlier day was suggested; but Mr. Daubeny saw reasons for postponing it for a fortnight. Mr. Daubeny's enemies were again very ferocious. It was all a trick. Mr. Daubeny had no right to continue Prime Minister a day after the decided expression of opinion as to unfitness which had been pronounced by the House of Commons. Men were waxing very wrath. Nevertheless, so much power remained in Mr. Daubeny's hand, and the election was delayed. That for Tankerville would not be held till the 20th of October. The whole House could not be chosen till the end of the month,—hardly by that time—and yet there was to be an autumn Session. The Ratlers and Bonteens were at any rate clear about the autumn Session. It was absolutely impossible that Mr. Daubeny should be allowed to remain in power over Christmas, and up to February.

Mr. Molescroft, whom Phineas saw in London, was not a comfortable counsellor. 'So you are going down to Tankerville?' he said.

'They seem to think I might as well try.'

'Quite right;—quite right. Somebody ought to try it, no doubt. It would be a disgrace to the whole party if Browborough were allowed to walk over. There isn't a borough in England more sure to return a Liberal than Tankerville if left to itself. And yet that lump of a legislator has sat there as a Tory for the last dozen years by dint of money and brass.'*

'You think we can unseat him?'

'I don't say that. He hasn't come to the end of his money, and as to his brass that is positively without end.'

'But surely he'll have some fear of consequences after what has been done?'

'None in the least. What has been done? Can you name a single Parliamentary aspirant who has been made to suffer?'

'They have suffered in character,' said Phineas. 'I should not like to have the things said of me that have been said of them.'

'I don't know a man of them who stands in a worse position among his own friends than he occupied before. And men of that sort don't want a good position among their enemies. They know they're safe. When the seat is in dispute everybody is savage enough; but when it is merely a question of punishing a man, what is the use of being savage? Who knows whose turn it may be next?'

'He'll play the old game, then?'

'Of course he'll play the old game,' said Mr. Molescroft. 'He doesn't know any other game. All the purists in England wouldn't teach him to think that a poor man ought not to sell his vote, and that a rich man oughtn't to buy it. You mean to go in for purity?'

'Certainly I do.'

'Browborough will think just as badly of you as you will of him. He'll hate you because he'll think you are trying to rob him of what he has honestly bought; but he'll hate you quite as much because you try to rob the borough. He'd tell you if you asked him that he doesn't want his seat for nothing, any more than he wants his house or his carriage-horses for nothing. To him you'll be a mean, low interloper. But you won't care about that.'

'Not in the least, if I can get the seat.'

'But I'm afraid you won't. He will be elected. You'll petition. He'll lose his seat. There will be a commission. And then the borough will be disfranchised. It's a fine career, but expensive; and then there is no reward beyond the self-satisfaction arising from a good action. However, Ruddles will do the best he can for you, and it certainly is possible that you may creep through.' This was very disheartening, but Barrington Erle assured our hero that such was Mr. Molescroft's usual way with candidates, and that it really meant little or nothing. At any rate, Phineas Finn was pledged to stand.

CHAPTER II

Harrington Hall

PHINEAS, on his first arrival in London, found a few of his old friends, men who were still delayed by business though the Session was over. He arrived on the 10th of August, which may be considered as the great day of the annual exodus, and he remembered how he, too, in former times had gone to Scotland to shoot grouse, and what he had done there besides shooting. He had been a welcome guest at Loughlinter, the magnificent seat of Mr. Kennedy, and indeed there had been that between him and Mr. Kennedy which ought to make him a welcome guest there still. But of Mr. Kennedy he had heard nothing directly since he had left London. From Mr. Kennedy's wife, Lady Laura, who had been his great friend, he had heard occasionally; but she was separated from her husband, and was living abroad with her father, the Earl of Brentford. Has it not been written in a former book* how this Lady Laura had been unhappy in her marriage, having wedded herself to a man whom she had never loved, because he was rich and powerful, and how this very Phineas had asked her to be his bride after she had accepted the rich man's hand? Thence had come great trouble, but nevertheless there had been that between Mr. Kennedy and our hero which made Phineas feel that he ought still to be welcomed as a guest should he show himself at the door of Loughlinter Castle. The idea came upon him simply because he found that almost every man for whom he inquired had just started, or was just starting, for the North; and he would have liked to go where others went. He asked a few questions as to Mr. Kennedy from Barrington Erle and others, who had known him, and was told that the man now lived quite alone. He still kept his seat in Parliament, but had hardly appeared during the last Session, and it was thought that he would not come forward again. Of his life in the country nothing was known. 'No one fishes his rivers, or shoots his moors, as far as I can learn,' said Barrington Erle. 'I suppose he looks after the sheep and says his prayers, and keeps his money together.'

'And there has been no attempt at a reconciliation?' Phineas asked.

'She went abroad to escape his attempts, and remains there in order that she may be safe. Of all hatreds that the world produces, a wife's hatred for her husband, when she does hate him, is the strongest.'

In September Finn was back in Ireland, and about the end of that month he made his first visit to Tankerville. He remained there for three or four days, and was terribly disgusted while staying at the 'Yellow' inn, to find that the people of the town would treat him as though he were rolling in wealth. He was soon tired of Tankerville, and as he could do nothing further, on the spot, till the time for canvassing should come on, about ten days previous to the election, he

returned to London, somewhat at a loss to know how to bestir himself. But in London he received a letter from another old friend, which decided him:—

'My dear Mr. Finn,' said the letter, 'of course you know that Oswald is now master of the Brake hounds. Upon my word, I think it is the place in the world for which he is most fit. He is a great martinet in the field, and works at it as though it were for his bread. We have been here looking after the kennels and getting up the horses since the beginning of August, and have been cub-hunting* ever so long. Oswald wants to know whether you won't come down to him till the election begins in earnest.

'We were so glad to hear that you were going to appear again. I have always known that it would be so. I have told Oswald scores of times that I was sure you would never be happy out of Parliament, and that your real home must be somewhere near the Treasury Chambers. You can't alter a man's nature. Oswald was born to be a master of hounds, and you were born to be a Secretary of State. He works the hardest and gets the least pay for it; but then, as he says, he does not run so great a risk of being turned out.

'We haven't much of a house, but we have plenty of room for you. As for the house, it was a matter of course, whether good or bad. It goes with the kennels, and I should as little think of having a choice as though I were one of the horses. We have very good stables, and such a stud! I can't tell you how many there are. In October it seems as though their name were legion. In March there is never anything for any body to ride on. I generally find then that mine are taken for the whips. Do come and take advantage of the flush. I can't tell you how glad we shall be to see you. Oswald ought to have written himself, but he says——; I won't tell you what he says. We shall take no refusal. You can have nothing to do before you are wanted at Tankerville.

'I was so sorry to hear of your great loss. I hardly know whether to mention it or to be silent in writing. If you were here of course I should speak of her. And I would rather renew your grief for a time than allow you to think that I am indifferent. Pray come to us.

'Yours ever most sincerely,

'VIOLET CHILTERN.

'Harrington Hall, Wednesday.'

Phineas Finn at once made up his mind that he would go to Harrington Hall. There was the prospect in this of an immediate return to some of the most charming pleasures of the old life, which was very grateful to him. It pleased him much that he should have been so thought of by this lady,—that she should have sought him out at once, at the moment of his reappearance. That she would have remembered him, he was quite sure, and that her husband, Lord Chiltern, should remember him also, was beyond a doubt. There had been passages in their joint lives which people cannot forget. But it might so well have been the case that they should not have cared to renew their acquaintance with him. As it was, they must have made close inquiry, and had sought him at the first day of his reappearance. The letter had reached him through the hands of Barrington Erle, who was a cousin of Lord Chiltern, and was at once answered as follows:—

'Fowler's Hotel, Jermyn Street,
'October 1st.

'My dear Lady Chiltern,

'I cannot tell you how much pleasure the very sight of your handwriting gave me. Yes, here I am again, trying my hand at the old game. They say that you can never cure a gambler or a politician; and, though I had very much to make me happy till that great blow came upon me, I believe that it is so. I am uneasy till I can see once more the Speaker's wig, and hear bitter things said of this "right honourable gentleman," and of that noble friend. I want to be once more in the midst of it; and as I have been left singularly desolate in the world, without a tie by which I am bound to aught but an honourable mode of living, I have determined to run the risk, and have thrown up the place which I held under Government. I am to stand for Tankerville, as you have heard, and I am told by those to whose tender mercies I have been confided by B. E. that I have not a chance of success.

'Your invitation is so tempting that I cannot refuse it. As you say, I have nothing to do till the play begins. I have issued my address, and must leave my name and my fame to be discussed by the Tankervillians till I make my appearance among them on the 10th of this month. Of course, I had heard that Chiltern has the Brake, and I have heard also that he is doing it uncommonly well. Tell him that I have hardly seen a hound since the memorable day on which I pulled him out from under his horse in the brook at Wissindine. I don't know whether I can ride a yard now. I will get to you on the 4th, and will remain if you will keep me till the 9th. If Chiltern can put me up on anything a little quieter than Bonebreaker, I'll go out steadily, and see how he does his cubbing. I may, perhaps, be justified in opining that Bonebreaker has before this left the establishment. If so I may, perhaps, find myself up to a little very light work.

'Remember me very kindly to him. Does he make a good nurse with the baby?

'Yours, always faithfully,

'PHINEAS FINN.

'I cannot tell you with what pleasure I look forward to seeing you both again.'

The next few days went very heavily with him. There had, indeed, been no real reason why he should not have gone to Harrington Hall at once, except that he did not wish to seem to be utterly homeless. And yet were he there, with his old friends, he would not scruple for a moment in owning that such was the case. He had fixed his day, however, and did remain in London till the 4th. Barrington Erle and Mr. Ratler he saw occasionally, for they were kept in town on the affairs of the election. The one was generally full of hope; but the other was no better than a Job's comforter.* 'I wouldn't advise you to expect too much at Tankerville, you know,' said Mr. Ratler.

'By no means,' said Phineas, who had always disliked Ratler, and had known himself to be disliked in return. 'I expect nothing.'

'Browborough understands such a place as Tankerville so well! He has been at it all his life. Money is no object to him, and he doesn't care a straw what anybody says of him. I don't think it's possible to unseat him.'

'We'll try at least,' said Phineas, upon whom, however, such remarks as these cast a gloom which he could not succeed in shaking off, though he could summon vigour sufficient to save him from showing the gloom. He knew very well that comfortable words would be spoken to him at Harrington Hall, and that then the gloom would go. The comforting words of his friends would mean quite as little as the discourtesies of Mr. Ratler. He understood that thoroughly, and felt that he ought to hold a stronger control over his own impulses. He must take the thing as it would come, and neither the flatterings of friends nor the threatenings of enemies could alter it; but he knew his own weakness, and confessed to himself that another week of life by himself at Fowler's Hotel, refreshed by occasional interviews with Mr. Ratler, would make him altogether unfit for the coming contest at Tankerville.

He reached Harrington Hall in the afternoon about four, and found Lady Chiltern alone. As soon as he saw her he told himself that she was not in the least altered since he had last been with her, and yet during the period she had undergone that great change which turns a girl into a mother. She had the baby with her when he came into the room, and at once greeted him as an old friend,—as a loved and loving friend who was to be made free at once to all the inmost privileges of real friendship, which are given to and are desired by so few. 'Yes, here we are again,' said Lady Chiltern, 'settled, as far as I suppose we ever shall be settled, for ever so many years to come. The place belongs to old Lord Gunthorpe, I fancy, but really I hardly know. I do know that we should give it up at once if we gave up the hounds, and that we can't be turned out as long as we have them. Doesn't it seem odd to have to depend on a lot of yelping dogs?'

'Only that the yelping dogs depend on you.'

'It's a kind of give and take, I suppose, like other things in the world. Of course, he's a beautiful baby. I had him in just that you might see him. I show Baby, and Oswald shows the hounds. We've nothing else to interest anybody. But nurse shall take him now. Come out and have a turn in the shrubbery before Oswald comes back. They're gone to-day as far as

Trumpeton Wood, out of which no fox was ever known to break, and they won't be home till six.'

'Who are "they"?' asked Phineas, as he took his hat.

'The "they" is only Adelaide Palliser. I don't think you ever knew her?'

'Never. Is she anything to the other Pallisers?'

'She is everything to them all; niece and grand-niece, and first cousin and grand-daughter. Her father was the fourth brother, and as she was one of six her share of the family wealth is small. Those Pallisers are very peculiar, and I doubt whether she ever saw the old duke. She has no father or mother, and lives when she is at home with a married sister, about seventy years older than herself, Mrs. Attenbury.'

'I remember Mrs. Attenbury.'

'Of course you do. Who does not? Adelaide was a child then, I suppose. Though I don't know why she should have been, as she calls herself one-and-twenty now. You'll think her pretty. I don't. But she is my great new friend, and I like her immensely. She rides to hounds, and talks Italian, and writes for the *Times*.'

'Writes for the *Times*!'

'I won't swear that she does, but she could. There's only one other thing about her. She's engaged to be married.'

'To whom?'

'I don't know that I shall answer that question, and indeed I'm not sure that she is engaged. But there's a man dying for her.'

'You must know, if she's your friend.'

'Of course I know; but there are ever so many ins and outs, and I ought not to have said a word about it. I shouldn't have done so to any one but you. And now we'll go in and have some tea, and go to bed.'

'Go to bed!'

'We always go to bed here before dinner on hunting days. When the cubbing began Oswald used to be up at three.'

'He doesn't get up at three now.'

'Nevertheless we go to bed. You needn't if you don't like,

and I'll stay with you if you choose till you dress for dinner. I did know so well that you'd come back to London, Mr. Finn. You are not a bit altered.'

'I feel to be changed in everything.'

'Why should you be altered? It's only two years. I am altered because of Baby. That does change a woman. Of course I'm thinking always of what he will do in the world; whether he'll be a master of hounds or a Cabinet Minister or a great farmer;—or perhaps a miserable spendthrift, who will let everything that his grandfathers and grandmothers have done for him go to the dogs.'

'Why do you think of anything so wretched, Lady Chiltern?'

'Who can help thinking? Men do do so. It seems to me that that is the line of most young men who come to their property early. Why should I dare to think that my boy should be better than others? But I do; and I fancy that he will be a great statesman. After all, Mr. Finn, that is the best thing that a man can be, unless it is given him to be a saint and a martyr and all that kind of thing,—which is not just what a mother looks for.'

'That would only be better than the spendthrift and gambler.'

'Hardly better you'll say, perhaps. How odd that is! We all profess to believe when we're told that this world should be used merely as a preparation for the next; and yet there is something so cold and comfortless in the theory that we do not relish the prospect even for our children. I fancy your people have more real belief in it than ours.'

Now Phineas Finn was a Roman Catholic. But the discussion was stopped by the noise of an arrival in the hall.

'There they are,' said Lady Chiltern; 'Oswald never comes in without a sound of trumpets to make him audible throughout the house.' Then she went to meet her husband, and Phineas followed her out of the drawing-room.

Lord Chiltern was as glad to see him as she had been, and in a very few minutes he found himself quite at home. In the hall he was introduced to Miss Palliser, but he was hardly able to see her as she stood there a moment in her hat and habit. There was ever so much said about the day's work. The earths had not been properly stopped, and Lord Chiltern had been very angry, and the owner of Trumpeton Wood, who was a great duke, had been much abused, and things had not gone altogether straight.

'Lord Chiltern was furious,' said Miss Palliser, laughing, 'and therefore, of course, I became furious too, and swore that it was an awful shame. Then they all swore that it was an awful shame, and everybody was furious. And you might hear one man saying to another all day long, "By George, this is too bad." But I never could quite make out what was amiss, and I'm sure the men didn't know.'

'What was it, Oswald?'

'Never mind now. One doesn't go to Trumpeton Wood expecting to be happy there. I've half a mind to swear I'll never draw it again.'

'I've been asking him what was the matter all the way home,' said Miss Palliser, 'but I don't think he knows himself.'

'Come upstairs, Phineas, and I'll show you your room,' said Lord Chiltern. 'It's not quite as comfortable as the old "Bull", but we make it do.'

Phineas, when he was alone, could not help standing for awhile with his back to the fire thinking of it all. He did already feel himself to be at home in that house, and his doing so was a contradiction to all the wisdom which he had been endeavouring to teach himself for the last two years. He had told himself over and over again that that life which he had lived in London had been, if not a dream, at any rate not more significant than a parenthesis in his days, which, as of course it had no bearing on those which had gone before, so neither would it influence those which were to follow. The dear friends of that period of feverish success would for the future be to him as—nothing. That was the lesson of wisdom which he had endeavoured to teach himself, and the facts of the last two years had seemed to show that the lesson was a true lesson. He had disappeared from among his former companions, and had heard almost nothing from them. From neither Lord Chiltern or his wife had he received any tidings. He had expected to receive none,—had known that in the common course of things none was to be expected. There were many others with whom he had been intimate—Barrington Erle, Laurence Fitzgibbon, Mr. Monk, a politician who had been in the Cabinet, and in consequence of whose political teaching he, Phineas Finn, had banished himself from the political world;—from none of these had he received a line till there came that letter summoning him back to the battle. There had never been a time during his late life in Dublin at which he had complained to himself that on this account his former friends had forgotten him. If they had not written to him, neither had he written to them. But on his first arrival in England he had, in the sadness of his solitude, told himself that he was forgotten. There would be no return, so he feared, of those pleasant intimacies which he now remembered so well, and which, as he remembered them, were so much more replete with unalloyed delights than they had ever been in their existing realities. And yet here he was, a welcome guest in Lord Chiltern's house, a welcome guest in Lady Chiltern's drawing-room, and quite as much at home with them as ever he had been in the old days.

Who is there that can write letters to all his friends, or would not find it dreary work to do so even in regard to those whom he really loves? When there is something palpable to be said, what a blessing is the penny post!* To one's wife, to one's child, one's mistress, one's steward if there be a steward; one's gamekeeper, if there be shooting forward; one's groom, if there be hunting; one's publisher, if there be a volume ready or money needed; or one's tailor occasionally, if a coat be required, a man is able to write. But what has a man to say to his friend,—or, for that matter, what has a woman? A Horace Walpole may write to a Mr. Mann* about all things under the sun, London gossip or transcendental philosophy, and if the Horace Walpole of the occasion can write well and will labour diligently at that vocation, his letters may be worth reading by his Mr. Mann, and by others; but, for the maintenance of love and friendship, continued correspondence between distant friends is naught. Distance in time and place, but especially in time, will diminish friendship. It is a rule of nature that it should be so, and thus the friendships which a man most fosters are those which he can best enjoy. If your friend leave you, and seek a residence in Patagonia, make a niche for him in your memory, and keep him there as warm as you may. Perchance he may return from Patagonia and the old joys may be repeated. But never think that those joys can be maintained by the assistance of ocean postage, let it be at never so cheap a rate. Phineas Finn had not thought this matter out very carefully, and now, after two years of absence, he was surprised to find that he was still had in remembrance by those who had never troubled themselves to write to him a line during his absence.

When he went down into the drawing-room he was surprised to find another old friend sitting there alone. 'Mr. Finn,' said the old lady, 'I hope I see you quite well. I am glad to meet you again. You find my niece much changed, I dare say?'

'Not in the least, Lady Baldock,' said Phineas, seizing the proffered hand of the dowager. In that hour of conversation, which they had had together, Lady Chiltern had said not a word to Phineas of her aunt, and now he felt himself to be almost discomposed by the meeting. 'Is your daughter here, Lady Baldock?'

Lady Baldock shook her head solemnly and sadly. 'Do not speak of her, Mr. Finn. It is too sad! We never mention her name now.' Phineas looked as sad as he knew how to look, but he said nothing. The lamentation of the mother did not seem to imply that the daughter was dead; and, from his remembrance of Augusta Boreham, he would have thought her to be the last woman in the world to run away with the coachman. At the moment there did not seem to be any other sufficient cause for so melancholy a wagging of that venerable head. He had been told to say nothing, and he could ask no questions; but Lady Baldock did not choose that he should be left to imagine things more terrible than the truth. 'She is lost to us for ever, Mr. Finn.'

'How very sad.'

'Sad, indeed! We don't know how she took it.'

'Took what, Lady Baldock?'

'I am sure it was nothing that she ever saw at home. If there is a thing I'm true to, it is the Protestant Established Church of England. Some nasty, low, lying, wheedling priest got hold of her, and now she's a nun, and calls herself—Sister Veronica John!' Lady Baldock threw great strength and unction into her description of the priest; but as soon as she had told her story a sudden thought struck her. 'Oh, laws! I quite forgot. I beg your pardon, Mr. Finn; but you're one of them!'

'Not a nun, Lady Baldock.' At that moment the door was opened, and Lord Chiltern came in, to the great relief of his wife's aunt.

CHAPTER III

Gerard Maule

'Why didn't you tell me?' said Phineas that night after Lady Baldock was gone to bed. The two men had taken off their dress coats, and had put on smoking caps,—Lord Chiltern, indeed, having clothed himself in a wonderful Chinese dressing-gown, and they were sitting round the fire in the smoking-room; but though they were thus employed and thus dressed the two younger ladies were still with them.

'How could I tell you everything in two minutes?' said Lady Chiltern.

'I'd have given a guinea to have heard her,' said Lord Chiltern, getting up and rubbing his hands as he walked about the room. 'Can't you fancy all that she'd say, and then her horror when she'd remember that Phineas was a Papist himself?'

'But what made Miss Boreham turn nun?'

'I fancy she found the penances lighter than they were at home,' said the lord. 'They couldn't well be heavier.'

'Dear old aunt!'

'Does she never go to see Sister Veronica?' asked Miss Palliser.

'She has been once,' said Lady Chiltern.

'And fumigated herself first so as to escape infection,' said the husband. 'You should hear Gerard Maule imitate her when she talks about the filthy priest.'

'And who is Gerard Maule?' Then Lady Chiltern looked at her friend, and Phineas was almost sure that Gerard Maule was the man who was dying for Adelaide Palliser.

'He's a great ally of mine,' said Lady Chiltern.

'He's a young fellow who thinks he can ride to hounds,' said Lord Chiltern, 'and who very often does succeed in riding over them.'

'That's not fair, Lord Chiltern,' said Miss Palliser.

'Just my idea of it,' replied the Master. 'I don't think it's at all fair. Because a man has plenty of horses, and nothing else to do, and rides twelve stone, and doesn't care how he's sworn at, he's always to be over the scent, and spoil every one's sport. I don't call it at all fair.'

'He's a very nice fellow, and a great friend of Oswald's. He is to be here to-morrow, and you'll like him very much. Won't he, Adelaide?'

'I don't know Mr. Finn's tastes quite so well as you do, Violet. But Mr. Maule is so harmless that no one can dislike him very much.'

'As for being harmless, I'm not so sure,' said Lady Chiltern. After that they all went to bed.

Phineas remained at Harrington Hall till the ninth, on which day he went to London so that he might be at Tankerville on the tenth. He rode Lord Chiltern's horses, and took an interest in the hounds, and nursed the baby. 'Now tell me what you think of Gerard Maule,' Lady Chiltern asked him, the day before he started.

'I presume that he is the young man that is dying for Miss Palliser.'

'You may answer my question, Mr. Finn, without making any such suggestion.'

'Not discreetly. Of course if he is to be made happy, I am bound at the present moment to say all good things of him. At such a crisis it would be wicked to tinge Miss Palliser's hopes with any hue less warm than rose colour.'

'Do you suppose that I tell everything that is said to me?'

'Not at all; but opinions do ooze out. I take him to be a good sort of a fellow; but why doesn't he talk a bit more?'

'That's just it.'

'And why does he pretend to do nothing? When he's out he rides hard; but at other times there's a ha-ha, lack a-daisical air about him which I hate. Why men assume it I never could understand. It can recommend them to nobody. A man can't suppose that he'll gain anything by pretending that he never reads, and never thinks, and never does anything, and never speaks, and doesn't care what he has for dinner, and, upon the whole, would just as soon lie in bed all day as get up. It isn't that he is really idle. He rides and eats, and does get up, and I daresay talks and thinks. It's simply a poor affectation.'

'That's your rose colour, is it?'

'You've promised secrecy, Lady Chiltern. I suppose he's well off?'

'He is an eldest son. The property is not large, and I'm afraid there's something wrong about it.'

'He has no profession?'

'None at all. He has an allowance of £800 a year, which in some sort of fashion is independent of his father. He has nothing on earth to do. Adelaide's whole fortune is four thousand pounds. If they were to marry what would become of them?'

'That wouldn't be enough to live on?'

'It ought to be enough,—as he must, I suppose, have the property some day,—if only he had something to do. What sort of a life would he lead?'

'I suppose he couldn't become a Master of Hounds?'

'That is ill-natured, Mr. Finn.'

'I did not mean it so. I did not indeed. You must know that I did not.'

'Of course Oswald had nothing to do, and, of course, there was a time when I wished that he should take to Parliament. No one knew all that better than you did. But he was very different from Mr. Maule.'

'Very different, indeed.'

'Oswald is a man full of energy, and with no touch of that affectation which you described. As it is, he does work hard. No man works harder. The learned people say that you should produce something, and I don't suppose that he produces much. But somebody must keep hounds, and nobody could do it better than he does.'

'You don't think that I mean to blame him?'

'I hope not.'

'Are he and his father on good terms now?'

'Oh, yes. His father wishes him to go to Saulsby, but he won't do that. He hates Saulsby.'

Saulsby was the country seat of the Earl of Brentford, the name of the property which must some day belong to this Lord Chiltern, and Phineas, as he heard this, remembered former days in which he had ridden about Saulsby Woods, and had thought them to be anything but hateful. 'Is Saulsby shut up?' he asked.

'Altogether, and so is the house in Portman Square. There never was anything more sad or desolate. You would find him altered, Mr. Finn. He is quite an old man now. He was here in the spring, for a week or two;—in England, that is; but he stayed at an hotel in London. He and Laura live at Dresden now, and a very sad time they must have.'

'Does she write?'

'Yes; and keeps up all her interest about politics. I have already told her that you are to stand for Tankerville. No one,—no other human being in the world will be so interested for you as she is. If any friend ever felt an interest almost selfish for a friend's welfare, she will feel such an interest for you. If you were to succeed it would give her a hope in life.' Phineas sat silent, drinking in the words that were said to him. Though they were true, or at least meant to be true, they were full of flattery. Why should this woman of whom they were speaking love him so dearly? She was nothing to him. She was highly born, greatly gifted, wealthy, and a married woman, whose character, as he well knew, was beyond the taint of suspicion, though she had been driven by the hard sullenness of her husband to refuse to live under his roof. Phineas Finn and Lady Laura Kennedy had not seen each other for two years, and when they had parted, though they had lived as friends, there had been no signs of still living friendship. True, indeed, she had written to him, but her letters had been short and cold, merely detailing certain circumstances of her outward life. Now he was told by this woman's dearest friend that his welfare was closer to her heart than any other interest!

'I daresay you often think of her?' said Lady Chiltern.

'Indeed, I do.'

'What virtues she used to ascribe to you! What sins she forgave you! How hard she fought for you! Now, though she can fight no more, she does not think of it all the less.'

'Poor Lady Laura!'

'Poor Laura, indeed! When one sees such shipwreck it makes a woman doubt whether she ought to marry at all.'

'And yet he was a good man. She always said so.'

'Men are so seldom really good. They are so little sympathetic. What man thinks of changing himself so as to suit his wife? And yet men expect that women shall put on altogether new characters when they are married, and girls think that they can do so. Look at this Mr. Maule, who is really over head and ears in love with Adelaide Palliser. She is full of hope and energy. He has none. And yet he has the effrontery to suppose that she will adapt herself to his way of living if he marries her.'

'Then they are to be married?'

'I suppose it will come to that. It always does if the man is in earnest. Girls will accept men simply because they think it ill-natured to return the compliment of an offer with a hearty "No."'

'I suppose she likes him?'

'Of course she does. A girl almost always likes a man who is in love with her,—unless indeed she positively dislikes him. But why should she like him? He is good-looking, is a gentleman, and not a fool. Is that enough to make such a girl as Adelaide Palliser think a man divine?'

'Is nobody to be accepted who is not credited with divinity?'

'The man should be a demigod, at least in respect to some part of his character. I can find nothing even demi-divine about Mr. Maule.'

'That's because you are not in love with him, Lady Chiltern.'

Six or seven very pleasant days Phineas Finn spent at Harrington Hall, and then he started alone, and very lonely, for Tankerville. But he admitted to himself that the pleasure which he had received during his visit was quite sufficient to qualify him in running any risk in an attempt to return to the kind of life which he had formerly led. But if he should fail at Tankerville what would become of him then?

CHAPTER IV
Tankerville

THE great Mr. Molescroft himself came over to Tankerville for the purpose of introducing our hero to the electors and to Mr. Ruddles, the local Liberal agent, who was to be employed. They met at the Lambton Arms, and there Phineas established himself, knowing well that he had before him ten days of unmitigated vexation and misery. Tankerville was a dirty, prosperous, ungainly town, which seemed to exude coal-dust or coal-mud at every pore. It was so well recognised as being dirty that people did not expect to meet each other with clean hands and faces. Linen was never white at Tankerville, and even ladies who sat in drawing-rooms were accustomed to the feel and taste and appearance of soot in all their daintiest recesses. We hear that at Oil City* the flavour of petroleum is hardly considered to be disagreeable, and so it was with the flavour of coal at Tankerville. And we know that at Oil City the flavour of petroleum must not be openly declared to be objectionable, and so it was with coal at Tankerville. At Tankerville coal was much loved, and was not thought to be dirty. Mr. Ruddles was very much begrimed himself, and some of the leading Liberal electors, upon whom Phineas Finn had already called, seemed to be saturated with the product of the district. It would not, however, in any event be his duty to live at Tankerville, and he had believed from the first moment of his entrance into the town that he would soon depart from it, and know it no more. He felt that the chance of his being elected was quite a forlorn hope, and could hardly understand why he had allowed himself to be embarrassed by so very unprofitable a speculation.

Phineas Finn had thrice before this been chosen to sit in Parliament—twice for the Irish borough of Loughshane, and once for the English borough of Loughton; but he had been so happy as hitherto to have known nothing of the miseries and occasional hopelessness of a contested election. At Loughton he had come forward as the nominee of the Earl of Brentford, and had been returned without any chance of failure by that nobleman's influence. At Loughshane things had nearly been as pleasant with him. He had almost been taught to think that nothing could be easier than getting into Parliament if only a man could live when he was there. But Loughton and Loughshane were gone,* with so many other comfortable things of old days, and now he found himself relegated to a borough to which, as it seemed to him, he was sent to fight, not that he might win, but because it was necessary to his party that the seat should not be allowed to be lost without fighting. He had had the pleasant things of parliamentary adventure, and now must undergo those which were unpleasant. No doubt he could have refused, but he had listened to the tempter, and could not now go back, though Mr. Ruddles was hardly more encouraging than Mr. Molescroft.

'Browborough has been at work for the last three days,' said Mr. Ruddles, in a tone of reproach. Mr. Ruddles had always thought that no amount of work could be too heavy for his candidates.

'Will that make much difference?' asked Mr. Molescroft.

'Well, it does. Of course, he has been among the colliers,—when we ought to have been before him.'

'I came when I was told,' said Phineas.

'I'd have telegraphed to you if I'd known where you were. But there's no help for spilt milk. We must get to work now,—that's all. I suppose you're for disestablishing the Church?'

'Not particularly,' said Phineas, who felt that with him, as a Roman Catholic, this was a delicate subject.

'We needn't go into that, need we?' said Mr. Molescroft, who, though a Liberal, was a good Churchman.

Mr. Ruddles was a Dissenter, but the very strong opinion

which Mr. Ruddles now expressed as to the necessity that the new candidate should take up the Church question did not spring at all from his own religious convictions. His present duty called upon him to have a Liberal candidate if possible returned for the borough with which he was connected, and not to disseminate the doctrines of his own sect. Nevertheless, his opinion was very strong. 'I think we must, Mr. Molescroft,' said he; 'I'm sure we must. Browborough has taken up the other side. He went to church last Sunday with the Mayor and two of the Aldermen, and I'm told he said all the responses louder than anybody else. He dined with the Vicar of Trinity on Monday. He has been very loud in denouncing Mr. Finn as a Roman Catholic, and has declared that everything will be up with the State if Tankerville returns a friend and supporter of the Pope. You'll find that the Church will be the cry here this election. You can't get anything by supporting it, but you may make a strong party by pledging yourself to disendowment.'

'Wouldn't local taxation do?' asked Mr. Molescroft, who indeed preferred almost any other reform to disendowment.

'I have made up my mind that we must have some check on municipal expenditure,' said Phineas.

'It won't do—not alone. If I understand the borough, the feeling at this election will altogether be about the Church. You see, Mr. Finn, your being a Roman Catholic gives them a handle, and they're already beginning to use it. They don't like Roman Catholics here; but if you can manage to give it a sort of Liberal turn,—as many of your constituents used to do, you know,—as though you disliked Church and State rather than cared for the Pope, may be it might act on our side rather than on theirs. Mr. Molescroft understands it all.'

'Oh, yes; I understand.'

Mr. Ruddles said a great deal more to the same effect, and though Mr. Molescroft did not express any acquiescence in these views, neither did he dissent. The candidate said but little at this interview, but turned the matter over in his mind. A seat in Parliament would be but a barren honour, and he could not afford to offer his services for barren honour. Honest political work he was anxious to do, but for what work he did he desired to be paid. The party to which he belonged had, as he knew, endeavoured to avoid the subject of the disendowment of the Church of England. It is the necessary nature of a political party in this country to avoid, as long as it can be avoided, the consideration of any question which involves a great change. There is a consciousness on the minds of leading politicians that the pressure from behind, forcing upon them great measures, drives them almost quicker than they can go, so that it becomes a necessity with them to resist rather than to aid the pressure which will certainly be at last effective by its own strength. The best carriage horses are those which can most steadily hold back against the coach as it trundles down the hill. All this Phineas knew, and was of opinion that the Barrington Erles and Ratlers of his party would not thank him for ventilating a measure which, however certain might be its coming, might well be postponed for a few years. Once already in his career he had chosen to be in advance of his party, and the consequences had been disastrous to him. On that occasion his feelings had been strong in regard to the measure upon which he broke away from his party; but, when he first thought of it, he did not care much about Church disendowment.

But he found that he must needs go as he was driven or else depart out of the place. He wrote a line to his friend Erle, not to ask advice, but to explain the circumstances. 'My only possible chance of success will lie in attacking the Church endowments. Of course I think they are bad, and of course I think that they must go. But I have never cared for the matter, and would have been very willing to leave it among those things which will arrange themselves. But I have no choice here.' And so he prepared himself to run his race on the course arranged for him by Mr. Ruddles. Mr. Molescroft, whose hours were precious, soon took his leave, and Phineas Finn was placarded about the town as the sworn foe to all Church endowments.

In the course of his canvass, and the commotions consequent upon it, he found that Mr. Ruddles was right. No other subject seemed at the moment to have any attraction in Tankerville. Mr. Browborough, whose life had not been passed in any strict obedience to the Ten Commandments, and whose religious observances had not hitherto interfered with either the pleasures or the duties of his life, repeated at every meeting which he attended, and almost to every elector whom he canvassed, the great Shibboleth* which he had now adopted—'The prosperity of England depends on the Church of her people.' He was not an orator. Indeed, it might be hard to find a man, who had for years been conversant with public life, less able to string a few words together for immediate use. Nor could he learn half-a-dozen sentences by rote. But he could stand up with unabashed brow and repeat with enduring audacity the same words a dozen times over—'The prosperity of England depends on the Church of her people.' Had he been asked whether the prosperity which he promised was temporal or spiritual in its nature, not only could he not have answered, but he would not in the least have understood the question. But the words as they came from his mouth had a weight which seemed to ensure their truth, and many men in Tankerville thought that Mr. Browborough was eloquent.

Phineas, on the other hand, made two or three great speeches every evening, and astonished even Mr. Ruddles by his oratory. He had accepted Mr. Ruddles's proposition with but lukewarm acquiescence, but in the handling of the matter he became zealous, fiery, and enthusiastic. He explained to his hearers with gracious acknowledgment that Church endowments had undoubtedly been most beneficent in past times. He spoke in the interests of no special creed. Whether in the so-called Popish days of Henry VIII and his ancestors, or in the so-called Protestant days that had followed, the state of society had required that spiritual teaching should be supplied from funds fixed and devoted to the purpose. The increasing intelligence and population of the country made this no longer desirable,—or, if desirable, no longer possible. Could these endowments be increased to meet the needs of the increasing millions? Was it not the fact that even among members of the Church of England they were altogether inefficient to supply the wants of our great towns? Did the people of Tankerville believe that the clergymen of London, of Liverpool, and of Manchester were paid by endowments? The arguments which had been efficacious in Ireland must be efficacious in England. He said this without reference to one creed or to another. He did believe in religious teaching. He had not a word to say against a Protestant Episcopal Church. But he thought, nay he was sure, that Church and State, as combined institutions, could no longer prevail in this country. If the people of Tankerville would return him to Parliament it should be his first object to put an end to this anomaly.

The Browboroughites were considerably astonished by his success. The colliers on this occasion did not seem to regard the clamour that was raised against Irish Papists. Much dirt was thrown and some heads were broken; but Phineas persevered. Mr. Ruddles was lost in admiration. They had never before had at Tankerville a man who could talk so well. Mr. Browborough without ceasing repeated his well-worn assurance, and it was received with the loudest exclamations of delight by his own party. The clergymen of the town and neighbourhood crowded round him and pursued him, and almost seemed to believe in him. They were at any rate fighting their battle as best they knew how to fight it. But the great body of the colliers listened to Phineas, and every collier was now a voter.* Then Mr. Ruddles, who had many eyes, began to perceive that the old game was to be played. 'There'll be money going to-morrow after all,' he whispered to Finn the evening before the election.

'I suppose you expected that.'

'I wasn't sure. They began by thinking they could do without it. They don't want to sacrifice the borough.'

'Nor do I, Mr. Ruddles.'

'But they'll sooner do that than lose the seat. A couple of dozen of men out of the Fallgate would make us safe.' Mr. Ruddles smiled as he said this.

And Phineas smiled as he answered, 'If any good can be done by talking to the men at the Fallgate, I'll talk to them by the hour together.'

'We've about done all that,' said Mr. Ruddles.

Then came the voting. Up to two o'clock the polling was so equal that the numbers at Mr. Browborough's committee room were always given in his favour, and those at the Liberal room in favour of Phineas Finn. At three o'clock Phineas was acknowledged to be ten ahead. He himself was surprised at his own success, and declared to himself that his old luck had not deserted him.

'They're giving £2 10*s.* a vote at the Fallgate this minute,' said Ruddles to him at a quarter-past three.

'We shall have to prove it.'

'We can do that, I think,' said Ruddles.

At four o'clock, when the poll was over, Browborough was declared to have won on the post by seven votes. He was that same evening declared by the Mayor to have been elected sitting member for the borough, and he again assured the people in his speech that the prosperity of England depends on the Church of her people.

'We shall carry the seat on a scrutiny as sure as eggs,' said Mr. Ruddles, who had been quite won by the gallant way in which Phineas had fought his battle.

CHAPTER V

Mr. Daubeny's Great Move

THE whole Liberal party was taken very much by surprise at the course which the election ran. Or perhaps it might be more proper to say that the parliamentary leaders of the party were surprised. It had not been recognised by them as necessary that the great question of Church and State should be generally discussed on this occasion. It was a matter of course that it should be discussed at some places, and by some men. Eager Dissenters would, of course, take advantage of the opportunity to press their views, and no doubt the entire abolition of the Irish Church as a State establishment* had taught Liberals to think and Conservatives to fear that the question would force itself forward at no very distant date. But it had not been expected to do so now. The general incompetence of a Ministry who could not command a majority on any measure was intended to be the strong point of the Liberal party, not only at the election, but at the meeting of Parliament. The Church question, which was necessarily felt by all statesmen to be of such magnitude as to dwarf every other, was not wanted as yet. It might remain in the background as the future standing-point for some great political struggle, in which it would be again necessary that every Liberal should fight, as though for life, with his teeth and nails.

Men who ten years since regarded almost with abhorrence, and certainly with distrust, the idea of disruption between Church and State in England, were no doubt learning to perceive that such disruption must come, and were reconciling themselves to it after that slow, silent, inargumentative fashion in which convictions force themselves among us. And from reconciliation to the idea some were advancing to enthusiasm on its behalf. 'It is only a question of time,' was now said by many who hardly remembered how devoted they had been to the Established Church of England a dozen years ago. But the fruit was not yet ripe, and the leaders of the Liberal party by no means desired that it should be plucked. They were, therefore, surprised, and but little pleased, when they found that the question was more discussed than any other on the hustings of enthusiastically political boroughs.

Barrington Erle was angry when he received the letter of Phineas Finn. He was at that moment staying with the Duke of St. Bungay, who was regarded by many as the only possible leader of the Liberal party, should Mr. Gresham for any reason fail them. Indeed the old Whigs, of whom Barrington Erle considered himself to be one, would have much preferred the Duke to Mr. Gresham, had it been possible to set Mr. Gresham aside. But Mr. Gresham was too strong to be set aside; and Erle and the Duke, with all their brethren, were minded to be thoroughly loyal to their leader. He was their leader, and not to be loyal was, in their minds, treachery. But occasionally they feared that the man would carry them whither they did not desire to go. In the meantime heavy things were spoken of our poor friend, Finn.

'After all, that man is an ass,' said Erle.

'If so, I believe you are altogether responsible for him,' said the Duke.

'Well, yes, in a measure; but not altogether. That, however, is a long story. He has many good gifts. He is clever, good-tempered, and one of the pleasantest fellows that ever lived. The women all like him.'

'So the Duchess tells me.'

'But he is not what I call loyal. He cannot keep himself from running after strange gods. What need had he to take up the Church question at Tankerville? The truth is, Duke, the thing is going to pieces. We get men into the House now who are clever, and all that sort of thing, and who force their way up, but who can't be made to understand that everybody should not want to be Prime Minister.' The Duke, who was now a Nestor*among politicians, though very green in his age, smiled as he heard remarks which had been familiar to him for the last forty years. He, too, liked his party, and was fond of loyal men; but he had learned at last that all loyalty must be built on a basis of self-advantage. Patriotism may exist without it, but that which Erle called loyalty in politics was simply devotion to the side which a man conceives to be his side, and which he cannot leave without danger to himself.

But if discontent was felt at the eagerness with which this subject was taken up at certain boroughs, and was adopted by men whose votes and general support would be essentially necessary to the would-be coming Liberal Government, absolute dismay was occasioned by a speech that was made at a certain county election. Mr. Daubeny had for many years been member for East Barsetshire, and was as sure of his seat as the Queen of her throne. No one would think of contesting Mr. Daubeny's right to sit for East Barsetshire, and no doubt he might have been returned without showing himself to the electors. But he did show himself to the electors; and, as a matter of course, made a speech on the occasion. It so happened that the day fixed for the election in this division of the county was quite at the close of this period of political excitement. When Mr. Daubeny addressed his friends in East Barsetshire the returns throughout the kingdom were nearly complete. No attention had been paid to this fact during the elections, but it was afterwards asserted that the arrangement had been made with a political purpose, and with a purpose which was politically dishonest. Mr. Daubeny, so said the angry Liberals, had not chosen to address his constituents till his speech at the hustings could have no effect on other counties. Otherwise,—so said the Liberals,—the whole Conservative party would have been called upon to disavow at the hustings the conclusion to which Mr. Daubeny hinted in East Barsetshire that he had arrived. The East Barsetshire men themselves,—so said the Liberals,—had been too crass to catch the meaning hidden under his ambiguous words; but those words, when read by the light of astute criticism, were found to contain an opinion that Church and State should be dissevered. 'By G——! he's going to take the bread out of our mouths again,' said Mr. Ratler.

The speech was certainly very ambiguous, and I am not sure that the East Barsetshire folk were so crass as they were accused of being, in not understanding it at once. The dreadful hint was wrapped up in many words, and formed but a small part of a very long oration. The bucolic mind of East Barsetshire took warm delight in the eloquence of the eminent personage who represented them, but was wont to extract more actual enjoyment from the music of his periods than from the strength of his arguments. When he would explain to them that he had discovered a new, or rather hitherto unknown, Conservative element in the character of his countrymen, which he could best utilise by changing everything in the Constitution, he manipulated his words with such grace, was so profound, so broad, and so exalted, was so brilliant in mingling a deep philosophy with the ordinary politics of the day, that the bucolic mind could only admire. It was a great honour to the electors of that agricultural county that they should be made the first recipients of these pearls, which were not wasted by being thrown before them. They were picked up by the gentlemen of the Press, and became the pearls, not of East Barsetshire, but of all England. On this occasion it was found that one pearl was very big, very rare, and worthy of great attention; but it was a black pearl, and was regarded by many as an abominable prodigy. 'The period of our history is one in which it becomes essential for us to renew those inquiries which have prevailed since man first woke to his destiny, as to the amount of connection which exists and which must exist between spiritual and simply human forms of government,—between our daily religion and our daily politics, between the Crown and the Mitre.' The East Barsetshire clergymen and the East Barsetshire farmers like to hear something of the mitre in political speeches at the hustings. The word sounds pleasantly in their ears, as appertaining to good old gracious times and good old gracious things. As honey falls fast from the mouth of the practised speaker, the less practised hearer is apt to catch more of the words than of the sense. The speech of Mr. Daubeny was taken all in good part by his assembled friends. But when it was read by the quidnuncs* on the following day it was found to contain so deep a meaning that it produced from Mr. Ratler's mouth those words of fear which have been already quoted.

Could it really be the case that the man intended to perform so audacious a trick of legerdemain as this for the preservation of his power, and that if he intended it he should have the power to carry it through? The renewal of inquiry as to the connection which exists between the Crown and the Mitre, when the bran was bolted,* could only mean the disestablishment of the Church. Mr. Ratler and his friends were not long in bolting the bran. Regarding the matter simply in its own light, without bringing to bear upon it the experience of the last half-century, Mr. Ratler would have thought his party strong enough to defy Mr. Daubeny utterly in such an attempt. The ordinary politician, looking at Mr. Daubeny's position as leader of the Conservative party, as a statesman depending on the support of the Church, as a Minister appointed to his present place for the express object of defending all that was left of old, and dear, and venerable in the Constitution, would have declared that Mr. Daubeny was committing political suicide, as to which future history would record a verdict of probably not temporary insanity. And when the speech was a week old this was said in many a respectable household through the country. Many a squire, many a parson, many a farmer was grieved for Mr. Daubeny when the words had been explained to him, who did not for a moment think that the words could be portentous as to the great Conservative party. But Mr. Ratler remembered Catholic emancipation, had himself been in the House when the Corn Laws were repealed, and had been nearly brokenhearted when household suffrage* had become the law of the land while a Conservative Cabinet and a Conservative Government were in possession of dominion in Israel.*

Mr. Bonteen was disposed to think that the trick was beyond the conjuring power even of Mr. Daubeny. 'After all, you know, there is the party,' he said to Mr. Ratler. Mr. Ratler's face was as good as a play, and if seen by that party would have struck that party with dismay and shame. The meaning of Mr. Ratler's face was plain enough. He thought so little of that party, on the score either of intelligence, honesty, or fidelity, as to imagine that it would consent to be led whithersoever Mr. Daubeny might choose to lead it. 'If they care about anything, it's about the Church,' said Mr. Bonteen.

'There's something they like a great deal better than the Church,' said Mr. Ratler. 'Indeed, there's only one thing they care about at all now. They've given up all the old things. It's very likely that if Daubeny were to ask them to vote for pulling down the Throne and establishing a Republic they'd all follow him into the lobby like sheep. They've been so knocked about by one treachery after another that they don't care now for anything beyond their places.'

'It's only a few of them get anything, after all.'

'Yes, they do. It isn't just so much a year they want, though those who have that won't like to part with it. But they like getting the counties, and the Garters, and the promotion in the army. They like their brothers to be made bishops, and their sisters like the Wardrobe and the Bedchamber. There isn't one of them that doesn't hang on somewhere,—or at least not many. Do you remember Peel's bill for the Corn Laws?'*

'There were fifty went against him then,' said Bonteen.

'And what are fifty? A man doesn't like to be one of fifty. It's too many for glory, and not enough for strength. There has come up among them a general feeling that it's just as well to let things slide,—as the Yankees say. They're down-hearted about it enough within their own houses, no doubt. But what can they do, if they hold back? Some stout old cavalier here and there may shut himself up in his own castle, and tell himself that the world around him may go to wrack and ruin, but that he will not help the evil work. Some are shutting themselves up. Look at old Quin, when they carried their Reform Bill. But men, as a rule, don't like to be shut up. How they reconcile it to their conscience,—that's what I can't understand.' Such was the wisdom, and such were the fears of

Mr. Ratler. Mr. Bonteen, however, could not bring himself to believe that the Arch-enemy would on this occasion be successful. 'It mayn't be too hot for him,' said Mr. Bonteen, when he reviewed the whole matter, 'but I think it'll be too heavy.'

They who had mounted higher than Mr. Ratler and Mr. Bonteen on the political ladder, but who had mounted on the same side, were no less astonished than their inferiors; and, perhaps, were equally disgusted, though they did not allow themselves to express their disgust as plainly. Mr. Gresham was staying in the country with his friend, Lord Cantrip, when the tidings reached them of Mr. Daubeny's speech to the electors of East Barsetshire. Mr. Gresham and Lord Cantrip had long sat in the same Cabinet, and were fast friends, understanding each other's views, and thoroughly trusting each other's loyalty. 'He means it,' said Lord Cantrip.

'He means to see if it be possible,' said the other. 'It is thrown out as a feeler to his own party.'

'I'll do him the justice of saying that he's not afraid of his party. If he means it, he means it altogether, and will not retract it, even though the party should refuse as a body to support him. I give him no other credit, but I give him that.'

Mr. Gresham paused for a few moments before he answered. 'I do not know,' said he, 'whether we are justified in thinking that one man will always be the same. Daubeny has once been very audacious, and he succeeded. But he had two things to help him,—a leader, who, though thoroughly trusted, was very idle, and an ill-defined question. When he had won his leader he had won his party. He has no such tower of strength now. And in the doing of this thing, if he means to do it, he must encounter the assured conviction of every man on his own side, both in the upper and lower House. When he told them that he would tap a Conservative element by reducing the suffrage they did not know whether to believe him or not. There might be something in it. It might be that they would thus resume a class of suffrage existing in former days, but which had fallen into abeyance, because not properly protected.

They could teach themselves to believe that it might be so, and those among them who found it necessary to free their souls did so teach themselves. I don't see how they are to free their souls when they are invited to put down the State establishment of the Church.'

'He'll find a way for them.'

'It's possible. I'm the last man in the world to contest the possibility, or even the expediency, of changes in political opinion. But I do not know whether it follows that because he was brave and successful once he must necessarily be brave and successful again. A man rides at some outrageous fence, and by the wonderful activity and obedient zeal of his horse is carried over it in safety. It does not follow that his horse will carry him over a house, or that he should be fool enough to ask the beast to do so.'

'He intends to ride at the house,' said Lord Cantrip; 'and he means it because others have talked of it. You saw the line which my rash young friend Finn took at Tankerville.'

'And all for nothing.'

'I am not so sure of that. They say he is like the rest. If Daubeny does carry the party with him, I suppose the days of the Church are numbered.'

'And what if they be?' Mr. Gresham almost sighed as he said this, although he intended to express a certain amount of satisfaction. 'What if they be? You know, and I know, that the thing has to be done. Whatever may be our own individual feelings, or even our present judgment on the subject,—as to which neither of us can perhaps say that his mind is not so made up that it may not soon be altered,—we know that the present union cannot remain. It is unfitted for that condition of humanity to which we are coming, and if so, the change must be for good. Why should not he do it as well as another? Or rather would not he do it better than another, if he can do it with less of animosity than we should rouse against us? If the blow would come softer from his hands than from ours, with less of a feeling of injury to those who dearly love the Church, should we not be glad that he should undertake the task?'

'Then you will not oppose him?'

'Ah;—there is much to be considered before we can say that. Though he may not be bound by his friends, we may be bound by ours. And then, though I can hint to you at a certain condition of mind, and can sympathise with you, feeling that such may become the condition of your mind, I cannot say that I should act upon it as an established conviction, or that I can expect that you will do so. If such be the political programme submitted to us when the House meets, then we must be prepared.'

Lord Cantrip also paused a moment before he answered, but he had his answer ready. 'I can frankly say that I should follow your leading, but that I should give my voice for opposition.'

'Your voice is always persuasive,' said Mr. Gresham.

But the consternation felt among Mr. Daubeny's friends was infinitely greater than that which fell among his enemies, when those wonderful words were read, discussed, criticised, and explained. It seemed to every clergyman in England that nothing short of disestablishment could be intended by them. And this was the man to whom they had all looked for protection! This was the bulwark of the Church, to whom they had trusted! This was the hero who had been so sound and so firm respecting the Irish Establishment, when evil counsels had been allowed to prevail in regard to that ill-used but still sacred vineyard! All friends of the Church had then whispered among themselves fearfully, and had, with sad looks and grievous forebodings, acknowledged that the thin edge*of the wedge had been driven into the very rock of the Establishment. The enemies of the Church were known to be powerful, numerous, and of course unscrupulous. But surely this Brutus would not raise a dagger against this Caesar! And yet, if not, what was the meaning of those words? And then men and women began to tell each other,—the men and women who are the very salt of the earth in this England of ours,—that their Brutus, in spite of his great qualities, had ever been mysterious, unintelligible, dangerous, and given to feats of conjuring. They had only been too submissive to their Brutus. Wonderful feats of conjuring they had endured, understanding nothing of the manner in which they were performed,—nothing of their probable results; but this feat of conjuring they would not endure. And so there were many meetings held about the country, though the time for combined action was very short.

Nothing more audacious than the speaking of those few words to the bucolic electors of East Barsetshire had ever been done in the political history of England. Cromwell was bold when he closed the Long Parliament. Shaftesbury was bold when he formed the plot for which Lord Russell and others suffered.* Walpole was bold when, in his lust for power, he discarded one political friend after another. And Peel was bold when he resolved to repeal the Corn Laws. But in none of these instances was the audacity displayed more wonderful than when Mr. Daubeny took upon himself to make known throughout the country his intention of abolishing the Church of England. For to such a declaration did those few words amount. He was now the recognised parliamentary leader of that party to which the Church of England was essentially dear. He had achieved his place by skill, rather than principle, —by the conviction on men's minds that he was necessary rather than that he was fit. But still, there he was; and, though he had alarmed many,—had, probably, alarmed all those who followed him by his eccentric and dangerous mode of carrying on the battle; though no Conservative regarded him as safe; yet on this question of the Church it had been believed that he was sound. What might be the special ideas of his own mind regarding ecclesiastical policy in general, it had not been thought necessary to consider. His utterances had been confusing, mysterious, and perhaps purposely unintelligible; but that was matter of little moment so long as he was prepared to defend the establishment of the Church of England as an institution adapted for English purposes. On that point it was believed that he was sound. To that mast it was supposed he had nailed his own colours and those of his party. In defending that fortress it was thought that he would be ready to fall, should the defence of it require a fall. It was because he was so far safe that he was there. And yet he spoke these words without consulting a single friend, or suggesting the propriety of his new scheme to a single supporter. And he knew what he was doing. This was the way in which he had thought it best to make known to his own followers, not only that he was about to abandon the old Institution, but that they must do so too!

As regarded East Barsetshire itself, he was returned, and fêted, and sent home with his ears stuffed with eulogy, before the bucolic mind had discovered his purpose. On so much he had probably calculated. But he had calculated also that after an interval of three or four days his secret would be known to all friends and enemies. On the day after his speech came the report of it in the newspapers; on the next day the leading articles, in which the world was told what it was that the Prime Minister had really said. Then, on the following day, the startled parsons, and the startled squires and farmers, and, above all, the startled peers and members of the Lower House, whose duty it was to vote as he should lead them, were all agog. Could it be that the newspapers were right in this meaning which they had attached to these words? On the day week after the election in East Barsetshire, a Cabinet Council was called in London, at which it would, of course, be Mr. Daubeny's duty to explain to his colleagues what it was that he did purpose to do.

In the meantime he saw a colleague or two.

'Let us look it straight in the face,' he said to a noble colleague; 'we must look it in the face before long.'

'But we need not hurry it forward.'

'There is a storm coming. We knew that before, and we heard the sound of it from every husting in the country. How shall we rule the storm so that it may pass over the land without devastating it? If we bring in a bill——'

'A bill for disestablishing the Church!' said the horror-stricken lord.

'If we bring in a bill, the purport of which shall be to moderate the ascendancy of the Church in accordance with the existing religious feelings of the population, we shall save much that otherwise must fall. If there must be a bill, would you rather that it should be modelled by us who love the Church, or by those who hate it?'

That lord was very wrath, and told the right honourable gentleman to his face that his duty to his party should have constrained him to silence on that subject till he had consulted his colleagues. In answer to this Mr. Daubeny said with much dignity that, should such be the opinion of his colleagues in general, he would at once abandon the high place which he held in their councils. But he trusted that it might be otherwise. He had felt himself bound to communicate his ideas to his constituents, and had known that in doing so some minds must be shocked. He trusted that he might be able to allay this feeling of dismay. As regarded this noble lord, he did succeed in lessening the dismay before the meeting was over, though he did not altogether allay it.

Another gentleman who was in the habit of sitting at Mr. Daubeny's elbow daily in the House of Commons was much gentler with him, both as to words and manner. 'It's a bold throw, but I'm afraid it won't come up sixes,' said the right honourable gentleman.

'Let it come up fives, then. It's the only chance we have; and if you think, as I do, that it is essentially necessary for the welfare of the country that we should remain where we are, we must run the risk.'

With another colleague, whose mind was really set on that which the Church is presumed to represent, he used another argument. 'I am convinced at any rate of this,' said Mr. Daubeny; 'that by sacrificing something of that ascendancy which the Establishment is supposed to give us, we can bring the Church, which we love, nearer to the wants of the people.' And so it came about that before the Cabinet met, every member of it knew what it was that was expected of him.

CHAPTER VI
Phineas and his old friends

PHINEAS FINN returned from Tankerville to London in much better spirits than those which had accompanied him on his journey thither. He was not elected; but then, before the election, he had come to believe that it was quite out of the question that he should be elected. And now he did think it probable that he should get the seat on a petition. A scrutiny used to be a very expensive business, but under the existing law,* made as the scrutiny would be in the borough itself, it would cost but little; and that little, should he be successful, would fall on the shoulders of Mr. Browborough. Should he knock off eight votes and lose none himself, he would be member for Tankerville. He knew that many votes had been given for Browborough which, if the truth were known of them, would be knocked off; and he did not know that the same could be said of any one of those by which he had been supported. But, unfortunately, the judge by whom all this would be decided might not reach Tankerville in his travels till after Christmas, perhaps not till after Easter; and in the meantime, what should he do with himself?

As for going back to Dublin, that was now out of the question. He had entered upon a feverish state of existence in which it was impossible that he should live in Ireland. Should he ultimately fail in regard to his seat he must—vanish out of the world. While he remained in his present condition he would not even endeavour to think how he might in such case best bestow himself. For the present he would remain within the region of politics, and live as near as he could to the whirl of the wheel of which the sound was so dear to him. Of one club he had always remained a member, and he had already been re-elected a member of the Reform.* So he took up his residence once more at the house of a certain Mr. and Mrs. Bunce, in Great Marlborough Street, with whom he had lodged when he first became a member of Parliament.

'So you're at the old game, Mr. Finn?' said his landlord.

'Yes; at the old game. I suppose it's the same with you?' Now Mr. Bunce had been a very violent politician, and used to rejoice in calling himself a Democrat.

'Pretty much the same, Mr. Finn. I don't see that things are much better than they used to be. They tell me at the People's Banner office that the lords have had as much to do with this election as with any that ever went before it.'

'Perhaps they don't know much about it at the People's Banner office. I thought Mr. Slide and the People's Banner had gone over to the other side, Bunce?'

'Mr. Slide is pretty wide-awake whatever side he's on. Not but what he's disgraced himself by what he's been and done now.' Mr. Slide in former days had been the editor of the People's Banner, and circumstances had arisen in consequence of which there had been some acquaintance between him and our hero. 'I see you was hammering away at the Church down at Tankerville.'

'I just said a word or two.'

'You was all right, there, Mr. Finn. I can't say as I ever saw very much in your religion; but what a man keeps in the way of religion for his own use is never nothing to me;—as what I keeps is nothing to him.'

'I'm afraid you don't keep much, Mr. Bunce.'

'And that's nothing to you, neither, is it, sir?'

'No, indeed.'

'But when we read of Churches as is called State Churches, —Churches as have bishops you and I have to pay for, as never goes into them——'

'But we don't pay the bishops, Mr. Bunce.'

'Oh yes, we do; because, if they wasn't paid, the money would come to us to do as we pleased with it. We proved all that when we pared them down a bit. What's an Ecclesiastical Commission? Only another name for a box to put the money into till you want to take it out again. When we hear of Churches such as these, as is not kept up by the people who uses them,—just as the theatres are, Mr. Finn, or the gin shops,—then I know there's a deal more to be done before honest men can come by their own. You're right enough, Mr. Finn, you are, as far as churches go, and you was right, too, when you cut and run off the Treasury Bench. I hope you ain't going to sit on that stool again.'

Mr. Bunce was a privileged person, and Mrs. Bunce made up for his apparent rudeness by her own affectionate cordiality. 'Deary me, and isn't it a thing for sore eyes to have you back again! I never expected this. But I'll do for you, Mr. Finn, just as I ever did in the old days; and it was I that was sorry when I heard of the poor young lady's death; so I was, Mr. Finn; well, then, I won't mention her name never again. But after all there's been betwixt you and us it wouldn't be natural to pass it by without one word; would it, Mr. Finn? Well, yes; he's just the same man as ever, without a ha'porth of difference. He's gone on paying that shilling to the Union every week of his life, just as he used to do; and never got so much out of it, not as a junketing*into the country. That he didn't. It makes me that sick sometimes when I think of where it's gone to, that I don't know how to bear it. Well, yes; that is true, Mr. Finn. There never was a man better at bringing home his money to his wife than Bunce, barring that shilling. If he'd drink it, which he never does, I think I'd bear it better than give it to that nasty Union. And young Jack writes as well as his father, pretty nigh, Mr. Finn, which is a comfort,' —Mr. Bunce was a journeyman scrivener at a law stationer's, —'and keeps his self; but he don't bring home his money, nor yet it can't be expected, Mr. Finn. I know what the young 'uns will do, and what they won't. And Mary Jane is quite handy about the house now,—only she do break things, which is an aggravation; and the hot water shall be always up at eight o'clock to a minute, if I bring it with my own hand, Mr. Finn.'

And so he was established once more in his old rooms in Great Marlborough Street; and as he sat back in the arm-chair, which he used to know so well, a hundred memories of former days crowded back upon him. Lord Chiltern for a few months had lived with him; and then there had arisen a quarrel, which he had for a time thought would dissolve his old life into ruin. Now Lord Chiltern was again his very intimate friend. And there had used to sit a needy money-lender whom he had been unable to banish. Alas! alas! how soon might he now require that money-lender's services! And then he recollected how he had left these rooms to go into others, grander and more appropriate to his life when he had filled high office under the State. Would there ever again come to him such cause for migration? And would he again be able to load the frame of the looking-glass over the fire with countless cards from Countesses and Ministers' wives? He had opened the oyster for himself once, though it had closed again with so sharp a snap when the point of his knife had been withdrawn. Would he be able to insert the point again between those two difficult shells? Would the Countesses once more be kind to him? Would drawing-rooms be opened to him, and sometimes opened to him and to no other? Then he thought of certain special drawing-rooms in which wonderful things had been said to him. Since that he had been a married man, and those special drawing-rooms and those wonderful words had in no degree actuated him in his choice of a wife. He had left all those things of his own free will, as though telling himself that there was a better life than they offered to him. But was he sure that he had found it to be better? He had certainly sighed for the gauds* which he had left. While his young wife was living he had kept his sighs down, so that she should not hear them; but he had been forced to acknowledge that his new life had been vapid and flavourless. Now he had been tempted back again to the old haunts. Would the Countesses' cards be showered upon him again?

One card, or rather note, had reached him while he was yet at Tankerville, reminding him of old days. It was from Mrs. Low, the wife of the barrister with whom he had worked when he had been a law student in London. She had asked him to come and dine with them after the old fashion in Baker Street, naming a day as to which she presumed that he would by that time have finished his affairs at Tankerville, intimating also that Mr. Low would then have finished his at North Broughton. Now Mr. Low had sat for North Broughton before Phineas left London, and his wife spoke of the seat as a certainty. Phineas could not keep himself from feeling that Mrs. Low intended to triumph over him; but, nevertheless, he accepted the invitation. They were very glad to see him, explaining that, as nobody was supposed to be in town, nobody had been asked to meet him. In former days he had been very intimate in that house, having received from both of them much kindness, mingled, perhaps, with some touch of severity on the part of the lady. But the ground for that was gone, and Mrs. Low was no longer painfully severe. A few words were said as to his great loss. Mrs. Low once raised her eyebrows in pretended surprise when Phineas explained that he had thrown up his place, and then they settled down on the question of the day. 'And so,' said Mrs. Low, 'you've begun to attack the Church?' It must be remembered that at this moment Mr. Daubeny had not as yet electrified the minds of East Barsetshire, and that, therefore, Mrs. Low was not disturbed. To Mrs. Low, Church and State was the very breath of her nostrils; and if her husband could not be said to live by means of the same atmosphere it was because the breath of his nostrils had been drawn chiefly in the Vice-Chancellor's Court in Lincoln's Inn. But he, no doubt, would be very much disturbed indeed should he ever be told that he was required, as an expectant member of Mr. Daubeny's party, to vote for the Disestablishment of the Church of England.

'You don't mean that I am guilty of throwing the first stone?' said Phineas.

'They have been throwing stones at the Temple since first it was built,' said Mrs. Low, with energy; 'but they have fallen off its polished shafts in dust and fragments.' I am afraid that Mrs. Low, when she allowed herself to speak thus energetically, entertained some confused idea that the Church of England and the Christian religion were one and the same thing, or, at least, that they had been brought into the world together.

'You haven't thrown the first stone,' said Mr. Low; 'but you have taken up the throwing at the first moment in which stones may be dangerous.'

'No stones can be dangerous,' said Mrs. Low.

'The idea of a State Church,' said Phineas, 'is opposed to my theory of political progress. What I hope is that my friends will not suppose that I attack the Protestant Church because I am a Roman Catholic. If I were a priest it would be my business to do so; but I am not a priest.'

Mr. Low gave his old friend a bottle of his best wine, and in all friendly observances treated him with due affection. But neither did he nor did his wife for a moment abstain from attacking their guest in respect to his speeches at Tankerville. It seemed, indeed, to Phineas that as Mrs. Low was buckled up in such triple armour that she feared nothing, she might have been less loud in expressing her abhorrence of the enemies of the Church. If she feared nothing, why should she scream so loudly? Between the two he was a good deal crushed and confounded, and Mrs. Low was very triumphant when she allowed him to escape from her hands at ten o'clock. But, at that moment, nothing had as yet been heard in Baker Street of Mr. Daubeny's proposition to the electors of East Barsetshire! Poor Mrs. Low! We can foresee that there is much grief in store for her, and some rocks ahead, too, in the political career of her husband.

Phineas was still in London, hanging about the clubs, doing nothing, discussing Mr. Daubeny's wonderful treachery with such men as came up to town, and waiting for the meeting of Parliament, when he received the following letter from Lady Laura Kennedy:—

'Dresden, November 18, ——

'MY DEAR MR. FINN,

'I have heard with great pleasure from my sister-in-law that you have been staying with them at Harrington Hall. It seems so like old days that you and Oswald and Violet should be together,—so much more natural than that you should be living in Dublin. I cannot conceive of you as living any other life than that of the House of Commons, Downing Street, and the clubs. Nor do I wish to do so. And when I hear of you at Harrington Hall I know that you are on your way to the other things.

'Do tell me what life is like with Oswald and Violet. Of course he never writes. He is one of those men who, on marrying, assume that they have at last got a person to do a duty which has always hitherto been neglected. Violet does write, but tells me little or nothing of themselves. Her letters are very nice, full of anecdote, well written,—letters that are fit to be kept and printed; but they are never family letters. She is inimitable in discussing the miseries of her own position as the wife of a Master of Hounds; but the miseries are as evidently fictitious as the art is real. She told me how poor dear Lady Baldock communicated to you her unhappiness about her daughter in a manner that made even me laugh; and would make thousands laugh in days to come were it ever to be published. But of her inside life, of her baby, or of her husband as a husband, she never says a word. You will have seen it all, and have enough of the feminine side of a man's character to be able to tell me how they are living. I am sure they are happy together, because Violet has more common sense than any woman I ever knew.

'And pray tell me about the affair at Tankerville. My cousin Barrington writes me word that you will certainly get the seat. He declares that Mr. Browborough is almost disposed not to fight the battle, though a man more disposed to fight never bribed an elector. But Barrington seems to think that you managed as well as you did by getting outside the traces,* as he calls it. We certainly did not think that you would come out strong against the Church. Don't suppose that I complain. For myself I hate to think of the coming severance; but if it must come, why not by your hands as well as by any other? It is hardly possible that you in your heart should love a Protestant ascendant Church. But, as Barrington says, a horse won't get oats unless he works steady between the traces.

'As to myself, what am I to say to you? I and my father live here a sad, sombre, solitary life, together. We have a large furnished house outside the town, with a pleasant view and a pretty garden. He does—nothing. He reads the English papers, and talks of English parties, is driven out, and eats his dinner, and sleeps. At home, as you know, not only did he take an active part in politics, but he was active also in the management of his own property. Now it seems to him to be almost too great a trouble to write a letter to his steward; and all this has come upon him because of me. He is here because he cannot bear that I should live alone. I have offered to return with him to Saulsby, thinking that Mr. Kennedy would trouble me no further,—or to remain here by myself; but he will consent to neither. In truth the burden of idleness has now fallen upon him so heavily that he cannot shake it off. He dreads that he may be called upon to do anything.

'To me it is all one tragedy. I cannot but think of things as they were two or three years since. My father and my husband were both in the Cabinet, and you, young as you were, stood but one step below it. Oswald was out in the cold. He was very poor. Papa thought all evil of him. Violet had refused him over and over again. He quarrelled with you, and all the world seemed against him. Then of a sudden you vanished, and we vanished. An ineffable misery fell upon me and upon my wretched husband. All our good things went from us at a blow. I and my poor father became as it were outcasts. But Oswald suddenly retricked his beams, and is flaming in the forehead of the morning sky.* He, I believe, has no more than he had deserved. He won his wife honestly;—did he not? And he has ever been honest. It is my pride to think I never gave him up. But the bitter part of my cup consists in this,—that as he has won what he has deserved, so have we. I complain of no injustice. Our castle was built upon the sand. Why should Mr. Kennedy have been a Cabinet Minister;—and why should I have been his wife? There is no one else of whom I can ask that question as I can of you, and no one else who can answer it as you can do.

'Of Mr. Kennedy it is singular how little I know, and how little I ever hear. There is no one whom I can ask to tell me of him. That he did not attend during the last Session I do know, and we presume that he has now abandoned his seat. I fear that his health is bad,—or perhaps, worse still, that his mind is affected by the gloom of his life. I suppose that he lives exclusively at Loughlinter. From time to time I am implored by him to return to my duty beneath his roof. He grounds his demand on no affection of his own, on no presumption that any affection can remain with me. He says no word of happiness. He offers no comfort. He does not attempt to persuade with promises of future care. He makes his claim simply on Holy Writ, and on the feeling of duty which thence ought to weigh upon me. He has never even told me that he loves me; but he is persistent in declaring that those whom God has joined together nothing human should separate. Since I have been here I have written to him once,—one sad, long, weary letter. Since that I am constrained to leave his letters unanswered.

'And now, my friend, could you not do for me a great kindness? For a while, till the inquiry be made at Tankerville, your time must be vacant. Cannot you come and see us? I have told Papa that I should ask you, and he would be delighted. I cannot explain to you what it would be to me to be able to talk again to one who knows all the errors and all the efforts of my past life as you do. Dresden is very cold in the winter. I do not know whether you would mind that. We are very particular about the rooms, but my father bears the temperature wonderfully well, though he complains. In March we move down south for a couple of months. Do come if you can.

'Most sincerely yours,

'LAURA KENNEDY.

'If you come, of course you will have yourself brought direct to us. If you can learn anything of Mr. Kennedy's life, and of his real condition, pray do. The faint rumours which reach me are painfully distressing.'

CHAPTER VII

Coming Home from Hunting

LADY CHILTERN was probably right when she declared that her husband must have been made to be a Master of Hounds,—presuming it to be granted that somebody must be Master of Hounds. Such necessity certainly does exist in this, the present condition of England. Hunting prevails; hunting men increase in numbers; foxes are preserved; farmers do not rebel; owners of coverts, even when they are not hunting men themselves, acknowledge the fact, and do not dare to maintain their pheasants at the expense of the much better-loved four-footed animal. Hounds are bred, and horses are trained specially to the work. A master of fox hounds is a necessity of the period. Allowing so much, we cannot but allow also that Lord Chiltern must have been made to fill the situation. He understood hunting, and, perhaps, there was nothing else requiring acute intelligence that he did understand. And he understood hunting, not only as a huntsman understands it,—in that branch of the science which refers simply to the judicious pursuit of the fox, being probably inferior to his own huntsman in that respect,—but he knew exactly what men should do, and what they should not. In regard to all those various interests with which he was brought in contact, he knew when to hold fast to his own claims, and when to make no claims at all. He was afraid of no one, but he was possessed of a sense of justice which induced him to acknowledge the rights of those around him. When he found that the earths were not stopped in Trumpeton Wood,—from which he judged that the keeper would complain that the hounds would not or could not kill any of the cubs found there,—he wrote in very round terms to the Duke who owned it. If His Grace did not want to have the wood drawn,* let him say so. If he did, let him have the earths stopped. But when that great question came up as to the Gartlow coverts—when that uncommonly disagreeable gentleman, Mr. Smith, of Gartlow, gave notice that the hounds should not be admitted into his place at all,—Lord Chiltern soon put the whole matter straight by taking part with the disagreeable gentleman. The disagreeable gentleman had been ill used. Men had ridden among his young laurels. If gentlemen who did hunt,—so said Lord Chiltern to his own supporters,—did not know how to conduct themselves in a matter of hunting, how was it to be expected that a gentleman who did not hunt should do so? On this occasion Lord Chiltern rated his own hunt so roundly that Mr. Smith and he were quite in a bond together, and the Gartlow coverts were re-opened. Now all the world knows that the Gartlow coverts, though small, are material as being in the very centre of the Brake country.

It is essential that a Master of Hounds should be somewhat feared by the men who ride with him. There should be much awe mixed with the love felt for him. He should be a man with whom other men will not care to argue; an irrational, cut and thrust, unscrupulous, but yet distinctly honest man; one who can be tyrannical, but will tyrannise only over the evil spirits; a man capable of intense cruelty to those alongside of him, but who will know whether his victim does in truth deserve scalping before he draws his knife. He should be savage and yet good-humoured; severe and yet forbearing; truculent and pleasant in the same moment. He should exercise unflinching authority, but should do so with the consciousness that he can support it only by his own popularity. His speech should be short, incisive, always to the point, but never founded on argument. His rules are based on no reason, and will never bear discussion. He must be the most candid of men, also the most close;—and yet never a hypocrite. He must condescend to no explanation, and yet must impress men with an assurance that his decisions will certainly be right. He must rule all as though no man's special welfare were of any account, and yet must administer all so as to offend none. Friends he must have, but not favourites. He must be self-sacrificing, diligent, eager, and watchful. He must be strong in health, strong in heart, strong in purpose, and strong in purse. He must be economical and yet lavish; generous as the wind and yet obdurate as the frost. He should be assured that of all human pursuits hunting is the best, and that of all living things a fox is the most valuable. He must so train his heart as to feel for the fox a mingled tenderness and cruelty which is inexplicable to ordinary men and women. His desire to preserve the brute and then to kill him should be equally intense and passionate. And he should do it all in accordance with a code of unwritten laws, which cannot be learnt without profound study. It may not perhaps be truly asserted that Lord Chiltern answered this description in every detail; but he combined so many of the qualities required that his wife showed her discernment when she declared that he seemed to have been made to be a Master of Hounds.

Early in that November he was riding home with Miss Palliser by his side, while the huntsmen and whips were trotting on with the hounds before him. 'You call that a good run, don't you?'

'No; I don't.'

'What was the matter with it? I declare it seems to me that something is always wrong. Men like hunting better than anything else, and yet I never find any man contented.'

'In the first place we didn't kill.'

'You know you're short of foxes at Gartlow,' said Miss Palliser, who, as is the manner with all hunting ladies, liked to show that she understood the affairs of the hunt.

'If I knew there were but one fox in a county, and I got upon that one fox, I would like to kill that one fox,—barring a vixen in March.'

'I thought it very nice. It was fast enough for anybody.'

'You might go as fast with a drag,* if that's all. I'll tell you something else. We should have killed him if Maule hadn't once ridden over the hounds when we came out of the little wood. I spoke very sharply to him.'

'I heard you, Lord Chiltern.'

'And I suppose you thought I was a brute.'

'Who? I? No, I didn't;—not particularly, you know. Men do say such things to each other!'

'He doesn't mind it, I fancy.'

'I suppose a man does not like to be told that directly he shows himself in a run the sport is all over and the hounds ought to be taken home.'

'Did I say that? I don't remember now what I said, but I know he made me angry. Come, let us trot on. They can take the hounds home without us.'

'Good night, Cox,' said Miss Palliser, as they passed by the pack. 'Poor Mr. Maule! I did pity him, and I do think he does care for it, though he is so impassive. He would be with us now, only he is chewing the cud of his unhappiness in solitude half a mile behind us.'

'That is hard upon you.'

'Hard upon me, Lord Chiltern! It is hard upon him, and, perhaps, upon you. Why should it be hard upon me?'

'Hard upon him, I should have said. Though why it shouldn't be the other way I don't know. He's a friend of yours.'

'Certainly.'

'And an especial friend, I suppose. As a matter of course Violet talks to me about you both.'

'No doubt she does. When once a woman is married she should be regarded as having thrown off her allegiance to her own sex. She is sure to be treacherous at any rate in one direction. Not that Lady Chiltern can tell anything of me that might not be told to all the world as far as I am concerned.'

'There is nothing in it, then?'

'Nothing at all.'

'Honour bright?'

'Oh,—honour as bright as it ever is in such matters as these.'

'I am sorry for that,—very sorry.'

'Why so, Lord Chiltern?'

'Because if you were engaged to him I thought that

perhaps you might have induced him to ride a little less forward.'

'Lord Chiltern,' said Miss Palliser, seriously; 'I will never again speak to you a word on any subject except hunting.'

At this moment Gerard Maule came up behind them, with a cigar in his mouth, apparently quite unconscious of any of that displeasure as to which Miss Palliser had supposed that he was chewing the cud in solitude. 'That was a goodish thing, Chiltern,' he said.

'Very good.'

'And the hounds hunted him well to the end.'

'Very well.'

'It's odd how the scent will die away at a moment. You see they couldn't carry on a field after we got out of the copse.'

'Not a field.'

'Considering all things I am glad we didn't kill him.'

'Uncommon glad,' said Lord Chiltern. Then they trotted on in silence a little way, and Maule again dropped behind. 'I'm blessed if he knows that I spoke to him, roughly,' said Chiltern. 'He's deaf, I think, when he chooses to be.'

'You're not sorry, Lord Chiltern.'

'Not in the least. Nothing will ever do any good. As for offending him, you might as well swear at a tree, and think to offend it. There's comfort in that, anyway. I wonder whether he'd talk to you if I went away?'

'I hope that you won't try the experiment.'

'I don't believe he would, or I'd go at once. I wonder whether you really do care for him?'

'Not in the least.'

'Or he for you.'

'Quite indifferent, I should say; but I can't answer for him, Lord Chiltern, quite as positively as I can for myself. You know, as things go, people have to play at caring for each other.'

'That's what we call flirting.'

'Just the reverse. Flirting I take to be the excitement of

love, without its reality, and without its ordinary result in marriage. This playing at caring has none of the excitement, but it often leads to the result, and sometimes ends in downright affection.'

'If Maule perseveres then you'll take him, and by-and-bye you'll come to like him.'

'In twenty years it might come to that, if we were always to live in the same house; but as he leaves Harrington tomorrow, and we may probably not meet each other for the next four years, I think the chance is small.'

Then Maule trotted up again, and after riding in silence with the other two for half an hour, he pulled out his case and lit a fresh cigar from the end of the old one, which he threw away. 'Have a baccy, Chiltern?' he said.

'No, thank you, I never smoke going home; my mind is too full. I've all that family behind to think of, and I'm generally out of sorts with the miseries of the day. I must say another word to Cox, or I should have to go to the kennels on my way home.' And so he dropped behind.

Gerard Maule smoked half his cigar before he spoke a word, and Miss Palliser was quite resolved that she would not open her mouth till he had spoken. 'I suppose he likes it?' he said at last.

'Who likes what, Mr. Maule?'

'Chiltern likes blowing fellows up.'

'It's a part of his business.'

'That's the way I look at it. But I should think it must be disagreeable. He takes such a deal of trouble about it. I heard him going on to-day to some one as though his whole soul depended on it.'

'He is very energetic.'

'Just so. I'm quite sure it's a mistake. What does a man ever get by it? Folks around you soon discount it till it goes for nothing.'

'I don't think energy goes for nothing, Mr. Maule.'

'A bull in a china shop is not a useful animal, nor is he ornamental, but there can be no doubt of his energy. The hare was full of energy, but he didn't win the race. The man who stands still is the man who keeps his ground.'

'You don't stand still when you're out hunting.'

'No;—I ride about, and Chiltern swears at me. Every man is a fool sometimes.'

'And your wisdom, perfect at all other times, breaks down in the hunting-field?'

'I don't in the least mind your chaffing. I know what you think of me just as well as though you told me.'

'What do I think of you?'

'That I'm a poor creature, generally half asleep, shallow-pated, slow-blooded, ignorant, useless, and unambitious.'

'Certainly unambitious, Mr. Maule.'

'And that word carries all the others. What's the good of ambition? There's the man they were talking about last night, —that Irishman.'

'Mr. Finn?'

'Yes; Phineas Finn. He is an ambitious fellow. He'll have to starve, according to what Chiltern was saying. I've sense enough to know I can't do any good.'

'You are sensible, I admit.'

'Very well, Miss Palliser. You can say just what you like, of course. You have that privilege.'

'I did not mean to say anything severe. I do admit that you are master of a certain philosophy, for which much may be said. But you are not to expect that I shall express an approval which I do not feel.'

'But I want you to approve it.'

'Ah!—there, I fear, I cannot oblige you.'

'I want you to approve it, though no one else may.'

'Though all else should do so, I cannot.'

'Then take the task of curing the sick one, and of strengthening the weak one, into your own hands. If you will teach, perhaps I may learn.'

'I have no mission for teaching, Mr. Maule.'

'You once said that,—that——'

'Do not be so ungenerous as to throw in my teeth what I

once said,—if I ever said a word that I would not now repeat.'

'I do not think that I am ungenerous, Miss Palliser.'

'I am sure you are not.'

'Nor am I self-confident. I am obliged to seek comfort from such scraps of encouragement as may have fallen in my way here and there. I once did think that you intended to love me.'

'Does love go by intentions?'

'I think so,—frequently with men, and much more so with girls.'

'It will never go so with me. I shall never intend to love any one. If I ever love any man it will be because I am made to do so, despite my intentions.'

'As a fortress is taken?'

'Well,—if you like to put it so. Only I claim this advantage, —that I can always get rid of my enemy when he bores me.'

'Am I boring you now?'

'I didn't say so. Here is Lord Chiltern again, and I know by the rattle of his horse's feet that something is the matter.'

Lord Chiltern came up full of wrath. One of the men's horses was thoroughly broken down, and, as the Master said, wasn't worth the saddle he carried. He didn't care a —— for the horse, but the man hadn't told him. 'At this rate there won't be anything to carry anybody by Christmas.'

'You'll have to buy some more,' said Gerard Maule.

'Buy some more!' said Lord Chiltern, turning round, and looking at the man. 'He talks of buying horses as he would sugar plums!' Then they trotted in at the gate, and in two minutes were at the hall door.

CHAPTER VIII
The Address

BEFORE the 11th of November, the day on which Parliament was to meet, the whole country was in a hubbub. Consternation and triumph were perhaps equally predominant, and equally strong. There were those who declared that now at length was Great Britain to be ruined in actual present truth; and those who asserted that, of a sudden, after a fashion so wholly unexpected as to be divine,—as great fires, great famines, and great wars are called divine,—a mighty hand had been stretched out to take away the remaining incubus of superstition, priestcraft, and bigotry under which England had hitherto been labouring. The proposed disestablishment of the State Church of England was, of course, the subject of this diversity of opinion.

And there was not only diversity, but with it great confusion. The political feelings of the country are, as a rule, so well marked that it is easy, as to almost every question, to separate the sheep from the goats. With but few exceptions one can tell where to look for the supporters and where for the opponents of one measure or of another. Meetings are called in this or in that public hall to assist or to combat the Minister of the day, and men know what they are about. But now it was not so. It was understood that Mr. Daubeny, the accredited leader of the Conservatives, was about to bring in the bill, but no one as yet knew who would support the bill. His own party, to a man,—without a single exception,—were certainly opposed to the measure in their minds. It must be so. It could not but be certain that they should hate it. Each individual sitting on the Conservative side in either House did most certainly within his own bosom cry Ichabod* when the fatal news reached his ears. But such private opinions and inward wailings need not, and probably would not, guide the body. Ichabod had been cried before, though probably never with such intensity of feeling. Disestablishment might be worse than Free Trade* or Household Suffrage, but was not more absolutely opposed to Conservative convictions than had been those great measures. And yet the party, as a party, had swallowed them both. To the first and lesser evil, a compact little body of staunch Commoners had stood forth in opposition,—but nothing had come of it to those true Britons beyond a feeling of living in the cold shade of exclusion. When the greater evil arrived, that of Household Suffrage,—a measure which twenty years since would hardly have been advocated by the advanced Liberals of the day,—the Conservatives had learned to acknowledge the folly of clinging to their own convictions, and had swallowed the dose without serious disruption of their ranks. Every man,—with but an exception or two,—took the measure up, some with faces so singularly distorted as to create true pity, some with an assumption of indifference, some with affected glee. But in the double process the party had become used to this mode of carrying on the public service. As poor old England must go to the dogs, as the doom had been pronounced against the country that it should be ruled by the folly of the many foolish, and not by the wisdom of the few wise, why should the few wise remain out in the cold,—seeing, as they did, that by so doing no good would be done to the country? Dissensions among their foes did, when properly used, give them power, —but such power they could only use by carrying measures which they themselves believed to be ruinous. But the ruin would be as certain should they abstain. Each individual might have gloried in standing aloof,—in hiding his face beneath his toga, and in remembering that Rome did once exist in her splendour. But a party cannot afford to hide its face in its toga. A party has to be practical. A party can only live by having its share of Garters, lord-lieutenants, bishops, and attorney-generals. Though the country were ruined, the party should be supported. Hitherto the party had been supported, and had latterly enjoyed almost its share of stars and Garters,—thanks to the individual skill and strategy of that great English political Von Moltke*Mr. Daubeny.

And now what would the party say about the disestablishment of the Church? Even a party must draw the line somewhere. It was bad to sacrifice things mundane; but this thing was the very Holy of Holies! Was nothing to be conserved by a Conservative party? What if Mr. Daubeny were to explain some day to the electors of East Barsetshire that an hereditary peerage was an absurdity? What if in some rural nook of his Bœotia*he should suggest in ambiguous language to the farmers that a Republic was the only form of Government capable of a logical defence? Duke had already said to Duke, and Earl to Earl, and Baronet to Baronet that there must be a line somewhere. Bishops as a rule say but little to each other, and now were afraid to say anything. The Church, which had been, which was, so truly beloved;—surely that must be beyond the line! And yet there crept through the very marrow of the party an agonising belief that Mr. Daubeny would carry the bulk of his party with him into the lobby of the House of Commons.

But if such was the dismay of the Conservatives, how shall any writer depict the consternation of the Liberals? If there be a feeling odious to the mind of a sober, hardworking man, it is the feeling that the bread he has earned is to be taken out of his mouth. The pay, the patronage, the powers, and the pleasure of Government were all due to the Liberals. 'God bless my soul,' said Mr. Ratler, who always saw things in a practical light, 'we have a larger fighting majority than any party has had since Lord Liverpool's time.* They have no right to attempt it. They are bound to go out.' 'There's nothing of honesty left in politics,' said Mr. Bonteen, declaring that he was sick of the life. Barrington Erle thought that the whole Liberal party should oppose the measure. Though they were Liberals they were not democrats; nor yet infidels. But when Barrington Erle said this, the great leaders of the Liberal party had not as yet decided on their ground of action.

There was much difficulty in reaching any decision. It had been asserted so often that the disestablishment of the Church was only a question of time, that the intelligence of the country had gradually so learned to regard it. Who had said so, men did not know and did not inquire;—but the words were spoken everywhere. Parsons with sad hearts,—men who in their own parishes were enthusiastic, pure, pious, and useful, —whispered them in the dead of the night to the wives of their bosoms. Bishops, who had become less pure by contact with the world at clubs, shrugged their shoulders and wagged their heads, and remembered comfortably the sanctity of vested interests. Statesmen listened to them with politeness, and did not deny that they were true. In the free intercourse of closest friendships the matter was discussed between ex-Secretaries of State. The Press teemed with the assertion that it was only a question of time. Some fervent, credulous friends predicted another century of life;—some hard-hearted logical opponents thought that twenty years would put an end to the anomaly:—a few stout enemies had sworn on the hustings with an anathema that the present Session should see the deposition from her high place of this eldest daughter of the woman of Babylon.* But none had expected the blow so soon as this; and none certainly had expected it from this hand.

But what should the Liberal party do? Ratler was for opposing Mr. Daubeny with all their force, without touching the merits of the case. It was no fitting work for Mr. Daubeny, and the suddenness of the proposition coming from such a quarter would justify them now and for ever, even though they themselves should disestablish everything before the Session were over. Barrington Erle, suffering under a real political conviction for once in his life, was desirous of a positive and chivalric defence of the Church. He believed in the twenty years. Mr. Bonteen shut himself up in disgust. Things were amiss; and, as he thought, the evil was due to want of party zeal on the part of his own leader, Mr. Gresham. He did not dare to say this, lest, when the house door should at last be opened, he might not be invited to enter with the others; but such was his conviction. 'If we were all a little less in the abstract, and a little more in the concrete, it would be better for us.' Laurence Fitzgibbon, when these words had been whispered to him by Mr. Bonteen, had hardly understood them; but it had been explained to him that his friend had meant 'men, not measures.'* When Parliament met, Mr. Gresham, the leader of the Liberal party, had not as yet expressed any desire to his general followers.

The Queen's Speech was read, and the one paragraph which seemed to possess any great public interest was almost a repetition of the words which Mr. Daubeny had spoken to the electors of East Barsetshire. 'It will probably be necessary for you to review the connection which still exists between, and which binds together, the Church and the State.' Mr. Daubeny's words had of course been more fluent, but the gist of the expression was the same. He had been quite in earnest when addressing his friends in the country. And though there had been but an interval of a few weeks, the Conservative party in the two Houses heard the paragraph read without surprise and without a murmur. Some said that the gentlemen on the Treasury Bench in the House of Commons did not look to be comfortable. Mr. Daubeny sat with his hat over his brow, mute, apparently impassive and unapproachable, during the reading of the Speech and the moving and seconding of the Address. The House was very full, and there was much murmuring on the side of the Opposition;—but from the Government benches hardly a sound was heard, as a young gentleman, from one of the Midland counties, in a deputy-lieutenant's uniform, who had hitherto been known for no particular ideas of his own, but had been believed to be at any rate true to the Church, explained, not in very clear language, that the time had at length come when the interests of religion demanded a wider support and a fuller sympathy than could be afforded under that system of Church endowment and State establishment for which the country had hitherto been so grateful, and for which the country had such boundless occasion for gratitude. Another gentleman, in the uniform of the Guards, seconded the Address, and declared that in nothing was the sagacity of a Legislature so necessary as in discerning the period in which that which had hitherto been good ceased to be serviceable. The status pupillaris* was mentioned, and it was understood that he had implied that England was now old enough to go on in matters of religion without a tutor in the shape of a State Church.

Who makes the speeches, absolutely puts together the words, which are uttered when the Address is moved and seconded? It can hardly be that lessons are prepared and sent to the noble lords and honourable gentlemen to be learned by heart like a school-boy's task. And yet, from their construction, style, and general tone,—from the platitudes which they contain as well as from the general safety and good sense of the remarks,—from the absence of any attempt to improve a great occasion by the fire of oratory, one cannot but be convinced that a very absolute control is exercised. The gorgeously apparelled speakers, who seem to have great latitude allowed them in the matter of clothing, have certainly very little in the matter of language. And then it always seems that either of the four might have made the speech of any of the others. It could not have been the case that the Hon. Colonel Mowbray Dick, the Member for West Bustard, had really elaborated out of his own head that theory of the status pupillaris. A better fellow, or a more popular officer, or a sweeter-tempered gentleman than Mowbray Dick does not exist; but he certainly never entertained advanced opinions respecting the religious education of his country. When he is at home with his family, he always goes to church, and there has been an end of it.

And then the fight began. The thunderbolts of opposition were unloosed, and the fires of political rancour blazed high. Mr. Gresham rose to his legs, and declared to all the world that which he had hitherto kept secret from his own party. It was known afterwards that in discussion with his own dearly-beloved political friend, Lord Cantrip, he had expressed his unbounded anger at the duplicity, greed for power, and want of patriotism displayed by his opponent; but he had acknowledged that the blow had come so quick and so unexpectedly that he thought it better to leave the matter to the House without instruction from himself. He now revelled in sarcasm, and before his speech was over raged into wrath. He would move an amendment to the Address for two reasons,—first because this was no moment for bringing before Parliament the question of the Church establishment, when as yet no well-considered opportunity of expressing itself on the subject had been afforded to the country, and secondly because any measure of reform on that matter should certainly not come to them from the right honourable gentleman opposite. As to the first objection, he should withhold his arguments till the bill suggested had been presented to them. It was in handling the second that he displayed his great power of invective. All those men who then sat in the House, and who on that night crowded the galleries, remember his tones as, turning to the dissenters who usually supported him, and pointing over the table to his opponents, he uttered that well-

worn quotation, *Quod minime reris,*—then he paused, and began again; *Quod minime reris,—Graiâ pandetur ab urbe.**
The power and inflexion of his voice at the word *Graiâ* were certainly very wonderful. He ended by moving an amendment to the Address, and asking for support equally from one side of the House as from the other.

When at length Mr. Daubeny moved his hat from his brow and rose to his legs he began by expressing his thankfulness that he had not been made a victim to the personal violence of the right honourable gentleman. He continued the same strain of badinage throughout,—in which he was thought to have been wrong, as it was a method of defence, or attack, for which his peculiar powers hardly suited him. As to any bill that was to be laid upon the table, he had not as yet produced it. He did not doubt that the dissenting interests of the country would welcome relief from an anomaly, let it come whence it might, even *Graiâ ab urbe,* and he waved his hand back to the clustering Conservatives who sat behind him. That the right honourable gentleman should be angry he could understand, as the return to power of the right honourable gentleman and his party had been anticipated, and he might almost say discounted as a certainty.

Then, when Mr. Daubeny sat down, the House was adjourned.

CHAPTER IX

The Debate

THE beginning of the battle as recorded in the last chapter took place on a Friday,—Friday, 11th November,—and consequently two entire days intervened before the debate could be renewed. There seemed to prevail an opinion during this interval that Mr. Gresham had been imprudent. It was acknowledged by all men that no finer speech than that delivered by him had ever been heard within the walls of that

House. It was acknowledged also that as regarded the question of oratory Mr. Daubeny had failed signally. But the strategy of the Minister was said to have been excellent, whereas that of the ex-Minister was very loudly condemned. There is nothing so prejudicial to a cause as temper. This man is declared to be unfit for any position of note, because he always shows temper. Anything can be done with another man,—he can be made to fit almost any hole,—because he has his temper under command. It may, indeed, be assumed that a man who loses his temper while he is speaking is endeavouring to speak the truth such as he believes it to be, and again it may be assumed that a man who speaks constantly without losing his temper is not always entitled to the same implicit faith. Whether or not this be a reason the more for preferring the calm and tranquil man may be doubted; but the calm and tranquil man is preferred for public services. We want practical results rather than truth. A clear head is worth more than an honest heart. In a matter of horseflesh of what use is it to have all manner of good gifts if your horse won't go whither you want him, and refuses to stop when you bid him? Mr. Gresham had been very indiscreet, and had especially sinned in opposing the Address without arrangements with his party.

And he made the matter worse by retreating within his own shell during the whole of that Saturday, Sunday, and Monday morning. Lord Cantrip was with him three or four times, and he saw both Mr. Palliser, who had been Chancellor of the Exchequer under him, and Mr. Ratler. But he went amidst no congregation of Liberals, and asked for no support. He told Ratler that he wished gentlemen to vote altogether in accordance with their opinions; and it came to be whispered in certain circles that he had resigned, or was resigning, or would resign, the leadership of his party. Men said that his passions were too much for him, and that he was destroyed by feelings of regret, and almost of remorse.

The Ministers held a Cabinet Council on the Monday morning, and it was supposed afterwards that that also had been stormy. Two gentlemen had certainly resigned their seats in the Government before the House met at four o'clock, and there were rumours abroad that others would do so if the suggested measure should be found really to amount to disestablishment. The rumours were, of course, worthy of no belief, as the transactions of the Cabinet are of necessity secret. Lord Drummond at the War Office, and Mr. Boffin from the Board of Trade, did, however, actually resign; and Mr. Boffin's explanations in the House were heard before the debate was resumed. Mr. Boffin had certainly not joined the present Ministry,—so he said,—with the view of destroying the Church. He had no other remark to make, and he was sure that the House would appreciate the course which had induced him to seat himself below the gangway. The House cheered very loudly, and Mr. Boffin was the hero of ten minutes. Mr. Daubeny detracted something from this triumph by the overstrained and perhaps ironic pathos with which he deplored the loss of his right honourable friend's services. Now this right honourable gentleman had never been specially serviceable.

But the wonder of the world arose from the fact that only two gentlemen out of the twenty or thirty who composed the Government did give up their places on this occasion. And this was a Conservative Government! With what a force of agony did all the Ratlers of the day repeat that inappropriate name! Conservatives! And yet they were ready to abandon the Church at the bidding of such a man as Mr. Daubeny! Ratler himself almost felt that he loved the Church. Only two resignations;—whereas it had been expected that the whole House would fall to pieces! Was it possible that these earls, that marquis, and the two dukes, and those staunch old Tory squires, should remain in a Government pledged to disestablish the Church? Was all the honesty, all the truth of the great party confined to the bosoms of Mr. Boffin and Lord Drummond? Doubtless they were all Esaus; but would they sell their great birthright for so very small a mess of pottage? The parsons in the country, and the little squires who but rarely come up to London, spoke of it all exactly as did the Ratlers. There were parishes in the country in which Mr. Boffin was canonised, though up to that date no Cabinet Minister could well have been less known to fame than was Mr. Boffin.

What would those Liberals do who would naturally rejoice in the disestablishment of the Church,—those members of the Lower House, who had always spoken of the ascendancy of Protestant episcopacy with the bitter acrimony of exclusion? After all, the success or failure of Mr. Daubeny must depend, not on his own party, but on them. It must always be so when measures of Reform are advocated by a Conservative Ministry. There will always be a number of untrained men ready to take the gift without looking at the giver. They have not expected relief from the hands of Greeks, but will take it when it comes from Greeks or Trojans. What would Mr. Turnbull say in this debate,—and what Mr. Monk? Mr. Turnbull was the people's tribune, of the day; Mr. Monk had also been a tribune, then a Minister, and now was again—something less than a tribune. But there were a few men in the House, and some out of it, who regarded Mr. Monk as the honestest and most patriotic politician of the day.

The debate was long and stormy, but was peculiarly memorable for the skill with which Mr. Daubeny's higher colleagues defended the steps they were about to take. The thing was to be done in the cause of religion. The whole line of defence was indicated by the gentlemen who moved and seconded the Address. An active, well-supported Church was the chief need of a prosperous and intelligent people. As to the endowments, there was some confusion of ideas; but nothing was to be done with them inappropriate to religion. Education would receive the bulk of what was left after existing interests had been amply guaranteed. There would be no doubt,—so said these gentlemen,—that ample funds for the support of an Episcopal Church would come from those wealthy members of the body to whom such a Church was dear. There seemed to be a conviction that clergymen under the new order of things would be much better off than under the old. As to the connection with the State, the time for it had clearly gone by. The Church, as a Church, would own increased power when it could appoint its own bishops, and be wholly dissevered from State patronage. It seemed to be almost a matter of surprise that really good Churchmen should have endured so long to be shackled by subservience to the State. Some of these gentlemen pleaded their cause so well that they almost made it appear that episcopal ascendancy would be restored in England by the disseverance of the Church and State.

Mr. Turnbull, who was himself a dissenter, was at last upon his legs, and then the Ratlers knew that the game was lost. It would be lost as far as it could be lost by a majority in that House on that motion; and it was by that majority or minority that Mr. Daubeny would be maintained in his high office or ejected from it. Mr. Turnbull began by declaring that he did not at all like Mr. Daubeny as a Minister of the Crown. He was not in the habit of attaching himself specially to any Minister of the Crown. Experience had taught him to doubt them all. Of all possible Ministers of the Crown at this period, Mr. Daubeny was he thought perhaps the worst, and the most dangerous. But the thing now offered was too good to be rejected, let it come from what quarter it would. Indeed, might it not be said of all the good things obtained for the people, of all really serviceable reforms, that they were gathered and garnered home in consequence of the squabbles of Ministers? When men wanted power, either to grasp at it or to retain it, then they offered bribes to the people. But in the taking of such bribes there was no dishonesty, and he should willingly take this bribe.

Mr. Monk spoke also. He would not, he said, feel himself justified in refusing the Address to the Crown proposed by Ministers, simply because that Address was founded on the proposition of a future reform, as to the expediency of which he had not for many years entertained a doubt. He could not allow it to be said of him that he had voted for the permanence of the Church establishment, and he must therefore support the Government. Then Ratler whispered a few words to his neighbour: 'I knew the way he'd run when Gresham insisted on poor old Mildmay's taking him into the Cabinet.' 'The whole thing has gone to the dogs,' said Bonteen. On the fourth night the House was divided, and Mr. Daubeny was the owner of a majority of fifteen.

Very many of the Liberal party expressed an opinion that the battle had been lost through the want of judgment evinced by Mr. Gresham. There was certainly no longer that sturdy adherence to their chief which is necessary for the solidarity of a party. Perhaps no leader of the House was ever more devoutly worshipped by a small number of adherents than was Mr. Gresham now; but such worship will not support power. Within the three days following the division the Ratlers had all put their heads together and had resolved that the Duke of St. Bungay was now the only man who could keep the party together. 'But who should lead our House?' asked Bonteen. Ratler sighed instead of answering. Things had come to that pass that Mr. Gresham was the only possible leader. And the leader of the House of Commons, on behalf of the Government, must be the chief man in the Government, let the so-called Prime Minister be who he may.

CHAPTER X

The deserted Husband

PHINEAS FINN had been in the gallery of the House throughout the debate, and was greatly grieved at Mr. Daubeny's success, though he himself had so strongly advocated the disestablishment of the Church in canvassing the electors of Tankerville. No doubt he had advocated the cause,—but he had done so as an advanced member of the Liberal party, and he regarded the proposition when coming from Mr. Daubeny as a horrible and abnormal birth. He, however, was only a looker-on,—could be no more than a looker-on for the existing short session. It had already been decided that the judge who was to try the case at Tankerville should visit that town early in January; and should it be decided on a scrutiny that the seat belonged to our hero, then he would enter upon his

privilege in the following Session without any further trouble to himself at Tankerville. Should this not be the case,—then the abyss of absolute vacuity would be open before him. He would have to make some disposition of himself, but he would be absolutely without an idea as to the how or where. He was in possession of funds to support himself for a year or two; but after that, and even during that time, all would be dark. If he should get his seat, then again the power of making an effort would at last be within his hands.

He had made up his mind to spend the Christmas with Lord Brentford and Lady Laura Kennedy at Dresden, and had already fixed the day of his arrival there. But this had been postponed by another invitation which had surprised him much, but which it had been impossible for him not to accept. It had come as follows:—

'November 9th, Loughlinter.

'Dear Sir,

'I am informed by letter from Dresden that you are in London on your way to that city with the view of spending some days with the Earl of Brentford. You will, of course, be once more thrown into the society of my wife, Lady Laura Kennedy.

'I have never understood, and certainly have never sanctioned, that breach of my wife's marriage vow which has led to her withdrawal from my roof. I never bade her go, and I have bidden her return. Whatever may be her feelings, or mine, her duty demands her presence here, and my duty calls upon me to receive her. This I am and always have been ready to do. Were the laws of Europe sufficiently explicit and intelligible I should force her to return to my house,—because she sins while she remains away, and I should sin were I to omit to use any means which the law might place in my hands for the due control of my own wife. I am very explicit to you although we have of late been strangers, because in former days you were closely acquainted with the condition of my family affairs.

'Since my wife left me I have had no means of communicating with her by the assistance of any common friend. Having heard that you are about to visit her at Dresden I feel a great desire to see you that I may be enabled to send by you a personal message. My health, which is now feeble, and the altered habits of my life render it almost impossible that I should proceed to London with this object, and I therefore ask it of your Christian charity that you should visit me here at Loughlinter. You, as a Roman Catholic, cannot but hold the bond of matrimony to be irrefragable. You cannot, at least, think that it should be set aside at the caprice of an excitable woman who is not able and never has been able to assign any reason for leaving the protection of her husband.

'I shall have much to say to you, and I trust you will come. I will not ask you to prolong your visit, as I have nothing to offer you in the way of amusement. My mother is with me; but otherwise I am alone. Since my wife left me I have not thought it even decent to entertain guests or to enjoy society. I have lived a widowed life. I cannot even offer you shooting, as I have no keepers on the mountains. There are fish in the river doubtless, for the gifts of God are given let men be ever so unworthy; but this, I believe, is not the month for fishermen. I ask you to come to me, not as a pleasure, but as a Christian duty.

'Yours truly,

'ROBERT KENNEDY.'

'Phineas Finn, Esq.'

As soon as he had read the letter Phineas felt that he had no alternative but to go. The visit would be very disagreeable, but it must be made. So he sent a line to Robert Kennedy naming a day; and wrote another to Lady Laura postponing his time at Dresden by a week, and explaining the cause of its postponement. As soon as the debate on the Address was over he started for Loughlinter.

A thousand memories crowded on his brain as he made the journey. Various circumstances had in his early life,—in that period of his life which had lately seemed to be cut off from the remainder of his days by so clear a line,—thrown him into close connection with this man, and with the man's wife. He had first gone to Loughlinter, not as Lady Laura's guest,—for Lady Laura had not then been married, or even engaged to be married,—but on her persuasion rather than on that of Mr. Kennedy. When there he had asked Lady Laura to be his own wife, and she had then told him that she was to become the wife of the owner of that domain. He remembered the blow as though it had been struck but yesterday, and yet the pain of the blow had not been long enduring. But though then rejected he had always been the chosen friend of the woman,—a friend chosen after an especial fashion. When he had loved another woman this friend had resented his defection with all a woman's jealousy. He had saved the husband's life, and had then become also the husband's friend, after that cold fashion which an obligation will create. Then the husband had been jealous, and dissension had come, and the ill-matched pair had been divided, with absolute ruin to both of them, as far as the material comforts and well-being of life were concerned. Then he, too, had been ejected, as it were, out of the world, and it had seemed to him as though Laura Standish and Robert Kennedy had been the inhabitants of another hemisphere. Now he was about to see them both again, both separately; and to become the medium of some communication between them. He knew, or thought that he knew, that no communication could avail anything.

It was dark night when he was driven up to the door of Loughlinter House in a fly* from the town of Callender. When he first made the journey, now some six or seven years since, he had done so with Mr. Ratler, and he remembered well that circumstance. He remembered also that on his arrival Lady Laura had scolded him for having travelled in such company. She had desired him to seek other friends,—friends higher in general estimation, and nobler in purpose. He had done so, partly at her instance, and with success. But Mr. Ratler was now somebody in the world, and he was nobody. And he remembered also how on that occasion he had been troubled in his mind in regard to a servant, not as yet knowing whether the usages of the world did or did not require that he should go so accompanied. He had taken the man, and had been thoroughly ashamed of himself for doing so. He had no servant now, no grandly developed luggage, no gun, no elaborate dress for the mountains. On that former occasion his heart had been very full when he reached Loughlinter, and his heart was full now. Then he had resolved to say a few words to Lady Laura, and he had hardly known how best to say them. Now he would be called upon to say a few to Lady Laura's husband, and the task would be almost as difficult.

The door was opened for him by an old servant in black,

who proposed at once to show him to his room. He looked round the vast hall, which, when he had before known it, was ever filled with signs of life, and felt at once that it was empty and deserted. It struck him as intolerably cold, and he saw that the huge fireplace was without a spark of fire. Dinner, the servant said, was prepared for half-past seven. Would Mr. Finn wish to dress? Of course he wished to dress. And as it was already past seven he hurried up stairs to his room. Here again everything was cold and wretched. There was no fire, and the man had left him with a single candle. There were candlesticks on the dressing-table, but they were empty. The man had suggested hot water, but the hot water did not come. In his poorest days he had never known discomfort such as this, and yet Mr. Kennedy was one of the richest commoners of Great Britain.

But he dressed, and made his way down stairs, not knowing where he should find his host or his host's mother. He recognised the different doors and knew the rooms within them, but they seemed inhospitably closed against him, and he went and stood in the cold hall. But the man was watching for him, and led him into a small parlour. Then it was explained to him that Mr. Kennedy's state of health did not admit of late dinners. He was to dine alone, and Mr. Kennedy would receive him after dinner. In a moment his cheeks became red, and a flash of wrath crossed his heart. Was he to be treated in this way by a man on whose behalf,—with no thought of his own comfort or pleasure,—he had made this long and abominable journey? Might it not be well for him to leave the house without seeing Mr. Kennedy at all? Then he remembered that he had heard it whispered that the man had become bewildered in his mind. He relented, therefore, and condescended to eat his dinner.

A very poor dinner it was. There was a morsel of flabby white fish, as to the nature of which Phineas was altogether in doubt, a beef steak as to the nature of which he was not at all in doubt, and a little crumpled-up tart which he thought the driver of the fly must have brought with him from the pastrycook's at Callender. There was some very hot sherry, but not much of it. And there was a bottle of claret, as to which Phineas, who was not usually particular in the matter of wine, persisted in declining to have anything to do with it after the first attempt. The gloomy old servant, who stuck to him during the repast, persisted in offering it, as though the credit of the hospitality of Loughlinter depended on it. There are so many men by whom the tenuis ratio saporum* has not been achieved, that the Caleb Balderstones* of those houses in which plenty does not flow are almost justified in hoping that goblets of Gladstone* may pass current. Phineas Finn was not a martyr to eating or drinking. He played with his fish without thinking much about it. He worked manfully at the steak. He gave another crumple to the tart, and left it without a pang. But when the old man urged him, for the third time, to take that pernicious draught with his cheese, he angrily demanded a glass of beer. The old man toddled out of the room, and on his return he proffered to him a diminutive glass of white spirit, which he called usquebaugh. Phineas, happy to get a little whisky, said nothing more about the beer, and so the dinner was over.

He rose so suddenly from his chair that the man did not dare to ask him whether he would not sit over his wine. A suggestion that way was indeed made, would he 'visit the laird out o' hand, or would he bide awee?' Phineas decided on visiting the laird out of hand, and was at once led across the hall, down a back passage which he had never before traversed, and introduced to the chamber which had ever been known as the 'laird's ain room.' Here Robert Kennedy rose to receive him.

Phineas knew the man's age well. He was still under fifty, but he looked as though he were seventy. He had always been thin, but he was thinner now than ever. He was very grey, and stooped so much, that though he came forward a step or two to greet his guest, it seemed as though he had not taken the trouble to raise himself to his proper height. 'You find me a much altered man,' he said. The change had been so great that it was impossible to deny it, and Phineas muttered something of regret that his host's health should be so bad. 'It is trouble of the mind,—not of the body, Mr. Finn. It is her doing,—her doing. Life is not to me a light thing, nor are the obligations of life light. When I married a wife, she became bone of my bone, and flesh of my flesh. Can I lose my bones and my flesh,—knowing that they are not with God but still subject elsewhere to the snares of the devil, and live as though I were a sound man? Had she died I could have borne it. I hope they have made you comfortable, Mr. Finn?'

'Oh, yes,' said Phineas.

'Not that Loughlinter can be comfortable now to any one. How can a man, whose wife has deserted him, entertain his guests? I am ashamed even to look a friend in the face, Mr. Finn.' As he said this he stretched forth his open hand as though to hide his countenance, and Phineas hardly knew whether the absurdity of the movement or the tragedy of the feeling struck him the more forcibly. 'What did I do that she should leave me? Did I strike her? Was I faithless? Had she not the half of all that was mine? Did I frighten her by hard words, or exact hard tasks? Did I not commune with her, telling her all my most inward purposes? In things of this world, and of that better world that is coming, was she not all in all to me? Did I not make her my very wife? Mr. Finn, do you know what made her go away?' He had asked perhaps a dozen questions. As to the eleven which came first it was evident that no answer was required; and they had been put with that pathetic dignity with which it is so easy to invest the interrogatory form of address. But to the last question it was intended that Phineas should give an answer, as Phineas presumed at once; and then it was asked with a wink of the eye, a low eager voice, and a sly twist of the face that were frightfully ludicrous. 'I suppose you do know,' said Mr. Kennedy, again working his eye, and thrusting his chin forward.

'I imagine that she was not happy.'

'Happy? What right had she to expect to be happy? Are we to believe that we should be happy here? Are we not told that we are to look for happiness there, and to hope for none below?' As he said this he stretched his left hand to the ceiling.

'But why shouldn't she have been happy? What did she want? Did she ever say anything against me, Mr. Finn?'

'Nothing but this,—that your temper and hers were incompatible.'

'I thought at one time that you advised her to go away?'

'Never!'

'She told you about it?'

'Not, if I remember, till she had made up her mind, and her father had consented to receive her. I had known, of course, that things were unpleasant.'

'How were they unpleasant? Why were they unpleasant? She wouldn't let you come and dine with me in London. I never knew why that was. When she did what was wrong, of course I had to tell her. Who else should tell her but her husband? If you had been her husband, and I only an acquaintance, then I might have said what I pleased. They rebel against the yoke because it is a yoke. And yet they accept the yoke, knowing it to be a yoke. It comes of the devil. You think a priest can put everything right.'

'No, I don't,' said Phineas.

'Nothing can put you right but the fear of God; and when a woman is too proud to ask for that, evils like these are sure to come. She would not go to church on Sunday afternoon, but had meetings of Belial* at her father's house instead.' Phineas well remembered those meetings of Belial, in which he with others had been wont to discuss the political prospects of the day. 'When she persisted in breaking the Lord's commandment, and defiling the Lord's day, I knew well what would come of it.'

'I am not sure, Mr. Kennedy, that a husband is justified in demanding that a wife shall think just as he thinks on matters of religion. If he is particular about it, he should find all that out before.'

'Particular! God's word is to be obeyed, I suppose?'

'But people doubt about God's word.'

'Then people will be damned,' said Mr. Kennedy, rising from his chair. 'And they will be damned.'

'A woman doesn't like to be told so.'

'I never told her so. I never said anything of the kind. I never spoke a hard word to her in my life. If her head did but ache, I hung over her with the tenderest solicitude. I refused her nothing. When I found that she was impatient I chose the shortest sermon for our Sunday evening's worship, to the great discomfort of my mother.' Phineas wondered whether this assertion as to the discomfort of old Mrs. Kennedy could possibly be true. Could it be that any human being really preferred a long sermon to a short one,—except the being who preached it or read it aloud? 'There was nothing that I did not do for her. I suppose you really do know why she went away, Mr. Finn?'

'I know nothing more than I have said.'

'I did think once that she was——'

'There was nothing more than I have said,' asserted Phineas sternly, fearing that the poor insane man was about to make some suggestion that would be terribly painful. 'She felt that she did not make you happy.'

'I did not want her to make me happy. I do not expect to be made happy. I wanted her to do her duty. You were in love with her once, Mr. Finn?'

'Yes, I was. I was in love with Lady Laura Standish.'

'Ah! Yes. There was no harm in that, of course; only when any thing of that kind happens, people had better keep out of each other's way afterwards. Not that I was ever jealous, you know.'

'I should hope not.'

'But I don't see why you should go all the way to Dresden to pay her a visit. What good can that do? I think you had much better stay where you are, Mr. Finn; I do indeed. It isn't a decent thing for a young unmarried man to go half across Europe to see a lady who is separated from her husband, and who was once in love with him;—I mean he was once in love with her. It's a very wicked thing, Mr. Finn, and I have to beg that you will not do it.'

Phineas felt that he had been grossly taken in. He had been

asked to come to Loughlinter in order that he might take a message from the husband to the wife, and now the husband made use of his compliance to forbid the visit on some grotesque score of jealousy. He knew that the man was mad, and that therefore he ought not to be angry; but the man was not too mad to require a rational answer, and had some method in his madness. 'Lady Laura Kennedy is living with her father,' said Phineas.

'Pshaw;—dotard!'

'Lady Laura Kennedy is living with her father,' repeated Phineas; 'and I am going to the house of the Earl of Brentford.'

'Who was it wrote and asked you?'

'The letter was from Lady Laura.'

'Yes;—from my wife. What right had my wife to write to you when she will not even answer my appeals? She is my wife;—my wife! In the presence of God she and I have been made one, and even man's ordinances have not dared to separate us. Mr. Finn, as the husband of Lady Laura Kennedy, I desire that you abstain from seeking her presence.' As he said this he rose from his chair, and took the poker in his hand. The chair in which he was sitting was placed upon the rug, and it might be that the fire required his attention. As he stood bending down, with the poker in his right hand, with his eye still fixed on his guest's face, his purpose was doubtful. The motion might be a threat, or simply have a useful domestic tendency. But Phineas, believing that the man was mad, rose from his seat and stood upon his guard. The point of the poker had undoubtedly been raised; but as Phineas stretched himself to his height, it fell gradually towards the fire, and at last was buried very gently among the coals. But he was never convinced that Mr. Kennedy had carried out the purpose with which he rose from his chair. 'After what has passed, you will no doubt abandon your purpose,' said Mr. Kennedy.

'I shall certainly go to Dresden,' said Phineas. 'If you have a message to send, I will take it.'

'Then you will be accursed among adulterers,' said the laird of Loughlinter. 'By such a one I will send no message. From the first moment that I saw you I knew you for a child of Apollyon.* But the sin was my own. Why did I ask to my house an idolater, one who pretends to believe that a crumb of bread is my God, a Papist, untrue alike to his country and to his Saviour? When she desired it of me I knew that I was wrong to yield. Yes;—it is you who have done it all, you, you, you;—and if she be a castaway, the weight of her soul will be doubly heavy on your own.'

To get out of the room, and then at the earliest possible hour of the morning out of the house, were now the objects to be attained. That his presence had had a peculiarly evil influence on Mr. Kennedy, Phineas could not doubt; as assuredly the unfortunate man would not have been left with mastery over his own actions had his usual condition been such as that which he now displayed. He had been told that 'poor Kennedy' was mad,—as we are often told of the madness of our friends when they cease for awhile to run in the common grooves of life. But the madman had now gone a long way out of the grooves;—so far, that he seemed to Phineas to be decidedly dangerous. 'I think I had better wish you good night,' he said.

'Look here, Mr. Finn.'

'Well?'

'I hope you won't go and make more mischief.'

'I shall not do that, certainly.'

'You won't tell her what I have said?'

'I shall tell her nothing to make her think that your opinion of her is less high than it ought to be.'

'Good night.'

'Good night,' said Phineas again; and then he left the room. It was as yet but nine o'clock, and he had no alternative but to go to bed. He found his way back into the hall, and from thence up to his own chamber. But there was no fire there, and the night was cold. He went to the window, and raised it for a moment, that he might hear the well-remembered sound of the Fall of Linter. Though the night was dark and wintry, a dismal damp November night, he would have crept out of the house and made his way up to the top of the brae, for the sake of auld lang syne, had he not feared that the inhospitable mansion would be permanently closed against him on his return. He rang the bell once or twice, and after a while the old serving man came to him. Could he have a cup of tea? The man shook his head, and feared that no boiling water could be procured at that late hour of the night. Could he have his breakfast the next morning at seven, and a conveyance to Callender at half-past seven? When the old man again shook his head, seeming to be dazed at the enormity of the demand, Phineas insisted that his request should be conveyed to the master of the house. As to the breakfast, he said he did not care about it, but the conveyance he must have. He did, in fact, obtain both, and left the house early on the following morning without again seeing Mr. Kennedy, and without having spoken a single word to Mr. Kennedy's mother. And so great was his hurry to get away from the place which had been so disagreeable to him, and which he thought might possibly become more so, that he did not even run across the sward that divided the gravel sweep from the foot of the waterfall.

CHAPTER XI

The truant wife

PHINEAS on his return to London wrote a line to Lady Chiltern in accordance with a promise which had been exacted from him. She was anxious to learn something as to the real condition of her husband's brother-in-law, and, when she heard that Phineas was going to Loughlinter, had begged that he would tell her the truth. 'He has become eccentric, gloomy, and very strange,' said Phineas. 'I do not believe that he is really mad, but his condition is such that I think no friend should recommend Lady Laura to return to him. He seems to have devoted himself to a gloomy religion,—and to the saving of money. I had but one interview with him, and that was essentially disagreeable.' Having remained two days in

London, and having participated, as far as those two days would allow him, in the general horror occasioned by the wickedness and success of Mr. Daubeny, he started for Dresden.

He found Lord Brentford living in a spacious house, with a huge garden round it, close upon the northern confines of the town. Dresden, taken altogether, is a clean cheerful city, and strikes the stranger on his first entrance as a place in which men are gregarious, busy, full of merriment, and pre-eminently social. Such is the happy appearance of but few towns either in the old or the new world, and is hardly more common in Germany than elsewhere. Leipsic is decidedly busy, but does not look to be social. Vienna is sufficiently gregarious, but its streets are melancholy. Munich is social, but lacks the hum of business. Frankfort is both practical and picturesque, but it is dirty, and apparently averse to mirth. Dresden has much to recommend it, and had Lord Brentford with his daughter come abroad in quest of comfortable easy social life, his choice would have been well made. But, as it was, any of the towns above named would have suited him as well as Dresden, for he saw no society, and cared nothing for the outward things of the world around him. He found Dresden to be very cold in the winter and very hot in the summer, and he liked neither heat nor cold; but he had made up his mind that all places, and indeed all things, are nearly equally disagreeable, and therefore he remained at Dresden, grumbling almost daily as to the climate and manners of the people.

Phineas, when he arrived at the hall door, almost doubted whether he had not been as wrong in visiting Lord Brentford as he had in going to Loughlinter. His friendship with the old Earl had been very fitful, and there had been quarrels quite as pronounced as the friendship. He had often been happy in the Earl's house, but the happiness had not sprung from any love for the man himself. How would it be with him if he found the Earl hardly more civil to him than the Earl's son-in-law had been? In former days the Earl had been a man quite capable of making himself disagreeable, and probably had not yet lost the power of doing so. Of all our capabilities this is the one which clings longest to us. He was thinking of all this when he found himself at the door of the Earl's house. He had travelled all night, and was very cold. At Leipsic there had been a nominal twenty minutes for refreshment, which the circumstances of the station had reduced to five. This had occurred very early in the morning, and had sufficed only to give him a bowl of coffee. It was now nearly ten, and breakfast had become a serious consideration with him. He almost doubted whether it would not have been better for him to have gone to an hotel in the first instance.

He soon found himself in the hall amidst a cluster of servants, among whom he recognised the face of a man from Saulsby. He had, however, little time allowed him for looking about. He was hardly in the house before Lady Laura Kennedy was in his arms. She had run forward, and before he could look into her face, she had put up her cheek to his lips and had taken both his hands. 'Oh, my friend,' she said; 'oh, my friend! How good you are to come to me! How good you are to come!' And then she led him into a large room, in which a table had been prepared for breakfast, close to an English-looking open fire. 'How cold you must be, and how hungry! Shall I have breakfast for you at once, or will you dress first? You are to be quite at home, you know; exactly as though we were brother and sister. You are not to stand on any ceremonies.' And again she took him by the hand. He had hardly looked her yet in the face, and he could not do so now because he knew that she was crying. 'Then I will show you to your room,' she said, when he had decided for a tub of water before breakfast. 'Yes, I will,—my own self. And I'd fetch the water for you, only I know it is there already. How long will you be? Half an hour? Very well. And you would like tea best, wouldn't you?'

'Certainly, I should like tea best.'

'I will make it for you. Papa never comes down till near two, and we shall have all the morning for talking. Oh, Phineas, it is such a pleasure to hear your voice again. You have been at Loughlinter?'

'Yes, I have been there.'

'How very good of you; but I won't ask a question now. You must put up with a stove here, as we have not open fires in the bed-rooms. I hope you will be comfortable. Don't be more than half an hour, as I shall be impatient.'

Though he was thus instigated to haste he stood a few minutes with his back to the warm stove that he might be enabled to think of it all. It was two years since he had seen this woman, and when they had parted there had been more between them of the remembrances of old friendship than of present affection. During the last few weeks of their intimacy she had made a point of telling him that she intended to separate herself from her husband; but she had done so as though it were a duty, and an arranged part of her own defence of her own conduct. And in the latter incidents of her London life,—that life with which he had been conversant,—she had generally been opposed to him, or, at any rate, had chosen to be divided from him. She had said severe things to him,—telling him that he was cold, heartless, and uninterested, never trying even to please him with that sort of praise which had once been so common with her in her intercourse with him, and which all men love to hear from the mouths of women. She had then been cold to him, though she would make wretched allusions to the time when he, at any rate, had not been cold to her. She had reproached him, and had at the same time turned away from him. She had repudiated him, first as a lover, then as a friend; and he had hitherto never been able to gauge the depth of the affection for him which had underlaid all her conduct. As he stood there thinking of it all, he began to understand it.

How natural had been her conduct on his arrival, and how like that of a genuine, true-hearted, honest woman! All her first thoughts had been for his little personal wants,—that he should be warmed, and fed, and made outwardly comfortable. Let sorrow be ever so deep, and love ever so true, a man will be cold who travels by winter, and hungry who has travelled by night. And a woman, who is a true, genuine woman, always takes delight in ministering to the natural wants of her friend. To see a man eat and drink, and wear his slippers, and sit at ease in his chair, is delightful to the feminine heart that loves. When I heard the other day that a girl had herself visited the room prepared for a man in her mother's house, then I knew that she loved him, though I had never before believed it. Phineas, as he stood there, was aware that this woman loved him dearly. She had embraced him, and given her face to him to kiss. She had clasped his hands, and clung to him, and had shown him plainly that in the midst of all her sorrow she could be made happy by his coming. But he was a man far too generous to take all this as meaning aught that it did not mean,—too generous, and intrinsically too manly. In his character there was much of weakness, much of vacillation, perhaps some deficiency of strength and purpose; but there was no touch of vanity. Women had loved him, and had told him so; and he had been made happy, and also wretched, by their love. But he had never taken pride, personally, to himself because they had loved him. It had been the accident of his life. Now he remembered chiefly that this woman had called herself his sister, and he was grateful.

Then he thought of her personal appearance. As yet he had hardly looked at her, but he felt that she had become old and worn, angular and hard-visaged. All this had no effect upon his feelings towards her, but filled him with ineffable regret. When he had first known her she had been a woman with a noble presence—not soft and feminine as had been Violet Effingham, but handsome and lustrous, with a healthy youth. In regard to age he and she were of the same standing. That he knew well. She had passed her thirty-second birthday, but that was all. He felt himself to be still a young man, but he could not think of her as of a young woman.

When he went down she had been listening for his footsteps, and met him at the door of the room. 'Now sit down,' she said, 'and be comfortable—if you can, with German surroundings. They are almost always late, and never give one any time. Everybody says so. The station at Leipsic is dreadful, I know. Good coffee is very well, but what is the use of good coffee if you have no time to drink it? You must eat our omelette. If there is one thing we can do better than you it is to make an omelette. Yes,—that is genuine German sausage. There is always some placed upon the table, but the Germans who come here never touch it themselves. You will have a cutlet, won't you? I breakfasted an hour ago, and more. I would not wait because then I thought I could talk to you better, and wait upon you. I did not think that anything would ever please me so much again as your coming has done. Oh, how much we shall have to say! Do you remember when we last parted;—when you were going back to Ireland?'

'I remember it well.'

'Ah me; as I look back upon it all, how strange it seems. I dare say you don't remember the first day I met you, at Mr. Mildmay's,—when I asked you to come to Portman Square because Barrington had said that you were clever?'

'I remember well going to Portman Square.'

'That was the beginning of it all. Oh dear, oh dear; when I think of it I find it so hard to see where I have been right, and where I have been wrong. If I had not been very wrong all this evil could not have come upon me.'

'Misfortune has not always been deserved.'

'I am sure it has been so with me. You can smoke here if you like.' This Phineas persistently refused to do. 'You may if you please. Papa never comes in here, and I don't mind it. You'll settle down in a day or two, and understand the extent of your liberties. Tell me first about Violet. She is happy?'

'Quite happy, I think.'

'I knew he would be good to her. But does she like the kind of life?'

'Oh, yes.'

'She has a baby, and therefore of course she is happy. She says he is the finest fellow in the world.'

'I dare say he is. They all seem to be contented with him, but they don't talk much about him.'

'No; they wouldn't. Had you a child you would have talked

about him, Phineas. I should have loved my baby better than all the world, but I should have been silent about him. With Violet of course her husband is the first object. It would certainly be so from her nature. And so Oswald is quite tame?'

'I don't know that he is very tame out hunting.'

'But to her?'

'I should think always. She, you know, is very clever.'

'So clever!'

'And would be sure to steer clear of all offence,' said Phineas, enthusiastically.

'While I could never for an hour avoid it. Did they say anything about the journey to Flanders?'

'Chiltern did, frequently. He made me strip my shoulder to show him the place where he hit me.'

'How like Oswald!'

'And he told me that he would have given one of his eyes to kill me, only Colepepper wouldn't let him go on. He half quarrelled with his second, but the man told him that I had not fired at him, and the thing must drop. "It's better as it is, you know," he said. And I agreed with him.'

'And how did Violet receive you?'

'Like an angel,—as she is.'

'Well, yes. I'll grant she is an angel now. I was angry with her once, you know. You men find so many angels in your travels. You have been honester than some. You have generally been off with the old angel before you were on with the new,—as far at least as I knew.'

'Is that meant for rebuke, Lady Laura?'

'No, my friend; no. That is all over. I said to myself when you told me that you would come, that I would not utter one ill-natured word. And I told myself more than that!'

'What more?'

'That you had never deserved it,—at least from me. But surely you were the most simple of men.'

'I dare say.'

'Men when they are true are simple. They are often false as hell, and then they are crafty as Lucifer. But the man who is true judges others by himself,—almost without reflection. A woman can be true as steel and cunning at the same time. How cunning was Violet, and yet she never deceived one of her lovers, even by a look. Did she?'

'She never deceived me,—if you mean that. She never cared a straw about me, and told me so to my face very plainly.'

'She did care,—many straws. But I think she always loved Oswald. She refused him again and again, because she thought it wrong to run a great risk, but I knew she would never marry any one else. How little Lady Baldock understood her. Fancy your meeting Lady Baldock at Oswald's house!'

'Fancy Augusta Boreham turning nun!'

'How exquisitely grotesque it must have been when she made her complaint to you.'

'I pitied her with all my heart.'

'Of course you did,—because you are so soft. And now, Phineas, we will put it off no longer. Tell me all that you have to tell me about him.'

CHAPTER XII
Königstein

PHINEAS FINN and Lady Laura Kennedy sat together discussing the affairs of the past till the servant told them that 'My Lord' was in the next room, and ready to receive Mr. Finn. 'You will find him much altered,' said Lady Laura, 'even more than I am.'

'I do not find you altered at all.'

'Yes, you do,—in appearance. I am a middle-aged woman, and conscious that I may use my privileges as such. But he has become quite an old man,—not in health so much as in manner. But he will be very glad to see you.' So saying she led him into a room, in which he found the Earl seated near the fireplace, and wrapped in furs. He got up to receive his guest, and Phineas saw at once that during the two years of his exile from England Lord Brentford had passed from manhood to senility. He almost tottered as he came forward, and he wrapped his coat around him with that air of studious self-preservation which belongs only to the infirm.

'It is very good of you to come and see me, Mr. Finn,' he said.

'Don't call him Mr. Finn, Papa. I call him Phineas.'

'Well, yes; that's all right, I dare say. It's a terrible long journey from London, isn't it, Mr. Finn?'

'Too long to be pleasant, my lord.'

'Pleasant! Oh, dear. There's no pleasantness about it. And so they've got an autumn session, have they? That's always a very stupid thing to do, unless they want money.'

'But there is a money bill which must be passed. That's Mr. Daubeny's excuse.'

'Ah, if they've a money bill of course it's all right. So you're in Parliament again?'

'I'm sorry to say I'm not.' Then Lady Laura explained to her father, probably for the third or fourth time, exactly what was their guest's position. 'Oh, a scrutiny. We didn't use to have any scrutinies at Loughton, did we? Ah, me; well, everything seems to be going to the dogs. I'm told they're attacking the Church now.' Lady Laura glanced at Phineas; but neither of them said a word. 'I don't quite understand it; but they tell me that the Tories are going to disestablish the Church. I'm very glad I'm out of it all. Things have come to such a pass that I don't see how a gentleman is to hold office now-a-days. Have you seen Chiltern lately?'

After a while, when Phineas had told the Earl all that there was to tell of his son and his grandson, and all of politics and of Parliament, Lady Laura suddenly interrupted them. 'You knew, Papa, that he was to see Mr. Kennedy. He has been to Loughlinter, and has seen him.'

'Oh, indeed!'

'He is quite assured that I could not with wisdom return to live with my husband.'

'It is a very grave decision to make,' said the Earl.

'But he has no doubt about it,' continued Lady Laura.

'Not a shadow of doubt,' said Phineas. 'I will not say that Mr. Kennedy is mad; but the condition of his mind is such in regard to Lady Laura that I do not think she could live with him in safety. He is crazed about religion.'

'Dear, dear, dear,' exclaimed the Earl.

'The gloom of his house is insupportable. And he does not pretend that he desires her to return that he and she may be happy together.'

'What for then?'

'That we might be unhappy together,' said Lady Laura.

'He repudiates all belief in happiness. He wishes her to return to him chiefly because it is right that a man and wife should live together.'

'So it is,' said the Earl.

'But not to the utter wretchedness of both of them,' said Lady Laura. 'He says,' and she pointed to Phineas, 'that were I there he would renew his accusation against me. He has not told me all. Perhaps he cannot tell me all. But I certainly will not return to Loughlinter.'

'Very well, my dear.'

'It is not very well, Papa; but, nevertheless, I will not return to Loughlinter. What I suffered there neither of you can understand.'

That afternoon Phineas went out alone to the galleries, but the next day she accompanied him, and showed him whatever of glory the town had to offer in its winter dress. They stood together before great masters, and together examined small gems. And then from day to day they were always in each other's company. He had promised to stay a month, and during that time he was petted and comforted to his heart's content. Lady Laura would have taken him into the Saxon Switzerland, in spite of the inclemency of the weather and her father's rebukes, had he not declared vehemently that he was happier remaining in the town. But she did succeed in carrying him off to the fortress of Königstein; and there as they wandered along the fortress constructed on that wonderful rock there occurred between them a conversation which he never forgot, and which it would not have been easy to forget. His own prospects had of course been frequently discussed. He had told her everything, down to the exact amount of money which he had to support him till he should again be enabled to earn an income, and had received assurances from her that everything would be just as it should be after a lapse of a few months. The Liberals would, as a matter of course, come in, and equally as a matter of course, Phineas would be in office. She spoke of this with such certainty that she almost convinced him. Having tempted him away from the safety of permanent income, the party could not do less than provide for him. If he could only secure a seat he would be safe; and it seemed that Tankerville would be a certain seat. This certainty he would not admit; but, nevertheless, he was comforted by his friend. When you have done the rashest thing in the world it is very pleasant to be told that no man of spirit could have acted otherwise. It was a matter of course that he should return to public life,—so said Lady Laura;—and doubly a matter of course when he found himself a widower without a child. 'Whether it be a bad life or a good life,' said Lady Laura, 'you and I understand equally well that no other life is worth having after it. We are like the actors, who cannot bear to be away from the gaslights when once they have lived amidst their glare.' As she said this they were leaning together over one of the parapets of the great fortress, and the sadness of the words struck him as they bore upon herself. She also had lived amidst the gaslights, and now she was self-banished into absolute obscurity. 'You could not have been content with your life in Dublin,' she said.

'Are you content with your life in Dresden?'

'Certainly not. We all like exercise; but the man who has had his leg cut off can't walk. Some can walk with safety; others only with a certain peril; and others cannot at all. You are in the second position, but I am in the last.'

'I do not see why you should not return.'

'And if I did what would come of it? In place of the seclusion of Dresden, there would be the seclusion of Portman Square or of Saulsby. Who would care to have me at their houses, or to come to mine? You know what a hazardous, chancy, short-lived thing is the fashion of a woman. With wealth, and wit, and social charm, and impudence, she may preserve it for some years, but when she has once lost it she can never recover it. I am as much lost to the people who did know me in London as though I had been buried for a century. A man makes himself really useful, but a woman can never do that.'

'All those general rules mean nothing,' said Phineas. 'I should try it.'

'No, Phineas. I know better than that. It would only be disappointment. I hardly think that after all you ever did understand when it was that I broke down utterly and marred my fortunes for ever.'

'I know the day that did it.'

'When I accepted him?'

'Of course it was. I know that, and so do you. There need be no secret between us.'

'There need be no secret between us certainly,—and on my part there shall be none. On my part there has been none.'

'Nor on mine.'

'There has been nothing for you to tell,—since you blurted out your short story of love that day over the waterfall, when I tried so hard to stop you.'

'How was I to be stopped then?'

'No; you were too simple. You came there with but one idea, and you could not change it on the spur of the moment. When I told you that I was engaged you could not swallow back the words that were not yet spoken. Ah, how well I remember it. But you are wrong, Phineas. It was not my engagement or my marriage that has made the world a blank for me.' A feeling came upon him which half-choked him, so that he could ask her no further question. 'You know that, Phineas.'

'It was your marriage,' he said, gruffly.

'It was, and has been, and still will be my strong, unalterable, unquenchable love for you. How could I behave to that other man with even seeming tenderness when my mind was always thinking of you, when my heart was always fixed upon you? But you have been so simple, so little given to vanity,'—she leaned upon his arm as she spoke,—'so pure and so manly, that you have not believed this, even when I told you. Has it not been so?'

'I do not wish to believe it now.'

'But you do believe it? You must and shall believe it. I ask for nothing in return. As my God is my judge, if I thought it possible that your heart should be to me as mine is to you, I could have put a pistol to my ear sooner than speak as I have spoken.' Though she paused for some word from him he could not utter a word. He remembered many things, but even to her in his present mood he could not allude to them;—how he had kissed her at the Falls, how she had bade him not come back to the house because his presence to her was insupportable; how she had again encouraged him to come, and had then forbidden him to accept even an invitation to dinner from her husband. And he remembered too the fierceness of her anger to him when he told her of his love for Violet Effingham. 'I must insist upon it,' she continued, 'that you shall take me now as I really am,—as your dearest friend, your sister, your mother, if you will. I know what I am. Were my husband not still living it would be the same. I should never under any circumstances marry again. I have passed the period of a woman's life when as a woman she is loved; but I have not outlived the power of loving. I shall fret about you, Phineas, like an old hen after her one chick; and though you turn out to be a duck, and get away into waters where I cannot follow you, I shall go cackling round the pond, and always have my eye upon you.' He was holding her now by the hand, but he could not speak for the tears were trickling down his cheeks. 'When I was young,' she continued, 'I did not credit myself with capacity for so much passion. I told myself that love after all should be a servant and not a master, and I married my husband fully intending to do my duty to him. Now we see what has come of it.'

'It has been his fault; not yours,' said Phineas.

'It was my fault,—mine; for I never loved him. Had you not told me what manner of man he was before? And I had believed you, though I denied it. And I knew when I went to Loughlinter that it was you whom I loved. And I knew too,—I almost knew that you would ask me to be your wife were not that other thing settled first. And I declared to myself that, in spite of both our hearts, it should not be so. I had no money then,—nor had you.'

'I would have worked for you.'

'Ah, yes; but you must not reproach me now, Phineas. I never deserted you as regarded your interests, though what little love you had for me was short-lived indeed. Nay; you are not accused, and shall not excuse yourself. You were right, —always right. When you had failed to win one woman your heart with a true natural spring went to another. And so entire had been the cure, that you went to the first woman with the tale of your love for the second.'

'To whom was I to go but to a friend?'

'You did come to a friend, and though I could not drive out of my heart the demon of jealousy, though I was cut to the very bone, I would have helped you had help been possible. Though it had been the fixed purpose of my life that Violet and Oswald should be man and wife, I would have helped you because that other purpose of serving you in all things had become more fixed. But it was to no good end that I sang your praises. Violet Effingham was not the girl to marry this man or that at the bidding of any one;—was she?'

'No, indeed.'

'It is of no use now talking of it; is it? But I want you to understand me from the beginning;—to understand all that was evil, and anything that was good. Since first I found that you were to me the dearest of human beings I have never once been untrue to your interests, though I have been unable not to be angry with you. Then came that wonderful episode in which you saved my husband's life.'

'Not his life.'

'Was it not singular that it should come from your hand? It seemed like Fate. I tried to use the accident, to make his friendship for you as thorough as my own. And then I was obliged to separate you, because,—because, after all I was so mere a woman that I could not bear to have you near me. I can bear it now.'

'Dear Laura!'

'Yes; as your sister. I think you cannot but love me a little when you know how entirely I am devoted to you. I can bear to have you near me now and think of you only as the hen thinks of her duckling. For a moment you are out of the pond, and I have gathered you under my wing. You understand?'

'I know that I am unworthy of what you say of me.'

'Worth has nothing to do with it,—has no bearing on it. I do not say that you are more worthy than all whom I have known. But when did worth create love? What I want is that you should believe me, and know that there is one bound to you who will never be unbound, one whom you can trust in all things,—one to whom you can confess that you have been wrong if you go wrong, and yet be sure that you will not lessen her regard. And with this feeling you must pretend to nothing more than friendship. You will love again, of course.'

'Oh, no.'

'Of course you will. I tried to blaze into power by a marriage, and I failed,—because I was a woman. A woman should marry only for love. You will do it yet, and will not fail. You may remember this too,—that I shall never be jealous again. You may tell me everything with safety. You will tell me everything?'

'If there be anything to tell, I will.'

'I will never stand between you and your wife,—though I would fain hope that she should know how true a friend I am. Now we have walked here till it is dark, and the sentry will think we are taking plans of the place. Are you cold?'

'I have not thought about the cold.'

'Nor have I. We will go down to the inn and warm ourselves before the train comes. I wonder why I should have brought you here to tell you my story. Oh, Phineas.' Then she threw herself into his arms, and he pressed her to his heart, and kissed first her forehead and then her lips. 'It shall never be so again,' she said. 'I will kill it out of my heart even though I should crucify my body. But it is not my love that I will kill. When you are happy I will be happy. When you prosper I will prosper. When you fail I will fail. When you rise,—as you will rise,—I will rise with you. But I will never again feel the pressure of your arm round my waist. Here is the gate, and the old guide. So, my friend, you see that we are not lost.' Then they walked down the very steep hill to the little town below the fortress, and there they remained till the evening train came from Prague, and took them back to Dresden.

Two days after this was the day fixed for Finn's departure. On the intermediate day the Earl begged for a few minutes' private conversation with him, and the two were closeted together for an hour. The Earl, in truth, had little or nothing to say. Things had so gone with him that he had hardly a will of his own left, and did simply that which his daughter directed him to do. He pretended to consult Phineas as to the expediency of his returning to Saulsby. Did Phineas think that his return would be of any use to the party? Phineas knew very well that the party would not recognise the difference whether the Earl lived at Dresden or in London. When a man has come to the end of his influence as the Earl had done he is as much a nothing in politics as though he had never risen above that quantity. The Earl had never risen very high, and even Phineas, with all his desire to be civil, could not say that the Earl's presence would materially serve the interests of the Liberal party. He made what most civil excuses he could, and suggested that if Lord Brentford should choose to return, Lady Laura would very willingly remain at Dresden alone. 'But why shouldn't she come too?' asked the Earl. And then, with the tardiness of old age, he proposed his little plan. 'Why should she not make an attempt to live once more with her husband?'

'She never will,' said Phineas.

'But think how much she loses,' said the Earl.

'I am quite sure she never will. And I am quite sure that she ought not to do so. The marriage was a misfortune. As it is they are better apart.' After that the Earl did not dare to say another word about his daughter; but discussed his son's affairs. Did not Phineas think that Chiltern might now be induced to go into Parliament? 'Nothing would make him do so,' said Phineas.

'But he might farm?'

'You see he has his hands full.'

'But other men keep hounds and farm too,' said the Earl.

'But Chiltern is not like other men. He gives his whole mind to it, and finds full employment. And then he is quite happy, and so is she. What more can you want for him? Everybody respects him.'

'That goes a very great way,' said the Earl. Then he thanked Phineas cordially, and felt that now as ever he had done his duty by his family.

There was no renewal of the passionate conversation which had taken place on the ramparts, but much of tenderness and of sympathy arose from it. Lady Laura took upon herself the tone and manners of an elder sister,—of a sister very much older than her brother,—and Phineas submitted to them not only gracefully but with delight to himself. He had not thanked her for her love when she expressed it, and he did not do so afterwards. But he accepted it, and bowed to it, and recognised it as constituting one of the future laws of his life. He was to do nothing of importance without her knowledge, and he was to be at her command should she at any time want assistance in England. 'I suppose I shall come back some day,' she said, as they were sitting together late on the evening before his departure.

'I cannot understand why you should not do so now. Your father wishes it.'

'He thinks he does; but were he told that he was to go to-morrow, or next summer, it would fret him. I am assured that Mr. Kennedy could demand my return,—by law.'

'He could not enforce it.'

'He would attempt it. I will not go back until he consents to my living apart from him. And, to tell the truth, I am better here for awhile. They say that the sick animals always creep somewhere under cover. I am a sick animal, and now that I have crept here I will remain till I am stronger. How terribly anxious you must be about Tankerville!'

'I am anxious.'

'You will telegraph to me at once? You will be sure to do that?'

'Of course I will, the moment I know my fate.'

'And if it goes against you?'

'Ah,—what then?'

'I shall at once write to Barrington Erle. I don't suppose he would do much now for his poor cousin, but he can at any rate say what can be done. I should bid you come here,—only that stupid people would say that you were my lover. I should not mind, only that he would hear it, and I am bound to save him from annoyance. Would you not go down to Oswald again?'

'With what object?'

'Because anything will be better than returning to Ireland. Why not go down and look after Saulsby? It would be a home, and you need not tie yourself to it. I will speak to Papa about that. But you will get the seat.'

'I think I shall,' said Phineas.

'Do;—pray do! If I could only get hold of that judge by the ears! Do you know what time it is? It is twelve, and your train starts at eight.' Then he arose to bid her adieu. 'No,' she said; 'I shall see you off.'

'Indeed you will not. It will be almost night when I leave this, and the frost is like iron.'

'Neither the night nor the frost will kill me. Do you think I will not give you your last breakfast? God bless you, dear.'

And on the following morning she did give him his breakfast by candle-light, and went down with him to the station.

The morning was black, and the frost was, as he had said, as hard as iron, but she was thoroughly good-humoured, and apparently happy. 'It has been so much to me to have you here, that I might tell you everything,' she said. 'You will understand me now.'

'I understand, but I know not how to believe,' he said.

'You do believe. You would be worse than a Jew if you did not believe me. But you understand also. I want you to marry, and you must tell her all the truth. If I can I will love her almost as much as I do you. And if I live to see them, I will love your children as dearly as I do you. Your children shall be my children;—or at least one of them shall be mine. You will tell me when it is to be.'

'If I ever intend such a thing, I will tell you.'

'Now, good-bye. I shall stand back there till the train starts, but do not you notice me. God bless you, Phineas.' She held his hand tight within her own for some seconds, and looked into his face with an unutterable love. Then she drew down her veil, and went and stood apart till the train had left the platform.

'He has gone, Papa,' Lady Laura said, as she stood afterwards by her father's bedside.

'Has he? Yes; I know he was to go, of course. I was very glad to see him, Laura.'

'So was I, Papa;—very glad indeed. Whatever happens to him, we must never lose sight of him again.'

'We shall hear of him, of course, if he is in the House.'

'Whether he is in the House or out of it we must hear of him. While we have aught he must never want.' The Earl stared at his daughter. The Earl was a man of large possessions, and did not as yet understand that he was to be called upon to share them with Phineas Finn. 'I know, Papa, you will never think ill of me.'

'Never, my dear.'

'I have sworn that I will be a sister to that man, and I will keep my oath.'

'I know you are a very good sister to Chiltern,' said the Earl. Lady Laura had at one time appropriated her whole fortune, which had been large, to the payment of her brother's debts. The money had been returned, and had gone to her husband. Lord Brentford now supposed that she intended at some future time to pay the debts of Phineas Finn.

CHAPTER XIII

'I have got the seat'

WHEN Phineas returned to London, the autumn Session, though it had been carried on so near to Christmas as to make many members very unhappy, had already been over for a fortnight. Mr. Daubeny had played his game with consummate skill to the last. He had brought in no bill, but had stated his intention of doing so early in the following Session. He had, he said, of course been aware from the first that it would have been quite impossible to carry such a measure as that proposed during the few weeks in which it had been possible for them to sit between the convening of Parliament and the Christmas holidays; but he thought that it was expedient that the proposition should be named to the House and ventilated as it had been, so that members on both sides might be induced to give their most studious attention to the subject before a measure, which must be so momentous, should be proposed to them. As had happened, the unforeseen division to which the House had been pressed on the Address had proved that the majority of the House was in favour of the great reform which it was the object of his ambition to complete. They were aware that they had been assembled at a somewhat unusual and inconvenient period of the year, because the service of the country had demanded that certain money bills should be passed. He, however, rejoiced greatly that this earliest opportunity had been afforded to him of explaining the intentions of the Government with which he had the honour of being connected. In answer to this there arose a perfect torrent of almost vituperative antagonism from the opposite side of the House. Did the Right Honourable gentleman dare to say that the question had been ventilated in the country, when it had never been broached by him or any of his followers till after the general election had been completed? Was it not notorious to the country that the first hint of it had been given when the Right Honourable gentleman was elected for East Barsetshire, and was it not equally notorious that that election had been so arranged that the marvellous proposition of the Right Honourable gentleman should not be known even to his own party till there remained no possibility of the expression of any condemnation from the hustings? It might be that the Right Honourable could so rule his own followers in that House as to carry them with him even in a matter so absolutely opposite to their own most cherished convictions. It certainly seemed that he had succeeded in doing so for the present. But would any one believe that he would have carried the country, had he dared to face the country with such a measure in his hands? Ventilation, indeed! He had not dared to ventilate his proposition. He had used this short Session in order that he might keep his clutch fastened on power, and in doing so was indifferent alike to the Constitution, to his party, and to the country. Harder words had never been spoken in the House than were uttered on this occasion. But the Minister was successful. He had been supported on the Address; and he went home to East Barsetshire at Christmas, perhaps with some little fear of the parsons around him; but with a full conviction that he would at least carry the second reading of his bill.

London was more than usually full and busy this year immediately after Christmas. It seemed as though it were admitted by all the Liberal party generally that the sadness of the occasion ought to rob the season of its usual festivities. Who could eat mince pies or think of Twelfth Night while so terribly wicked a scheme was in progress for keeping the real majority out in the cold? It was the injustice of the thing that rankled so deeply,—that, and a sense of inferiority to the cleverness displayed by Mr. Daubeny! It was as when a player is checkmated by some audacious combination of two pawns and a knight, such being all the remaining forces of the victorious adversary, when the beaten man has two castles and a queen upon the board. It was, indeed, worse than this,—for the adversary had appropriated to his own use the castles and the queen of the unhappy vanquished one. This Church Reform was the legitimate property of the Liberals, and had not been as yet used by them only because they had felt it right to keep in the background for some future great occasion so great and so valuable a piece of ordnance. It was theirs so safely that they could afford to bide their time. And then,—so they all said, and so some of them believed,—the country was not ready for so great a measure. It must come; but there must be tenderness in the mode of producing it. The parsons must be respected, and the great Church-of-England feeling of the people must be considered with affectionate regard. Even the most rabid Dissenter would hardly wish to see a structure so nearly divine attacked and destroyed by rude hands. With grave and slow and sober earnestness, with loving touches and soft caressing manipulation let the beautiful old Church be laid to its rest, as something too exquisite, too lovely, too refined for the present rough manners of the world! Such were the ideas as to Church Reform of the leading Liberals of the day; and now this man, without even a majority to back him, this audacious Cagliostro* among statesmen, this destructive leader of all declared Conservatives, had come forward without a moment's warning, and pretended that he would do the thing out of hand! Men knew that it had to be done. The country had begun to perceive that the old Establishment must fall; and, knowing this, would not the Liberal backbone of Great Britain perceive the enormity of this Cagliostro's wickedness,—and rise against him and bury him beneath its scorn as it ought to do? This was the feeling that made a real Christmas impossible to Messrs. Ratler and Bonteen.

'The one thing incredible to me,' said Mr. Ratler, 'is that

Englishmen should be so mean.' He was alluding to the Conservatives who had shown their intention of supporting Mr. Daubeny, and whom he accused of doing so, simply with a view to power and patronage, without any regard to their own consistency or to the welfare of the country. Mr. Ratler probably did not correctly read the minds of the men whom he was accusing, and did not perceive, as he should have done with his experience, how little there was among them of concerted action. To defend the Church was a duty to each of them; but then, so also was it a duty to support his party. And each one could see his way to the one duty, whereas the other was vague, and too probably ultimately impossible. If it were proper to throw off the incubus of this conjuror's authority, surely some wise, and great, and bold man would get up and so declare. Some junto of wise men of the party would settle that he should be deposed. But where were they to look for the wise and bold men? where even for the junto? Of whom did the party consist?—Of honest, chivalrous, and enthusiastic men, but mainly of men who were idle, and unable to take upon their own shoulders the responsibility of real work. Their leaders had been selected from the outside,—clever, eager, pushing men, but of late had been hardly selected from among themselves. As used to be the case with Italian Powers, they entrusted their cause to mercenary foreign generals, soldiers of fortune, who carried their good swords whither they were wanted; and, as of old, the leaders were ever ready to fight, but would themselves declare what should be and what should not be the *casus belli*.* There was not so much meanness as Mr. Ratler supposed in the Conservative ranks, but very much more unhappiness. Would it not be better to go home and live at the family park all the year round, and hunt, and attend Quarter Sessions,* and be able to declare morning and evening with a clear conscience that the country was going to the dogs? Such was the mental working of many a Conservative who supported Mr. Daubeny on this occasion.

At the instance of Lady Laura, Phineas called upon the

Duke of St. Bungay soon after his return, and was very kindly received by his Grace. In former days, when there were Whigs instead of Liberals, it was almost a rule of political life that all leading Whigs should be uncles, brothers-in-law, or cousins to each other. This was pleasant and gave great consistency to the party; but the system has now gone out of vogue. There remain of it, however, some traces, so that among the nobler born Liberals of the day there is still a good deal of agreeable family connection. In this way the St. Bungay Fitz-Howards were related to the Mildmays and Standishes, and such a man as Barrington Erle was sure to be cousin to all of them. Lady Laura had thus only sent her friend to a relation of her own, and as the Duke and Phineas had been in the same Government, his Grace was glad enough to receive the returning aspirant. Of course there was something said at first as to the life of the Earl at Dresden. The Duke recollected the occasion of such banishment, and shook his head; and attempted to look unhappy when the wretched condition of Mr. Kennedy was reported to him. But he was essentially a happy man, and shook off the gloom at once when Phineas spoke of politics. 'So you are coming back to us, Mr. Finn?'

'They tell me I may perhaps get the seat.'

'I am heartily glad, for you were very useful. I remember how Cantrip almost cried when he told me you were going to leave him. He had been rather put upon, I fancy, before.'

'There was perhaps something in that, your Grace.'

'There will be nothing to return to now beyond barren honours.'

'Not for a while.'

'Not for a long while,' said the Duke;—'for a long while, that is, as candidates for office regard time. Mr. Daubeny will be safe for this Session at least. I doubt whether he will really attempt to carry his measure this year. He will bring it forward, and after the late division he must get his second reading. He will then break down gracefully in Committee, and declare that the importance of the interests concerned demands further inquiry. It wasn't a thing to be done in one year.'

'Why should he do it at all?' asked Phineas.

'That's what everybody asks, but the answer seems to be so plain! Because he can do it, and we can't. He will get from our side much support, and we should get none from his.'

'There is something to me sickening in their dishonesty,' said Phineas energetically.

'The country has the advantage; and I don't know that they are dishonest. Ought we to come to a deadlock in legislation in order that parties might fight out their battle till one had killed the other?'

'I don't think a man should support a measure which he believes to be destructive.'

'He doesn't believe it to be destructive. The belief is theoretic,—or not even quite that. It is hardly more than romantic. As long as acres are dear, and he can retain those belonging to him, the country gentleman will never really believe his country to be in danger. It is the same with commerce. As long as the Three per Cents. do not really mean Four per Cent.,—I may say as long as they don't mean Five per Cent., —the country will be rich, though every one should swear that it be ruined.'

'I'm very glad, at the same time, that I don't call myself a Conservative,' said Phineas.

'That shows how disinterested you are, as you certainly would be in office. Good-bye. Come and see the Duchess when she comes to town. And if you've nothing better to do, give us a day or two at Longroyston at Easter.' Now Longroyston was the Duke's well-known country seat, at which Whig hospitality had been dispensed with a lavish hand for two centuries.

On the 20th January Phineas travelled down to Tankerville again in obedience to a summons served upon him at the instance of the judge who was to try his petition against Browborough. It was the special and somewhat unusual nature of this petition that the complainants not only sought to oust the sitting member, but also to give the seat to the late unsuccessful candidate. There was to be a scrutiny, by which, if it should be successful, so great a number of votes would be deducted from those polled on behalf of the unfortunate Mr. Browborough as to leave a majority for his opponent, with the additional disagreeable obligation upon him of paying the cost of the transaction by which he would thus lose his seat. Mr. Browborough, no doubt, looked upon the whole thing with the greatest disgust. He thought that a battle when once won should be regarded as over till the occasion should come for another battle. He had spent his money like a gentleman, and hated these mean ways. No one could ever say that he had ever petitioned. That was his way of looking at it. That Shibboleth of his as to the prospects of England and the Church of her people had, no doubt, made the House less agreeable to him during the last Short session than usual; but he had stuck to his party, and voted with Mr. Daubeny on the Address,—the obligation for such vote having inconveniently pressed itself upon him before the presentation of the petition had been formally completed. He had always stuck to his party. It was the pride of his life that he had been true and consistent. He also was summoned to Tankerville, and he was forced to go, although he knew that the Shibboleth would be thrown in his teeth.

Mr. Browborough spent two or three very uncomfortable days at Tankerville, whereas Phineas was triumphant. There were worse things in store for poor Mr. Browborough than his repudiated Shibboleth, or even than his lost seat. Mr. Ruddles, acting with wondrous energy, succeeded in knocking off the necessary votes, and succeeded also in proving that these votes were void by reason of gross bribery. He astonished Phineas by the cool effrontery with which he took credit to himself for not having purchased votes in the Fallgate on the Liberal side, but Phineas was too wise to remind him that he himself had hinted at one time that it would be well to lay out a little money in that way. No one at the present moment was more clear than was Ruddles as to the necessity of purity at elections. Not a penny had been misspent by the Finnites. A vote or two from their score was knocked off on grounds which did not touch the candidate or his agents. One man had personated* a vote, but this appeared to have been done at the instigation of some very cunning Browborough partisan. Another man had been wrongly described. This, however, amounted to nothing. Phineas Finn was seated for the borough, and the judge declared his purpose of recommending the House of Commons to issue a commission with reference to the expediency of instituting a prosecution. Mr. Browborough left the town in great disgust, not without various publicly expressed intimations from his opponents that the prosperity of England depended on the Church of her people. Phineas was gloriously entertained by the Liberals of the borough, and then informed that as so much had been done for him it was hoped that he would now open his pockets on behalf of the charities of the town. 'Gentlemen,' said Phineas, to one or two of the leading Liberals, 'it is as well that you should know at once that I am a very poor man.' The leading Liberals made wry faces, but Phineas was member for the borough.

The moment that the decision was announced, Phineas, shaking off for the time his congratulatory friends, hurried to the post-office and sent his message to Lady Laura Standish* at Dresden: 'I have got the seat.' He was almost ashamed of himself as the telegraph boy looked up at him when he gave in the words, but this was a task which he could not have entrusted to any one else. He almost thought that this was in truth the proudest and happiest moment of his life. She would so thoroughly enjoy his triumph, would receive from it such great and unselfish joy, that he almost wished that he could have taken the message himself. Surely had he done so there would have been fit occasion for another embrace.

He was again a member of the British House of Commons, —was again in possession of that privilege for which he had never ceased to sigh since the moment in which he lost it. A drunkard or a gambler may be weaned from his ways, but not a politician. To have been in the House and not to be there was, to such a one as Phineas Finn, necessarily, a state of discontent. But now he had worked his way up again, and he was determined that no fears for the future should harass him. He would give his heart and soul to the work while his money lasted. It would surely last him for the Session. He was all alone in the world, and would trust to the chapter of accidents for the future.

'I never knew a fellow with such luck as yours,' said Barrington Erle to him, on his return to London. 'A seat always drops into your mouth when the circumstances seem to be most forlorn.'

'I have been lucky, certainly.'

'My cousin, Laura Kennedy, has been writing to me about you.'

'I went over to see them, you know.'

'So I heard. She talks some nonsense about the Earl being willing to do anything for you. What could the Earl do? He has no more influence in the Loughton borough than I have. All that kind of thing is clean done for,—with one or two exceptions. We got much better men while it lasted than we do now.'

'I should doubt that.'

'We did;—much truer men,—men who went straighter. By the bye, Phineas, we must have no tricks on this Church matter. We mean to do all we can to throw out the second reading.'

'You know what I said at the hustings.'

'D—— the hustings. I know what Browborough said, and Browborough voted like a man with his party. You were against the Church at the hustings, and he was for it. You will vote just the other way. There will be a little confusion, but the people of Tankerville will never remember the particulars.'

'I don't know that I can do that.'

'By heavens, if you don't, you shall never more be officer of ours*;—though Laura Kennedy should cry her eyes out.'

CHAPTER XIV
Trumpeton Wood

IN the meantime the hunting season was going on in the Brake country with chequered success. There had arisen the great Trumpeton Wood question, about which the sporting world was doomed to hear so much for the next twelve months,—and Lord Chiltern was in an unhappy state of mind. Trumpeton Wood belonged to that old friend of ours, the Duke of Omnium,* who had now almost fallen into second childhood. It was quite out of the question that the Duke should himself interfere in such a matter, or know anything about it; but Lord Chiltern, with headstrong resolution, had persisted in writing to the Duke himself. Foxes had always hitherto been preserved in Trumpeton Wood, and the earths had always been stopped on receipt of due notice by the keepers. During the cubbing season there had arisen quarrels.

The keepers complained that no effort was made to kill the foxes. Lord Chiltern swore that the earths were not stopped. Then there came tidings of a terrible calamity. A dying fox, with a trap to its pad, was found in the outskirts of the Wood; and Lord Chiltern wrote to the Duke. He drew the Wood in regular course before any answer could be received,—and three of his hounds picked up poison, and died beneath his eyes. He wrote to the Duke again,—a cutting letter; and then came from the Duke's man of business, Mr. Fothergill, a very short reply, which Lord Chiltern regarded as an insult. Hitherto the affair had not got into the sporting papers, and was simply a matter of angry discussion at every meet in the neighbouring counties. Lord Chiltern was very full of wrath, and always looked as though he desired to avenge those poor hounds on the Duke and all belonging to him. To a Master of Hounds the poisoning of one of his pack is murder of the deepest dye. There probably never was a Master who in his heart of hearts would not think it right that a detected culprit should be hung for such an offence. And most Masters would go further than this, and declare that in the absence of such detection the owner of the covert in which the poison had been picked up should be held to be responsible. In this instance the condition of ownership was unfortunate. The Duke himself was old, feeble, and almost imbecile. He had never been eminent as a sportsman; but, in a not energetic manner, he had endeavoured to do his duty by the country. His heir, Plantagenet Palliser, was simply a statesman, who, as regarded himself, had never a day to spare for amusement; and who, in reference to sport, had unfortunate fantastic notions that pheasants and rabbits destroyed crops, and that foxes were injurious to old women's poultry. He, however, was not the owner, and had refused to interfere. There had been family quarrels too, adverse to the sporting interests of the younger Palliser scions, so that the shooting of this wood had drifted into the hands of Mr. Fothergill and his friends. Now, Lord Chiltern had settled it in his own mind that the hounds had been poisoned, if not in compliance with Mr Fothergill's orders, at any rate in furtherance of his wishes, and, could he have had his way, he certainly would have sent Mr. Fothergill to the gallows. Now, Miss Palliser, who was still staying at Lord Chiltern's house, was niece to the old Duke, and first cousin to the heir. 'They are nothing to me,' she said once, when Lord Chiltern had attempted to apologise for the abuse he was heaping on her relatives. 'I haven't seen the Duke since I was a little child, and I shouldn't know my cousin were I to meet him.'

'So much the more gracious is your condition,' said Lady Chiltern,—'at any rate in Oswald's estimation.'

'I know them, and once spent a couple of days at Matching with them,' said Lord Chiltern. 'The Duke is an old fool, who always gave himself greater airs than any other man in England,—and as far as I can see, with less to excuse them. As for Planty Pall, he and I belong so essentially to different orders of things, that we can hardly be reckoned as being both men.'

'And which is the man, Lord Chiltern?'

'Whichever you please, my dear; only not both. Doggett was over there yesterday, and found three separate traps.'

'What did he do with the traps?' said Lady Chiltern.

'I wasn't fool enough to ask him, but I don't in the least doubt that he threw them into the water—or that he'd throw Palliser there too if he could get hold of him. As for taking the hounds to Trumpeton again, I wouldn't do it if there were not another covert in the country.'

'Then leave it so, and have done with it,' said his wife. 'I wouldn't fret as you do for what another man did with his own property, for all the foxes in England.'

'That is because you understand nothing of hunting, my dear. A man's property is his own in one sense, but isn't his own in another. A man can't do what he likes with his coverts.'

'He can cut them down.'

'But he can't let another pack hunt them, and he can't hunt them himself. If he's in a hunting county he is bound to preserve foxes.'

'What binds him, Oswald? A man can't be bound without a penalty.'

'I should think it penalty enough for everybody to hate me. What are you going to do about Phineas Finn?'

'I have asked him to come on the 1st and stay till Parliament meets.'

'And is that woman coming?'

'There are two or three women coming.'

'She with the German name, whom you made me dine with in Park Lane?'

'Madame Max Goesler is coming. She brings her own horses, and they will stand at Doggett's.'

'They can't stand here, for there is not a stall.'

'I am so sorry that my poor little fellow should incommode you,' said Miss Palliser.

'You're a licensed offender,—though, upon my honour, I don't know whether I ought to give a feed of oats to any one having a connection with Trumpeton Wood. And what is Phineas to ride?'

'He shall ride my horses,' said Lady Chiltern, whose present condition in life rendered hunting inopportune to her.

'Neither of them would carry him a mile. He wants about as good an animal as you can put him upon. I don't know what I'm to do. It's all very well for Laura to say that he must be mounted.'

'You wouldn't refuse to give Mr. Finn a mount!' said Lady Chiltern, almost with dismay.

'I'd give him my right hand to ride, only it wouldn't carry him. I can't make horses. Harry brought home that brown mare on Tuesday with an overreach that she won't get over this season. What the deuce they do with their horses to knock them about so, I can't understand. I've killed horses in my time, and ridden them to a stand-still, but I never bruised them and battered them about as these fellows do.'

'Then I'd better write to Mr. Finn, and tell him,' said Lady Chiltern, very gravely.

'Oh, Phineas Finn!' said Lord Chiltern; 'oh, Phineas Finn!

what a pity it was that you and I didn't see the matter out when we stood opposite to each other on the sands at Blankenberg!'

'Oswald,' said his wife, getting up, and putting her arm over his shoulder, 'you know you would give your best horse to Mr. Finn, as long as he chose to stay here, though you rode upon a donkey yourself.'

'I know that if I didn't, you would,' said Lord Chiltern. And so the matter was settled.

At night, when they were alone together, there was further discussion as to the visitors who were coming to Harrington Hall. 'Is Gerard Maule to come back?' asked the husband.

'I have asked him. He left his horses at Doggett's, you know.'

'I didn't know.'

'I certainly told you, Oswald. Do you object to his coming? You can't really mean that you care about his riding?'

'It isn't that. You must have some whipping post, and he's as good as another. But he shilly-shallies about that girl. I hate all that stuff like poison.'

'All men are not so—abrupt shall I say?—as you were.'

'I had something to say, and I said it. When I had said it a dozen times, I got to have it believed. He doesn't say it as though he meant to have it believed.'

'You were always in earnest, Oswald.'

'I was.'

'To the extent of the three minutes which you allowed yourself. It sufficed, however;—did it not? You are glad you persevered?'

'What fools women are.'

'Never mind that. Say you are glad. I like you to tell me so. Let me be a fool if I will.'

'What made you so obstinate?'

'I don't know. I never could tell. It wasn't that I didn't dote upon you, and think about you, and feel quite sure that there never could be any other one than you.'

'I've no doubt it was all right;—only you very nearly made

me shoot a fellow, and now I've got to find horses for him. I wonder whether he could ride Dandolo?'

'Don't put him up on anything very hard.'

'Why not? His wife is dead, and he hasn't got a child, nor yet an acre of property. I don't know who is entitled to break his neck if he is not. And Dandolo is as good a horse as there is in the stable, if you can once get him to go. Mind, I have to start to-morrow at nine, for it's all eighteen miles.' And so the Master of the Brake Hounds took himself to his repose.

Lady Laura Kennedy had written to Barrington Erle respecting her friend's political interests, and to her sister-in-law, Lady Chiltern, as to his social comfort. She could not bear to think that he should be left alone in London till Parliament should meet, and had therefore appealed to Lady Chiltern as to the memory of many past events. The appeal had been unnecessary and superfluous. It cannot be said that Phineas and his affairs were matters of as close an interest to Lady Chiltern as to Lady Laura. If any woman loved her husband beyond all things Lord Chiltern's wife did, and ever had done so. But there had been a tenderness in regard to the young Irish Member of Parliament, which Violet Effingham had in old days shared with Lady Laura, and which made her now think that all good things should be done for him. She believed him to be addicted to hunting, and therefore horses must be provided for him. He was a widower, and she remembered of old that he was fond of pretty women, and she knew that in coming days he might probably want money;—and therefore she had asked Madame Max Goesler to spend a fortnight at Harrington Hall. Madame Max Goesler and Phineas Finn had been acquainted before, as Lady Chiltern was well aware. But perhaps Lady Chiltern, when she summoned Madame Max into the country, did not know how close the acquaintance had been.

Madame Max came a couple of days before Phineas, and was taken out hunting on the morning after her arrival. She was a lady who could ride to hounds,—and who, indeed, could do nearly anything to which she set her mind. She was dark, thin, healthy, good-looking, clever, ambitious, rich, unsatisfied, perhaps unscrupulous,—but not without a conscience. As has been told in a former portion of this chronicle, she could always seem to be happy with her companion of the day, and yet there was ever present a gnawing desire to do something more and something better than she had as yet achieved. Of course, as he took her to the meet, Lord Chiltern told her his grievance respecting Trumpeton Wood. 'But, my dear Lord Chiltern, you must not abuse the Duke of Omnium to me.'

'Why not to you?'

'He and I are sworn friends.'

'He's a hundred years old.'

'And why shouldn't I have a friend a hundred years old? And as for Mr. Palliser, he knows no more of your foxes than I know of his taxes. Why don't you write to Lady Glencora? She understands everything.'

'Is she a friend of yours, too?'

'My particular friend. She and I, you know, look after the poor dear Duke between us.'

'I can understand why she should sacrifice herself.'

'But not why I do. I can't explain it myself; but so it has come to pass, and I must not hear the Duke abused. May I write to Lady Glencora about it?'

'Certainly,—if you please; but not as giving her any message from me. Her uncle's property is mismanaged most damnably. If you choose to tell her that I say so you can. I'm not going to ask anything as a favour. I never do ask favours. But the Duke or Planty Palliser among them should do one of two things. They should either stand by the hunting, or they should let it alone;—and they should say what they mean. I like to know my friends, and I like to know my enemies.'

'I am sure the Duke is not your enemy, Lord Chiltern.'

'These Pallisers have always been running with the hare and hunting with the hounds. They are great aristocrats, and yet are always going in for the people. I'm told that Planty

Pall calls fox-hunting barbarous. Why doesn't he say so out loud, and stub up Trumpeton Wood and grow corn?'

'Perhaps he will when Trumpeton Wood belongs to him.'

'I should like that much better than poisoning hounds and trapping foxes.' When they got to the meet, conclaves of men might be seen gathered together here and there, and in each conclave they were telling something new or something old as to the iniquities perpetrated at Trumpeton Wood.

On that evening before dinner Madame Goesler was told by her hostess that Phineas Finn was expected on the following day. The communication was made quite as a matter of course; but Lady Chiltern had chosen a time in which the lights were shaded, and the room was dark. Adelaide Palliser was present, as was also a certain Lady Baldock,—not that Lady Baldock who had abused all Papists to poor Phineas, but her son's wife. They were drinking tea together over the fire, and the dim lights were removed from the circle. This, no doubt, was simply an accident; but the gloom served Madame Goesler during one moment of embarrassment. 'An old friend of yours is coming here to-morrow,' said Lady Chiltern.

'An old friend of mine! Shall I call my friend he or she?'

'You remember Mr. Finn?'

That was the moment in which Madame Goesler rejoiced that no strong glare of light fell upon her face. But she was a woman who would not long leave herself subject to any such embarrassment. 'Surely,' she said, confining herself at first to the single word.

'He is coming here. He is a great friend of mine.'

'He always was a good friend of yours, Lady Chiltern.'

'And of yours, too, Madame Max. A sort of general friend, I think, was Mr. Finn in the old days. I hope you will be glad to see him.'

'Oh, dear, yes.'

'I thought him very nice,' said Adelaide Palliser.

'I remember mamma saying, before she was mamma, you know,' said Lady Baldock, 'that Mr. Finn was very nice indeed, only he was a Papist, and only he had got no money, and only he would fall in love with everybody. Does he go on falling in love with people, Violet?'

'Never with married women, my dear. He has had a wife himself since that, Madame Goesler, and the poor thing died.'

'And now here he is beginning all over again,' said Lady Baldock.

'And as pleasant as ever,' said her cousin. 'You know he has done all manner of things for our family. He picked Oswald up once after one of those terrible hunting accidents; and he saved Mr. Kennedy when men were murdering him.'

'That was questionable kindness,' said Lady Baldock.

'And he sat for Lord Brentford's borough.'

'How good of him!' said Miss Palliser.

'And he has done all manner of things,' said Lady Chiltern.

'Didn't he once fight a duel?' asked Madame Goesler.

'That was the grandest thing of all,' said his friend, 'for he didn't shoot somebody whom perhaps he might have shot had he been as bloodthirsty as somebody else. And now he has come back to Parliament, and all that kind of thing, and he's coming here to hunt. I hope you'll be glad to see him, Madame Goesler.'

'I shall be very glad to see him,' said Madame Goesler, slowly; 'I heard about his success at that town, and I knew that I should meet him somewhere.'

CHAPTER XV

'How well you knew!'

IT was necessary also that some communication should be made to Phineas, so that he might not come across Madame Goesler unawares. Lady Chiltern was more alive to that necessity than she had been to the other, and felt that the gentleman, if not warned of what was to take place, would be much more likely than the lady to be awkward at the trying moment. Madame Goesler would in any circumstances be sure to recover her self-possession very quickly, even were she to lose it for a moment; but so much could hardly be said for the social powers of Phineas Finn. Lady Chiltern therefore contrived to see him alone for a moment on his arrival. 'Who do you think is here?'

'Lady Laura has not come!'

'Indeed, no; I wish she had. An old friend, but not so old as Laura!'

'I cannot guess;—not Lord Fawn?'

'Lord Fawn! What would Lord Fawn do here? Don't you know that Lord Fawn goes nowhere since his last matrimonial trouble? It's a friend of yours, not of mine.'

'Madame Goesler?' whispered Phineas.

'How well you knew when I said it was a friend of yours. Madame Goesler is here,—not altered in the least.'

'Madame Goesler!'

'Does it annoy you?'

'Oh, no. Why should it annoy me?'

'You never quarrelled with her?'

'Never!'

'There is no reason why you should not meet her?'

'None at all;—only I was surprised. Did she know that I was coming?'

'I told her yesterday. I hope that I have not done wrong or made things unpleasant. I knew that you used to be friends.'

'And as friends we parted, Lady Chiltern.' He had nothing more to say in the matter; nor had she. He could not tell the story of what had taken place between himself and the lady, and she could not keep herself from surmising that something had taken place, which, had she known it, would have prevented her from bringing the two together at Harrington.

Madame Goesler, when she was dressing, acknowledged to herself that she had a task before her which would require all her tact and all her courage. She certainly would not have accepted Lady Chiltern's invitation had she known that she would encounter Phineas Finn at the house. She had twenty-four hours to think of it, and at one time had almost made up her mind that some sudden business should recall her to London. Of course, her motive would be suspected. Of course Lady Chiltern would connect her departure with the man's arrival. But even that, bad as it would be, might be preferable to the meeting! What a fool had she been,—so she accused herself,—in not foreseeing that such an accident might happen, knowing as she did that Phineas Finn had reappeared in the political world, and that he and the Chiltern people had ever been fast friends! As she had thought about it, lying awake at night, she had told herself that she must certainly be recalled back to London by business. She would telegraph up to town, raising a question about any trifle, and on receipt of the answer she could be off with something of an excuse. The shame of running away from the man seemed to be a worse evil than the shame of meeting him. She had in truth done nothing to disgrace herself. In her desire to save a man whom she had loved from the ruin which she thought had threatened him, she had—offered him her hand. She had made the offer, and he had refused it! That was all. No; she would not be driven to confess to herself that she had ever fled from the face of man or woman. This man would be again in London, and she could not always fly. It would be only necessary that she should maintain her own composure, and the misery of the meeting would pass away after the first few minutes. One consolation was assured to her. She thoroughly believed in the man,—feeling certain that he had not betrayed her, and would not betray her. But now, as the time for the meeting drew near, as she stood for a moment before the glass,—pretending to look at herself in order that her maid might not remark her uneasiness, she found that her courage, great as it was, hardly sufficed her. She almost plotted some scheme of a headache, by which she might be enabled not to show herself till after dinner. 'I am so blind that I can hardly see out of my eyes,' she said to the maid, actually beginning the scheme. The woman assumed a look of painful solicitude, and declared that 'Madame did not look quite her best.' 'I suppose I shall shake it off,' said Madame Goesler; and then she descended the stairs.

The condition of Phineas Finn was almost as bad, but he had a much less protracted period of anticipation than that with which the lady was tormented. He was sent up to dress for dinner with the knowledge that in half an hour he would find himself in the same room with Madame Goesler. There could be no question of his running away, no possibility even of his escaping by a headache. But it may be doubted whether his dismay was not even more than hers. She knew that she could teach herself to use no other than fitting words; but he was almost sure that he would break down if he attempted to speak to her. She would be safe from blushing, but he would assuredly become as red as a turkey-cock's comb up to the roots of his hair. Her blood would be under control, but his would be coursing hither and thither through his veins, so as to make him utterly unable to rule himself. Nevertheless, he also plucked up his courage and descended, reaching the drawing-room before Madame Goesler had entered it. Chiltern was going on about Trumpeton Wood to Lord Baldock, and was renewing his fury against all the Pallisers, while Adelaide stood by and laughed. Gerard Maule was lounging on a chair, wondering that any man could expend such energy on such a subject. Lady Chiltern was explaining the merits of the case to Lady Baldock,—who knew nothing about hunting; and the other guests were listening with eager attention. A certain Mr. Spooner, who rode hard and did nothing else, and who acted as an unacknowledged assistant-master under Lord Chiltern,—there is such a man in every hunt,—acted as chorus, and indicated, chiefly with dumb show, the strong points of the case.

'Finn, how are you?' said Lord Chiltern, stretching out his left hand. 'Glad to have you back again, and congratulate you about the seat. It was put down in red herrings, and we found nearly a dozen of them afterwards,—enough to kill half the pack.'

'Picked up nine,' said Mr. Spooner.

'Children might have picked them up quite as well,—and eaten them,' said Lady Chiltern.

'They didn't care about that,' continued the Master. 'And now they've wires and traps over the whole place. Palliser's a friend of yours—isn't he, Finn?'

'Of course I knew him,—when I was in office.'

'I don't know what he may be in office, but he's an uncommon bad sort of fellow to have in a county.'

'Shameful!' said Mr. Spooner, lifting up both his hands.

'This is my first cousin, you know,' whispered Adelaide, to Lady Baldock.

'If he were my own brother, or my grandmother, I should say the same,' continued the angry lord. 'We must have a meeting about it, and let the world know it,—that's all.' At this moment the door was again opened, and Madame Goesler entered the room.

When one wants to be natural, of necessity one becomes the reverse of natural. A clever actor,—or more frequently a clever actress,—will assume the appearance; but the very fact of the assumption renders the reality impossible. Lady Chiltern was generally very clever in the arrangement of all little social difficulties, and, had she thought less about it, might probably have managed the present affair in an easy and graceful manner. But the thing had weighed upon her mind, and she had decided that it would be expedient that she should say something when those two old friends first met each other again in her drawing-room. 'Madame Max,' she said, 'you remember Mr. Finn.' Lord Chiltern for a moment stopped the torrent of his abuse. Lord Baldock made a little effort to look uninterested, but quite in vain. Mr. Spooner stood on one side. Lady Baldock stared with all her eyes,—with some feeling of instinct that there would be something to see; and Gerard Maule, rising from the sofa, joined the circle. It seemed as though Lady Chiltern's words had caused the formation of a ring in the midst of which Phineas and Madame Goesler were to renew their acquaintance.

'Very well indeed,' said Madame Max, putting out her hand and looking full into our hero's face with her sweetest smile. 'And I hope Mr. Finn will not have forgotten me.' She did it admirably—so well that surely she need not have thought of running away.

But poor Phineas was not happy. 'I shall never forget you,' said he; and then that unavoidable blush suffused his face, and the blood began to career through his veins.

'I am so glad you are in Parliament again,' said Madame Max.

'Yes;—I've got in again, after a struggle. Are you still living in Park Lane?'

'Oh, yes;—and shall be most happy to see you.' Then she seated herself,—as did also Lady Chiltern by her side. 'I see the poor Duke's iniquities are still under discussion. I hope Lord Chiltern recognises the great happiness of having a grievance. It would be a pity that so great a blessing should be thrown away upon him.' For the moment Madame Max had got through her difficulty, and, indeed, had done so altogether till the moment should come in which she should find herself alone with Phineas. But he slunk back from the gathering before the fire, and stood solitary and silent till dinner was announced. It became his fate to take an old woman into dinner who was not very clearsighted. 'Did you know that lady before?' she asked.

'Oh, yes; I knew her two or three years ago in London.'

'Do you think she is pretty?'

'Certainly.'

'All the men say so, but I never can see it. They have been saying ever so long that the old Duke of Omnium means to marry her on his deathbed, but I don't suppose there can be anything in it.'

'Why should he put it off for so very inopportune an occasion?' asked Phineas.

CHAPTER XVI

Copperhouse Cross and Broughton Spinnies

AFTER all, the thing had not been so very bad. With a little courage and hardihood we can survive very great catastrophes, and go through them even without broken bones. Phineas, when he got up to his room, found that he had spent the evening in company with Madame Goesler, and had not suffered materially, except at the very first moment of the meeting. He had not said a word to the lady, except such as were spoken in mixed conversation with her and others; but they had been together, and no bones had been broken. It could not be that his old intimacy should be renewed, but he could now encounter her in society, as the Fates might direct, without a renewal of that feeling of dismay which had been so heavy on him.

He was about to undress when there came a knock at the door, and his host entered the room. 'What do you mean to do about smoking?' Lord Chiltern asked.

'Nothing at all.'

'There's a fire in the smoking-room, but I'm tired, and I want to go to bed. Baldock doesn't smoke. Gerard Maule is smoking in his own room, I take it. You'll probably find Spooner at this moment established somewhere in the back slums, having a pipe with old Doggett, and planning retribution. You can join them if you please.'

'Not to-night, I think. They wouldn't trust me,—and I should spoil their plans.'

'They certainly wouldn't trust you,—or any other human being. You don't mind a horse that baulks a little, do you?'

'I'm not going to hunt, Chiltern.'

'Yes, you are. I've got it all arranged. Don't you be a fool, and make us all uncomfortable. Everybody rides here;—every man, woman, and child about the place. You shall have one of the best horses I've got;—only you must be particular about your spurs.'

'Indeed, I'd rather not. The truth is, I can't afford to ride my own horses, and therefore I'd rather not ride my friends'.'

'That's all gammon. When Violet wrote she told you you'd be expected to come out. Your old flame, Madame Max, will be there, and I tell you she has a very pretty idea of keeping to hounds. Only Dandolo has that little defect.'

'Is Dandolo the horse?'

'Yes;—Dandolo is the horse. He's up to a stone over your weight, and can do any mortal thing within a horse's compass. Cox won't ride him because he baulks, and so he has come into my stable. If you'll only let him know that you're on his back, and have got a pair of spurs on your heels with rowels* in them, he'll take you anywhere. Good-night, old fellow. You can smoke if you choose, you know.'

Phineas had resolved that he would not hunt; but, nevertheless, he had brought boots with him, and breeches, fancying that if he did not he would be forced out without those comfortable appurtenances. But there came across his heart a feeling that he had reached a time of life in which it was no longer comfortable for him to live as a poor man with men who were rich. It had been his lot to do so when he was younger, and there had been some pleasure in it; but now he would rather live alone and dwell upon the memories of the past. He, too, might have been rich, and have had horses at command, had he chosen to sacrifice himself for money.

On the next morning they started in a huge waggonette for Copperhouse Cross,—a meet that was suspiciously near to the Duke's fatal wood. Spooner had explained to Phineas over night that they never did draw Trumpeton Wood on Copperhouse Cross days, and that under no possible circumstances would Chiltern now draw Trumpeton Wood. But there is no saying where a fox may run. At this time of the year, just the beginning of February, dog-foxes from the big woods were very apt to be away from home, and when found would go straight for their own earths. It was very possible that they might find themselves in Trumpeton Wood, and then certainly there would be a row. Spooner shrugged his shoulders, and shook his head, and seemed to insinuate that Lord Chiltern would certainly do something very dreadful to the Duke or to the Duke's heir if any law of venery should again be found to have been broken on this occasion.

The distance to Copperhouse Cross was twelve miles, and Phineas found himself placed in the carriage next to Madame Goesler. It had not been done of fixed design; but when a party of six are seated in a carriage, the chances are that one given person will be next to or opposite to any other given person. Madame Max had remembered this, and had prepared herself, but Phineas was taken aback when he found how close was his neighbourhood to the lady. 'Get in, Phineas,' said his lordship. Gerard Maule had already seated himself next to Miss Palliser, and Phineas had no alternative but to take the place next to Madame Max.

'I didn't know that you rode to hounds?' said Phineas.

'Oh, yes; I have done so for years. When we met it was always in London, Mr. Finn; and people there never know what other people do. Have you heard of this terrible affair about the Duke?'

'Oh, dear, yes.'

'Poor Duke! He and I have seen a great deal of each other since,—since the days when you and I used to meet. He knows nothing about all this, and the worst of it is, he is not in a condition to be told.'

'Lady Glencora could put it all right.'

'I'll tell Lady Glencora, of course,' said Madame Max. 'It seems so odd in this country that the owner of a property does not seem at all to have any exclusive right to it. I suppose the Duke could shut up the wood if he liked.'

'But they poisoned the hounds.'

'Nobody supposes the Duke did that,—or even the Duke's servants, I should think. But Lord Chiltern will hear us if we don't take care.'

'I've heard every word you've been saying,' exclaimed Lord Chiltern.

'Has it been traced to any one?'

'No,—not traced, I suppose.'

'What then, Lord Chiltern? You may speak out to me. When I'm wrong I like to be told so.'

'Then you're wrong now,' said Lord Chiltern, 'if you take the part of the Duke or of any of his people. He is bound to find foxes for the Brake hunt. It is almost a part of his title deeds. Instead of doing so he has had them destroyed.'

'It's as bad as voting against the Church establishment,' said Madame Goesler.

There was a very large meet at Copperhouse Cross, and both Madame Goesler and Phineas Finn found many old acquaintances there. As Phineas had formerly sat in the House for five years, and had been in office, and had never made himself objectionable either to his friends or adversaries, he had been widely known. He now found half a dozen men who were always members of Parliament,—men who seem, though commoners, to have been born legislators,—who all spoke to him as though his being member for Tankerville and hunting with the Brake hounds were equally matters of course.* They knew him, but they knew nothing of the break in his life. Or if they remembered that he had not been seen about the House for the last two or three years they remembered also that accidents do happen to some men. It will occur now and again that a regular denizen of Westminster will get a fall in the political hunting-field, and have to remain about the world for a year or two without a seat. That Phineas had lately triumphed over Browborough at Tankerville was known, the event having been so recent; and men congratulated him, talking of poor Browborough,—whose heavy figure had been familiar to them for many a year,—but by no means recognising that the event of which they spoke had been, as it were, life and death to their friend. Roby was there, who was at this moment Mr. Daubeny's head whip and patronage secretary. If any one should have felt acutely the exclusion of

Mr. Browborough from the House,—any one beyond the sufferer himself,—it should have been Mr. Roby; but he made himself quite pleasant, and even condescended to be jocose upon the occasion. 'So you've beat poor Browborough in his own borough,' said Mr. Roby.

'I've beat him,' said Phineas; 'but not, I hope, in a borough of his own.'

'He's been there for the last fifteen years. Poor old fellow! He's awfully cut up about this Church Question. I shouldn't have thought he'd have taken anything so much to heart. There are worse fellows than Browborough, let me tell you. What's all this I hear about the Duke poisoning the foxes?' But the crowd had begun to move, and Phineas was not called upon to answer the question.

Copperhouse Cross in the Brake Hunt was a very popular meet. It was easily reached by a train from London, was in the centre of an essentially hunting country, was near to two or three good coverts, and was in itself a pretty spot. Two roads intersected each other on the middle of Copperhouse Common, which, as all the world knows, lies just on the outskirts of Copperhouse Forest. A steep winding hill leads down from the Wood to the Cross, and there is no such thing within sight as an enclosure. At the foot of the hill, running under the wooden bridge, straggles the Copperhouse Brook,—so called by the hunting men of the present day, though men who know the country of old, or rather the county, will tell you that it is properly called the river Cobber, and that the spacious old farm buildings above were once known as the Cobber Manor House. He would be a vain man who would now try to change the name, as Copperhouse Cross has been printed in all the lists of hunting meets for at least the last thirty years; and the Ordnance map has utterly rejected the two b's. Along one of the cross-roads there was a broad extent of common, some seven or eight hundred yards in length, on which have been erected the butts*used by those well-known defenders of their country, the Copperhouse Volunteer Rifles; and just below the bridge the sluggish water becomes a little lake, having probably at some time been artificially widened, and there is a little island and a decoy for ducks. On the present occasion carriages were drawn up on all the roads, and horses were clustered on each side of the brook, and the hounds sat stately on their haunches where riflemen usually kneel to fire, and there was a hum of merry voices, and the bright colouring of pink coats, and the sheen of ladies' hunting toilettes, and that mingled look of business and amusement which is so peculiar to our national sports. Two hundred men and women had come there for the chance of a run after a fox,—for a chance against which the odds are more than two to one at every hunting day,—for a chance as to which the odds are twenty to one against the success of the individuals collected; and yet, for every horseman and every horsewoman there, not less than £5 a head will have been spent for this one day's amusement. When we give a guinea for a stall at the opera we think that we pay a large sum; but we are fairly sure of having our music. When you go to Copperhouse Cross you are by no means sure of your opera.

Why is it that when men and women congregate, though the men may beat the women in numbers by ten to one, and though they certainly speak the louder, the concrete sound that meets the ears of any outside listener is always a sound of women's voices? At Copperhouse Cross almost every one was talking, but the feeling left upon the senses was that of an amalgam of feminine laughter, feminine affectation, and feminine eagerness. Perhaps at Copperhouse Cross the determined perseverance with which Lady Gertrude Fitzaskerley addressed herself to Lord Chiltern, to Cox the huntsman, to the two whips, and at last to Mr. Spooner, may have specially led to the remark on this occasion. Lady Chiltern was very short with her, not loving Lady Gertrude. Cox bestowed upon her two 'my lady's,' and then turned from her to some peccant hound. But Spooner was partly gratified, and partly incapable, and underwent a long course of questions about the Duke and the poisoning. Lady Gertrude, whose father seemed to have owned half the coverts in Ireland, had never before heard of such enormity. She suggested a round robin* and would not be at all ashamed to put her own name to it. 'Oh, for the matter of that,' said Spooner, 'Chiltern can be round enough himself without any robin.' 'He can't be too round,' said Lady Gertrude, with a very serious aspect.

At last they moved away, and Phineas found himself riding by the side of Madame Goesler. It was natural that he should do so, as he had come with her. Maule had, of course, remained with Miss Palliser, and Chiltern and Spooner had taken themselves to their respective duties. Phineas might have avoided her, but in doing so he would have seemed to avoid her. She accepted his presence apparently as a matter of course, and betrayed by her words and manner no memory of past scenes. It was not customary with them to draw the forest, which indeed, as it now stood, was a forest only in name, and they trotted off to a gorse a mile and a half distant. This they drew blank,—then another gorse also blank,—and two or three little fringes of wood, such as there are in every country, and through which huntsmen run their hounds, conscious that no fox will lie there. At one o'clock they had not found, and the hilarity of the really hunting men as they ate their sandwiches and lit their cigars was on the decrease. The ladies talked more than ever, Lady Gertrude's voice was heard above them all, and Lord Chiltern trotted on close behind his hounds in obdurate silence. When things were going bad with him no one in the field dared to speak to him.

Phineas had never seen his horse till he reached the meet, and there found a fine-looking, very strong, bay animal, with shoulders like the top of a hay-stack, short-backed, short-legged, with enormous quarters, and a wicked-looking eye. 'He ought to be strong,' said Phineas to the groom. 'Oh, sir; strong ain't no word for him,' said the groom; ''e can carry a 'ouse.' 'I don't know whether he's fast?' inquired Phineas. 'He's fast enough for any 'ounds, sir,' said the man with that tone of assurance which always carries conviction. 'And he can jump?' 'He can jump!' continued the groom; 'no 'orse in my lord's stables can't beat him.' 'But he won't?' said Phineas.

'It's only sometimes, sir, and then the best thing is to stick him at it till he do. He'll go, he will, like a shot at last; and then he's right for the day.' Hunting men will know that all this was not quite comfortable. When you ride your own horse, and know his special defects, you know also how far that defect extends, and what real prospect you have of overcoming it. If he be slow through the mud, you keep a good deal on the road in heavy weather, and resolve that the present is not an occasion for distinguishing yourself. If he be bad at timber, you creep through a hedge. If he pulls, you get as far from the crowd as may be. You gauge your misfortune, and make your little calculation as to the best mode of remedying the evil. But when you are told that your friend's horse is perfect,—only that he does this or that,—there comes a weight on your mind from which you are unable to release it. You cannot discount your trouble at any percentage. It may amount to absolute ruin, as far as that day is concerned; and in such a circumstance you always look forward to the worst. When the groom had done his description, Phineas Finn would almost have preferred a day's canvass at Tankerville under Mr. Ruddles's authority to his present position.

When the hounds entered Broughton Spinnies, Phineas and Madame Goesler were still together. He had not been riding actually at her side all the morning. Many men and two or three ladies had been talking to her. But he had never been far from her in the ruck, and now he was again close by her horse's head. Broughton Spinnies were in truth a series of small woods, running one into another almost without intermission, never thick, and of no breadth. There was always a litter or two of cubs at the place, and in no part of the Brake country was greater care taken in the way of preservation and encouragement to interesting vixens; but the lying was bad; there was little or no real covert; and foxes were very apt to travel and get away into those big woods belonging to the Duke,—where, as the Brake sportsmen now believed, they would almost surely come to an untimely end. 'If we draw this blank I don't know what we are to do,' said Mr. Spooner, addressing himself to Madame Goesler with lachrymose anxiety.

'Have you nothing else to draw?' asked Phineas.

'In the common course of things we should take Muggery Gorse, and so on to Trumpeton Wood. But Muggery is on the Duke's land, and Chiltern is in such a fix! He won't go there unless he can't help it. Muggery Gorse is only a mile this side of the big wood.'

'And foxes of course go to the big wood?' asked Madame Max.

'Not always. They often come here,—and as they can't hang here, we have the whole country before us. We get as good runs from Muggery as from any covert in the country. But Chiltern won't go there to-day unless the hounds show a line. By George, that's a fox! That's Dido. That's a find!' And Spooner galloped away, as though Dido could do nothing with the fox she had found unless he was there to help her.

Spooner was quite right, as he generally was on such occasions. He knew the hounds even by voice, and knew what hound he could believe. Most hounds will lie occasionally, but Dido never lied. And there were many besides Spooner who believed in Dido. The whole pack rushed to her music, though the body of them would have remained utterly unmoved at the voice of any less reverenced and less trustworthy colleague. The whole wood was at once in commotion,—men and women riding hither and thither, not in accordance with any judgment; but as they saw or thought they saw others riding who were supposed to have judgment. To get away well is so very much! And to get away well is often so very difficult! There are so many things of which the horseman is bound to think in that moment. Which way does the wind blow? And then, though a fox will not long run up wind, he will break covert up wind, as often as not. From which of the various rides can you find a fair exit into the open country, without a chance of breaking your neck before the run begins? When you hear some wild halloa, informing you that one fox has gone in the direction exactly opposite to that in which the hounds are hunting, are you sure that the noise is not made about a second fox? On all these matters you are bound to make up your mind without losing a moment; and if you make up your mind wrongly the five pounds you have invested in that day's amusement will have been spent for nothing. Phineas and Madame Goesler were in the very centre of the wood when Spooner rushed away from them down one of the rides on hearing Dido's voice; and at that time they were in a crowd. Almost immediately the fox was seen to cross another ride, and a body of horsemen rushed away in that direction, knowing that the covert was small, and there the animal must soon leave the wood. Then there was a shout of 'Away!' repeated over and over again, and Lord Chiltern, running up like a flash of lightning, and passing our two friends, galloped down a third ride to the right of the others. Phineas at once followed the master of the pack, and Madame Goesler followed Phineas. Men were still riding hither and thither; and a farmer, meeting them, with his horse turned back towards the centre of the wood which they were leaving, halloaed out as they passed that there was no way out at the bottom. They met another man in pink, who screamed out something as to 'the devil of a bank down there.' Chiltern, however, was still going on, and our hero had not the heart to stop his horse in its gallop and turn back from the direction in which the hounds were running. At that moment he hardly remembered the presence of Madame Goesler, but he did remember every word that had been said to him about Dandolo. He did not in the least doubt but that Chiltern had chosen his direction rightly, and that if he were once out of the wood he would find himself with the hounds; but what if this brute should refuse to take him out of the wood? That Dandolo was very fast he soon became aware, for he gained upon his friend before him as they neared the fence. And then he saw what there was before him. A new broad ditch had been cut, with the express object of preventing egress or ingress at that point; and a great bank had been constructed with the clay. In all probability there might be another ditch on the other side. Chiltern, however, had clearly made up his mind about it. The horse he was riding went at it gallantly, cleared the first ditch, balanced himself for half a moment on the bank, and then, with a fresh spring, got into the field beyond. The tail hounds were running past outside the covert, and the master had placed himself exactly right for the work in hand. How excellent would be the condition of Finn if only Dandolo would do just as Chiltern's horse had done before him!

And Phineas almost began to hope that it might be so. The horse was going very well, and very willingly. His head was stretched out, he was pulling, not more, however, than pleasantly, and he seemed to be as anxious as his rider. But there was a little twitch about his ears which his rider did not like, and then it was impossible not to remember that awful warning given by the groom, 'It's only sometimes, sir.' And after what fashion should Phineas ride him at the obstacle? He did not like to strike a horse that seemed to be going well, and was unwilling, as are all good riders, to use his heels. So he spoke to him, and proposed to lift him at the ditch. To the very edge the horse galloped,—too fast, indeed, if he meant to take the bank as Chiltern's horse had done,—and then stopping himself so suddenly that he must have shaken every joint in his body, he planted his fore feet on the very brink, and there he stood, with his head down, quivering in every muscle. Phineas Finn, following naturally the momentum which had been given to him, went over the brute's neck head-foremost into the ditch. Madame Max was immediately off her horse. 'Oh, Mr. Finn, are you hurt?'

But Phineas, happily, was not hurt. He was shaken and dirty, but not so shaken, and not so dirty, but that he was on his legs in a minute, imploring his companion not to mind him but go on. 'Going on doesn't seem to be so easy,' said Madame Goesler, looking at the ditch as she held her horse in her hand. But to go back in such circumstances is a terrible disaster. It amounts to complete defeat; and is tantamount to a confession that you must go home, because you are unable to ride to hounds. A man, when he is compelled to do this, is almost driven to resolve at the spur of the moment that he will give up hunting for the rest of his life. And if one thing be more essential than any other to the horseman in general, it is that he, and not the animal which he rides, shall be the master. 'The best thing is to stick him at it till he do,' the groom had said; and Phineas resolved to be guided by the groom.

But his first duty was to attend on Madame Goesler. With very little assistance she was again in her saddle, and she at once declared herself certain that her horse could take the fence. Phineas again instantly jumped into his saddle, and turning Dandolo again at the ditch, rammed the rowels into the horse's sides. But Dandolo would not jump yet. He stood with his fore feet on the brink, and when Phineas with his whip struck him severely over the shoulders, he went down into the ditch on all fours, and then scrambled back again to his former position. 'What an infernal brute!' said Phineas, gnashing his teeth.

'He is a little obstinate, Mr. Finn; I wonder whether he'd jump if I gave him a lead.' But Phineas was again making the attempt, urging the horse with spurs, whip, and voice. He had brought himself now to that condition in which a man is utterly reckless as to falling himself,—or even to the kind of fall he may get,—if he can only force his animal to make the attempt. But Dandolo would not make the attempt. With ears down and head outstretched, he either stuck obstinately on the brink, or allowed himself to be forced again and again into the ditch. 'Let me try it once, Mr. Finn,' said Madame Goesler in her quiet way.

She was riding a small horse, very nearly thoroughbred, and known as a perfect hunter by those who habitually saw Madame Goesler ride. No doubt he would have taken the fence readily enough had his rider followed immediately after Lord Chiltern; but Dandolo had baulked at the fence nearly a dozen times, and evil communications will corrupt good manners. Without any show of violence, but still with persistent determination, Madame Goesler's horse also declined to jump. She put him at it again and again, and he would make no slightest attempt to do his business. Phineas raging, fuming, out of breath, miserably unhappy, shaking his reins, plying his whip, rattling himself about in the saddle, and banging his legs against the horse's sides, again and again plunged away at the obstacle. But it was all to no purpose. Dandolo was constantly in the ditch, sometimes lying with his side against the bank, and had now been so hustled and driven that, had he been on the other side, he would have had no breath left to carry his rider, even in the ruck of the hunt. In the meantime the hounds and the leading horsemen were far away,—never more to be seen on that day by either Phineas Finn or Madame Max Goesler. For a while, during the frantic efforts that were made, an occasional tardy horseman was viewed galloping along outside the covert, following the tracks of those who had gone before. But before the frantic efforts had been abandoned as utterly useless every vestige of the morning's work had left the neighbourhood of Broughton Spinnies, except these two unfortunate ones. At last it was necessary that the defeat should be acknowledged. 'We're beaten, Madame Goesler,' said Phineas, almost in tears.

'Altogether beaten, Mr. Finn.'

'I've a good mind to swear that I'll never come out hunting again.'

'Swear what you like, if it will relieve you, only don't think of keeping such an oath. I've known you before this to be depressed by circumstances quite as distressing as these, and to be certain that all hope was over;—but yet you have recovered.' This was the only allusion she had yet made to their former acquaintance. 'And now we must think of getting out of the wood.'

'I haven't the slightest idea of the direction of anything.'

'Nor have I; but as we clearly can't get out this way we might as well try the other. Come along. We shall find somebody to put us in the right road. For my part I'm glad it is no worse. I thought at one time that you were going to break your neck.' They rode on for a few minutes in silence, and then she spoke again. 'Is it not odd, Mr. Finn, that after all that has come and gone you and I should find ourselves riding about Broughton Spinnies together?'

CHAPTER XVII

Madame Goesler's story

'AFTER all that has come and gone, is it not odd that you and I should find ourselves riding about Broughton Spinnies together?' That was the question which Madame Goesler asked Phineas Finn when they had both agreed that it was impossible to jump over the bank out of the wood, and it was, of course, necessary that some answer should be given to it.

'When I saw you last in London,' said Phineas, with a voice that was gruff, and a manner that was abrupt, 'I certainly did not think that we should meet again so soon.'

'No;—I left you as though I had grounds for quarrelling; but there was no quarrel. I wrote to you, and tried to explain that.'

'You did;—and though my answer was necessarily short, I was very grateful.'

'And here you are back among us; and it does seem so odd. Lady Chiltern never told me that I was to meet you.'

'Nor did she tell me.'

'It is better so, for otherwise I should not have come, and then, perhaps, you would have been all alone in your discomfiture at the bank.'

'That would have been very bad.'

'You see I can be quite frank with you, Mr. Finn. I am heartily glad to see you, but I should not have come had I been told. And when I did see you, it was quite improbable that we should be thrown together as we are now,—was it not? Ah;—here is a man, and he can tell us the way back to Copperhouse Cross. But I suppose we had better ask for Harrington Hall at once.'

The man knew nothing at all about Harrington Hall, and very little about Copperhouse; but he did direct them on to the road, and they found that they were about sixteen miles from Lord Chiltern's house. The hounds had gone away in the direction of Trumpeton Wood, and it was agreed that it would be useless to follow them. The waggonette had been left at an inn about two miles from Copperhouse Cross, but they resolved to abandon that and to ride direct to Harrington Hall. It was now nearly three o'clock, and they would not be subjected to the shame which falls upon sportsmen who are seen riding home very early in the day. To get oneself lost before twelve, and then to come home, is a very degrading thing; but at any time after two you may be supposed to have ridden the run of the season, and to be returning after an excellent day's work.

Then Madame Goesler began to talk about herself, and to give a short history of her life during the last two-and-a-half years. She did this in a frank natural manner, continuing her tale in a low voice, as though it were almost a matter of course that she should make the recital to so old a friend. And Phineas soon began to feel that it was natural that she should do so. 'It was just before you left us,' she said, 'that the Duke took to coming to my house.' The duke spoken of was the Duke of Omnium, and Phineas well remembered to have heard some rumours about the Duke and Madame Max. It had been hinted to him that the Duke wanted to marry the lady, but that rumour he had never believed. The reader, if he has duly studied the history of the age, will know that the Duke did make an offer to Madame Goesler,* pressing it with all his eloquence, but that Madame Goesler, on mature consideration, thought it best to decline to become a duchess. Of all this, however, the reader who understands Madame Goesler's character will be quite sure that she did not say a word to Phineas Finn. Since the business had been completed she had spoken of it to no one but to Lady Glencora Palliser, who had forced herself into a knowledge of all the circumstances while they were being acted.

'I met the Duke once at Matching,' said Phineas.

'I remember it well. I was there, and first made the Duke's acquaintance on that occasion. I don't know how it was that we became intimate;—but we did, and then I formed a sort of friendship with Lady Glencora; and somehow it has come about that we have been a great deal together since.'

'I suppose you like Lady Glencora?'

'Very much indeed,—and the Duke, too. The truth is, Mr. Finn, that let one boast as one may of one's independence,—and I very often do boast of mine to myself,—one is inclined to do more for a Duke of Omnium than for a Mr. Jones.'

'The Dukes have more to offer than the Joneses;—I don't mean in the way of wealth only, but of what one enjoys most in society generally.'

'I suppose they have. At any rate, I am glad that you should make some excuse for me. But I do like the man. He is gracious and noble in his bearing. He is now very old, and sinking fast into the grave; but even the wreck is noble.'

'I don't know that he ever did much,' said Phineas.

'I don't know that he ever did anything according to your idea of doing. There must be some men who do nothing.'

'But a man with his wealth and rank has opportunities so great! Look at his nephew!'

'No doubt Mr. Palliser is a great man. He never has a moment to speak to his wife or to anybody else; and is always thinking so much about the country that I doubt if he knows anything about his own affairs. Of course he is a man of a different stamp,—and of a higher stamp, if you will. But I have an idea that such characters as those of the present Duke are necessary to the maintenance of a great aristocracy. He has had the power of making the world believe in him simply because he has been rich and a duke. His nephew, when he comes to the title, will never receive a tithe of the respect that has been paid to this old fainéant.'*

'But he will achieve much more than ten times the reputation,' said Phineas.

'I won't compare them, nor will I argue; but I like the Duke. Nay;—I love him. During the last two years I have allowed the whole fashion of my life to be remodelled by this intimacy. You knew what were my habits. I have only been in Vienna for one week since I last saw you, and I have spent months and months at Matching.'

'What do you do there?'

'Read to him;—talk to him;—give him his food, and do all that in me lies to make his life bearable. Last year, when it was thought necessary that very distinguished people should be entertained at the great family castle,—in Barsetshire, you know——'

'I have heard of the place.'

'A regular treaty or agreement was drawn up. Conditions were sealed and signed. One condition was that both Lady Glencora and I should be there. We put our heads together to try to avoid this; as, of course, the Prince would not want to see me particularly,—and it was altogether so grand an affair that things had to be weighed. But the Duke was inexorable. Lady Glencora at such a time would have other things to do, and I must be there, or Gatherum Castle should not be opened. I suggested whether I could not remain in the background and look after the Duke as a kind of upper nurse, —but Lady Glencora said it would not do.'

'Why should you subject yourself to such indignity?'

'Simply from love of the man. But you see I was not subjected. For two days I wore my jewels beneath royal eyes,—eyes that will sooner or later belong to absolute majesty. It was an awful bore, and I ought to have been at Vienna. You ask me why I did it. The fact is that things sometimes become too strong for one, even when there is no real power of constraint. For years past I have been used to have my own way, but when there came a question of the entertainment of royalty I found myself reduced to blind obedience. I had to go to Gatherum Castle, to the absolute neglect of my business; and I went.'

'Do you still keep it up?'

'Oh, dear, yes. He is at Matching now, and I doubt whether he will ever leave it again. I shall go there from here as a matter of course, and relieve guard with Lady Glencora.'

'I don't see what you get for it all.'

'Get;—what should I get? You don't believe in friendship, then?'

'Certainly I do;—but this friendship is so unequal. I can hardly understand that it should have grown from personal liking on your side.'

'I think it has,' said Madame Goesler, slowly. 'You see, Mr. Finn, that you as a young man can hardly understand how natural it is that a young woman,—if I may call myself young,—should minister to an old man.'

'But there should be some bond to the old man.'

'There is a bond.'

'You must not be angry with me,' said Phineas.

'I am not in the least angry.'

'I should not venture to express any opinion, of course,—only that you ask me.'

'I do ask you, and you are quite welcome to express your opinion. And were it not expressed, I should know what you thought just the same. I have wondered at it myself sometimes,—that I should have become as it were engulfed in this new life, almost without will of my own. And when he dies, how shall I return to the other life? Of course I have the house in Park Lane still, but my very maid talks of Matching as my home.'

'How will it be when he has gone?'

'Ah,—how indeed? Lady Glencora and I will have to curtsey to each other, and there will be an end of it. She will be a duchess then, and I shall no longer be wanted.'

'But even if you were wanted——?'

'Oh, of course. It must last the Duke's time, and last no longer. It would not be a healthy kind of life were it not that I do my very best to make the evening of his days pleasant for him, and in that way to be of some service in the world. It has done me good to think that I have in some small degree sacrificed myself. Let me see;—we are to turn here to the left. That goes to Copperhouse Cross, no doubt. Is it not odd that I should have told you all this history?'

'Just because this brute would not jump over the fence.'

'I dare say I should have told you, even if he had jumped over; but certainly this has been a great opportunity. Do you tell your friend Lord Chiltern not to abuse the poor Duke any more before me. I dare say our host is all right in what he says; but I don't like it. You'll come and see me in London, Mr. Finn?'

'But you'll be at Matching?'

'I do get a few days at home sometimes. You see I have escaped for the present,—or otherwise you and I would not have come to grief together in Broughton Spinnies.'

Soon after this they were overtaken by others who were returning home, and who had been more fortunate than they in getting away with the hounds. The fox had gone straight for Trumpeton Wood, not daring to try the gorse on the way, and then had been run to ground. Chiltern was again in a towering passion, as the earths, he said, had been purposely left open. But on this matter the men who had overtaken our friends were both of opinion that Chiltern was wrong. He had allowed it to be understood that he would not draw Trumpeton Wood, and he had therefore no right to expect that the earths should be stopped. But there were and had been various opinions on this difficult point, as the laws of hunting are complex, recondite, numerous, traditional, and not always perfectly understood. Perhaps the day may arrive in which they shall be codified under the care of some great and laborious master of hounds.

'And they did nothing more?' asked Phineas.

'Yes;—they chopped another fox before they left the place,—so that in point of fact they have drawn Trumpeton. But they didn't mean it.'

When Madame Max Goesler and Phineas had reached Harrington Hall they were able to give their own story of the day's sport to Lady Chiltern, as the remainder of the party had not as yet returned.

CHAPTER XVIII
Spooner of Spoon Hall

ADELAIDE PALLISER was a tall, fair girl, exquisitely made, with every feminine grace of motion, highly born, and carrying always the warranty of her birth in her appearance; but with no special loveliness of face. Let not any reader suppose that therefore she was plain. She possessed much more than a sufficiency of charm to justify her friends in claiming her as a beauty, and the demand had been generally allowed by public opinion. Adelaide Palliser was always spoken of as a girl to be admired; but she was not one whose countenance would strike with special admiration any beholder who did not know her. Her eyes were pleasant and bright, and, being in truth green, might, perhaps with propriety, be described as grey. Her nose was well formed. Her mouth was, perhaps, too small. Her teeth were perfect. Her chin was somewhat too long, and was on this account the defective feature of her face. Her hair was brown and plentiful; but in no way peculiar. No doubt she wore a chignon;* but if so she wore it with the special view of being in no degree remarkable in reference to her head-dress. Such as she was,—beauty or no beauty—her own mind on the subject was made up, and she had resolved long since that the gift of personal loveliness had not been bestowed upon her. And yet after a fashion she was proud of her own appearance. She knew that she looked like a lady, and she knew also that she had all that command of herself which health and strength can give to a woman when she is without feminine affectation.

Lady Chiltern, in describing her to Phineas Finn, had said that she talked Italian, and wrote for the *Times*. The former assertion was, no doubt, true, as Miss Palliser had passed some years of her childhood in Florence; but the latter statement was made probably with reference to her capability rather than her performance. Lady Chiltern intended to imply that Miss Palliser was so much better educated than young ladies in general that she was able to express herself intelligibly in her own language. She had been well educated, and would, no doubt, have done the *Times* credit had the *Times* chosen to employ her.

She was the youngest daughter of the youngest brother of the existing Duke of Omnium, and the first cousin, therefore, of Mr. Plantagenet Palliser, who was the eldest son of the second brother. And as her mother had been a Bavilard there could be no better blood. But Adelaide had been brought up so far away from the lofty Pallisers and lofty Bavilards as almost to have lost the flavour of her birth. Her father and mother had died when she was an infant, and she had gone to the custody of a much older half-sister, Mrs. Atterbury, whose mother had been not a Bavilard, but a Brown. And Mr. Atterbury was a mere nobody, a rich, erudite, highly-accomplished gentleman, whose father had made his money at the bar, and whose grandfather had been a country clergyman. Mrs. Atterbury, with her husband, was still living at Florence; but Adelaide Palliser had quarrelled with Florence life, and had gladly consented to make a long visit to her friend Lady Chiltern.

In Florence she had met Gerard Maule, and the acquaintance had not been viewed with favour by the Atterburys. Mrs. Atterbury knew the history of the Maule family, and declared to her sister that no good could come from any intimacy. Old Mr. Maule, she said, was disreputable. Mrs. Maule, the mother,—who, according to Mr. Atterbury, had been the only worthy member of the family,—was long since dead. Gerard Maule's sister had gone away with an Irish cousin, and they were now living in India on the professional income of a captain in a foot regiment. Gerard Maule's younger brother had gone utterly to the dogs, and nobody knew anything about him. Maule Abbey, the family seat in Herefordshire, was,—so said Mrs. Atterbury,—absolutely in ruins. The furniture, as all the world knew, had been sold by the squire's creditors under the sheriff's order ten years ago, and not a chair or a table had been put into the house since that time. The property, which was small,—£2,000 a year at the outside,—was, no doubt, entailed on the eldest son; and Gerard, fortunately, had a small fortune of his own, independent of his father. But then he was also a spendthrift,—so said Mrs. Atterbury,—keeping a stable full of horses, for which he could not afford to pay; and he was, moreover, the most insufferably idle man who ever wandered about the world without any visible occupation for his hours. 'But he hunts,' said Adelaide. 'Do you call that an occupation?' asked Mrs. Atterbury with scorn. Now Mrs. Atterbury painted pictures, copied Madonnas, composed sonatas, corresponded with learned men in Rome, Berlin, and Boston, had been the intimate friend of Cavour, had paid a visit to Garibaldi on his island with the view of explaining to him the real condition of Italy,—and was supposed to understand Bismarck.* Was it possible that a woman who so filled her own life should accept hunting as a creditable employment for a young man, when it was admitted to be his sole employment? And, moreover, she desired that her sister Adelaide should marry a certain Count Brudi, who, according to her belief, had more advanced ideas about things in general than any other living human being. Adelaide Palliser had determined that she would not marry Count Brudi; had, indeed, almost determined that she would marry Gerard Maule, and had left her brother-in-law's house in Florence after something like a quarrel. Mrs. Atterbury had declined to authorise the visit to Harrington Hall, and then Adelaide had pleaded her age and independence. She was her own mistress if she so chose to call herself, and would not, at any rate, remain in Florence at the present moment to receive the attentions of Signor Brudi. Of the previous winter she had passed three months with some relatives in England, and there she had learned to ride to hounds, had first met Gerard Maule, and had made acquaintance with Lady Chiltern. Gerard Maule had wandered to Italy after her, appearing at Florence in his desultory way, having no definite purpose, not even that of asking Adelaide to be his wife,—but still pursuing her, as though he wanted her without knowing what he wanted. In the course of the Spring, however, he had proposed, and had been almost accepted. But Adelaide, though she would not yield to her sister, had been frightened. She knew that she loved the man, and she swore to herself a thousand times that she would not be dictated to by her sister;—but was she prepared to accept the fate which would at once be hers were she now to marry Gerard Maule? What could she do with a man who had no ideas of his own as to what he ought to do with himself?

Lady Chiltern was in favour of the marriage. The fortune, she said, was as much as Adelaide was entitled to expect, the man was a gentleman, was tainted by no vices, and was truly in love. 'You had better let them fight it out somewhere else,' Lord Chiltern had said when his wife proposed that the invitation to Gerard Maule should be renewed; but Lady Chiltern had known that if 'fought out' at all, it must be fought out at Harrington Hall. 'We have asked him to come back,' she said to Adelaide, 'in order that you may make up your mind. If he chooses to come, it will show that he is in earnest; and then you must take him, or make him understand that he is not to be taken.' Gerard Maule had chosen to come; but Adelaide Palliser had not as yet quite made up her mind.

Perhaps there is nothing so generally remarkable in the conduct of young ladies in the phase of life of which we are now speaking as the facility,—it may almost be said audacity, —with which they do make up their minds. A young man seeks a young woman's hand in marriage, because she has waltzed stoutly with him, and talked pleasantly between the dances;—and the young woman gives it, almost with gratitude. As to the young man, the readiness of his action is less marvellous than hers. He means to be master, and, by the very nature of the joint life they propose to lead, must take her to his sphere of life, not bind himself to hers. If he worked before he will work still. If he was idle before he will be idle still; and he probably does in some sort make a calculation and strike a balance between his means and the proposed additional burden of a wife and children. But she, knowing nothing, takes a monstrous leap in the dark, in which everything is to be changed, and in which everything is trusted to chance. Miss Palliser, however, differing in this from the majority of her friends and acquaintances, frightened, perhaps by those representations of her sister to which she would not altogether yield, had paused, and was still pausing. 'Where should we go and live if I did marry him?' she said to Lady Chiltern.

'I suppose he has an opinion of his own on that subject?'

'Not in the least, I should think.'

'Has he never said anything about it?'

'Oh dear no. Matters have not got so far as that at all;—nor would they ever, out of his own head. If we were married and taken away to the train he would only ask what place he should take the tickets for when he got to the station.'

'Couldn't you manage to live at Maule Abbey?'

'Perhaps we might; only there is no furniture, and, as I am told, only half a roof.'

'It does seem to be absurd that you two should not make up your mind, just as other people do,' said Lady Chiltern. 'Of course he is not a rich man, but you have known that all along.'

'It is not a question of wealth or poverty, but of an utterly lack-a-daisical indifference to everything in the world.'

'He is not indifferent to you.'

'That is the marvellous part of it,' said Miss Palliser.

This was said on the evening of the famous day at Broughton Spinnies, and late on that night Lord Chiltern predicted to his wife that another episode was about to occur in the life of their friend.

'What do you think Spooner has just asked me?'

'Permission to fight the Duke, or Mr. Palliser?'

'No,—it's nothing about the hunting. He wants to know if you'd mind his staying here three or four days longer.'

'What a very odd request!'

'It is odd, because he was to have gone to-morrow. I suppose there's no objection.'

'Of course not if you like to have him.'

'I don't like it a bit,' said Lord Chiltern; 'but I couldn't turn him out. And I know what it means.'

'What does it mean?'

'You haven't observed anything?'

'I have observed nothing in Mr. Spooner, except an awe-struck horror at the trapping of a fox.'

'He's going to propose to Adelaide Palliser.'

'Oswald! You are not in earnest.'

'I believe he is. He would have told me if he thought I could give him the slightest encouragement. You can't very well turn him out now.'

'He'll get an answer that he won't like if he does,' said Lady Chiltern.

Miss Palliser had ridden well on that day, and so had Gerard Maule. That Mr. Spooner should ride well to hounds was quite a matter of course. It was the business of his life to do so, and he did it with great judgment. He hated Maule's style of riding, considering it to be flashy, injurious to hunting, and unsportsmanlike; and now he had come to hate the man. He had, of course, perceived how close were the attentions paid by Mr. Maule to Miss Palliser, and he thought that he perceived that Miss Palliser did not accept them with thorough satisfaction. On his way back to Harrington Hall he made some inquiries, and was taught to believe that Mr. Maule was not a man of very high standing in the world. Mr. Spooner himself had a very pretty property of his own,—which was all his own. There was no doubt about his furniture, or about the roof at Spoon Hall. He was Spooner of Spoon Hall, and had been High Sheriff for his county. He was not so young as he once had been;—but he was still a young man, only just turned forty, and was his own master in everything. He could read, and he always looked at the country newspaper; but a book was a thing that he couldn't bear to handle. He didn't think he had ever seen a girl sit a horse better than Adelaide Palliser sat hers, and a girl who rode as she did would probably like a man addicted to hunting. Mr.

Spooner knew that he understood hunting, whereas that fellow Maule cared for nothing but jumping over flights of rails. He asked a few questions that evening of Phineas Finn respecting Gerard Maule, but did not get much information. 'I don't know where he lives;' said Phineas; 'I never saw him till I met him here.'

'Don't you think he seems sweet upon that girl?'

'I shouldn't wonder if he is.'

'She's an uncommonly clean-built young woman, isn't she?' said Mr. Spooner; 'but it seems to me she don't care much for Master Maule. Did you see how he was riding to-day?'

'I didn't see anything, Mr. Spooner.'

'No, no; you didn't get away. I wish he'd been with you. But she went uncommon well.' After that he made his request to Lord Chiltern, and Lord Chiltern, with a foresight quite unusual to him, predicted the coming event to his wife.

There was shooting on the following day, and Gerard Maule and Mr. Spooner were both out. Lunch was sent down to the covert side, and the ladies walked down and joined the sportsmen. On this occasion Mr. Spooner's assiduity was remarkable, and seemed to be accepted with kindly grace. Adelaide even asked a question about Trumpeton Wood, and expressed an opinion that her cousin was quite wrong because he did not take the matter up. 'You know it's the keepers do it all,' said Mr. Spooner, shaking his head with an appearance of great wisdom. 'You never can have foxes unless you keep your keepers well in hand. If they drew the Spoon Hall coverts blank I'd dismiss my man the next day.'

'It mightn't be his fault.'

'He knows my mind, and he'll take care that there are foxes. They've been at my stick covert* three times this year, and put a brace out each time. A leash* went from it last Monday week. When a man really means a thing, Miss Palliser, he can pretty nearly always do it.' Miss Palliser replied with a smile that she thought that to be true, and Mr. Spooner was not slow at perceiving that this afforded good encouragement to him in regard to that matter which was now weighing most heavily upon his mind.

On the next day there was hunting again, and Phineas was mounted on a horse more amenable to persuasion than old Dandolo. There was a fair run in the morning, and both Phineas and Madame Max were carried well. The remarkable event in the day, however, was the riding of Dandolo in the afternoon by Lord Chiltern himself. He had determined that the horse should go out, and had sworn that he would ride him over a fence if he remained there making the attempt all night. For two weary hours he did remain, with a groom behind him, spurring the brute against a thick hedge, with a ditch at the other side of it, and at the end of the two hours he succeeded. The horse at last made a buck leap and went over with a loud grunt. On his way home Lord Chiltern sold the horse to a farmer for fifteen pounds;—and that was the end of Dandolo as far as the Harrington Hall stables were concerned. This took place on the Friday, the 8th of February. It was understood that Mr. Spooner was to return to Spoon Hall on Saturday, and on Monday, the 11th, Phineas was to go to London. On the 12th the Session would begin, and he would once more take his seat in Parliament.

'I give you my word and honour, Lady Chiltern,' Gerard Maule said to his hostess, 'I believe that oaf of a man is making up to Adelaide.' Mr. Maule had not been reticent about his love towards Lady Chiltern, and came to her habitually in all his troubles.

'Chiltern has told me the same thing.'

'No!'

'Why shouldn't he see it, as well as you? But I wouldn't believe it.'

'Upon my word I believe it's true. But, Lady Chiltern——'

'Well, Mr. Maule.'

'You know her so well.'

'Adelaide, you mean?'

'You understand her thoroughly. There can't be anything in it; is there?'

'How anything?'

'She can't really—like him?'

'Mr. Maule, if I were to tell her that you had asked such a question as that I don't believe that she'd ever speak a word to you again; and it would serve you right. Didn't you call him an oaf?'

'I did.'

'And how long has she known him?

'I don't believe she ever spoke to him before yesterday.'

'And yet you think that she will be ready to accept this oaf as her husband to-morrow! Do you call that respect?'

'Girls do such wonderful strange things. What an impudent ass he must be!'

'I don't see that at all. He may be an ass and yet not impudent, or impudent and yet not an ass. Of course he has a right to speak his mind,—and she will have a right to speak hers.'

CHAPTER XIX

Something out of the way

THE Brake hounds went out four days a week, Monday, Wednesday, Friday, and Saturday; but the hunting party on this Saturday was very small. None of the ladies joined in it, and when Lord Chiltern came down to breakfast at half-past eight he met no one but Gerard Maule. 'Where's Spooner?' he asked. But neither Maule nor the servant could answer the question. Mr. Spooner was a man who never missed a day from the beginning of cubbing to the end of the season, and who, when April came, could give you an account of the death of every fox killed. Chiltern cracked his eggs, and said nothing more for the moment, but Gerard Maule had his suspicions. 'He must be coming,' said Maule; 'suppose you send up to him.' The servant was sent, and came down with Mr. Spooner's compliments. Mr. Spooner didn't mean to hunt to-day. He had something of a headache. He would see Lord Chiltern at the meet on Monday.

Maule immediately declared that neither would he hunt; but Lord Chiltern looked at him, and he hesitated. 'I don't care about your knowing,' said Gerard.

'Oh,—I know. Don't you be an ass.'

'I don't see why I should give him an opportunity.'

'You're to go and pull your boots and breeches off because he has not put his on, and everybody is to be told of it! Why shouldn't he have an opportunity, as you call it? If the opportunity can do him any good, you may afford to be very indifferent.'

'It's a piece of d—— impertinence,' said Maule, with most unusual energy.

'Do you finish your breakfast, and come and get into the trap. We've twenty miles to go. You can ask Spooner on Monday how he spent his morning.'

At ten o'clock the ladies came down to breakfast, and the whole party were assembled. 'Mr. Spooner!' said Lady Chiltern to that gentleman, who was the last to enter the room. 'This is a marvel!' He was dressed in a dark-blue frock-coat, with a coloured silk handkerchief round his neck, and had brushed his hair down close to his head. He looked quite unlike himself, and would hardly have been known by those who had never seen him out of the hunting field. In his dress clothes of an evening, or in his shooting coat, he was still himself. But in the garb he wore on the present occasion he was quite unlike Spooner of Spoon Hall, whose only pride in regard to clothes had hitherto been that he possessed more pairs of breeches than any other man in the county. It was ascertained afterwards, when the circumstances came to be investigated, that he had sent a man all the way across to Spoon Hall for that coat and the coloured neck-handkerchief on the previous day; and some one, most maliciously, told the story abroad. Lady Chiltern, however, always declared that her secrecy on the matter had always been inviolable.

'Yes, Lady Chiltern; yes,' said Mr. Spooner, as he took a

seat at the table; 'wonders never cease, do they?' He had prepared himself even for this moment, and had determined to show Miss Palliser that he could be sprightly and engaging even without his hunting habiliments.

'What will Lord Chiltern do without you?' one of the ladies asked.

'He'll have to do his best.'

'He'll never kill a fox,' said Miss Palliser.

'Oh, yes; he knows what he's about. I was so fond of my pillow this morning that I thought I'd let the hunting slide for once. A man should not make a toil of his pleasure.'

Lady Chiltern knew all about it, but Adelaide Palliser knew nothing. Madame Goesler, when she observed the light-blue necktie, at once suspected the execution of some great intention. Phineas was absorbed in his observation of the difference in the man. In his pink coat he always looked as though he had been born to wear it, but his appearance was now that of an amateur actor got up in a miscellaneous middle-age costume. He was sprightly, but the effort was painfully visible. Lady Baldock said something afterwards, very ill-natured, about a hog in armour, and old Mrs. Burnaby spoke the truth when she declared that all the comfort of her tea and toast was sacrificed to Mr. Spooner's frock coat. But what was to be done with him when breakfast was over? For a while he was fixed upon poor Phineas, with whom he walked across to the stables. He seemed to feel that he could hardly hope to pounce upon his prey at once, and that he must bide his time.

Out of the full heart the mouth speaks. 'Nice girl, Miss Palliser,' he said to Phineas, forgetting that he had expressed himself nearly in the same way to the same man on a former occasion.

'Very nice, indeed. It seems to me that you are sweet upon her yourself.'

'Who? I! Oh, no—I don't think of those sort of things. I suppose I shall marry some day. I've a house fit for a lady to-morrow, from top to bottom, linen and all. And my property's my own.'

'That's a comfort.'

'I believe you. There isn't a mortgage on an acre of it, and that's what very few men can say. As for Miss Palliser, I don't know that a man could do better; only I don't think much of those things. If ever I do pop the question, I shall do it on the spur of the moment. There'll be no preparation with me, nor yet any beating about the bush. "Would it suit your views, my dear, to be Mrs. Spooner?" that's about the long and the short of it. A clean-made little mare, isn't she?' This last observation did not refer to Adelaide Palliser, but to an animal standing in Lord Chiltern's stables. 'He bought her from Charlie Dickers for a twenty pound note last April. The mare hadn't a leg to stand upon. Charlie had been stagging with her for the last two months, and knocked her all to pieces. She's a screw,* of course, but there isn't anything carries Chiltern so well. There's nothing like a good screw. A man'll often go with two hundred and fifty guineas between his legs, supposed to be all there because the animal's sound, and yet he don't know his work. If you like schooling a young 'un, that's all very well. I used to be fond of it myself; but I've come to feel that being carried to hounds without much thinking about it is the cream of hunting, after all. I wonder what the ladies are at? Shall we go back and see?' Then they turned to the house, and Mr. Spooner began to be a little fidgety. 'Do they sit altogether mostly all the morning?'

'I fancy they do.'

'I suppose there's some way of dividing them. They tell me you know all about women. If you want to get one to yourself, how do you manage it?'

'In perpetuity, do you mean, Mr. Spooner?'

'Any way;—in the morning, you know.'

'Just to say a few words to her?'

'Exactly that;—just to say a few words. I don't mind asking you, because you've done this kind of thing before.'

'I should watch my opportunity,' said Phineas, remembering a period of his life in which he had watched much and had found it very difficult to get an opportunity.

'But I must go after lunch,' said Mr. Spooner; 'I'm expected home to dinner, and I don't know much whether they'll like me to stop over Sunday.'

'If you were to tell Lady Chiltern——'

'I was to have gone on Thursday, you know. You won't tell anybody?'

'Oh dear no.'

'I think I shall propose to that girl. I've about made up my mind to do it, only a fellow can't call her out before half a dozen of them. Couldn't you get Lady C. to trot her out into the garden? You and she are as thick as thieves.'

'I should think Miss Palliser was rather difficult to be managed.'

Phineas declined to interfere, taking upon himself to assure Mr. Spooner that attempts to arrange matters in that way never succeeded. He went in and settled himself to the work of answering correspondents at Tankerville, while Mr. Spooner hung about the drawing-room, hoping that circumstances and time might favour him. It is to be feared that he made himself extremely disagreeable to poor Lady Chiltern, to whom he was intending to open his heart could he only find an opportunity for so much as that. But Lady Chiltern was determined not to have his confidence, and at last withdrew from the scene in order that she might not be entrapped. Before lunch had come all the party knew what was to happen,—except Adelaide herself. She, too, perceived that something was in the wind, that there was some stir, some discomfort, some secret affair forward, or some event expected which made them all uneasy;—and she did connect it with the presence of Mr. Spooner. But, in pitiable ignorance of the facts that were clear enough to everybody else, she went on watching and wondering, with a half-formed idea that the house would be more pleasant as soon as Mr. Spooner should have taken his departure. He was to go after lunch. But on such occasions there is, of course, a latitude, and 'after lunch' may be stretched at any rate to the five o'clock tea. At three o'clock Mr. Spooner was still hanging about. Madame Goesler and Phineas, with an openly declared intention of friendly intercourse, had gone out to walk together. Lord and Lady Baldock were on horseback. Two or three old ladies hung over the fire and gossiped. Lady Chiltern had retired to her baby;—when on a sudden Adelaide Palliser declared her intention of walking into the village. 'Might I accompany you, Miss Palliser?' said Mr. Spooner; 'I want a walk above all things.' He was very brave, and persevered though it was manifest that the lady did not desire his company. Adelaide said something about an old woman whom she intended to visit; whereupon Mr. Spooner declared that visiting old women was the delight of his life. He would undertake to give half a sovereign to the old woman if Miss Palliser would allow him to come. He was very brave, and persevered in such a fashion that he carried his point. Lady Chiltern from her nursery window saw them start through the shrubbery together.

'I have been waiting for this opportunity all the morning,' said Mr. Spooner, gallantly.

But in spite of his gallantry, and although she had known, almost from breakfast time, that he had been waiting for something, still she did not suspect his purpose. It has been said that Mr. Spooner was still young, being barely over forty years of age; but he had unfortunately appeared to be old to Miss Palliser. To himself it seemed as though the fountains of youth were still running through all his veins. Though he had given up schooling young horses, he could ride as hard as ever. He could shoot all day. He could take 'his whack of wine,' as he called it, sit up smoking half the night, and be on horseback the next morning after an early breakfast without the slightest feeling of fatigue. He was a red-faced little man, with broad shoulders, clean shaven, with small eyes, and a nose on which incipient pimples began to show themselves. To himself and the comrades of his life he was almost as young as he had ever been; but the young ladies of the county called him Old Spooner, and regarded him as a permanent assistant unpaid huntsman to the Brake hounds. It was not within the compass of Miss Palliser's imagination to conceive that this man should intend to propose himself to her as her lover.

'I have been waiting for this opportunity all the morning,' said Mr. Spooner. Adelaide Palliser turned round and looked at him, still understanding nothing. Ride at any fence hard enough, and the chances are you'll get over. The harder you ride the heavier the fall, if you get a fall; but the greater the chance of your getting over. This had been a precept in the life of Mr. Spooner, verified by much experience, and he had resolved that he would be guided by it on this occasion. 'Ever since I first saw you, Miss Palliser, I have been so much taken by you that,—that,—in point of fact, I love you better than all the women in the world I ever saw; and will you,—will you be Mrs. Spooner?'

He had at any rate ridden hard at his fence. There had been no craning,—no looking about for an easy place, no hesitation as he brought his horse up to it. No man ever rode straighter than he did on this occasion. Adelaide stopped short on the path, and he stood opposite to her, with his fingers inserted between the closed buttons of his frock-coat. 'Mr. Spooner!' exclaimed Adelaide.

'I am quite in earnest, Miss Palliser; no man ever was more in earnest. I can offer you a comfortable well-furnished home, an undivided heart, a good settlement, and no embarrassment on the property. I'm fond of a country life myself, but I'll adapt myself to you in everything reasonable.'

'You are mistaken, Mr. Spooner; you are indeed.'

'How mistaken?'

'I mean that it is altogether out of the question. You have surprised me so much that I couldn't stop you sooner; but pray do not speak of it again.'

'It is a little sudden, but what is a man to do? If you will only think of it——'

'I can't think of it at all. There is no need for thinking. Really, Mr. Spooner, I can't go on with you. If you wouldn't mind turning back I'll walk into the village by myself.'

Mr. Spooner, however, did not seem inclined to obey this injunction, and stood his ground, and, when she moved on, walked on beside her. 'I must insist on being left alone,' she said.

'I haven't done anything out of the way,' said the lover.

'I think it's very much out of the way. I have hardly ever spoken to you before. If you will only leave me now there shall not be a word more said about it.'

But Mr. Spooner was a man of spirit. 'I'm not in the least ashamed of what I've done,' he said.

'But you might as well go away, when it can't be of any use.'

'I don't know why it shouldn't be of use. Miss Palliser, I'm a man of good property. My great-great-grandfather lived at Spoon Hall, and we've been there ever since. My mother was one of the Platters of Platter House. I don't see that I've done anything out of the way. As for shilly-shallying, and hanging about, I never knew any good come from it. Don't let us quarrel, Miss Palliser. Say that you'll take a week to think of it.'

'But I won't think of it at all; and I won't go on walking with you. If you'll go one way, Mr. Spooner, I'll go the other.'

Then Mr. Spooner waxed angry. 'Why am I to be treated with disdain?' he said.

'I don't want to treat you with disdain. I only want you to go away.'

'You seem to think that I'm something,—something altogether beneath you.'

And so in truth she did. Miss Palliser had never analysed her own feelings and emotions about the Spooners whom she met in society; but she probably conceived that there were people in the world who, from certain accidents, were accustomed to sit at dinner with her, but who were no more fitted for her intimacy than were the servants who waited upon her. Such people were to her little more than the tables and chairs with which she was brought in contact. They were persons with whom it seemed to her to be impossible that she should have anything in common,—who were her inferiors, as completely as were the menials around her. Why she should thus despise Mr. Spooner, while in her heart of hearts she loved Gerard Maule, it would be difficult to explain. It was not simply an affair of age,—nor of good looks, nor altogether of education. Gerard Maule was by no means wonderfully erudite. They were both addicted to hunting. Neither of them did anything useful. In that respect Mr. Spooner stood the higher, as he managed his own property successfully. But Gerard Maule so wore his clothes, and so carried his limbs, and so pronounced his words that he was to be regarded as one entitled to make love to any lady; whereas poor Mr. Spooner was not justified in proposing to marry any woman much more gifted than his own housemaid. Such, at least, were Adelaide Palliser's ideas. 'I don't think anything of the kind,' she said, 'only I want you to go away. I shall go back to the house, and I hope you won't accompany me. If you do, I shall turn the other way.' Whereupon she did retire at once, and he was left standing in the path.

There was a seat there, and he sat down for a moment to think of it all. Should he persevere in his suit, or should he rejoice that he had escaped from such an ill-conditioned minx? He remembered that he had read, in his younger days, that lovers in novels generally do persevere, and that they are almost always successful at last. In affairs of the heart, such perseverance was, he thought, the correct thing. But in this instance the conduct of the lady had not given him the slightest encouragement. When a horse balked with him at a fence, it was his habit to force the animal till he jumped it,—as the groom had recommended Phineas to do. But when he had encountered a decided fall, it was not sensible practice to ride the horse at the same place again. There was probably some occult cause for failure. He could not but own that he had been thrown on the present occasion,—and upon the whole, he thought, that he had better give it up. He found his way back to the house, put up his things, and got away to Spoon Hall in time for dinner, without seeing Lady Chiltern or any of her guests.

'What has become of Mr. Spooner?' Maule asked, as soon as he returned to Harrington Hall.

'Nobody knows,' said Lady Chiltern, 'but I believe he has gone.'

'Has anything happened?'

'I have heard no tidings; but, if you ask for my opinion, I think something has happened. A certain lady seems to have been ruffled, and a certain gentleman has disappeared. I am inclined to think that a few unsuccessful words have been spoken.' Gerard Maule saw that there was a smile in her eye, and he was satisfied.

'My dear, what did Mr. Spooner say to you during his walk?' This question was asked by the ill-natured old lady in the presence of nearly all the party.

'We were talking of hunting,' said Adelaide.

'And did the poor old woman get her half-sovereign?'

'No;—he forgot that. We did not go into the village at all. I was tired and came back.'

'Poor old woman;—and poor Mr. Spooner!'

Everybody in the house knew what had occurred, as Mr. Spooner's discretion in the conduct of this affair had not been equal to his valour; but Miss Palliser never confessed openly, and almost taught herself to believe that the man had been mad or dreaming during that special hour.

CHAPTER XX
Phineas again in London

PHINEAS, on his return to London, before he had taken his seat in the House, received the following letter from Lady Laura Kennedy:—

'Dresden, Feb. 8, 1870.*

'DEAR FRIEND,—

'I thought that perhaps you would have written to me from Harrington. Violet has told me of the meeting between you and Madame Goesler, and says that the old friendship seems to have been perfectly re-established. She used to think once that there might be more than friendship, but I never quite believed that. She tells me that Chiltern is quarrelling with the Pallisers. You ought not to let him quarrel with people. I know that he would listen to you. He always did.

'I write now especially because I have just received so dreadful a letter from Mr. Kennedy! I would send it you were it not that there are in it a few words which on his behalf I shrink from showing even to you. It is full of threats. He begins by quotations from the Scriptures, and from the Prayer-Book, to show that a wife has no right to leave her husband,—and then he goes on to the law. One knows all that of course. And then he asks whether he ever ill-used me? Was he ever false to me? Do I think, that were I to choose to submit the matter to the iniquitous practices of the present Divorce Court,* I could prove anything against him by which even that low earthly judge would be justified in taking from him his marital authority? And if not,—have I no conscience? Can I reconcile it to myself to make his life utterly desolate and wretched simply because duties which I took upon myself at my marriage have become distasteful to me?

'These questions would be very hard to answer, were there not other questions that I could ask. Of course I was wrong to marry him. I know that now, and I repent my sin in sack-cloth and ashes. But I did not leave him after I married him till he had brought against me horrid accusations,—accusations which a woman could not bear, which, if he believed them himself, must have made it impossible for him to live with me. Could any wife live with a husband who declared to her face that he believed that she had a lover? And in this very letter he says that which almost repeats the accusation. He has asked me how I can have dared to receive you, and desires me never either to see you or to wish to see you again. And yet he sent for you to Loughlinter before you came, in order that you might act as a friend between us. How could I possibly return to a man whose power of judgment has so absolutely left him?

'I have a conscience in the matter, a conscience that is very far from being at ease. I have done wrong, and have shipwrecked every hope in this world. No woman was ever more severely punished. My life is a burden to me, and I may truly say that I look for no peace this side the grave. I am conscious, too, of continued sin,—a sin unlike other sins,—not to be avoided, of daily occurrence, a sin which weighs me to the ground. But I should not sin the less were I to return to him. Of course he can plead his marriage. The thing is done. But it can't be right that a woman should pretend to love a man whom she loathes. I couldn't live with him. If it were simply to go and die, so that his pride would be gratified by my return, I would do it; but I should not die. There would come some horrid scene, and I should be no more a wife to him than I am while living here.

'He now threatens me with publicity. He declares that unless I return to him he will put into some of the papers a statement of the whole case. Of course this would be very bad. To be obscure and untalked of is all the comfort that now remains to me. And he might say things that would be prejudicial to others,—especially to you. Could this in any way be prevented? I suppose the papers would publish anything; and you know how greedily people will read slander about those whose names are in any way remarkable. In my heart I believe he is insane; but it is very hard that one's privacy should be at the mercy of a madman. He says that he can get an order from the Court of Queen's Bench which will oblige the judges in Saxony to send me back to England in the custody of the police, but that I do not believe. I had the opinion of Sir Gregory Grogram before I came away, and he told me that it was not so. I do not fear his power over my person, while I remain here, but that the matter should be dragged forward before the public.

'I have not answered him yet, nor have I shown his letter to Papa. I hardly liked to tell you when you were here, but I almost fear to talk to Papa about it. He never urges me to go back, but I know that he wishes that I should do so. He has ideas about money, which seem singular to me, knowing, as I do, how very generous he has been himself. When I married, my fortune, as you knew, had been just used in paying Chiltern's debts. Mr. Kennedy had declared himself to be quite indifferent about it, though the sum was large. The whole thing was explained to him, and he was satisfied. Before a year was over he complained to Papa, and then Papa and Chiltern together raised the money,—£40,000,—and it was paid to Mr. Kennedy. He has written more than once to Papa's lawyer to say that, though the money is altogether useless to him, he will not return a penny of it, because by doing so he would seem to abandon his rights. Nobody has asked him to return it. Nobody has asked him to defray a penny on my account since I left him. But Papa continues to say that the money should not be lost to the family. I cannot, however, return to such a husband for the sake of £40,000. Papa is very angry about the money, because he says that if it had been paid in the usual way at my marriage, settlements would have been required that it should come back to the family after Mr. Kennedy's death in the event of my having no child. But, as it is now, the money would go to his estate after my death. I don't understand why it should be so, but Papa is always harping upon it, and declaring that Mr. Kennedy's pretended generosity has robbed us all. Papa thinks that were I to return this could be arranged; but I could not go back to him for such a reason. What does it matter? Chiltern and Violet will have enough; and of what use would it be to such a one as I am to have a sum of money to leave behind me? I should leave it to your children, Phineas, and not to Chiltern's.

'He bids me neither see you nor write to you,—but how can I obey a man whom I believe to be mad? And when I will not obey him in the greater matter by returning to him it would be absurd were I to attempt to obey him in smaller details. I don't suppose I shall see you very often. His letter has, at any rate, made me feel that it would be impossible for me to return to England, and it is not likely that you will soon come here again. I will not even ask you to do so, though your presence gave a brightness to my life for a few days which nothing else could have produced. But when the lamp for a while burns with special brightness there always comes afterwards a corresponding dullness. I had to pay for your visit, and for the comfort of my confession to you at Königstein. I was determined that you should know it all; but, having told you, I do not want to see you again. As for writing, he shall not deprive me of the consolation,—nor I trust will you.

'Do you think that I should answer his letter, or will it be better that I should show it to Papa? I am very averse to doing this, as I have explained to you; but I would do so if I thought that Mr. Kennedy really intended to act upon his threats. I will not conceal from you that it would go nigh to kill me if my name were dragged through the papers. Can anything be done to prevent it? If he were known to be mad of course the papers would not publish his statements; but I suppose that if he were to send a letter from Loughlinter with his name to it they would print it. It would be very, very cruel.

'God bless you. I need not say how faithfully I am

'Your friend,

'L. K.'

This letter was addressed to Phineas at his club, and there he received it on the evening before the meeting of Parliament. He sat up for nearly an hour thinking of it after he read it. He must answer it at once. That was a matter of course. But he could give her no advice that would be of any service to her. He was, indeed, of all men the least fitted to give her counsel in her present emergency. It seemed to him that as she was safe from any attack on her person, she need only remain at Dresden, answering his letter by what softest negatives she could use. It was clear to him that in his present condition she could take no steps whatever in regard to the money. That must be left to his conscience, to time, and to chance. As to the threat of publicity, the probability, he thought, was that it would lead to nothing. He doubted whether any respectable newspaper would insert such a statement as that suggested. Were it published, the evil must be borne. No diligence on her part, or on the part of her lawyers, could prevent it.

But what had she meant when she wrote of continual sin, sin not to be avoided, of sin repeated daily which nevertheless weighed her to the ground? Was it expected of him that he should answer that portion of her letter? It amounted to a passionate renewal of that declaration of affection for himself which she had made at Königstein, and which had pervaded her whole life since some period antecedent to her wretched marriage. Phineas, as he thought of it, tried to analyse the nature of such a love. He also, in those old days, had loved her, and had at once resolved that he must tell her so, though his hopes of success had been poor indeed. He had taken the first opportunity, and had declared his purpose. She, with the imperturbable serenity of a matured kind-hearted woman, had patted him on the back, as it were, as she told him of her existing engagement with Mr. Kennedy. Could it be that at that moment she could have loved him as she now said she did, and that she should have been so cold, so calm, and so kind; while, at that very moment, this coldness, calmness, and kindness was but a thin crust over so strong a passion? How different had been his own love! He had been neither calm nor kind. He had felt himself for a day or two to be so terribly knocked about that the world was nothing to him. For a month or two he had regarded himself as a man peculiarly circumstanced,—marked for misfortune and for a solitary life. Then he had retricked his beams;* and before twelve months were passed had almost forgotten his love. He knew now, or thought that he knew,—that the continued indulgence of a hopeless passion was a folly opposed to the very instincts of man and woman,—a weakness showing want of fibre and of muscle in the character. But here was a woman who could calmly conceal her passion in its early days and marry a man whom she did not love in spite of it, who could make her heart, her feelings, and all her feminine delicacy subordinate to material considerations, and nevertheless could not rid herself of her passion in the course of years, although she felt its existence to be an intolerable burden on her conscience. On which side lay strength of character and on which side weakness? Was he strong or was she?

And he tried to examine his own feelings in regard to her. The thing was so long ago that she was to him as some aunt, or sister, so much the elder as to be almost venerable. He acknowledged to himself a feeling which made it incumbent upon him to spend himself in her service, could he serve her by any work of his. He was,—or would be, devoted to her. He owed her a never-dying gratitude. But were she free to marry again to-morrow, he knew that he could not marry her. She herself had said the same thing. She had said that she would be his sister. She had specially required of him that he should make known to her his wife, should he ever marry again. She had declared that she was incapable of further jealousy;—and yet she now told him of daily sin of which her conscience could not assoil itself.

'Phineas,' said a voice close to his ears, 'are you repenting your sins?'

'Oh, certainly;—what sins?'

It was Barrington Erle. 'You know that we are going to do nothing to-morrow,' continued he.

'So I am told.'

'We shall let the Address pass almost without a word. Gresham will simply express his determination to oppose the Church Bill to the knife. He means to be very plain-spoken about it. Whatever may be the merits of the Bill, it must be regarded as an unconstitutional effort to retain power in the hands of the minority, coming from such hands as those of Mr. Daubeny. I take it he will go at length into the question of majorities, and show how inexpedient it is on behalf of the nation that any Ministry should remain in power who cannot command a majority in the House on ordinary questions. I don't know whether he will do that to-morrow or at the second reading of the Bill.'

'I quite agree with him.'

'Of course you do. Everybody agrees with him. No gentleman can have a doubt on the subject. Personally, I hate the idea of Church Reform. Dear old Mildmay, who taught me all I know, hates it too. But Mr. Gresham is the head of our party now, and much as I may differ from him on many things, I am bound to follow him. If he proposes Church Reform in my time, or anything else, I shall support him.'

'I know those are your ideas.'

'Of course they are. There are no other ideas on which things can be made to work. Were it not that men get drilled into it by the force of circumstances any government in this country would be impossible. Were it not so, what should we come to? The Queen would find herself justified in keeping in any set of Ministers who could get her favour, and ambitious men would prevail without any support from the country. The Queen must submit to dictation from some quarter.'

'She must submit to advice, certainly.'

'Don't cavil at a word when you know it to be true,' said Barrington, energetically. 'The constitution of the country requires that she should submit to dictation. Can it come safely from any other quarter than that of a majority of the House of Commons?'

'I think not.'

'We are all agreed about that. Not a single man in either House would dare to deny it. And if it be so, what man in his senses can think of running counter to the party which he believes to be right in its general views? A man so burthened with scruples as to be unable to act in this way should keep himself aloof from public life. Such a one cannot serve the country in Parliament, though he may possibly do so with pen and ink in his closet.'

'I wonder then that you should have asked me to come forward again after what I did about the Irish land question,' said Phineas.

'A first fault may be forgiven when the sinner has in other respects been useful. The long and the short of it is that you must vote with us against Daubeny's bill. Browborough sees it plainly enough. He supported his chief in the teeth of all his protestations at Tankerville.'

'I am not Browborough.'

'Nor half so good a man if you desert us,' said Barrington Erle, with anger.

'I say nothing about that. He has his ideas of duty, and I have mine. But I will go so far as this. I have not yet made up my mind. I shall ask advice; but you must not quarrel with me if I say that I must seek it from some one who is less distinctly a partisan than you are.'

'From Monk?'

'Yes;—from Mr. Monk. I do think it will be bad for the country that this measure should come from the hands of Mr. Daubeny.'

'Then why the d—— should you support it, and oppose your own party at the same time? After that you can't do it. Well, Ratler, my guide and philosopher, how is it going to be?'

Mr. Ratler had joined them, but was still standing before the seat they occupied, not condescending to sit down in amicable intercourse with a man as to whom he did not yet know whether to regard him as a friend or foe. 'We shall be very quiet for the next month or six weeks,' said Ratler.

'And then?' asked Phineas.

'Well, then it will depend on what may be the number of a few insane men who never ought to have seats in the House.'

'Such as Mr. Monk and Mr. Turnbull?' Now it was well known that both those gentlemen, who were recognised as leading men, were strong Radicals, and it was supposed that they both would support any bill, come whence it might, which would separate Church and State.

'Such as Mr. Monk,' said Ratler. 'I will grant that Turnbull may be an exception. It is his business to go in for everything in the way of agitation, and he at any rate is consistent. But when a man has once been in office,—why then——'

'When he has taken the shilling?' said Phineas.

'Just so. I confess I do not like a deserter.'

'Phineas will be all right,' said Barrington Erle.

'I hope so,' said Mr. Ratler, as he passed on.

'Ratler and I run very much in the same groove,' said Barrington, 'but I fancy there is some little difference in the motive power.'

'Ratler wants place.'

'And so do I.'

'He wants it just as most men want professional success,' said Phineas. 'But if I understand your object, it is chiefly the maintenance of the old-established political power of the Whigs. You believe in families?'

'I do believe in the patriotism of certain families. I believe that the Mildmays, FitzHowards, and Pallisers have for some centuries brought up their children to regard the well-being of their country as their highest personal interest, and that such teaching has been generally efficacious. Of course, there have been failures. Every child won't learn its lesson however well it may be taught. But the school in which good training is most practised will, as a rule, turn out the best scholars. In this way I believe in families. You have come in for some of the teaching, and I expect to see you a scholar yet.'

The House met on the following day, and the Address was moved and seconded; but there was no debate. There was not even a full House. The same ceremony had taken place so short a time previously, that the whole affair was flat and uninteresting. It was understood that nothing would in fact be done. Mr. Gresham, as leader of his side of the House, confined himself to asserting that he should give his firmest opposition to the proposed measure, which was, it seemed, so popular with the gentlemen who sat on the other side, and who supported the so-called Conservative Government of the day. His reasons for doing so had been stated very lately, and must unfortunately be repeated very soon, and he would not, therefore, now trouble the House with them. He did not on this occasion explain his ideas as to majorities, and the Address was carried by seven o'clock in the evening. Mr. Daubeny named a day a month hence for the first reading of his bill, and was asked the cause of the delay by some member on a back bench. 'Because it cannot be ready sooner,' said Mr. Daubeny. 'When the honourable gentleman has achieved a position which will throw upon him the responsibility of bringing forward some great measure for the benefit of his country, he will probably find it expedient to devote some little time to details. If he do not, he will be less anxious to avoid attack than I am.' A Minister can always give a reason; and, if he be clever, he can generally when doing so punish the man who asks for it. The punishing of an influential enemy is an indiscretion; but an obscure questioner may often be crushed with good effect.

Mr. Monk's advice to Phineas was both simple and agreeable. He intended to support Mr. Gresham, and of course counselled his friend to do the same.

'But you supported Mr. Daubeny on the Address before Christmas,' said Phineas.

'And shall therefore be bound to explain why I oppose him now;—but the task will not be difficult. The Queen's speech to Parliament was in my judgment right, and therefore I concurred in the Address. But I certainly cannot trust Mr. Daubeny with Church Reform. I do not know that many will make the same distinction, but I shall do so.'

Phineas soon found himself sitting in the House as though he had never left it. His absence had not been long enough to make the place feel strange to him. He was on his legs before a fortnight was over asking some question of some Minister, and of course insinuating as he did so that the Minister in question had been guilty of some enormity of omission or commission. It all came back upon him as though he had been born to the very manner. And as it became known to the Ratlers that he meant to vote right on the great coming question,—to vote right and to speak right in spite of his doings at Tankerville,—everybody was civil to him. Mr. Bonteen did express an opinion to Mr. Ratler that it was quite impossible that Phineas Finn should ever again accept office, as of course the Tankervillians would never replace him in his seat after manifest apostasy to his pledge; but Mr. Ratler seemed to think very little of that. 'They won't remember, Lord bless you;—and then he's one of those fellows that always get in somewhere. He's not a man I particularly like; but you'll always see him in the House;—up and down, you know. When a fellow begins early, and has got it in him, it's hard to shake him off.' And thus even Mr. Ratler was civil to our hero.

Lady Laura Kennedy's letter had, of course, been answered, —not without very great difficulty. 'My dear Laura,' he had begun,—for the first time in his life. She had told him to treat her as a brother would do, and he thought it best to comply with her instructions. But beyond that, till he declared himself at the end to be hers affectionately, he made no further protestation of affection. He made no allusion to that sin which weighed so heavily on her, but answered all her questions. He advised her to remain at Dresden. He assured her that no power could be used to enforce her return. He expressed his belief that Mr. Kennedy would abstain from making any public statement, but suggested that if any were made the answering of it should be left to the family lawyer. In regard to the money, he thought it impossible that any step should be taken. He then told her all there was to tell of Lord and Lady Chiltern, and something also of himself. When the letter was written he found that it was cold and almost constrained. To his own ears it did not sound like the hearty letter of a generous friend. It savoured of the caution with which it had been prepared. But what could he do? Would he not sin against her and increase her difficulties if he addressed her with warm affection? Were he to say a word that ought not to be addressed to any woman he might do her an irreparable injury; and yet the tone of his own letter was odious to him.

CHAPTER XXI

Mr. Maule, Senior

THE life of Mr. Maurice Maule, of Maule Abbey, the father of Gerard Maule, had certainly not been prosperous. He had from his boyhood enjoyed a reputation for cleverness, and at school had done great things,—winning prizes, spouting speeches on Speech days, playing in elevens, and looking always handsome. He had been one of those show boys of which two or three are generally to be found at our great schools, and all manner of good things had been prophesied on his behalf. He had been in love before he was eighteen, and very nearly succeeded in running away with the young lady before he went to college. His father had died when he was an infant, so that at twenty-one he was thought to be in possession of comfortable wealth. At Oxford he was considered to have got into a good set,—men of fashion who were also given to talking of books,—who spent money, read poetry, and had opinions of their own respecting the Tracts and Mr. Newman.* He took his degree, and then started himself in the world upon that career which is of all the most difficult to follow with respect and self-comfort. He proposed to himself the life of an idle man with a moderate income,—a life which should be luxurious, refined, and graceful, but to which should be attached the burden of no necessary occupation. His small estate gave him but little to do, as he would not farm any portion of his own acres. He became a magistrate in his county; but he would not interest himself with the price of a good yoke of bullocks, as did Mr. Justice Shallow,*—nor did he ever care how a score of ewes went at any fair. There is no harder life than this. Here and there we may find a man who has so trained himself that day after day he can devote his mind without compulsion to healthy pursuits, who can induce himself to work, though work be not required from him for any ostensible object, who can save himself from the curse of misusing his time, though he has for it no defined and necessary use; but such men are few, and are made of better metal than was Mr. Maule. He became an idler, a man of luxury, and then a spendthrift. He was now hardly beyond middle life, and he assumed for himself the character of a man of taste. He loved music, and pictures, and books, and pretty women. He loved also good eating and drinking; but conceived of himself that in his love for them he was an artist, and not a glutton. He had married early, and his wife had died soon. He had not given himself up with any special zeal to the education of his children, nor to the preservation of his property. The result of his indifference has been told in a previous chapter. His house was deserted, and his children were scattered about the world. His eldest son, having means of his own, was living an idle, desultory life, hardly with prospects of better success than had attended his father.

Mr. Maule was now something about fifty-five years of age, and almost considered himself young. He lived in chambers on a flat in Westminster, and belonged to two excellent clubs. He had not been near his property for the last ten years, and as he was addicted to no country sport there were ten weeks in the year which were terrible to him. From the middle of August to the end of October for him there was no whist, no society,—it may almost be said no dinner. He had tried going to the seaside; he had tried going to Paris; he had endeavoured to enjoy Switzerland and the Italian lakes;—but all had failed, and he had acknowledged to himself that this sad period of the year must always be endured without relaxation, and without comfort.

Of his children he now took but little notice. His daughter was married and in India. His younger son had disappeared, and the father was perhaps thankful that he was thus saved from trouble. With his elder son he did maintain some amicable intercourse, but it was very slight in its nature. They never corresponded unless the one had something special to say to the other. They had no recognised ground for meeting. They did not belong to the same clubs. They did not live in the same circles. They did not follow the same pursuits. They were interested in the same property;—but, as on that subject there had been something approaching to a quarrel, and as neither looked for assistance from the other, they were now silent on the matter. The father believed himself to be a poorer man than his son, and was very sore on the subject; but he had nothing beyond a life interest in his property, and there remained to him a certain amount of prudence which induced him to abstain from eating more of his pudding,—lest absolute starvation and the poorhouse should befall him. There still remained to him the power of spending some five or six hundred a year, and upon this practice had taught him to live with a very considerable amount of self-indulgence. He dined out a great deal, and was known everywhere as Mr. Maule of Maule Abbey.

He was a slight, bright-eyed, grey-haired, good-looking

man, who had once been very handsome. He had married, let us say for love;—probably very much by chance. He had ill-used his wife, and had continued a long-continued liaison with a complaisant friend. This had lasted some twenty years of his life, and had been to him an intolerable burden. He had come to see the necessity of employing his good looks, his conversational powers, and his excellent manners on a second marriage which might be lucrative; but the complaisant lady had stood in his way. Perhaps there had been a little cowardice on his part; but at any rate he had hitherto failed. The season for such a mode of relief was not, however, as yet clean gone with him, and he was still on the look out. There are women always in the market ready to buy for themselves the right to hang on the arm of a real gentleman. That Mr. Maurice Maule was a real gentleman no judge in such matters had ever doubted.

On a certain morning just at the end of February Mr. Maule was sitting in his library,—so-called,—eating his breakfast, at about twelve o'clock; and at his side there lay a note from his son Gerard. Gerard had written to say that he would call on that morning, and the promised visit somewhat disturbed the father's comfort. He was in his dressing-gown and slippers, and had his newspaper in his hand. When his newspaper and breakfast should be finished,—as they would be certainly at the same moment,—there were in store for him two cigarettes, and perhaps some new French novel which had just reached him. They would last him till two o'clock. Then he would dress and saunter out in his great coat, made luxurious with furs. He would see a picture, or perhaps some china-vase, of which news had reached him, and would talk of them as though he might be a possible buyer. Everybody knew that he never bought anything;—but he was a man whose opinion on such matters was worth having. Then he would call on some lady whose acquaintance at the moment might be of service to him;—for that idea of blazing once more out into the world on a wife's fortune was always present to him. At about five he would saunter into his club, and play a rubber in a gentle unexcited manner till seven. He never played for high points, and would never be enticed into any bet beyond the limits of his club stakes. Were he to lose £10 or £20 at a sitting his arrangements would be greatly disturbed, and his comfort seriously affected. But he played well, taking pains with his game, and some who knew him well declared that his whist was worth a hundred a year to him. Then he would dress and generally dine in society. He was known as a good diner out, though in what his excellence consisted they who entertained him might find it difficult to say. He was not witty, nor did he deal in anecdotes. He spoke with a low voice, never addressing himself to any but his neighbour, and even to his neighbour saying but little. But he looked like a gentleman, was well dressed, and never awkward. After dinner he would occasionally play another rubber; but twelve o'clock always saw him back into his own rooms. No one knew better than Mr. Maule that the continual bloom of lasting summer which he affected requires great accuracy in living. Late hours, nocturnal cigars, and midnight drinkings, pleasurable though they may be, consume too quickly the free-flowing lamps of youth, and are fatal at once to the husbanded candle-ends of age.

But such as his days were, every minute of them was precious to him. He possessed the rare merit of making a property of his time and not a burden. He had so shuffled off his duties that he had now rarely anything to do that was positively disagreeable. He had been a spendthrift; but his creditors, though perhaps never satisfied, had been quieted. He did not now deal with reluctant and hard-tasked tenants, but with punctual, though inimical, trustees, who paid to him with charming regularity that portion of his income which he was allowed to spend. But that he was still tormented with the ambition of a splendid marriage it might be said of him that he was completely at his ease. Now, as he lit his cigarette, he would have been thoroughly comfortable, were it not that he was threatened with disturbance by his son. Why should his son wish to see him, and thus break in upon him at the most charming hour of the day? Of course his son would not come to him without having some business in hand which must be disagreeable. He had not the least desire to see his son,—and yet, as they were on amicable terms, he could not deny himself after the receipt of his son's note. Just at one, as he finished his first cigarette, Gerard was announced.

'Well, Gerard!'

'Well, father,—how are you? You are looking as fresh as paint, sir.'

'Thanks for the compliment, if you mean one. I am pretty well. I thought you were hunting somewhere.'

'So I am; but I have just come up to town to see you. I find you have been smoking;—may I light a cigar?'

'I never do smoke cigars here, Gerard. I'll offer you a cigarette.' The cigarette was reluctantly offered, and accepted with a shrug. 'But you didn't come here merely to smoke, I dare say.'

'Certainly not, sir. We do not often trouble each other, father; but there are things about which I suppose we had better speak. I'm going to be married!'

'To be married!' The tone in which Mr. Maule, senior, repeated the words was much the same as might be used by any ordinary father if his son expressed an intention of going into the shoe-black business.

'Yes, sir. It's a kind of thing men do sometimes.'

'No doubt;—and it's a kind of thing that they sometimes repent of having done.'

'Let us hope for the best. It is too late at any rate to think about that, and as it is to be done, I have come to tell you.'

'Very well. I suppose you are right to tell me. Of course you know that I can do nothing for you; and I don't suppose that you can do anything for me. As far as your own welfare goes, if she has a large fortune,——'

'She has no fortune.'

'No fortune!'

'Two or three thousand pounds perhaps.'

'Then I look upon it as an act of simple madness, and can

only say that as such I shall treat it. I have nothing in my power, and therefore I can neither do you good or harm; but I will not hear any particulars, and I can only advise you to break it off, let the trouble be what it may.'

'I certainly shall not do that, sir.'

'Then I have nothing more to say. Don't ask me to be present, and don't ask me to see her.'

'You haven't heard her name yet.'

'I do not care one straw what her name is.'

'It is Adelaide Palliser.'

'Adelaide Muggins would be exactly the same thing to me. My dear Gerard, I have lived too long in the world to believe that men can coin into money the noble blood of well-born wives. Twenty thousand pounds is worth more than all the blood of all the Howards;* and a wife even with twenty thousand pounds would make you a poor, embarrassed, and half-famished man.'

'Then I suppose I shall be whole famished, as she certainly has not got a quarter of that sum.'

'No doubt you will.'

'Yet, sir, married men with families have lived on my income.'

'And on less than a quarter of it. The very respectable man who brushes my clothes no doubt does so. But then you see he has been brought up in that way. I suppose that you as a bachelor put by every year at least half your income?'

'I never put by a shilling, sir. Indeed, I owe a few hundred pounds.'

'And yet you expect to keep a house over your head, and an expensive wife and family, with lady's maid, nurses, cook, footman, and grooms, on a sum which has been hitherto insufficient for your own wants! I didn't think you were such an idiot, my boy.'

'Thank you, sir.'

'What will her dress cost?'

'I have not the slightest idea.'

'I dare say not. Probably she is a horsewoman. As far as I

know anything of your life that is the sphere in which you will have made the lady's acquaintance.'

'She does ride.'

'No doubt, and so do you; and it will be very easy to say whither you will ride together if you are fools enough to get married. I can only advise you to do nothing of the kind. Is there anything else?'

There was much more to be said if Gerard could succeed in forcing his father to hear him. Mr. Maule, who had hitherto been standing, seated himself as he asked that last question, and took up the book which had been prepared for his morning's delectation. It was evidently his intention that his son should leave him. The news had been communicated to him, and he had said all that he could say on the subject. He had at once determined to confine himself to a general view of the matter, and to avoid details,—which might be personal to himself. But Gerard had been specially required to force his father into details. Had he been left to himself he would certainly have thought that the conversation had gone far enough. He was inclined, almost as well as his father, to avoid present discomfort. But when Miss Palliser had suddenly,—almost suddenly,—accepted him; and when he had found himself describing the prospects of his life in her presence and in that of Lady Chiltern, the question of the Maule Abbey inheritance had of necessity been discussed. At Maule Abbey there might be found a home for the married couple, and,—so thought Lady Chiltern,—the only fitting home. Mr. Maule, the father, certainly did not desire to live there. Probably arrangements might be made for repairing the house and furnishing it with Adelaide's money. Then, if Gerard Maule would be prudent, and give up hunting, and farm a little himself,—and if Adelaide would do her own housekeeping and dress upon forty pounds a year, and if they would both live an exemplary, model, energetic, and strictly economical life, both ends might be made to meet. Adelaide had been quite enthusiastic as to the forty pounds, and had suggested that she would do it for thirty. The housekeeping was a matter of course, and the more so as a leg of mutton roast or boiled would be the beginning and the end of it. To Adelaide the discussion had been exciting and pleasurable, and she had been quite in earnest when looking forward to a new life at Maule Abbey. After all there could be no such great difficulty for a young married couple to live on £800 a year, with a house and garden of their own. There would be no carriage and no man servant till,—till old Mr. Maule was dead. The suggestion as to the ultimate and desirable haven was wrapped up in ambiguous words. 'The property must be yours some day,' suggested Lady Chiltern. 'If I outlive my father.' 'We take that for granted; and then, you know——' So Lady Chiltern went on, dilating upon a future state of squirearchal bliss and rural independence. Adelaide was enthusiastic; but Gerard Maule, —after he had assented to the abandonment of his hunting, much as a man assents to being hung when the antecedents of his life have put any option in the matter out of his power,— had sat silent and almost moody while the joys of his coming life were described to him. Lady Chiltern, however, had been urgent in pointing out to him that the scheme of living at Maule Abbey could not be carried on without his father's assistance. They all knew that Mr. Maule himself could not be affected by the matter, and they also knew that he had but very little power in reference to the property. But the plan could not be matured without some sanction from him. Therefore there was still much more to be said when the father had completed the exposition of his views on marriage in general. 'I wanted to speak to you about the property,' said Gerard. He had been specially enjoined to be staunch in bringing his father to the point.

'And what about the property?'

'Of course my marriage will not affect your interests.'

'I should say not. It would be very odd if it did. As it is, your income is much larger than mine.'

'I don't know how that is, sir; but I suppose you will not refuse to give me a helping hand if you can do so without disturbance to your own comfort.'

'In what sort of way? Don't you think anything of that kind can be managed better by the lawyer? If there is a thing I hate, it is business.'

Gerard remembering his promise to Lady Chiltern did persevere, though the perseverance went much against the grain with him. 'We thought, sir, that if you would consent we might live at Maule Abbey.'

'Oh;—you did; did you?'

'Is there any objection?'

'Simply the fact that it is my house, and not yours.'

'It belongs, I suppose, to the property; and as——'

'As what?' asked the father, turning upon the son with sharp angry eyes, and with something of real animation in his face.

Gerard was very awkward in conveying his meaning to his father. 'And as,' he continued,—'as it must come to me, I suppose, some day, and it will be the proper sort of thing that we should live there then, I thought that you would agree that if we went and lived there now it would be a good sort of thing to do.'

'That was your idea?'

'We talked it over with our friend, Lady Chiltern.'

'Indeed! I am so much obliged to your friend, Lady Chiltern, for the interest she takes in my affairs. Pray make my compliments to Lady Chiltern, and tell her at the same time that, though no doubt I have one foot in the grave, I should like to keep my house for the other foot, though too probably I may never be able to drag it so far as Maule Abbey.'

'But you don't think of living there.'

'My dear boy, if you will inquire among any friends you may happen to know who understand the world better than Lady Chiltern seems to do, they will tell you that a son should not suggest to his father the abandonment of the family property, because the father may—probably—soon—be conveniently got rid of under ground.'

'There was no thought of such a thing,' said Gerard.

'It isn't decent. I say that with all due deference to Lady

Chiltern's better judgment. It's not the kind of thing that men do. I care less about it than most men, but even I object to such a proposition when it is made so openly. No doubt I am old.' This assertion Mr. Maule made in a weak, quavering voice, which showed that had his intention been that way turned in his youth, he might probably have earned his bread on the stage.

'Nobody thought of your being old, sir.'

'I shan't last long, of course. I am a poor feeble creature. But while I do live, I should prefer not to be turned out of my own house,—if Lady Chiltern could be induced to consent to such an arrangement. My doctor seems to think that I might linger on for a year or two,—with great care.'

'Father, you know I was thinking of nothing of the kind.'

'We won't act the king and the prince* any further, if you please. The prince protested very well, and, if I remember right, the father pretended to believe him. In my weak state you have rather upset me. If you have no objection I would choose to be left to recover myself a little.'

'And is that all that you will say to me?'

'Good heavens;—what more can you want? I will not—consent—to give up—my house at Maule Abbey for your use,—as long as I live. Will that do? And if you choose to marry a wife and starve, I won't think that any reason why I should starve too. Will that do? And your friend, Lady Chiltern, may—go—and be d——d. Will that do?'

'Good morning, sir.'

'Good morning, Gerard.' So the interview was over, and Gerard Maule left the room. The father, as soon as he was alone, immediately lit another cigarette, took up his French novel, and went to work as though he was determined to be happy and comfortable again without losing a moment. But he found this to be beyond his power. He had been really disturbed, and could not easily compose himself. The cigarette was almost at once chucked into the fire, and the little volume was laid on one side. Mr. Maule rose almost impetuously from his chair, and stood with his back to the fire, contemplating the proposition that had been made to him.

It was actually true that he had been offended by the very faint idea of death which had been suggested to him by his son. Though he was a man bearing no palpable signs of decay, in excellent health, with good digestion,—who might live to be ninety,—he did not like to be warned that his heir would come after him. The claim which had been put forward to Maule Abbey by his son had rested on the fact that when he should die the place must belong to his son;—and the fact was unpleasant to him. Lady Chiltern had spoken of him behind his back as being mortal, and in doing so had been guilty of an impertinence. Maule Abbey, no doubt, was a ruined old house, in which he never thought of living,—which was not let to a tenant by the creditors of his estate, only because its condition was unfit for tenancy. But now Mr. Maule began to think whether he might not possibly give the lie to these people who were compassing his death, by returning to the halls of his ancestors, if not in the bloom of youth, still in the pride of age. Why should he not live at Maule Abbey if this successful marriage could be effected? He almost knew himself well enough to be aware that a month at Maule Abbey would destroy him; but it is the proper thing for a man of fashion to have a place of his own, and he had always been alive to the glory of being Mr. Maule of Maule Abbey. In preparing the way for the marriage that was to come he must be so known. To be spoken of as the father of Maule of Maule Abbey would have been fatal to him. To be the father of a married son at all was disagreeable, and therefore when the communication was made to him he had managed to be very unpleasant. As for giving up Maule Abbey,—! He fretted and fumed as he thought of the proposition through the hour which should have been to him an hour of enjoyment; and his anger grew hot against his son as he remembered all that he was losing. At last, however, he composed himself sufficiently to put on with becoming care his luxurious furred great coat, and then he sallied forth in quest of the lady.

CHAPTER XXII

'Purity of morals, Finn'

MR. QUINTUS SLIDE was now, as formerly, the editor of the People's Banner, but a change had come over the spirit of his dream. His newspaper was still the People's Banner, and Mr. Slide still professed to protect the existing rights of the people, and to demand new rights for the people. But he did so as a Conservative. He had watched the progress of things, and had perceived that duty called upon him to be the organ of Mr. Daubeny. This duty he performed with great zeal, and with an assumption of consistency and infallibility which was charming. No doubt the somewhat difficult task of veering round without inconsistency, and without flaw to his infallibility, was eased by Mr. Daubeny's newly-declared views on Church matters. The People's Banner could still be a genuine People's Banner in reference to ecclesiastical policy. And as that was now the subject mainly discussed by the newspapers, the change made was almost entirely confined to the lauding of Mr. Daubeny instead of Mr. Turnbull. Some other slight touches were no doubt necessary. Mr. Daubeny was the head of the Conservative party in the kingdom, and though Mr. Slide himself might be of all men in the kingdom the most democratic, or even the most destructive, still it was essential that Mr. Daubeny's organ should support the Conservative party all round. It became Mr. Slide's duty to speak of men as heaven-born patriots whom he had designated a month or two since as bloated aristocrats and leeches fattened on the blood of the people. Of course remarks were made by his brethren of the press,—remarks which were intended to be very unpleasant. One evening newspaper took the trouble to divide a column of its own into double columns, printing on one side of the inserted line remarks made by the People's Banner in September respecting the Duke of ——, and the Marquis of ——, and Sir —— ——, which were certainly very harsh; and on the other side remarks equally laudatory as to the characters of the same titled politicians. But a journalist, with the tact and experience of Mr. Quintus Slide, knew his business too well to allow himself to be harassed by any such small stratagem as that. He did not pause to defend himself, but boldly attacked the meanness, the duplicity, the immorality, the grammar, the paper, the type, and the wife of the editor of the evening newspaper. In the storm of wind in which he rowed it was unnecessary for him to defend his own conduct. 'And then,' said he at the close of a very virulent and successful article, 'the hirelings of —— dare to accuse me of inconsistency!' The readers of the People's Banner all thought that their editor had beaten his adversary out of the field.

Mr. Quintus Slide was certainly well adapted for his work. He could edit his paper with a clear appreciation of the kind of matter which would best conduce to its success, and he could write telling leading articles himself. He was indefatigable, unscrupulous, and devoted to his paper. Perhaps his great value was shown most clearly in his distinct appreciation of the low line of public virtue with which his readers would be satisfied. A highly-wrought moral strain would he knew well create either disgust or ridicule. 'If there is any beastliness I 'ate it is 'igh-faluting,' he has been heard to say to his underlings. The sentiment was the same as that conveyed in the 'Point de zèle' of Talleyrand.* 'Let's 'ave no d——d nonsense,' he said on another occasion, when striking out from a leading article a passage in praise of the patriotism of a certain public man. 'Mr. Gresham is as good as another man, no doubt; what we want to know is whether he's along with us.' Mr. Gresham was not along with Mr. Slide at present, and Mr. Slide found it very easy to speak ill of Mr. Gresham.

Mr. Slide one Sunday morning called at the house of Mr. Bunce in Great Marlborough Street, and asked for Phineas Finn. Mr. Slide and Mr. Bunce had an old acquaintance with each other, and the editor was not ashamed to exchange a few friendly words with the law-scrivener before he was shown up to the member of Parliament. Mr. Bunce was an outspoken, eager, and honest politician,—with very little accurate knowledge of the political conditions by which he was surrounded, but with a strong belief in the merits of his own class. He was a sober, hardworking man, and he hated all men who were not sober and hardworking. He was quite clear in his mind that all nobility should be put down, and that all property in land should be taken away from men who were enabled by such property to live in idleness. What should be done with the land when so taken away was a question which he had not yet learnt to answer. At the present moment he was accustomed to say very hard words of Mr. Slide behind his back, because of the change which had been effected in the People's Banner, and he certainly was not the man to shrink from asserting in a person's presence aught that he said in his absence. 'Well, Mr. Conservative Slide,' he said, stepping into the little back parlour, in which the editor was left while Mrs. Bunce went up to learn whether the member of Parliament would receive his visitor.

'None of your chaff, Bunce.'

'We have enough of your chaff, anyhow; don't we, Mr. Slide? I still sees the Banner, Mr. Slide,—most days; just for the joke of it.'

'As long as you take it, Bunce, I don't care what the reason is.'

'I suppose a heditor's about the same as a Cabinet Minister. You've got to keep your place;—that's about it, Mr. Slide.'

'We've got to tell the people who's true to 'em. Do you believe that Gresham 'd ever have brought in a Bill for doing away with the Church? Never;—not if he'd been Prime Minister till doomsday. What you want is progress.'

'That's about it, Mr. Slide.'

'And where are you to get it? Did you ever hear that a rose by any other name 'd smell as sweet? If you can get progress from the Conservatives, and you want progress, why not go to the Conservatives for it? Who repealed the corn laws? Who gave us 'ousehold suffrage?'

'I think I've been told all that before, Mr. Slide; them things weren't given by no manner of means, as I look at it. We just went in and took 'em. It was hall a haccident whether it was Cobden or Peel, Gladstone or Disraeli, as was the servants we employed to do our work. But Liberal is Liberal, and Conservative is Conservative. What are you, Mr. Slide, to-day?'

'If you'd talk of things, Bunce, which you understand, you would not talk quite so much nonsense.'

At this moment Mrs. Bunce entered the room, perhaps preventing a quarrel, and offered to usher Mr. Slide up to the young member's room. Phineas had not at first been willing to receive the gentleman, remembering that when they had last met the intercourse had not been pleasant,—but he knew that enmities are foolish things, and that it did not become him to perpetuate a quarrel with such a man as Mr. Quintus Slide. 'I remember him very well, Mrs. Bunce.'

'I know you didn't like him, Sir.'

'Not particularly.'

'No more don't I. No more don't Bunce. He's one of them as 'd say a'most anything for a plate of soup and a glass of wine. That's what Bunce says.'

'It won't hurt me to see him.'

'No, sir; it won't hurt you. It would be a pity indeed if the likes of him could hurt the likes of you.' And so Mr. Quintus Slide was shown up into the room.

The first greeting was very affectionate, at any rate on the part of the editor. He grasped the young member's hand, congratulated him on his seat, and began his work as though he had never been all but kicked out of that very same room by its present occupant. 'Now you want to know what I'm come about; don't you?'

'No doubt I shall hear in good time, Mr. Slide.'

'It's an important matter;—and so you'll say when you do hear. And it's one in which I don't know whether you'll be able to see your way quite clear.'

'I'll do my best, if it concerns me.'

'It does.' So saying Mr. Slide, who had seated himself in an arm-chair by the fireside opposite to Phineas, crossed his legs, folded his arms on his breast, put his head a little on one side, and sat for a few moments in silence, with his eyes fixed on his companion's face. 'It does concern you, or I shouldn't be here. Do you know Mr. Kennedy,—the Right Honourable Robert Kennedy, of Loughlinter, in Scotland?'

'I do know Mr. Kennedy.'

'And do you know Lady Laura Kennedy, his wife?'

'Certainly I do.'

'So I supposed. And do you know the Earl of Brentford, who is, I take it, father to the lady in question?'

'Of course I do. You know that I do.' For there had been a time in which Phineas had been subjected to the severest censure which the People's Banner could inflict upon him, because of his adherence to Lord Brentford, and the vials of wrath had been poured out by the hands of Mr. Quintus Slide himself.

'Very well. It does not signify what I know or what I don't. Those preliminary questions I have been obliged to ask as my justification for coming to you on the present occasion. Mr. Kennedy has I believe been greatly wronged.'

'I am not prepared to talk about Mr. Kennedy's affairs,' said Phineas gravely.

'But unfortunately he is prepared to talk about them. That's the rub. He has been ill-used, and he has come to the People's Banner for redress. Will you have the kindness to cast your eye down that slip?' Whereupon the editor handed to Phineas a long scrap of printed paper, amounting to about a column and a half of the People's Banner, containing a letter to the editor dated from Loughlinter, and signed Robert Kennedy at full length.

'You don't mean to say that you're going to publish this,' said Phineas before he had read it.

'Why not?'

'The man is a madman.'

'There's nothing in the world easier than calling a man

mad. It's what we do to dogs when we want to hang them. I believe Mr. Kennedy has the management of his own property. He is not too mad for that. But just cast your eye down and read it.'

Phineas did cast his eye down, and read the whole letter;—nor as he read it could he bring himself to believe that the writer of it would be judged to be mad from its contents. Mr. Kennedy had told the whole story of his wrongs, and had told it well,—with piteous truthfulness, as far as he himself knew and understood the truth. The letter was almost simple in its wailing record of his own desolation. With a marvellous absence of reticence he had given the names of all persons concerned. He spoke of his wife as having been, and being, under the influence of Mr. Phineas Finn;—spoke of his own former friendship for that gentleman, who had once saved his life when he fell among thieves, and then accused Phineas of treachery in betraying that friendship. He spoke with bitter agony of the injury done him by the Earl, his wife's father, in affording a home to his wife, when her proper home was at Loughlinter. And then declared himself willing to take the sinning woman back to his bosom. 'That she had sinned is certain,' he said; 'I do not believe she has sinned as some sin; but, whatever be her sin, it is for a man to forgive as he hopes for forgiveness.' He expatiated on the absolute and almost divine right which it was intended that a husband should exercise over his wife, and quoted both the Old and New Testament in proof of his assertions. And then he went on to say that he appealed to public sympathy, through the public press, because, owing to some gross insufficiency in the laws of extradition, he could not call upon the magistracy of a foreign country to restore to him his erring wife. But he thought that public opinion, if loudly expressed, would have an effect both upon her and upon her father, which his private words could not produce. 'I wonder very greatly that you should put such a letter as that into type,' said Phineas when he had read it all.

'Why shouldn't we put it into type?'

'You don't mean to say that you'll publish it.'

'Why shouldn't we publish it?'

'It's a private quarrel between a man and his wife. What on earth have the public got to do with that?'

'Private quarrels between gentlemen and ladies have been public affairs for a long time past. You must know that very well.'

'When they come into court they are.'

'In court and out of court! The morale of our aristocracy,—what you call the Upper Ten,*—would be at a low ebb indeed if the public press didn't act as their guardians. Do you think that if the Duke of —— beats his wife black and blue, nothing is to be said about it unless the Duchess brings her husband into court? Did you ever know of a separation among the Upper Ten, that wasn't handled by the press one way or the other? It's my belief that there isn't a peer among 'em all as would live with his wife constant, if it was not for the press;—only some of the very old ones, who couldn't help themselves.'

'And you call yourself a Conservative?'

'Never mind what I call myself. That has nothing to do with what we're about now. You see that letter, Finn. There is nothing little or dirty about us. We go in for morals and purity of life, and we mean to do our duty by the public without fear or favour. Your name is mentioned there in a manner that you won't quite like, and I think I am acting uncommon kind by you in showing it to you before we publish it.' Phineas, who still held the slip in his hand, sat silent thinking of the matter. He hated the man. He could not endure the feeling of being called Finn by him without showing his resentment. As regarded himself, he was thoroughly well inclined to kick Mr. Slide and his Banner into the street. But he was bound to think first of Lady Laura. Such a publication as this, which was now threatened, was the misfortune which the poor woman dreaded more than any other. He, personally, had certainly been faultless in the matter. He had never addressed a word of love to Mr. Kennedy's wife since the moment in which she had told him that she was engaged to marry the Laird of

Loughlinter. Were the letter to be published he could answer it, he thought, in such a manner as to defend himself and her without damage to either. But on her behalf he was bound to prevent this publicity if it could be prevented;—and he was bound also, for her sake, to allow himself to be called Finn by this most obnoxious editor. 'In the ordinary course of things, Finn, it will come out to-morrow morning,' said the obnoxious editor.

'Every word of it is untrue,' said Phineas.

'You say that, of course.'

'And I should at once declare myself willing to make such a statement on oath. It is a libel of the grossest kind, and of course there would be a prosecution. Both Lord Brentford and I would be driven to that.'

'We should be quite indifferent. Mr. Kennedy would hold us harmless. We're straightforward. My showing it to you would prove that.'

'What is it you want, Mr. Slide?'

'Want! You don't suppose we want anything. If you think that the columns of the People's Banner are to be bought, you must have opinions respecting the press of the day which make me pity you as one grovelling in the very dust. The daily press of London is pure and immaculate. That is, the morning papers are. Want, indeed! What do you think I want?'

'I have not the remotest idea.'

'Purity of morals, Finn;—punishment for the guilty;—defence for the innocent;—support for the weak;—safety for the oppressed;—and a rod of iron for the oppressors!'

'But that is a libel.'

'It's very heavy on the old Earl, and upon you, and upon Lady Laura;—isn't it?'

'It's a libel,—as you know. You tell me that purity of morals can be supported by such a publication as this! Had you meant to go on with it, you would hardly have shown it to me.'

'You're in the wrong box there, Finn. Now I'll tell you

what we'll do,—on behalf of what I call real purity. We'll delay the publication if you'll undertake that the lady shall go back to her husband.'

'The lady is not in my hands.'

'She's under your influence. You were with her over at Dresden not much more than a month ago. She'd go sharp enough if you told her.'

'You never made a greater mistake in your life.'

'Say that you'll try.'

'I certainly will not do so.'

'Then it goes in to-morrow,' said Mr. Quintus Slide, stretching out his hand and taking back the slip.

'What on earth is your object?'

'Morals! Morals! We shall be able to say that we've done our best to promote domestic virtue and secure forgiveness for an erring wife. You've no notion, Finn, in your mind of what will soon be the hextent of the duties, privileges, and hinfluences of the daily press;—the daily morning press, that is; for I look on those little evening scraps as just so much paper and ink wasted. You won't interfere, then?'

'Yes, I will;—if you'll give me time. Where is Mr. Kennedy?'

'What has that to do with it? Do you write over to Lady Laura and the old lord and tell them that if she'll undertake to be at Loughlinter within a month this shall be suppressed. Will you do that?'

'Let me first see Mr. Kennedy.'

Mr. Slide thought a while over that matter. 'Well,' said he at last, 'you can see Kennedy if you will. He came up to town four or five days ago, and he's staying at an hotel in Judd Street.'

'An hotel in Judd Street?'

'Yes;—Macpherson's in Judd Street. I suppose he likes to keep among the Scotch. I don't think he ever goes out of the house, and he's waiting in London till this thing is published.'

'I will go and see him,' said Phineas.

'I shouldn't wonder if he murdered you;—but that's between you and him.'

'Just so.'

'And I shall hear from you?'

'Yes,' said Phineas, hesitating as he made the promise. 'Yes, you shall hear from me.'

'We've got our duty to do, and we mean to do it. If we see that we can induce the lady to go back to her husband, we shall habstain from publishing, and virtue will be its own reward. I needn't tell you that such a letter as that would sell a great many copies, Finn.' Then, at last, Mr. Slide arose and departed.

CHAPTER XXIII
Macpherson's Hotel

PHINEAS, when he was left alone, found himself greatly at a loss as to what he had better do. He had pledged himself to see Mr. Kennedy, and was not much afraid of encountering personal violence at the hands of that gentleman. But he could think of nothing which he could with advantage say to Mr. Kennedy. He knew that Lady Laura would not return to her husband. Much as she dreaded such exposure as was now threatened, she would not return to Loughlinter to avoid even that. He could not hold out any such hope to Mr. Kennedy;—and without doing so how could he stop the publication? He thought of getting an injunction from the Vice-Chancellor;—but it was now Sunday, and he had understood that the publication would appear on the morrow, unless stopped by some note from himself. He thought of finding some attorney, and taking him to Mr. Kennedy; but he knew that Mr. Kennedy would be deterred by no attorney. Then he thought of Mr. Low. He would see Mr. Kennedy first, and then go to Mr. Low's house.

Judd Street*runs into the New Road near the great stations

of the Midland and Northern Railways, and is a highly respectable street. But it can hardly be called fashionable, as is Piccadilly; or central, as is Charing Cross; or commercial, as is the neighbourhood of St. Paul's. Men seeking the shelter of an hotel in Judd Street most probably prefer decent and respectable obscurity to other advantages. It was some such feeling, no doubt, joined to the fact that the landlord had originally come from the neighbourhood of Loughlinter, which had taken Mr. Kennedy to Macpherson's Hotel. Phineas, when he called at about three o'clock on Sunday afternoon, was at once informed by Mrs. Macpherson that Mr. Kennedy was 'nae doubt at hame, but was nae willing to see folk on the Saaboth.' Phineas pleaded the extreme necessity of his business, alleging that Mr. Kennedy himself would regard its nature as a sufficient justification for such Sabbath-breaking,—and sent up his card. Then there came down a message to him. Could not Mr. Finn postpone his visit to the following morning? But Phineas declared that it could not be postponed. Circumstances, which he would explain to Mr. Kennedy, made it impossible. At last he was desired to walk up stairs, though Mrs. Macpherson, as she showed him the way, evidently thought that her house was profaned by such wickedness.

Macpherson in preparing his house had not run into that extravagance of architecture which has lately become so common in our hotels. It was simply an ordinary house, with the words 'Macpherson's Hotel' painted on a semi-circular board over the doorway. The front parlour had been converted into a bar, and in the back parlour the Macphersons lived. The staircase was narrow and dirty, and in the front drawing-room,—with the chamber behind for his bedroom,—Mr. Kennedy was installed. Mr. Macpherson probably did not expect any customers beyond those friendly Scots who came up to London from his own side of the Highlands. Mrs. Macpherson, as she opened the door, was silent and almost mysterious. Such a breach of the law might perhaps be justified by circumstances of which she knew nothing, but should receive no sanction from her which she could avoid. So she did not even whisper the name.

Mr. Kennedy, as Phineas entered, slowly rose from his chair, putting down the Bible which had been in his hands. He did not speak at once, but looked at his visitor over the spectacles which he wore. Phineas thought that he was even more haggard in appearance and aged than when they two had met hardly three months since at Loughlinter. There was no shaking of hands, and hardly any pretence at greeting. Mr. Kennedy simply bowed his head, and allowed his visitor to begin the conversation.

'I should not have come to you on such a day as this, Mr. Kennedy——'

'It is a day very unfitted for the affairs of the world,' said Mr. Kennedy.

'Had not the matter been most pressing in regard both to time and its own importance.'

'So the woman told me, and therefore I have consented to see you.'

'You know a man of the name of —— Slide, Mr. Kennedy?' Mr. Kennedy shook his head. 'You know the editor of the People's Banner?' Again he shook his head. 'You have, at any rate, written a letter for publication to that newspaper.'

'Need I consult you as to what I write?'

'But he,—the editor,—has consulted me.'

'I can have nothing to do with that.'

'This Mr. Slide, the editor of the People's Banner, has just been with me, having in his hand a printed letter from you, which,—you will excuse me, Mr. Kennedy,—is very libellous.'

'I will bear the responsibility of that.'

'But you would not wish to publish falsehood about your wife, or even about me.'

'Falsehood! sir; how dare you use that word to me? Is it false to say that she has left my house? Is it false to say that she is my wife, and cannot desert me, as she has done, without breaking her vows, and disregarding the laws both of God and man? Am I false when I say that I gave her no cause? Am I false when I offer to take her back, let her faults be what they may have been? Am I false when I say that her father acts illegally in detaining her? False! False in your teeth! Falsehood is villany, and it is not I that am the villain.'

'You have joined my name in the accusation.'

'Because you are her paramour. I know you now;—viper that was warmed in my bosom! Will you look me in the face and tell me that, had it not been for you, she would not have strayed from me?' To this Phineas could make no answer. 'Is it not true that when she went with me to the altar you had been her lover?'

'I was her lover no longer, when she once told me that she was to be your wife.'

'Has she never spoken to you of love since? Did she not warn you from the house in her faint struggle after virtue? Did she not whistle you back again when she found the struggle too much for her? When I asked you to the house, she bade you not come. When I desired that you might never darken my eyes again, did she not seek you? With whom was she walking on the villa grounds by the river banks when she resolved that she would leave all her duties and desert me? Will you dare to say that you were not then in her confidence? With whom was she talking when she had the effrontery to come and meet me at the house of the Prime Minister, which I was bound to attend? Have you not been with her this very winter in her foreign home?'

'Of course I have,—and you sent her a message by me.'

'I sent no message. I deny it. I refused to be an accomplice in your double guilt. I laid my command upon you that you should not visit my wife in my absence, and you disobeyed, and you are an adulterer. Who are you that you are to come for ever between me and my wife?'

'I never injured you in thought or deed. I come to you now because I have seen a printed letter which contains a gross libel upon myself.'

'It is printed then?' he asked, in an eager tone.

'It is printed; but it need not, therefore, be published. It is a libel, and should not be published. I shall be forced to seek redress at law. You cannot hope to regain your wife by publishing false accusations against her.'

'They are true. I can prove every word that I have written. She dare not come here, and submit herself to the laws of her country. She is a renegade from the law, and you abet her in her sin. But it is not vengeance that I seek. "Vengeance is mine, saith the Lord."'

'It looks like vengeance, Mr. Kennedy.'

'Is it for you to teach me how I shall bear myself in this time of my great trouble?' Then suddenly he changed; his voice falling from one of haughty defiance to a low, mean, bargaining whisper. 'But I'll tell you what I'll do. If you will say that she shall come back again I'll have it cancelled, and pay all the expenses.'

'I cannot bring her back to you.'

'She'll come if you tell her. If you'll let them understand that she must come they'll give way. You can try it at any rate.'

'I shall do nothing of the kind. Why should I ask her to submit herself to misery?'

'Misery! What misery? Why should she be miserable? Must a woman need be miserable because she lives with her husband? You hear me say that I will forgive everything. Even she will not doubt me when I say so, because I have never lied to her. Let her come back to me, and she shall live in peace and quiet, and hear no word of reproach.'

'I can have nothing to do with it, Mr. Kennedy.'

'Then, sir, you shall abide my wrath.' With that he sprang quickly round, grasping at something which lay upon a shelf near him, and Phineas saw that he was armed with a pistol. Phineas, who had hitherto been seated, leaped to his legs; but the pistol in a moment was at his head, and the madman pulled at the trigger. But the mechanism of the instrument required that some bolt should be loosed before the hammer would fall upon the nipple, and the unhandy wretch for an instant fumbled over the work so that Phineas, still facing his enemy, had time to leap backwards towards the door. But Kennedy, though he was awkward, still succeeded in firing before our friend could leave the room. Phineas heard the thud of the bullet, and knew that it must have passed near his head. He was not struck, however; and the man, frightened at his own deed, abstained from the second shot, or loitered long enough in his remorse to enable his prey to escape. With three or four steps Phineas leaped down the stairs, and, finding the front door closed, took shelter within Mrs. Macpherson's bar. 'The man is mad,' he said; 'did you not hear the shot?' The woman was too frightened to reply, but stood trembling, holding Phineas by the arm. There was nobody in the house, she said, but she and the two lasses. 'Nae doobt the Laird's by ordinaire,' she said at last. She had known of the pistol; but had not dared to have it removed. She and Macpherson had only feared that he would hurt himself,—and had at last agreed, as day after day passed without any injury from the weapon, to let the thing remain unnoticed. She had heard the shot, and had been sure that one of the two men above would have been killed.

Phineas was now in great doubt as to what duty was required of him. His first difficulty consisted in this,—that his hat was still in Mr. Kennedy's room, and that Mrs. Macpherson altogether refused to go and fetch it. While they were still discussing this, and Phineas had not as yet resolved whether he would first get a policeman or go at once to Mr. Low, the bell from the room was rung furiously. 'It's the Laird,' said Mrs. Macpherson, 'and if naebody waits on him he'll surely be shooting ane of us.' The two girls were now outside the bar shaking in their shoes, and evidently unwilling to face the danger. At last the door of the room above was opened, and our hero's hat was sent rolling down the stairs.

It was clear to Phineas that the man was so mad as to be not even aware of the act he had perpetrated. 'He'll do nothing more with the pistol,' he said, 'unless he should attempt to destroy himself.' At last it was determined that one of the girls should be sent to fetch Macpherson home from the Scotch Church, and that no application should be made at once to the police. It seemed that the Macphersons knew the circumstances of their guest's family, and that there was a cousin of his in London who was the only one with whom he seemed to have any near connection. The thing that had occurred was to be told to this cousin, and Phineas left his address, so that if it should be thought necessary he might be called upon to give his account of the affair. Then, in his perturbation of spirit, he asked for a glass of brandy; and having swallowed it, was about to take his leave. 'The brandy wull be saxpence, sir,' said Mrs. Macpherson, as she wiped the tears from her eyes.

Having paid for his refreshment, Phineas got into a cab, and had himself driven to Mr. Low's house. He had escaped from his peril, and now again it became his strongest object to stop the publication of the letter which Slide had shown him. But as he sat in the cab he could not hinder himself from shuddering at the danger which had been so near to him. He remembered his sensation as he first saw the glimmer of the barrel of the pistol, and then became aware of the man's first futile attempt, and afterwards saw the flash and heard the hammer fall at the same moment. He had once stood up to be fired at in a duel, and had been struck by the ball. But nothing in that encounter had made him feel sick and faint through every muscle as he had felt just now. As he sat in the cab he was aware that but for the spirits he had swallowed he would be altogether overcome, and he doubted even now whether he would be able to tell his story to Mr. Low. Luckily perhaps for him neither Mr. Low nor his wife were at home. They were out together, but were expected in between five and six. Phineas declared his purpose of waiting for them, and requested that Mr. Low might be asked to join him in the dining-room immediately on his return. In this way an hour was allowed him, and he endeavoured to compose himself. Still, even at the end of the hour, his heart was beating so violently that he could hardly control the motion of his own limbs. 'Low, I have been shot at by a madman,' he said, as soon as his friend entered the room. He had determined to be calm, and to speak much more of the document in the editor's hands than of the attempt which had been made on his own life; but he had been utterly unable to repress the exclamation.

'Shot at?'

'Yes; by Robert Kennedy; the man who was Chancellor of the Duchy;—almost within a yard of my head.' Then he sat down and burst out into a fit of convulsive laughter.

The story about the pistol was soon told, and Mr. Low was of opinion that Phineas should not have left the place without calling in policemen and giving an account to them of the transaction. 'But I had something else on my mind,' said Phineas, 'which made it necessary that I should see you at once;—something more important even than this madman's attack upon me. He has written a most foul-mouthed attack upon his wife, which is already in print, and will I fear be published to-morrow morning.' Then he told the story of the letter. 'Slide no doubt will be at the People's Banner office to-night, and I can see him there. Perhaps when I tell him what has occurred he will consent to drop the publication altogether.'

But in this view of the matter Mr. Low did not agree with his visitor. He argued the case with a deliberation which to Phineas in his present state of mind was almost painful. If the whole story of what had occurred were told to Quintus Slide, that worthy protector of morals and caterer for the amusement of the public would, Mr. Low thought, at once publish the letter and give a statement of the occurrence at Macpherson's Hotel. There would be nothing to hinder him from so profitable a proceeding, as he would know that no one would stir on behalf of Lady Laura in the matter of the libel, when the tragedy of Mr. Kennedy's madness should have been made known. The publication would be as safe as attractive. But if Phineas should abstain from going to him at all, the same calculation which had induced him to show the letter would induce him to postpone the publication, at any rate for another twenty-four hours. 'He means to make capital out of his virtue; and he won't give that up for the sake of being a day in advance. In the meantime we will get an injunction from the Vice-Chancellor to stop the publication.'

'Can we do that in one day?'

'I think we can. Chancery isn't what it used to be,' said Mr. Low, with a sigh. 'I'll tell you what I'll do. I'll go this very moment to Pickering.' Mr. Pickering at this time was one of the three Vice-Chancellors. 'It isn't exactly the proper thing for counsel to call on a judge on a Sunday afternoon with the direct intention of influencing his judgment for the following morning; but this is a case in which a point may be strained. When such a paper as the People's Banner gets hold of a letter from a madman, which if published would destroy the happiness of a whole family, one shouldn't stick at a trifle. Pickering is just the man to take a common-sense view of the matter. You'll have to make an affidavit in the morning, and we can get the injunction served before two or three o'clock. Mr. Septimus Slope* or whatever his name is, won't dare to publish it after that. Of course, if it comes out to-morrow morning, we shall have been too late; but this will be our best chance.' So Mr. Low got his hat and umbrella, and started for the Vice-Chancellor's house. 'And I tell you what, Phineas;—do you stay and dine here. You are so flurried by all this, that you are not fit to go anywhere else.'

'I am flurried.'

'Of course you are. Never mind about dressing. Do you go up and tell Georgiana all about it;—and have dinner put off half an hour. I must hunt Pickering up, if I don't find him at home.' Then Phineas did go upstairs and tell Georgiana—otherwise Mrs. Low—the whole story. Mrs. Low was deeply affected, declaring her opinion very strongly as to the horrible condition of things, when madmen could go about with pistols, and without anybody to take care against them. But as to Lady Laura Kennedy, she seemed to think that the poor husband had great cause of complaint, and that Lady Laura ought to be punished. Wives, she thought, should never leave their husbands on any pretext; and, as far as she had heard the story, there had been no pretext at all in the case. Her sympathies were clearly with the madman, though she was quite ready to acknowledge that any and every step should be taken which might be adverse to Mr. Quintus Slide.

CHAPTER XXIV

Madame Goesler is sent for

WHEN the elder Mr. Maule had sufficiently recovered from the perturbation of mind and body into which he had been thrown by the ill-timed and ill-worded proposition of his son to enable him to resume the accustomed tenour of his life, he arrayed himself in his morning winter costume, and went forth in quest of a lady. So much was told some few chapters back, but the name of the lady was not then disclosed. Starting from Victoria Street, Westminster, he walked slowly across St. James's Park and the Green Park till he came out in Piccadilly, near the bottom of Park Lane. As he went up the Lane he looked at his boots, at his gloves, and at his trousers, and saw that nothing was unduly soiled. The morning air was clear and frosty, and had enabled him to dispense with the costly comfort of a cab. Mr. Maule hated cabs in the morning,—preferring never to move beyond the tether of his short daily constitutional walk. A cab for going out to dinner was a necessity;—but his income would not stand two or three cabs a day. Consequently he never went north of Oxford Street, or east of the theatres, or beyond Eccleston Square towards the river. The regions of South Kensington and New Brompton were a trouble to him, as he found it impossible to lay down a limit in that direction which would not exclude him from things which he fain would not exclude. There are dinners given at South Kensington which such a man as Mr. Maule cannot afford not to eat. In Park Lane he knocked at the door of a very small house,—a house that might almost be called tiny by comparison of its dimensions with those around it, and then asked for Madame Goesler. Madame Goesler had that morning gone into the country. Mr. Maule in his blandest manner expressed some surprise, having understood that she had not long since returned from Harrington Hall. To this the servant assented, but went on to explain that she had been in town only a day or two when she was summoned down to Matching by a telegram. It was believed, the man said, that the Duke of Omnium was poorly. 'Oh! indeed;—I am sorry to hear that,' said Mr. Maule, with a wry face. Then, with steps perhaps a little less careful, he walked back across the park to his club. On taking up the evening paper he at once saw a paragraph stating that the Duke of Omnium's condition to-day was much the same as yesterday; but that he had passed a quiet night. That very distinguished but now aged physician, Sir Omicron Pie, was still staying at Matching Priory. 'So old Omnium is going off the hooks at last,' said Mr. Maule to a club acquaintance.

The club acquaintance was in Parliament, and looked at the matter from a strictly parliamentary point of view. 'Yes, indeed. It has given a deal of trouble.'

Mr. Maule was not parliamentary, and did not understand. 'Why trouble,—except to himself? He'll leave his Garter and strawberry-leaves* and all his acres behind him.'

'What is Gresham to do about the Exchequer when he comes in? I don't know whom he's to send there. They talk of Bonteen, but Bonteen hasn't half weight enough. They'll offer it to Monk, but Monk 'll never take office again.'

'Ah, yes. Planty Pall was Chancellor of the Exchequer. I suppose he must give that up now?'

The parliamentary acquaintance looked up at the unparliamentary man with that mingled disgust and pity which parliamentary gentlemen and ladies always entertain for those who have not devoted their minds to the constitutional forms of the country. 'The Chancellor of the Exchequer can't very well sit in the House of Lords, and Palliser can't very well help becoming Duke of Omnium. I don't know whether he can take the decimal coinage question with him, but I fear not. They don't like it at all in the city.'

'I believe I'll go and play a rubber of whist,' said Mr. Maule. He played his whist, and lost thirty points without showing the slightest displeasure, either by the tone of his voice or by any grimace of his countenance. And yet the money which passed from his hands was material to him. But he was great at such efforts as these, and he understood well the fluctuations of the whist table. The half-crowns which he had paid were only so much invested capital.

He dined at his club this evening, and joined tables with another acquaintance who was not parliamentary. Mr. Parkinson Seymour was a man much of his own stamp, who cared not one straw as to any difficulty which the Prime Minister might feel in filling the office of Chancellor of the Exchequer. There were men by dozens ready and willing, and no doubt able,—or at any rate, one as able as the other,—to manage the taxes of the country. But the blue riband and the Lord Lieutenancy of Barsetshire were important things,—which would now be in the gift of Mr. Daubeny; and Lady Glencora would at last be a duchess,—with much effect on Society, either good or bad. And Planty Pall would be a duke, with very much less capability, as Mr. Parkinson Seymour thought, for filling that great office, than that which the man had displayed who was now supposed to be dying at Matching. 'He has been a fine old fellow,' said Mr. Parkinson Seymour.

'Very much so. There ain't many of that stamp left.'

'I don't know one,' continued the gentleman, with enthusiasm. 'They all go in for something now, just as Jones goes in for being a bank clerk. They are politicians, or gamblers, or, by heaven, tradesmen, as some of them are. The Earl of Tydvil and Lord Merthyr are in partnership together working their own mines,—by the Lord, with a regular deed of partnership, just like two cheesemongers. The Marquis of Maltanops has a share in a bitter beer house at Burton. And the Duke of Discount, who married old Ballance's daughter, and is brother-in-law to young George Advance, retains his interest in the house in Lombard Street. I know it for a fact.'

'Old Omnium was above that kind of thing,' said Mr. Maule.

'Lord bless you;—quite another sort of man. There is nothing left like it now. With a princely income I don't suppose he ever put by a shilling in his life. I've heard it said that he couldn't afford to marry, living in the manner in which he chose to live. And he understood what dignity meant. None of them understand that now. Dukes are as common as dogs in the streets, and a marquis thinks no more of himself than a market-gardener. I'm very sorry the old duke should go. The nephew may be very good at figures, but he isn't fit to fill his uncle's shoes. As for Lady Glencora, no doubt as things go now she's very popular, but she's more like a dairy-maid than a duchess to my way of thinking.'

There was not a club in London, and hardly a drawing-room in which something was not said that day in consequence of the two bulletins which had appeared as to the condition of the old Duke;—and in no club and in no drawing-room was a verdict given against the dying man. It was acknowledged everywhere that he had played his part in a noble and even in a princely manner, that he had used with a becoming grace the rich things that had been given him, and that he had deserved well of his country. And yet, perhaps, no man who had lived during the same period, or any portion of the period, had done less, or had devoted himself more entirely to the consumption of good things without the slightest idea of producing anything in return! But he had looked like a duke, and known how to set a high price on his own presence.

To Mr. Maule the threatened demise of this great man was not without a peculiar interest. His acquaintance with Madame Goesler had not been of long standing, nor even as yet had it reached a close intimacy. During the last London season he had been introduced to her, and had dined twice at her house. He endeavoured to make himself agreeable to her, and he flattered himself that he had succeeded. It may be said of him generally, that he had the gift of making himself pleasant to women. When last she had parted from him with a smile, repeating the last few words of some good story which he had told her, the idea struck him that she after all might perhaps be the woman. He made his inquiries, and had learned that there was not a shadow of a doubt as to her wealth,—or even to her power of disposing of that wealth as she pleased. So he wrote to her a pretty little note, in which he gave to her the history of that good story, how it originated with a certain Cardinal, and might be found in certain memoirs;*—which did not, however, bear the best reputation in the world. Madame Goesler answered his note very graciously, thanking him for the reference, but declaring that the information given was already so sufficient that she need prosecute the inquiry no further. Mr. Maule smiled as he declared to himself that those memoirs would certainly be in Madame Goesler's hands before many days were over. Had his intimacy been a little more advanced he would have sent the volume to her.

But he also learned that there was some romance in the lady's life which connected her with the Duke of Omnium. He was diligent in seeking information, and became assured that there could be no chance for himself, or for any man, as long as the Duke was alive. Some hinted that there had been a private marriage,—a marriage, however, which Madame Goesler had bound herself by solemn oaths never to disclose. Others surmised that she was the Duke's daughter. Hints were, of course, thrown out as to a connection of another kind, —but with no great vigour, as it was admitted on all hands that Lady Glencora, the Duke's niece by marriage, and the mother of the Duke's future heir, was Madame Goesler's great friend. That there was a mystery was a fact very gratifying to the world at large; and perhaps, upon the whole, the more gratifying in that nothing had occurred to throw a gleam of light upon the matter since the fact of the intimacy had become generally known. Mr. Maule was aware, however, that there could be no success for him as long as the Duke lived. Whatever might be the nature of the alliance, it was too strong to admit of any other while it lasted. But the Duke was a very old,—or, at least, a very infirm man. And now the Duke was dying. Of course it was only a chance. Mr. Maule knew the world too well to lay out any great portion of his hopes on a prospect so doubtful. But it was worth a struggle, and he would so struggle that he might enjoy success, should success come, without laying himself open to the pangs of disappointment. Mr. Maule hated to be unhappy or uncomfortable, and therefore never allowed any aspiration to proceed to such length as to be inconvenient to his feelings should it not be gratified.

In the meantime Madame Max Goesler had been sent for, and had hurried off to Matching almost without a moment's preparation. As she sat in the train, thinking of it, tears absolutely filled her eyes. 'Poor dear old man,' she said to herself; and yet the poor dear old man had simply been a trouble to her, adding a most disagreeable task to her life, and one which she was not called on to perform by any sense of duty. 'How is he?' she said anxiously, when she met Lady Glencora in the hall at Matching. The two women kissed each other as though they had been almost sisters since their birth. 'He is a little better now, but he was very uneasy when we telegraphed this morning. He asked for you twice, and then we thought it better to send.'

'Oh, of course it was best,' said Madame Goesler.

CHAPTER XXV

'I would do it now'

THOUGH it was rumoured all over London that the Duke of Omnium was dying, his Grace had been dressed and taken out of his bed-chamber into a sitting-room, when Madame Goesler was brought into his presence by Lady Glencora Palliser. He was reclining in a great arm-chair, with his legs propped up on cushions, and a respectable old lady in a black silk gown and a very smart cap was attending to his wants. The respectable old lady took her departure when the younger ladies entered the room, whispering a word of instruction to

Lady Glencora as she went. 'His Grace should have his broth at half-past four, my lady, and a glass and a half of champagne. His Grace won't drink his wine out of a tumbler, so perhaps your ladyship won't mind giving it him at twice.'

'Marie has come,' said Lady Glencora.

'I knew she would come,' said the old man, turning his head round slowly on the back of his chair. 'I knew she would be good to me to the last.' And he laid his withered hand on the arm of his chair, so that the woman whose presence gratified him might take it within hers and comfort him.

'Of course I have come,' said Madame Goesler, standing close by him and putting her left arm very lightly on his shoulder. It was all that she could do for him, but it was in order that she might do this that she had been summoned from London to his side. He was wan and worn and pale,—a man evidently dying, the oil of whose lamp was all burned out; but still as he turned his eyes up to the woman's face there was a remnant of that look of graceful fainéant nobility which had always distinguished him. He had never done any good, but he had always carried himself like a duke, and like a duke he carried himself to the end.

'He is decidedly better than he was this morning,' said Lady Glencora.

'It is pretty nearly all over, my dear. Sit down, Marie. Did they give you anything after your journey?'

'I could not wait, Duke.'

'I'll get her some tea,' said Lady Glencora. 'Yes, I will. I'll do it myself. I know he wants to say a word to you alone.' This she added in a whisper.

But sick people hear everything, and the Duke did hear the whisper. 'Yes, my dear;—she is quite right. I am glad to have you for a minute alone. Do you love me, Marie?'

It was a foolish question to be asked by a dying old man of a young woman who was in no way connected with him, and whom he had never seen till some three or four years since. But it was asked with feverish anxiety, and it required an answer. 'You know I love you, Duke. Why else should I be here?'

'It is a pity you did not take the coronet when I offered it you.'

'Nay, Duke, it was no pity. Had I done so, you could not have had us both.'

'I should have wanted only you.'

'And I should have stood aloof,—in despair to think that

I was separating you from those with whom your Grace is bound up so closely. We have ever been dear friends since that.'

'Yes;—we have been dear friends. But——' Then he closed his eyes, and put his long thin fingers across his face, and lay back awhile in silence, still holding her by the other hand. 'Kiss me, Marie,' he said at last; and she stooped over him and kissed his forehead. 'I would do it now if I thought it would serve you.' She only shook her head and pressed his hand closely. 'I would; I would. Such things have been done, my dear.'

'Such a thing shall never be done by me, Duke.'

They remained seated side by side, the one holding the other by the hand, but without uttering another word, till Lady Glencora returned bringing a cup of tea and a morsel of toast in her own hand. Madame Goesler, as she took it, could not help thinking how it might have been with her had she accepted the coronet which had been offered. In that case she might have been a duchess herself, but assuredly she would not have been waited upon by a future duchess. As it was, there was no one in that family who had not cause to be grateful to her. When the Duke had sipped a spoonful of his broth, and swallowed his allowance of wine, they both left him, and the respectable old lady with the smart cap was summoned back to her position. 'I suppose he whispered something very gracious to you,' Lady Glencora said when they were alone.

'Very gracious.'

'And you were gracious to him,—I hope.'

'I meant to be.'

'I'm sure you did. Poor old man! If you had done what he asked you I wonder whether his affection would have lasted as it has done.'

'Certainly not, Lady Glen. He would have known that I had injured him.'

'I declare I think you are the wisest woman I ever met, Madame Max. I am sure you are the most discreet. If I had always been as wise as you are!'

'You always have been wise.'

'Well,—never mind. Some people fall on their feet like cats; but you are one of those who never fall at all. Others tumble about in the most unfortunate way, without any great fault of their own. Think of that poor Lady Laura.'

'Yes, indeed.'

'I suppose it's true about Mr. Kennedy. You've heard of it of course in London.' But as it happened Madame Goesler had not heard the story. 'I got it from Barrington Erle, who always writes to me if anything happens. Mr. Kennedy has fired a pistol at the head of Phineas Finn.'

'At Phineas Finn!'

'Yes, indeed. Mr. Finn went to him at some hotel in London. No one knows what it was about; but Mr. Kennedy went off in a fit of jealousy, and fired a pistol at him.'

'He did not hit him?'

'It seems not. Mr. Finn is one of those Irish gentlemen who always seem to be under some special protection. The ball went through his whiskers and didn't hurt him.'

'And what has become of Mr. Kennedy?'

'Nothing, it seems. Nobody sent for the police, and he has been allowed to go back to Scotland,—as though a man were permitted by special Act of Parliament to try to murder his wife's lover. It would be a bad law, because it would cause such a deal of bloodshed.'

'But he is not Lady Laura's lover,' said Madame Goesler, gravely.

'That would make the law difficult, because who is to say whether a man is or is not a woman's lover?'

'I don't think there was ever anything of that kind.'

'They were always together, but I dare say it was Platonic. I believe these kind of things generally are Platonic. And as for Lady Laura;—heavens and earth!—I suppose it must have been Platonic. What did the Duke say to you?'

'He bade me kiss him.'

'Poor dear old man. He never ceases to speak of you when you are away, and I do believe he could not have gone in peace without seeing you. I doubt whether in all his life he ever loved any one as he loves you. We dine at half-past seven, dear: and you had better just go into his room for a moment as you come down. There isn't a soul here except Sir Omicron Pie, and Plantagenet, and two of the other nephews,—whom, by the bye, he has refused to see. Old Lady Hartletop wanted to come.'

'And you wouldn't have her?'

'I couldn't have refused. I shouldn't have dared. But the Duke would not hear of it. He made me write to say that he was too weak to see any but his nearest relatives. Then he made me send for you, my dear;—and now he won't see the relatives. What shall we do if Lady Hartletop turns up? I'm living in fear of it. You'll have to be shut up out of sight somewhere if that should happen.'

During the next two or three days the Duke was neither

much better nor much worse. Bulletins appeared in the newspapers, though no one at Matching knew from whence they came. Sir Omicron Pie, who, having retired from general practice, was enabled to devote his time to the 'dear Duke,' protested that he had no hand in sending them out. He declared to Lady Glencora every morning that it was only a question of time. 'The vital spark is on the spring,' said Sir Omicron, waving a gesture heavenward with his hand. For three days Mr. Palliser was at Matching, and he duly visited his uncle twice a day. But not a syllable was ever said between them beyond the ordinary words of compliments. Mr. Palliser spent his time with his private secretary, working out endless sums and toiling for unapproachable results in reference to decimal coinage. To him his uncle's death would be a great blow, as in his eyes to be Chancellor of the Exchequer was much more than to be Duke of Omnium. For herself Lady Glencora was nearly equally indifferent, though she did in her heart of hearts wish that her son should go to Eton with the title of Lord Silverbridge.

On the third morning the Duke suddenly asked a question of Madame Goesler. The two were again sitting near to each other, and the Duke was again holding her hand; but Lady Glencora was also in the room. 'Have you not been staying with Lord Chiltern?'

'Yes, Duke.'

'He is a friend of yours.'

'I used to know his wife before they were married.'

'Why does he go on writing me letters about a wood?' This he asked in a wailing voice, as though he were almost weeping. 'I know nothing of Lord Chiltern. Why does he write to me about the wood? I wish he wouldn't write to me.'

'He does not know that you are ill, Duke. By-the-bye, I promised to speak to Lady Glencora about it. He says that foxes are poisoned at Trumpeton Wood.'

'I don't believe a word of it,' said the Duke. 'No one would poison foxes in my wood. I wish you'd see about it, Glencora. Plantagenet will never attend to anything. But he shouldn't write to me. He ought to know better than to write letters to me. I will not have people writing letters to me. Why don't they write to Fothergill?' and then the Duke began in truth to whimper.

'I'll put it all right,' said Lady Glencora.

'I wish you would. I don't like them to say there are no foxes; and Plantagenet never will attend to anything.' The wife had long since ceased to take the husband's part when accusations such as this were brought against him. Nothing could make Mr. Palliser think it worth his while to give up any shred of his time to such a matter as the preservation of foxes.

On the fourth day the catastrophe happened which Lady Glencora had feared. A fly with a pair of horses from the Matching Road station was driven up to the door of the Priory, and Lady Hartletop was announced. 'I knew it,' said Lady Glencora, slapping her hand down on the table in the room in which she was sitting with Madame Goesler. Unfortunately the old lady was shown into the room before Madame Goesler could escape, and they passed each other on the threshold. The Dowager Marchioness of Hartletop was a very stout old lady, now perhaps nearer to seventy than sixty-five years of age, who for many years had been the intimate friend of the Duke of Omnium. In latter days, during which she had seen but little of the Duke himself, she had heard of Madame Max Goesler, but she had never met that lady. Nevertheless, she knew the rival friend at a glance. Some instinct told her that that woman with the black brow and the dark curls was Madame Goesler. In these days the Marchioness was given to waddling rather than to walking, but she waddled past the foreign female,—as she had often called Madame Max,—with a dignified though duck-like step. Lady Hartletop was a bold woman; and it must be supposed that she had some heart within her or she would hardly have made such a journey with such a purpose. 'Dear Lady Hartletop,' said Lady Glencora, 'I am so sorry that you should have had this trouble.'

'I must see him,' said Lady Hartletop. Lady Glencora put

both her hands together piteously, as though deprecating her visitor's wrath. 'I must insist on seeing him.'

'Sir Omicron has refused permission to any one to visit him.'

'I shall not go till I've seen him. Who was that lady?'

'A friend of mine,' said Lady Glencora, drawing herself up.

'She is——, Madame Goesler.'

'That is her name, Lady Hartletop. She is my most intimate friend.'

'Does she see the Duke?'

Lady Glencora, when expressing her fear that the woman would come to Matching, had confessed that she was afraid of Lady Hartletop. And a feeling of dismay—almost of awe—had fallen upon her on hearing the Marchioness announced. But when she found herself thus cross-examined, she resolved that she would be bold. Nothing on earth should induce her to open the door of the Duke's room to Lady Hartletop, nor would she scruple to tell the truth about Madame Goesler. 'Yes,' she said, 'Madame Goesler does see the Duke.'

'And I am to be excluded!'

'My dear Lady Hartletop, what can I do? The Duke for some time past has been accustomed to the presence of my friend, and therefore her presence now is no disturbance. Surely that can be understood.'

'I should not disturb him.'

'He would be inexpressibly excited were he to know that you were even in the house. And I could not take it upon myself to tell him.'

Then Lady Hartletop threw herself upon a sofa, and began to weep piteously. 'I have known him for more than forty years,' she moaned, through her choking tears. Lady Glencora's heart was softened, and she was kind and womanly; but she would not give way about the Duke. It would, as she knew, have been useless, as the Duke had declared that he would see no one except his eldest nephew, his nephew's wife, and Madame Goesler.

That evening was very dreadful to all of them at Matching, —except to the Duke, who was never told of Lady Hartletop's perseverance. The poor old woman could not be sent away on that afternoon, and was therefore forced to dine with Mr. Palliser. He, however, was warned by his wife to say nothing in the lady's presence about his uncle, and he received her as he would receive any other chance guest at his wife's table. But the presence of Madame Goesler made the chief difficulty. She herself was desirous of disappearing for that evening, but Lady Glencora would not permit it. 'She has seen you, my dear, and asked about you. If you hide yourself, she'll say all sorts of things.' An introduction was therefore necessary, and Lady Hartletop's manner was grotesquely grand. She dropped a very low curtsey, and made a very long face, but she did not say a word. In the evening the Marchioness sat close to Lady Glencora, whispering many things about the Duke; and condescending at last to a final entreaty that she might be permitted to see him on the following morning. 'There is Sir Omicron,' said Lady Glencora, turning round to the little doctor. But Lady Hartletop was too proud to appeal to Sir Omicron, who, as a matter of course, would support the orders of Lady Glencora. On the next morning Madame Goesler did not appear at the breakfast-table, and at eleven Lady Hartletop was taken back to the train in Lady Glencora's carriage. She had submitted herself to discomfort, indignity, fatigue, and disappointment; and it had all been done for love. With her broad face, and her double chin, and her heavy jowl, and the beard that was growing round her lips, she did not look like a romantic woman; but, in spite of appearances, romance and a duck-like waddle may go together. The memory of those forty years had been strong upon her, and her heart was heavy because she could not see that old man once again. Men will love to the last, but they love what is fresh and new. A woman's love can live on the recollection of the past, and cling to what is old and ugly. 'What an episode!' said Lady Glencora, when the unwelcome visitor was gone;—'but it's odd how much less dreadful things are than you think they will be. I was frightened when I heard her name; but you see we've got through it without much harm.'

A week passed by, and still the Duke was living. But now he was too weak to be moved from one room to another, and Madame Goesler passed two hours each day sitting by his bedside. He would lie with his hand out upon the coverlid, and she would put hers upon it; but very few words passed between them. He grumbled again about the Trumpeton Woods, and Lord Chiltern's interference, and complained of his nephew's indifference. As to himself and his own condition, he seemed to be, at any rate, without discomfort, and was certainly free from fear. A clergyman attended him, and gave him the sacrament. He took it,—as the champagne prescribed by Sir Omicron, or the few mouthfuls of chicken broth which were administered to him by the old lady with the smart cap; but it may be doubted whether he thought much more of the one remedy than of the other. He knew that he had lived, and that the thing was done. His courage never failed him. As to the future, he neither feared much nor hoped much; but was, unconsciously, supported by a general trust in the goodness and the greatness of the God who had made him what he was. 'It is nearly done now, Marie,' he said to Madame Goesler one evening. She only pressed his hand in answer. His condition was too well understood between them to allow of her speaking to him of any possible recovery. 'It has been a great comfort to me that I have known you,' he said.

'Oh no!'

'A great comfort;—only I wish it had been sooner. I could have talked to you about things which I never did talk of to any one. I wonder why I should have been a duke, and another man a servant.'

'God Almighty ordained such difference.'

'I'm afraid I have not done it well;—but I have tried; indeed I have tried.' Then she told him he had ever lived as a great nobleman ought to live. And, after a fashion, she herself believed what she was saying. Nevertheless, her nature was much nobler than his; and she knew that no man should dare to live idly as the Duke had lived.

CHAPTER XXVI
The Duke's Will

ON the ninth day after Madame Goesler's arrival the Duke died, and Lady Glencora Palliser became Duchess of Omnium. But the change probably was much greater to Mr. Palliser than to his wife. It would seem to be impossible to imagine a greater change than had come upon him. As to rank, he was raised from that of a simple commoner to the very top of the tree. He was made master of almost unlimited wealth, Garters, and lord-lieutenancies; and all the added grandeurs which come from high influence when joined to high rank were sure to be his. But he was no more moved by these things than would have been a god, or a block of wood. His uncle was dead; but his uncle had been an old man, and his grief on that score was moderate. As soon as his uncle's body had been laid in the family vault at Gatherum, men would call him Duke of Omnium; and then he could never sit again in the House of Commons. It was in that light, and in that light only, that he regarded the matter. To his uncle it had been everything to be Duke of Omnium. To Plantagenet Palliser it was less than nothing. He had lived among men and women with titles all his life, himself untitled, but regarded by them as one of themselves, till the thing, in his estimation, had come to seem almost nothing. One man walked out of a room before another man; and he, as Chancellor of the Exchequer, had, during a part of his career, walked out of most rooms before most men. But he cared not at all whether he walked out first or last,—and for him there was nothing else in it. It was a toy that would perhaps please his wife, but he doubted even whether she would not cease to be Lady Glencora with regret. In himself this thing that had happened had absolutely crushed him. He had won for himself by his own aptitudes and his own industry one special position in the empire,—and that position, and that alone, was incompatible with the rank which he was obliged to assume! His case was very hard, and he felt it;—but he made no complaint to human ears. 'I suppose you must give up the Exchequer,' his wife said to him. He shook his head, and made no reply. Even to her he could not explain his feelings.

I think, too, that she did regret the change in her name, though she was by no means indifferent to the rank. As Lady Glencora she had made a reputation which might very possibly fall away from her as Duchess of Omnium. Fame is a skittish jade, more fickle even than Fortune, and apt to shy, and bolt, and plunge away on very trifling causes. As Lady Glencora Palliser she was known to every one, and had always done exactly as she had pleased. The world in which she lived had submitted to her fantasies, and had placed her on a pedestal from which, as Lady Glencora, nothing could have moved her. She was by no means sure that the same pedestal would be able to carry the Duchess of Omnium. She must begin again, and such beginnings are dangerous. As Lady Glencora she had almost taken upon herself to create a rivalry in society to certain very distinguished, and indeed illustrious, people. There were only two houses in London, she used to say, to which she never went. The 'never' was not quite true;—but there had been something in it. She doubted whether as Duchess of Omnium she could go on with this. She must lay down her mischief, and abandon her eccentricity, and in some degree act like other duchesses. 'The poor old man,' she said to Madame Goesler; 'I wish he could have gone on living a little longer.' At this time the two ladies were alone together at Matching. Mr. Palliser, with the cousins, had gone to Gatherum, whither also had been sent all that remained of the late Duke, in order that fitting funeral obsequies might be celebrated over the great family vault.

'He would hardly have wished it himself, I think.'

'One never knows,—and as far as one can look into futurity one has no idea what would be one's own feelings. I suppose he did enjoy life.'

'Hardly, for the last twelve months,' said Madame Goesler.

'I think he did. He was happy when you were about him; and he interested himself about things. Do you remember how much he used to think of Lady Eustace and her diamonds? When I first knew him he was too magnificent to care about anything.'

'I suppose his nature was the same.'

'Yes, my dear; his nature was the same, but he was strong enough to restrain his nature, and wise enough to know that his magnificence was incompatible with ordinary interests. As he got to be older he broke down, and took up with mere mortal gossip. But I think it must have made him happier.'

'He showed his weakness in coming to me,' said Madame Goesler, laughing.

'Of course he did;—not in liking your society, but in wanting to give you his name. I have often wondered what kind of things he used to say to that old Lady Hartletop. That was in his full grandeur, and he never condescended to speak much then. I used to think him so hard; but I suppose he was only acting his part. I used to call him the Grand Lama to Plantagenet when we were first married,—before Planty was born. I shall always call him Silverbridge now instead of Planty.'

'I would let others do that.'

'Of course I was joking; but others will, and he will be spoilt. I wonder whether he will live to be a Grand Lama or a popular Minister. There cannot be two positions further apart. My husband, no doubt, thinks a good deal of himself as a statesman and a clever politician,—at least I suppose he does; but he has not the slightest reverence for himself as a nobleman. If the dear old Duke were hobbling along Piccadilly, he was conscious that Piccadilly was graced by his presence, and never moved without being aware that people looked at him, and whispered to each other,—There goes the Duke of Omnium. Plantagenet considers himself inferior to a sweeper while on the crossing, and never feels any pride of place unless he is sitting on the Treasury Bench with his hat over his eyes.'

'He'll never sit on the Treasury Bench again.'

'No;—poor dear. He's an Othello now with a vengeance, for his occupation is gone. I spoke to him about your friend and the foxes, and he told me to write to Mr. Fothergill. I will as soon as it's decent. I fancy a new duchess shouldn't write letters about foxes till the old Duke is buried. I wonder what sort of a will he'll have made. There's nothing I care twopence for except his pearls. No man in England had such a collection of precious stones. They'd been yours, my dear, if you had consented to be Mrs. O.'

The Duke was buried and the will was read, and Plantagenet Palliser was addressed as Duke of Omnium by all the tenantry and retainers of the family in the great hall of Gatherum Castle. Mr. Fothergill, who had upon occasion in former days* been driven by his duty to remonstrate with the heir, was all submission. Planty Pall had come to the throne, and half a county was ready to worship him. But he did not know how to endure worship, and the half county declared that he was stern and proud, and more haughty even than his uncle. At every 'Grace' that was flung at him he winced and was miserable, and declared to himself that he should never become accustomed to his new life. So he sat all alone, and meditated how he might best reconcile the forty-eight farthings which go to a shilling with that thorough-going useful decimal, fifty.

But his meditations did not prevent him from writing to his wife, and on the following morning, Lady Glencora,—as she shall be called now for the last time,—received a letter from him which disturbed her a good deal. She was in her room when it was brought to her, and for an hour after reading it hardly knew how to see her guest and friend, Madame Goesler. The passage in the letter which produced this dismay was as follows:—'He has left to Madame Goesler twenty thousand pounds and all his jewels. The money may be very well, but I think he has been wrong about the jewellery. As to myself I do not care a straw, but you will be sorry; and then people will talk. The lawyers will, of course, write to her, but I suppose you had better tell her. They seem to think that the stones are worth a great deal of money; but I have long learned never to believe any statement that is made to me. They are all here, and I suppose she will have to send some authorised person to have them packed. There is a regular inventory, of which a copy shall be sent to her by post as soon as it can be prepared.' Now it must be owned that the duchess did begrudge her friend the duke's collection of pearls and diamonds.

About noon they met. 'My dear,' she said, 'you had better hear your good fortune at once. Read that,—just that side. Plantagenet is wrong in saying that I shall regret it. I don't care a bit about it. If I want a ring or a brooch he can buy me one. But I never did care about such things, and I don't now. The money is all just as it should be.' Madame Goesler read the passage, and the blood mounted up into her face. She read it very slowly, and when she had finished reading it she was for a moment or two at a loss for her words to express herself. 'You had better send one of Garnett's people,' said the Duchess, naming the house of a distinguished jeweller and goldsmith in London.

'It will hardly need,' said Madame Goesler.

'You had better be careful. There is no knowing what they are worth. He spent half his income on them, I believe, during part of his life.' There was a roughness about the Duchess of which she was herself conscious, but which she could not restrain, though she knew that it betrayed her chagrin.

Madame Goesler came gently up to her and touched her arm caressingly. 'Do you remember,' said Madame Goesler, 'a small ring with a black diamond,—I suppose it was a diamond,—which he always wore?'

'I remember that he always did wear such a ring.'

'I should like to have that,' said Madame Goesler.

'You have them all,—everything. He makes no distinction.'

'I should like to have that, Lady Glen,—for the sake of the hand that wore it. But, as God is great above us, I will never take aught else that has belonged to the Duke.'

'Not take them!'

'Not a gem; not a stone; not a shilling.'

'But you must.'

'I rather think that I can be under no such obligation,' she said, laughing. 'Will you write to Mr. Palliser,—or I should say, to the Duke,—to-night, and tell him that my mind is absolutely made up?'

'I certainly shall not do that.'

'Then I must. As it is, I shall have pleasant memories of his Grace. According to my ability I have endeavoured to be good to him, and I have no stain on my conscience because of his friendship. If I took his money and his jewels,—or rather your money and your jewels,—do you think I could say as much?'

'Everybody takes what anybody leaves them by will.'

'I will be an exception to the rule, Lady Glen. Don't you think that your friendship is more to me than all the diamonds in London?'

'You shall have both, my dear,' said the Duchess,—quite in earnest in her promise. Madame Goesler shook her head.

'Nobody ever repudiates legacies. The Queen would take the jewels if they were left to her.'

'I am not the Queen. I have to be more careful what I do than any queen. I will take nothing under the Duke's will. I will ask a boon which I have already named, and if it be given me as a gift by the Duke's heir, I will wear it till I die. You will write to Mr. Palliser?'

'I couldn't do it,' said the Duchess.

'Then I will write myself.' And she did write, and of all the rich things which the Duke of Omnium had left to her, she took nothing but the little ring with the black stone which he had always worn on his finger.

CHAPTER XXVII

An Editor's Wrath

On that Sunday evening in London Mr. Low was successful in finding the Vice-Chancellor, and the great judge smiled and nodded, listened to the story, and acknowledged that the circumstances were very peculiar. He thought that an injunction to restrain the publication might be given at once upon Mr. Finn's affidavit; and that the peculiar circumstances justified the peculiarity of Mr. Low's application. Whether he would have said as much had the facts concerned the families of Mr. Joseph Smith and his son-in-law Mr. John Jones, instead of the Earl of Brentford and the Right Honourable Robert Kennedy, some readers will perhaps doubt, and may doubt also whether an application coming from some newly-fledged barrister would have been received as graciously as that made by Mr. Low, Q.C. and M.P.,—who would probably himself soon sit on some lofty legal bench. On the following morning Phineas and Mr. Low,—and no doubt also Mr. Vice-Chancellor Pickering,—obtained early copies of the People's Banner, and were delighted to find that Mr. Kennedy's letter did not appear in it. Mr. Low had made his calculation rightly. The editor, considering that he would gain more by having the young member of Parliament and the Standish family, as it were, in his hands than by the publication of a certain libellous letter, had resolved to put the document back for at least twenty-four hours, even though the young member neither came nor wrote as he had promised. The letter did not appear, and before ten o'clock Phineas Finn had made his affidavit in a dingy little room behind the Vice-Chancellor's Court. The injunction was at once issued, and was of such potency that should any editor dare to publish any paper therein prohibited, that editor and that editor's newspaper would assuredly be crumpled up in a manner very disagreeable, if not altogether destructive. Editors of newspapers are self-willed, arrogant, and stiff-necked, a race of men who believe much in themselves and little in anything else, with no feelings of reverence or respect for matters which are august enough to other men;—but an injunction from a Court of Chancery is a power which even an editor respects. At about noon Vice-Chancellor Pickering's injunction was served at the office of the People's Banner in Quartpot Alley, Fleet Street. It was done in duplicate,—or perhaps in triplicate,—so that there should be no evasion; and all manner of crumpling was threatened in the event of any touch of disobedience. All this happened on Monday, March the first, while the poor dying Duke was waiting impatiently for the arrival of his friend at Matching. Phineas was busy all the morning till it was time that he should go down to the House. For as soon as he could leave Mr. Low's chambers in Lincoln's Inn he had gone to Judd Street, to inquire as to the condition of the man who had tried to murder him. He there saw Mr. Kennedy's cousin, and received an assurance from that gentleman that Robert Kennedy should be taken down at once to Loughlinter. Up to that moment not a word had been said to the police as to what had been done. No more notice had been taken of the attempt to murder than might have been necessary had Mr. Kennedy thrown a clothes-brush at his visitor's head. There was the little hole in the post of the door with the bullet in it, just six feet above the ground; and there was the pistol, with five chambers still loaded, which Macpherson had cunningly secured on his return from church, and given over to the cousin that same evening. There was certainly no want of evidence, but nobody was disposed to use it.

At noon the injunction was served in Quartpot Alley, and was put into Mr. Slide's hands on his arrival at the office at three o'clock. That gentleman's duties required his attendance from three till five in the afternoon, and then again from nine in the evening till any hour in the morning at which he might be able to complete the People's Banner for that day's use. He had been angry with Phineas when the Sunday night passed without a visit or letter at the office, as a promise had been made that there should be either a visit or a letter; but he had felt sure, as he walked into the city from his suburban residence at Camden Town, that he would now find some communication on the great subject. The matter was one of most serious importance. Such a letter as that which was in his possession would no doubt create much surprise, and receive no ordinary attention. A People's Banner could hardly ask for a better bit of good fortune than the privilege of first publishing such a letter. It would no doubt be copied into every London paper, and into hundreds of provincial papers, and every journal so copying it would be bound to declare that it was taken from the columns of the People's Banner. It was, indeed, addressed 'To the Editor of the People's Banner' in the printed slip which Mr. Slide had shown to Phineas Finn, though Kennedy himself had not prefixed to it any such direction. And the letter, in the hands of Quintus Slide, would not simply have been a letter. It might have been groundwork for, perhaps, some half-dozen leading articles, all of a most attractive kind. Mr. Slide's high moral tone upon such an occasion would have been qualified to do good to every British matron, and to add virtues to the Bench of Bishops. All this he had postponed with some inadequately defined idea that he could do better with the property in his hands by putting himself into personal communication with the persons concerned. If he could manage to reconcile such a husband to such a wife,—or even to be conspicuous in an attempt to do so; and if he could make the old Earl and the young Member of Parliament feel that he had spared them by abstaining from the publication, the results might be very beneficial. His conception of the matter had been somewhat hazy, and he had certainly made a mistake. But, as he walked from his home to Quartpot Alley, he little dreamed of the treachery with which he had been treated. 'Has Phineas Finn been here?' he asked as he took his accustomed seat within a small closet, that might be best described as a glass cage. Around him lay the debris of many past newspapers, and the germs of many future publications. To all the world except himself it would have been a chaos, but to him, with his experience, it was admirable order. No; Mr. Finn had not been there. And then, as he was searching among the letters for one from the Member for Tankerville, the injunction was thrust into his hands. To say that he was aghast is but a poor form of speech for the expression of his emotion.

He had been 'done'—'sold,'—absolutely robbed by that wretchedly-false Irishman whom he had trusted with all the confidence of a candid nature and an open heart! He had been most treacherously misused! Treachery was no adequate word for the injury inflicted on him. The more potent is a man, the less accustomed to endure injustice, and the more his power to inflict it,—the greater is the sting and the greater the astonishment when he himself is made to suffer. Newspaper editors sport daily with the names of men of whom they do not hesitate to publish almost the severest words that can be uttered;—but let an editor be himself attacked, even without his name, and he thinks that the thunderbolts of heaven should fall upon the offender. Let his manners, his truth, his judgment, his honesty, or even his consistency be questioned, and thunderbolts are forthcoming, though they may not be from heaven. There should certainly be a thunderbolt or two now, but Mr. Slide did not at first quite see how they were to be forged.

He read the injunction again and again. As far as the document went he knew its force, and recognised the necessity of obedience. He might, perhaps, be able to use the information contained in the letter from Mr. Kennedy, so as to harass Phineas and Lady Laura and the Earl, but he was at once aware that it must not be published. An editor is bound to avoid the meshes of the law, which are always infinitely more costly to companies, or things, or institutions, than they are to individuals. Of fighting with Chancery he had no notion; but it should go hard with him if he did not have a fight with Phineas Finn. And then there arose another cause for deep sorrow. A paragraph was shown to him in a morning paper of that day which must, he thought, refer to Mr. Kennedy and Phineas Finn. 'A rumour has reached us that a member of Parliament, calling yesterday afternoon upon a right honourable gentleman, a member of a late Government, at his hotel, was shot at by the latter in his sitting room. Whether the rumour be true or not we have no means of saying, and therefore abstain from publishing names. We are informed that the gentleman who used the pistol was out of his mind. The bullet did not take effect.' How cruel it was that such information should have reached the hands of a rival, and not fallen in the way of the People's Banner! And what a pity that the bullet should have been wasted! The paragraph must certainly refer to Phineas Finn and Kennedy. Finn, a Member of Parliament, had been sent by Slide himself to call upon Kennedy, a member of the late Government, at Kennedy's hotel. And the paragraph must be true. He himself had warned Finn that there would be danger in the visit. He had even prophesied murder, —and murder had been attempted! The whole transaction had been, as it were, the very goods and chattels of the People's Banner, and the paper had been shamefully robbed of its property. Mr. Slide hardly doubted that Phineas Finn had himself sent the paragraph to an adverse paper, with the express view of adding to the injury inflicted upon the Banner. That day Mr. Slide hardly did his work effectively within his glass cage, so much was his mind affected, and at five o'clock, when he left his office, instead of going at once home to Mrs. Slide at Camden Town, he took an omnibus, and went down to Westminster. He would at once confront the traitor who had deceived him.

It must be acknowledged on behalf of this editor that he did in truth believe that he had been hindered from doing good. The whole practice of his life had taught him to be confident that the editor of a newspaper must be the best possible judge, —indeed the only possible good judge,—whether any statement or story should or should not be published. Not altogether without a conscience, and intensely conscious of such conscience as did constrain him, Mr. Quintus Slide imagined that no law of libel, no injunction from any Vice-Chancellor, no outward power or pressure whatever was needed to keep his energies within their proper limits. He and his newspaper formed together a simply beneficent institution, any interference with which must of necessity be an injury to the public. Everything done at the office of the People's Banner was done in the interest of the People,—and, even though individuals might occasionally be made to suffer by the severity with which their names were handled in its columns, the general result was good. What are the sufferings of the few to the advantage of the many? If there be fault in high places, it is proper that it be exposed. If there be fraud, adulteries, gambling, and lasciviousness,—or even quarrels and indiscretions among those whose names are known, let every detail be laid open to the light, so that the people may have a warning. That such details will make a paper 'pay' Mr. Slide knew also; but it is not only in Mr. Slide's path of life that the bias of a man's mind may lead him to find that virtue and profit are compatible. An unprofitable newspaper cannot long continue its existence, and, while existing, cannot be widely beneficial. It is the circulation, the profitable circulation,—of forty, fifty, sixty, or a hundred thousand copies through all the arteries and veins of the public body which is beneficent. And how can such circulation be effected unless the taste of the public be consulted? Mr. Quintus Slide, as he walked up Westminster

Hall, in search of that wicked member of Parliament, did not at all doubt the goodness of his cause. He could not contest the Vice-Chancellor's injunction, but he was firm in his opinion that the Vice-Chancellor's injunction had inflicted an evil on the public at large, and he was unhappy within himself in that the power and majesty and goodness of the press should still be hampered by ignorance, prejudice, and favour for the great. He was quite sure that no injunction would have been granted in favour of Mr. Joseph Smith and Mr. John Jones.

He went boldly up to one of the policemen who sit guarding the door of the lobby of our House of Commons, and asked for Mr. Finn. The Cerberus*on the left was not sure whether Mr. Finn was in the House, but would send in a card if Mr. Slide would stand on one side. For the next quarter of an hour Mr. Slide heard no more of his message, and then applied again to the Cerberus. The Cerberus shook his head, and again desired the applicant to stand on one side. He had done all that in him lay. The other watchful Cerberus standing on the right, observing that the intruder was not accommodated with any member, intimated to him the propriety of standing back in one of the corners. Our editor turned round upon the man as though he would bite him;—but he did stand back, meditating an article on the gross want of attention to the public shown in the lobby of the House of Commons. Is it possible that any editor should endure any inconvenience without meditating an article? But the judicious editor thinks twice of such things. Our editor was still in his wrath when he saw his prey come forth from the House with a card,—no doubt his own card. He leaped forward in spite of the policeman, in spite of any Cerberus, and seized Phineas by the arm. 'I want just to have a few words,' he said. He made an effort to repress his wrath, knowing that the whole world would be against him should he exhibit any violence of indignation on that spot; but Phineas could see it all in the fire of his eye.

'Certainly,' said Phineas, retiring to the side of the lobby, with a conviction that the distance between him and the House was already sufficient.

'Can't you come down into Westminster Hall?'

'I should only have to come up again. You can say what you've got to say here.'

'I've got a great deal to say. I never was so badly treated in my life;—never.' He could not quite repress his voice, and he saw that a policeman looked at him. Phineas saw it also.

'Because we have hindered you from publishing an untrue and very slanderous letter about a lady!'

'You promised me that you'd come to me yesterday.'

'I think not. I think I said that you should hear from me,—and you did.'

'You call that truth,—and honesty!'

'Certainly I do. Of course it was my first duty to stop the publication of the letter.'

'You haven't done that yet.'

'I've done my best to stop it. If you have nothing more to say I'll wish you good evening.'

'I've a deal more to say. You were shot at, weren't you?'

'I have no desire to make any communication to you on anything that has occurred, Mr. Slide. If I stayed with you all the afternoon I could tell you nothing more. Good evening.'

'I'll crush you,' said Quintus Slide, in a stage whisper; 'I will, as sure as my name is Slide.'

Phineas looked at him and retired into the House, whither Quintus Slide could not follow him, and the editor of the People's Banner was left alone in his anger.

'How a cock can crow on his own dunghill!' That was Mr. Slide's first feeling, as with a painful sense of diminished consequence he retraced his steps through the outer lobbies and down into Westminster Hall. He had been browbeaten by Phineas Finn, simply because Phineas had been able to retreat within those happy doors. He knew that to the eyes of all the policemen and strangers assembled Phineas Finn had been a hero, a Parliamentary hero, and he had been some poor outsider,—to be ejected at once should he make himself disagreeable to the Members. Nevertheless, had he not all the columns of the People's Banner in his pocket? Was he not great in the Fourth Estate,—much greater than Phineas Finn in his estate? Could he not thunder every night so that an audience to be counted by hundreds of thousands should hear his thunder;—whereas this poor Member of Parliament must struggle night after night for an opportunity of speaking; and could then only speak to benches half deserted; or to a few Members half asleep,—unless the Press should choose to convert his words into thunderbolts. Who could doubt for a moment with which lay the greater power? And yet this wretched Irishman, who had wriggled himself into Parliament on a petition, getting the better of a good, downright English John Bull by a quibble, had treated him with scorn,—the wretched Irishman being for the moment like a cock on his own dunghill. Quintus Slide was not slow to tell himself that he also had an elevation of his own, from which he could make himself audible. In former days he had forgiven Phineas Finn more than once. If he ever forgave Phineas Finn again might his right hand forget its cunning, and never again draw blood or tear a scalp.

CHAPTER XXVIII

The first Thunderbolt

It was not till after Mr. Slide had left him that Phineas wrote the following letter to Lady Laura:—

'*House of Commons*, 1*st March*, 18—

'My dear Friend,

'I have a long story to tell, which I fear I shall find difficult in the telling; but it is so necessary that you should know the facts that I must go through with it as best I may. It will give you very great pain; but the result as regards your own position will not I think be injurious to you.

'Yesterday, Sunday, a man came to me who edits a newspaper, and whom I once knew. You will remember when I used to tell you in Portman Square of the amenities and angers of Mr. Slide,—the man who wanted to sit for Loughton. He is the editor. He brought me a long letter from Mr. Kennedy himself, intended for publication, and which was already printed, giving an elaborate and, I may say, a most cruelly untrue account of your quarrel. I read the letter, but of course cannot remember the words. Nor if I could remember them should I repeat them. They contained all the old charges with which you are familiar, and which your unfortunate husband now desired to publish in consummation of his threats. Why Mr. Slide should have brought me the paper before publishing it I can hardly understand. But he did so;—and told me that Mr. Kennedy was in town. We have managed among us to obtain a legal warrant for preventing the publication of the letter, and I think I may say that it will not see the light.

'When Mr. Slide left me I called on Mr. Kennedy, whom I found in a miserable little hotel, in Judd Street, kept by Scotch people named Macpherson. They had come from the neighbourhood of Loughlinter, and knew Mr. Kennedy well. This was yesterday afternoon, Sunday, and I found some difficulty in making my way into his presence. My object was to induce him to withdraw the letter;—for at that time I doubted whether the law could interfere quickly enough to prevent the publication.

'I found your husband in a very sad condition. What he said or what I said I forget; but he was as usual intensely anxious that you should return to him. I need not hesitate now to say that he is certainly mad. After a while, when I expressed my assured opinion that you would not go back to Loughlinter, he suddenly turned round, grasped a revolver, and fired at my head. How I got out of the room I don't quite remember. Had he repeated the shot, which he might have done over and over again, he must have hit me. As it was I escaped, and blundered down the stairs to Mrs. Macpherson's room.

'They whom I have consulted in the matter, namely, Barrington Erle and my particular friend, Mr. Low,—to whom I went for legal assistance in stopping the publication,—seem to think that I should have at once sent for the police, and given Mr. Kennedy in charge. But I did not do so, and hitherto the police have, I believe, no knowledge of what occurred. A paragraph appeared in one of the morning papers to-day, giving almost an accurate account of the matter, but mentioning neither the place nor any of the names. No doubt it will be repeated in all the papers, and the names will soon be known. But the result will be simply a general conviction as to the insanity of poor Mr. Kennedy,—as to which they who know him have had for a long time but little doubt.

'The Macphersons seem to have been very anxious to screen their guest. At any other hotel no doubt the landlord would have sent for the police;—but in this case the attempt was kept quite secret. They did send for George Kennedy, a cousin of your husband's, whom I think you know, and whom I saw this morning. He assures me that Robert Kennedy is quite aware of the wickedness of the attempt he made, and that he is plunged in deep remorse. He is to be taken down to Loughlinter to-morrow, and is,—so says his cousin,—as tractable as a child. What George Kennedy means to do, I cannot say; but for myself, as I did not send for the police at the moment, as I am told I ought to have done, I shall now do nothing. I don't know that a man is subject to punishment because he does not make complaint. I suppose I have a right to regard it all as an accident if I please.

'But for you this must be very important. That Mr. Kennedy is insane there cannot now, I think, be a doubt; and therefore the question of your returning to him,—as far as there has been any question,—is absolutely settled. None of your friends would be justified in allowing you to return. He is undoubtedly mad, and has done an act which is not murderous only on that conclusion. This settles the question so perfectly that you could, no doubt, reside in England now without danger. Mr. Kennedy himself would feel that he could take no steps to enforce your return after what he did yesterday. Indeed, if you could bring yourself to face the publicity, you could, I imagine, obtain a legal separation which would give you again the control of your own fortune. I feel myself bound to mention this; but I give you no advice. You will no doubt explain all the circumstances to your father.

'I think I have now told you everything that I need tell you. The thing only happened yesterday, and I have been all the morning busy, getting the injunction, and seeing Mr. George Kennedy. Just before I began this letter that horrible editor was with me again, threatening me with all the penalties which an editor can inflict. To tell the truth, I do feel confused among them all, and still fancy that I hear the click of the pistol. That newspaper paragraph says that the ball went through my whiskers, which was certainly not the case;—but a foot or two off is quite near enough for a pistol ball.

'The Duke of Omnium is dying, and I have heard to-day that Madame Goesler, our old friend, has been sent for to Matching. She and I renewed our acquaintance the other day at Harrington.

'God bless you.

'Your most sincere friend,

'PHINEAS FINN.

'Do not let my news oppress you. The firing of the pistol is a thing done and over without evil results. The state of Mr. Kennedy's mind is what we have long suspected; and, melancholy though it be, should contain for you at any rate this consolation,—that the accusations made against you would not have been made had his mind been unclouded.'

Twice while Finn was writing this letter was he rung into the House for a division, and once it was suggested to him to say a few words of angry opposition to the Government on some not important subject under discussion. Since the beginning of the Session hardly a night had passed without some verbal sparring, and very frequently the limits of parliamentary decorum had been almost surpassed. Never within the memory of living politicians had political rancour been so sharp, and the feeling of injury so keen, both on the one side and on the other. The taunts thrown at the Conservatives, in reference to the Church, had been almost unendurable,—and the more so because the strong expressions of feeling from their own party throughout the country were against them. Their own convictions also were against them. And there had for a while been almost a determination through the party to deny their leader and disclaim the bill. But a feeling of duty to the party had prevailed, and this had not been done. It had not been done; but the not doing of it was a sore burden on the half-broken shoulders of many a man who sat gloomily on the benches behind Mr. Daubeny. Men goaded as they were, by their opponents, by their natural friends, and by their own consciences, could not bear it in silence, and very bitter things were said in return. Mr. Gresham was accused of a degrading lust for power. No other feeling could prompt him to oppose with a factious acrimony never before exhibited in that House,—so said some wretched Conservative with broken back and broken heart,—a measure which he himself would only be too willing to carry were he allowed the privilege of passing over to the other side of the House for the purpose. In these encounters, Phineas Finn had already exhibited his prowess, and, in spite of his declarations at Tankerville, had become prominent as an opponent to Mr. Daubeny's bill. He had, of course, himself been taunted, and held up in the House to the execration of his own constituents; but he had enjoyed his fight, and had remembered how his friend Mr. Monk had once told him that the pleasure lay all on the side of opposition. But on this evening he declined to speak. 'I suppose you have hardly recovered from Kennedy's pistol,' said Mr. Ratler, who had, of course, heard the whole story. 'That, and the whole affair together have upset me,' said Phineas. 'Fitzgibbon will do it for you; he's in the House.' And so it happened that on that occasion the Honourable Laurence Fitzgibbon made a very effective speech against the Government.

On the next morning from the columns of the People's Banner was hurled the first of those thunderbolts with which it was the purpose of Mr. Slide absolutely to destroy the political and social life of Phineas Finn. He would not miss his aim as Mr. Kennedy had done. He would strike such blows that no constituency should ever venture to return Mr. Finn again to Parliament; and he thought that he could also so strike his blows that no mighty nobleman, no distinguished commoner, no lady of rank should again care to entertain the miscreant and feed him with the dainties of fashion. The first thunderbolt was as follows:—

'We abstained yesterday from alluding to a circumstance which occurred at a small hotel in Judd Street on Sunday afternoon, and which, as we observe, was mentioned by one of our contemporaries. The names, however, were not given, although the persons implicated were indicated. We can see no reason why the names should be concealed. Indeed, as both the gentlemen concerned have been guilty of very great criminality, we think that we are bound to tell the whole story, —and this the more especially as certain circumstances have in a very peculiar manner placed us in possession of the facts.

'It is no secret that for the last two years Lady Laura Kennedy has been separated from her husband, the Honourable Robert Kennedy, who, in the last administration, under Mr. Mildmay, held the office of Chancellor of the Duchy of Lancaster; and we believe as little a secret that Mr. Kennedy has been very persistent in endeavouring to recall his wife to her home. With equal persistence she has refused to obey, and we have in our hands the clearest possible evidence that Mr. Kennedy has attributed her obstinate refusal to influence exercised over her by Mr. Phineas Finn, who three years since was her father's nominee for the then existing borough of Loughton, and who lately succeeded in ousting poor Mr. Browborough from his seat for Tankerville by his impetuous promises to support that very measure of Church Reform which he is now opposing with that venom which makes him valuable to his party. Whether Mr. Phineas Finn will ever sit in another Parliament we cannot, of course, say, but we think we can at least assure him that he will never again sit for Tankerville.

'On last Sunday afternoon Mr. Finn, knowing well the feeling with which he is regarded by Mr. Kennedy, outraged all decency by calling upon that gentleman, whose address he obtained from our office. What took place between them no one knows, and, probably, no one ever will know. But the interview was ended by Mr. Kennedy firing a pistol at Mr. Finn's head. That he should have done so without the grossest provocation no one will believe. That Mr. Finn had gone to the husband to interfere with him respecting his wife is an undoubted fact,—a fact which, if necessary, we are in a position to prove. That such interference must have been most heartrending every one will admit. This intruder, who had thrust himself upon the unfortunate husband on the Sabbath afternoon, was the very man whom the husband accuses of having robbed him of the company and comfort of his wife. But we cannot, on that account, absolve Mr. Kennedy of the criminality of his act. It should be for a jury to decide what view should be taken of that act, and to say how far the outrageous provocation offered should be allowed to palliate the offence. But hitherto the matter has not reached the police. Mr. Finn was not struck, and managed to escape from the room. It was his manifest duty as one of the community, and more especially so as a member of Parliament, to have reported all the circumstances at once to the police. This was not done by him, nor by the persons who keep the hotel. That Mr. Finn should have reasons of his own for keeping the whole affair secret, and for screening the attempt at murder, is clear enough. What inducements have been used with the people of the house we cannot, of course, say. But we understand that Mr. Kennedy has been allowed to leave London without molestation.

'Such is the true story of what occurred on Sunday afternoon in Judd Street, and, knowing what we do, we think ourselves justified in calling upon Major Mackintosh to take the case into his own hands.' Now Major Mackintosh was at this time the head of the London constabulary. 'It is quite out of the question that such a transaction should take place in the heart of London at three o'clock on a Sunday afternoon, and be allowed to pass without notice. We intend to keep as little of what we know from the public as possible, and do not hesitate to acknowledge that we are debarred by an injunction of the Vice-Chancellor from publishing a certain document which would throw the clearest light upon the whole circumstance. As soon as possible after the shot was fired Mr. Finn went to work, and, as we think, by misrepresentations, obtained the injunction early on yesterday morning. We feel sure that it would not have been granted had the transaction in Judd Street been at the time known to the Vice-Chancellor in all its enormity. Our hands are, of course, tied. The document in question is still with us, but it is sacred. When called upon to show it by any proper authority we shall be ready; but, knowing what we do know, we should not be justified in allowing the matter to sleep. In the meantime we call upon those whose duty it is to preserve the public peace to take the steps necessary for bringing the delinquents to justice.

'The effect upon Mr. Finn, we should say, must be his immediate withdrawal from public life. For the last year or two he has held some subordinate but permanent place in Ireland, which he has given up on the rumour that the party to which he has attached himself is likely to return to office. That he is a seeker after office is notorious. That any possible Government should now employ him, even as a tide-waiter, is quite out of the question; and it is equally out of the question that he should be again returned to Parliament, were he to resign his seat on accepting office. As it is, we believe, notorious that this gentlemen cannot maintain the position which he holds without being paid for his services, it is reasonable to suppose that his friends will recommend him to retire, and seek his living in some obscure, and, let us hope, honest profession.' Mr. Slide, when his thunderbolt was prepared, read it over with delight, but still with some fear as to probable results. It was expedient that he should avoid a prosecution for libel, and essential that he should not offend the majesty of the Vice-Chancellor's injunction. Was he sure that he was safe in each direction? As to the libel, he could not tell himself that he was certainly safe. He was saying very hard things both of Lady Laura and of Phineas Finn, and sailing very near the wind. But neither of those persons would probably be willing to prosecute; and, should he be prosecuted, he would then, at any rate, be able to give in Mr. Kennedy's letter as evidence in his own defence. He really did believe that what he was doing was all done in the cause of morality. It was the business of such a paper as that which he conducted to run some risk in defending morals, and exposing distinguished culprits on behalf of the public. And then, without some such risk, how could Phineas Finn be adequately punished for the atrocious treachery of which he had been guilty? As to the Chancellor's order, Mr. Slide thought that he had managed that matter very completely. No doubt he had acted in direct opposition to the spirit of the injunction, but legal orders are read by the letter, and not by the spirit. It was open to him to publish anything he pleased respecting Mr. Kennedy and his wife, subject, of course, to the general laws of the land in regard to libel. The Vice-Chancellor's special order to him referred simply to a particular document, and from that document he had not quoted a word, though he had contrived to repeat all the bitter things which it contained, with much added venom of his own. He felt secure of being safe from any active anger on the part of the Vice-Chancellor.

The article was printed and published. The reader will perceive that it was full of lies. It began with a lie in that statement that 'we abstained yesterday from alluding to circumstances' which had been unknown to the writer when his yesterday's paper was published. The indignant reference to poor Finn's want of delicacy in forcing himself upon Mr. Kennedy on the Sabbath afternoon, was, of course, a tissue of lies. The visit had been made almost at the instigation of the editor himself. The paper from beginning to end was full of falsehood and malice, and had been written with the express intention of creating prejudice against the man who had offended the writer. But Mr. Slide did not know that he was lying, and did not know that he was malicious. The weapon which he used was one to which his hand was accustomed, and he had been led by practice to believe that the use of such weapons by one in his position was not only fair, but also beneficial to the public. Had anybody suggested to him that he was stabbing his enemy in the dark, he would have averred that he was doing nothing of the kind, because the anonymous accusation of sinners in high rank was, on behalf of the public, the special duty of writers and editors attached to the public press. Mr. Slide's blood was running high with virtuous indignation against our hero as he inserted those last cruel words as to the choice of an obscure but honest profession.

Phineas Finn read the article before he sat down to breakfast on the following morning, and the dagger went right into his bosom. Every word told upon him. With a jaunty laugh within his own sleeve he had assured himself that he was safe against any wound which could be inflicted on him from the columns of the People's Banner. He had been sure that he would be attacked, and thought that he was armed to bear it. But the thin blade penetrated every joint of his harness, and every particle of the poison curdled in his blood. He was hurt about Lady Laura; he was hurt about his borough of Tankerville; he was hurt by the charges against him of having outraged delicacy; he was hurt by being handed over to the tender mercies of Major Mackintosh; he was hurt by the craft with which the Vice-Chancellor's injunction had been evaded; but he was especially hurt by the allusions to his own poverty. It was necessary that he should earn his bread, and no doubt he was a seeker after place. But he did not wish to obtain wages without working for them; and he did not see why the work and wages of a public office should be less honourable than those of any other profession. To him, with his ideas, there was no profession so honourable, as certainly there were none which demanded greater sacrifices or were more precarious. And he did believe that such an article as that would have the effect of shutting against him the gates of that dangerous Paradise which he desired to enter. He had no great claim upon his party; and, in giving away the good things of office, the giver is only too prone to recognise any objections against an individual which may seem to relieve him from the necessity of bestowing aught in that direction. Phineas felt that he would almost be ashamed to show his face at the clubs or in the House. He must do so as a matter of course, but he knew that he could not do so without confessing by his visage that he had been deeply wounded by the attack in the People's Banner.

He went in the first instance to Mr. Low, and was almost surprised that Mr. Low should not have yet even have heard that such an attack had been made. He had almost felt, as he walked to Lincoln's Inn, that everybody had looked at him, and that passers-by in the street had declared to each other that he was the unfortunate one who had been doomed by the editor of the People's Banner to seek some obscure way of earning his bread. Mr. Low took the paper, read, or probably only half read, the article, and then threw the sheet aside as worthless. 'What ought I to do?'

'Nothing at all.'

'One's first desire would be to beat him to a jelly.'

'Of all courses that would be the worst, and would most certainly conduce to his triumph.'

'Just so;—I only allude to the pleasure one would have, but which one has to deny oneself. I don't know whether he has laid himself open for libel.'

'I should think not. I have only just glanced at it, and therefore can't give an opinion; but I should think you would not dream of such a thing. Your object is to screen Lady Laura's name.'

'I have to think of that first.'

'It may be necessary that steps should be taken to defend her character. If an accusation be made with such publicity as to enforce belief if not denied, the denial must be made, and may probably be best made by an action for libel. But that must be done by her or her friends,—but certainly not by you.'

'He has laughed at the Vice-Chancellor's injunction.'

'I don't think that you can interfere. If, as you believe, Mr. Kennedy be insane, that fact will probably soon be proved, and will have the effect of clearing Lady Laura's character. A wife may be excused for leaving a mad husband.'

'And you think I should do nothing?'

'I don't see what you can do. You have encountered a chimney sweeper, and of course you get some of the soot. What you do do, and what you do not do, must depend at any rate on the wishes of Lady Laura Kennedy and her father. It is a matter in which you must make yourself subordinate to them.'

Fuming and fretting, and yet recognising the truth of Mr. Low's words, Phineas left the chambers, and went down to his club. It was a Wednesday, and the House was to sit in the morning; but before he went to the House he put himself in the way of certain of his associates in order that he might hear what would be said, and learn if possible what was thought. Nobody seemed to treat the accusations in the newspaper as very serious, though all around him congratulated him on his escape from Mr. Kennedy's pistol. 'I suppose the poor man really is mad,' said Lord Cantrip, whom he met on the steps of one of the clubs.

'No doubt, I should say.'

'I can't understand why you didn't go to the police.'

'I had hoped the thing would not become public,' said Phineas.

'Everything becomes public;—everything of that kind. It is very hard upon poor Lady Laura.'

'That is the worst of it, Lord Cantrip.'

'If I were her father I should bring her to England, and demand a separation in a regular and legal way. That is what he should do now in her behalf. She would then have an opportunity of clearing her character from imputations which, to a certain extent, will affect it, even though they come from a madman, and from the very scum of the press.'

'You have read that article?'

'Yes;—I saw it but a minute ago.'

'I need not tell you that there is not the faintest ground in the world for the imputation made against Lady Laura there.'

'I am sure that there is none;—and therefore it is that I tell you my opinion so plainly. I think that Lord Brentford should be advised to bring Lady Laura to England, and to put down the charges openly in Court. It might be done either by an application to the Divorce Court for a separation, or by an action against the newspaper for libel. I do not know Lord Brentford quite well enough to intrude upon him with a letter, but I have no objection whatever to having my name mentioned to him. He and I and you and poor Mr. Kennedy sat together in the same Government, and I think that Lord Brentford would trust my friendship so far.' Phineas thanked him, and assured him that what he had said should be conveyed to Lord Brentford.

CHAPTER XXIX

The Spooner Correspondence

It will be remembered that Adelaide Palliser had accepted the hand of Mr. Maule, junior, and that she and Lady Chiltern between them had despatched him up to London on an embassy to his father, in which he failed very signally. It had been originally Lady Chiltern's idea that the proper home for the young couple would be the ancestral hall, which must be theirs some day, and in which, with exceeding prudence, they might be able to live as Maules of Maule Abbey upon the very limited income which would belong to them. How slight were the grounds for imputing such stern prudence to Gerard Maule both the ladies felt;—but it had become essential to do something; the young people were engaged to each other, and a manner of life must be suggested, discussed, and as far as possible arranged. Lady Chiltern was useful at such work, having a practical turn of mind, and understanding well the condition of life for which it was necessary that her friend should prepare herself. The lover was not vicious, he neither drank nor gambled, nor ran himself hopelessly in debt. He was good-humoured and tractable, and docile enough when nothing disagreeable was asked from him. He would have, he said, no objection to live at Maule Abbey if Adelaide liked it. He didn't believe much in farming, but would consent at Adelaide's request to be the owner of bullocks. He was quite ready to give up hunting, having already taught himself to think that the very few good runs in a season were hardly worth the trouble of getting up before daylight all the winter. He went forth, therefore, on his embassy, and we know how he failed. Another lover would have communicated the disastrous tidings at once to the lady; but Gerard Maule waited a week before he did so, and then told his story in half-a-dozen words. 'The governor cut up rough about Maule Abbey, and will not hear of it. He generally does cut up rough.'

'But he must be made to hear of it,' said Lady Chiltern. Two days afterwards the news reached Harrington of the death of the Duke of Omnium. A letter of an official nature reached Adelaide from Mr. Fothergill, in which the writer explained that he had been desired by Mr. Palliser to communicate to her and the relatives the sad tidings. 'So the poor old man has gone at last,' said Lady Chiltern, with that affectation of funereal gravity which is common to all of us.

'Poor old Duke!' said Adelaide. 'I have been hearing of him as a sort of bugbear all my life. I don't think I ever saw him but once, and then he gave me a kiss and a pair of earrings. He never paid any attention to us at all, but we were taught to think that Providence had been very good to us in making the Duke our uncle.'

'He was very rich?'

'Horribly rich, I have always heard.'

'Won't he leave you something? It would be very nice now that you are engaged to find that he has given you five thousand pounds.'

'Very nice indeed;—but there is not a chance of it. It has always been known that everything is to go to the heir. Papa had his fortune and spent it. He and his brother were never friends, and though the Duke did once give me a kiss I imagine that he forgot my existence immediately afterwards.'

'So the Duke of Omnium is dead,' said Lord Chiltern when he came home that evening.

'Adelaide has had a letter to tell her so this afternoon.'

'Mr. Fothergill wrote to me,' said Adelaide;—'the man who is so wicked about the foxes.'

'I don't care a straw about Mr. Fothergill; and now my mouth is closed against your uncle. But it's quite frightful to think that a Duke of Omnium must die like anybody else.'

'The Duke is dead;—long live the Duke,' said Lady Chiltern. 'I wonder how Mr. Palliser will like it.'

'Men always do like it, I suppose,' said Adelaide.

'Women do,' said Lord Chiltern. 'Lady Glencora will be delighted to reign,—though I can hardly fancy her by any other name. By the bye, Adelaide, I have got a letter for you.'

'A letter for me, Lord Chiltern!'

'Well,—yes; I suppose I had better give it you. It is not addressed to you, but you must answer it.'

'What on earth is it?'

'I think I can guess,' said Lady Chiltern, laughing. She had guessed rightly, but Adelaide Palliser was still altogether in the dark when Lord Chiltern took a letter from his pocket and handed it to her. As he did so he left the room, and his wife followed him. 'I shall be upstairs, Adelaide, if you want advice,' said Lady Chiltern.

The letter was from Mr. Spooner. He had left Harrington Hall after the uncourteous reception which had been accorded to him by Miss Palliser in deep disgust, resolving that he would never again speak to her, and almost resolving that Spoon Hall should never have a mistress in his time. But with his wine after dinner his courage came back to him, and he began to reflect once more that it is not the habit of young ladies to accept their lovers at the first offer. There was living with Mr. Spooner at this time a very attached friend, whom he usually consulted in all emergencies, and to whom on this occasion he opened his heart. Mr. Edward Spooner, commonly called Ned by all who knew him, and not unfrequently so addressed by those who did not, was a distant cousin of the Squire's, who unfortunately had no particular income of his own. For the last ten years he had lived at Spoon Hall, and had certainly earned his bread. The Squire had achieved a certain credit for success as a country gentleman. Nothing about his place was out of order. His own farming, which was extensive, succeeded. His bullocks and sheep won prizes. His horses were always useful and healthy. His tenants were solvent, if not satisfied, and he himself did not owe a shilling. Now many people in the neighbourhood attributed all this to the judicious care of Mr. Edward Spooner, whose eye was never off the place, and whose discretion was equal to his zeal. In giving the Squire his due, one must acknowledge that he recognised the merits of his cousin, and trusted him in everything. That night, as soon as the customary bottle of claret had succeeded the absolutely normal bottle of port after dinner, Mr. Spooner of Spoon Hall opened his heart to his cousin.

'I shall have to walk, then,' said Ned.

'Not if I know it,' said the Squire. 'You don't suppose I'm going to let any woman have the command of Spoon Hall?'

'They do command,—inside, you know.'

'No woman shall ever turn you out of this house, Ned.'

'I'm not thinking of myself, Tom,' said the cousin. 'Of course you'll marry some day, and of course I must take my chance. I don't see why it shouldn't be Miss Palliser as well as another.'

'The jade almost made me angry.'

'I suppose that's the way with most of 'em. "Ludit exultim metuitque tangi".' For Ned Spooner had himself preserved some few tattered shreds of learning from his school days. 'You don't remember about the filly?'

'Yes I do; very well,' said the Squire.

'"Nuptiarum expers"* That's what it is, I suppose. Try it again.' The advice on the part of the cousin was genuine and unselfish. That Mr. Spooner of Spoon Hall should be rejected by a young lady without any fortune seemed to him to be impossible. At any rate it is the duty of a man in such circumstances to persevere. As far as Ned knew the world, ladies always required to be asked a second or a third time. And then no harm can come from such perseverance. 'She can't break your bones, Tom.'

There was much honesty displayed on this occasion. The Squire, when he was thus instigated to persevere, did his best to describe the manner in which he had been rejected. His powers of description were not very great, but he did not conceal anything wilfully. 'She was as hard as nails, you know.'

'I don't know that that means much. Horace's filly kicked a few, no doubt.'

'She told me that if I'd go one way, she'd go the other!'

'They always say about the hardest things that come to their tongues. They don't curse and swear as we do, or there'd be no bearing them. If you really like her——'

'She's such a well-built creature! There's a look of blood about her I don't see in any of 'em. That sort of breeding is what one wants to get through the mud with.'

Then it was that the cousin recommended a letter to Lord Chiltern. Lord Chiltern was at the present moment to be regarded as the lady's guardian, and was the lover's intimate friend. A direct proposal had already been made to the young lady, and this should now be repeated to the gentleman who for the time stood in the position of her father. The Squire for a while hesitated, declaring that he was averse to make his secret known to Lord Chiltern. 'One doesn't want every fellow in the country to know it,' he said. But in answer to this the cousin was very explicit. There could be but little doubt that Lord Chiltern knew the secret already; and he would certainly be rather induced to keep it as a secret than to divulge it if it were communicated to him officially. And what other step could the Squire take? It would not be likely that he should be asked again to Harrington Hall with the express view of repeating his offer. The cousin was quite of opinion that a written proposition should be made; and on that very night the cousin himself wrote out a letter for the Squire to copy in the morning. On the morning the Squire copied the letter,—not without additions of his own, as to which he had very many words with his discreet cousin,—and in a formal manner handed it to Lord Chiltern towards the afternoon of that day, having devoted his whole morning to the finding of a proper opportunity for doing so. Lord Chiltern had read the letter, and had, as we see, delivered it to Adelaide Palliser. 'That's another proposal from Mr. Spooner,' Lady Chiltern said, as soon as they were alone.

'Exactly that.'

'I knew he'd go on with it. Men are such fools.'

'I don't see that he's a fool at all;' said Lord Chiltern, almost in anger. 'Why shouldn't he ask a girl to be his wife? He's a rich man, and she hasn't got a farthing.'

'You might say the same of a butcher, Oswald.'

'Mr. Spooner is a gentleman.'

'You do not mean to say that he's fit to marry such a girl as Adelaide Palliser?'

'I don't know what makes fitness. He's got a red nose, and if she don't like a red nose,—that's unfitness. Gerard Maule's nose isn't red, and I dare say therefore he's fitter. Only, unfortunately, he has no money.'

'Adelaide Palliser would no more think of marrying Mr. Spooner than you would have thought of marrying the cook.'

'If I had liked the cook I should have asked her, and I don't see why Mr. Spooner shouldn't ask Miss Palliser. She needn't take him.'

In the meantime Miss Palliser was reading the following letter:—

'*Spoon Hall*, 11*th March*, 18—.

'MY DEAR LORD CHILTERN,—

'I venture to suppose that at present you are acting as the guardian of Miss Palliser, who has been staying at your house all the winter. If I am wrong in this I hope you will pardon me, and consent to act in that capacity for this occasion. I entertain feelings of the greatest admiration and warmest affection for the young lady I have named, which I ventured to express when I had the pleasure of staying at Harrington Hall in the early part of last month. I cannot boast that I was received on that occasion with much favour; but I know that I am not very good at talking, and we are told in all the books that no man has a right to expect to be taken at the first time of asking. Perhaps Miss Palliser will allow me, through you, to request her to consider my proposal with more deliberation than was allowed to me before, when I spoke to her perhaps with injudicious hurry.' So far the Squire adopted his cousin's words without alteration.

'I am the owner of my own property,—which is more than everybody can say. My income is nearly £4,000 a year. I shall be willing to make any proper settlement that may be recommended by the lawyers,—though I am strongly of opinion that an estate shouldn't be crippled for the sake of the widow. As to refurnishing the old house, and all that, I'll do anything that Miss Palliser may please. She knows my taste about hunting, and I know hers, so that there need not be any difference of opinion on that score.

'Miss Palliser can't suspect me of any interested motives. I come forward because I think she is the most charming girl I ever saw, and because I love her with all my heart. I haven't got very much to say for myself, but if she'll consent to be the mistress of Spoon Hall, she shall have all that the heart of a woman can desire.

'Pray believe me,
'My dear Lord Chiltern,
'Yours very sincerely,
'THOMAS PLATTER SPOONER.

'As I believe that Miss Palliser is fond of books, it may be well to tell her that there is an uncommon good library at Spoon Hall. I shall have no objection to go abroad for the honeymoon for three or four months in the summer.'

The postscript was the Squire's own, and was inserted in opposition to the cousin's judgment. 'She won't come for the sake of the books,' said the cousin. But the Squire thought that the attractions should be piled up. 'I wouldn't talk of the honeymoon till I'd got her to come round a little,' said the cousin. The Squire thought that the cousin was falsely delicate, and pleaded that all girls like to be taken abroad when they're married. The second half of the body of the letter was very much disfigured by the Squire's petulance; so that the modesty with which he commenced was almost put to the blush by a touch of arrogance in the conclusion. That sentence in which the Squire declared that an estate ought not to be crippled for the sake of the widow was very much questioned by the cousin. 'Such a word as "widow" never ought to go into such a letter as this.' But the Squire protested that he would not be mealy-mouthed. 'She can bear to think of it, I'll go bail; and why shouldn't she hear about what she can think about?' 'Don't talk about furniture yet, Tom,' the cousin said; but the Squire was obstinate, and the cousin became hopeless. That word about loving her with all his heart was the cousin's own, but what followed, as to her being mistress of Spoon Hall, was altogether opposed to his judgment. 'She'll be proud enough of Spoon Hall if she comes here,' said the Squire. 'I'd let her come first,' said the cousin.

We all know that the phraseology of the letter was of no importance whatever. When it was received the lady was engaged to another man; and she regarded Mr. Spooner of Spoon Hall as being guilty of unpardonable impudence in approaching her at all.

'A red-faced vulgar old man, who looks as if he did nothing but drink,' she said to Lady Chiltern.

'He does you no harm, my dear.'

'But he does do harm. He makes things very uncomfortable. He has no business to think it possible. People will suppose that I gave him encouragement.'

'I used to have lovers coming to me year after year,—the same people,—whom I don't think I ever encouraged; but I never felt angry with them.'

'But you didn't have Mr. Spooner.'

'Mr. Spooner didn't know me in those days, or there is no saying what might have happened.' Then Lady Chiltern argued the matter on views directly opposite to those which she had put forward when discussing the matter with her husband. 'I always think that any man who is privileged to sit down to table with you is privileged to ask. There are disparities of course which may make the privilege questionable, —disparities of age, rank, and means.'

'And of tastes,' said Adelaide.

'I don't know about that.—A poet doesn't want to marry a poetess, nor a philosopher a philosopheress. A man may make himself a fool by putting himself in the way of certain refusal; but I take it the broad rule is that a man may fall in love with any lady who habitually sits in his company.'

'I don't agree with you at all. What would be said if the curate at Long Royston were to propose to one of the Fitz-Howard girls?'

'The Duchess would probably ask the Duke to make the young man a bishop out of hand, and the Duke would have to spend a morning in explaining to her the changes which have come over the making of bishops since she was young. There is no other rule that you can lay down, and I think that girls should understand that they have to fight their battles subject to that law. It's very easy to say, "No."'

'But a man won't take "No."'

'And it's lucky for us sometimes that they don't,' said Lady Chiltern, remembering certain passages in her early life.

The answer was written that night by Lord Chiltern after much consultation. As to the nature of the answer,—that it should be a positive refusal,—of course there could be no doubt; but then arose a question whether a reason should be given, or whether the refusal should be simply a refusal. At last it was decided that a reason should be given, and the letter ran as follows:—

'My dear Mr. Spooner,

'I am commissioned to inform you that Miss Palliser is engaged to be married to Mr. Gerard Maule.

'Yours faithfully,

'Chiltern.'

The young lady had consented to be thus explicit because it had been already determined that no secret should be kept as to her future prospects.

'He is one of those poverty-stricken wheedling fellows that one meets about the world every day,' said the Squire to his cousin—'a fellow that rides horses that he can't pay for, and owes some poor devil of a tailor for the breeches that he sits in. They eat, and drink, and get along heaven only knows how. But they're sure to come to smash at last. Girls are such fools nowadays.'

'I don't think there has ever been much difference in that,' said the cousin.

'Because a man greases his whiskers, and colours his hair, and paints his eyebrows, and wears kid gloves, by

George, they'll go through fire and water after him. He'll never marry her.'

'So much the better for her.'

'But I hate such d—— impudence. What right has a man to come forward in that way who hasn't got a house over his head, or the means of getting one? Old Maule is so hard up that he can barely get a dinner at his club in London. What I wonder at is that Lady Chiltern shouldn't know better.'

CHAPTER XXX

Regrets

MADAME GOESLER remained at Matching till after the return of Mr. Palliser—or, as we must now call him, the Duke of Omnium—from Gatherum Castle, and was therefore able to fight her own battle with him respecting the gems and the money which had been left her. He brought to her with his own hands the single ring which she had requested, and placed it on her finger. 'The goldsmith will soon make that all right,' she said, when it was found to be much too large for the largest finger on which she could wear a ring. 'A bit shall be taken out, but I will not have it reset.'

'You got the lawyer's letter and the inventory, Madame Goesler?'

'Yes, indeed. What surprises me is that the dear old man should never have spoken of so magnificent a collection of gems.'

'Orders have been given that they shall be packed.'

'They may be packed or unpacked, of course, as your Grace pleases, but pray do not connect me with the packing.'

'You must be connected with it.'

'But I wish not to be connected with it, Duke. I have written to the lawyer to renounce the legacy, and, if your Grace persists, I must employ a lawyer of my own to renounce them after some legal form. Pray do not let the case be sent to me, or there will be so much trouble, and we shall have another great jewel robbery?* I won't take it in, and I won't have the money, and I will have my own way. Lady Glen will tell you that I can be very obstinate when I please.'

Lady Glencora had told him so already. She had been quite sure that her friend would persist in her determination as to the legacy, and had thought that her husband should simply accept Madame Goesler's assurances to that effect. But a man who had been Chancellor of the Exchequer could not deal with money, or even with jewels, so lightly. He assured his wife that such an arrangement was quite out of the question. He remarked that property was property, by which he meant to intimate that the real owner of substantial wealth could not be allowed to disembarrass himself of his responsibilities or strip himself of his privileges by a few generous but idle words. The late Duke's will was a very serious thing, and it seemed to the heir that this abandoning of a legacy bequeathed by the Duke was a making light of the Duke's last act and deed. To refuse money in such circumstances was almost like refusing rain from heaven, or warmth from the sun. It could not be done. The things were her property, and though she might, of course, chuck them into the street, they would no less be hers. 'But I won't have them, Duke,' said Madame Goesler; and the late Chancellor of the Exchequer found that no proposition made by him in the House had ever been received with a firmer opposition. His wife told him that nothing he could say would be of any avail, and rather ridiculed his idea of the solemnity of wills. 'You can't make a person take a thing because you write it down on a thick bit of paper, any more than if you gave it her across a table. I understand it all, of course. She means to show that she didn't want anything from the Duke. As she refused the name and title, she won't have the money and jewels. You can't make her take them, and I'm quite sure you can't talk her over.' The young Duke was not persuaded, but had to give the battle up,—at any rate, for the present.

On the 19th of March Madame Goesler returned to London, having been at Matching Priory for more than three weeks.

On her journey back to Park Lane many thoughts crowded on her mind. Had she, upon the whole, done well in reference to the Duke of Omnium? The last three years of her life had been sacrificed to an old man with whom she had not in truth possessed aught in common. She had persuaded herself that there had existed a warm friendship between them;—but of what nature could have been a friendship with one whom she had not known till he had been in his dotage? What words of the Duke's speaking had she ever heard with pleasure, except certain terms of affection which had been half mawkish and half senile? She had told Phineas Finn, while riding home with him from Broughton Spinnies, that she had clung to the Duke because she loved him, but what had there been to produce such love? The Duke had begun his acquaintance with her by insulting her,—and had then offered to make her his wife. This,—which would have conferred upon her some tangible advantages, such as rank, and wealth, and a great name,—she had refused, thinking that the price to be paid for them was too high, and that life might even yet have something better in store for her. After that she had permitted herself to become, after a fashion, head nurse to the old man, and in that pursuit had wasted three years of what remained to her of her youth. People, at any rate, should not say of her that she had accepted payment for the three years' service by taking a casket of jewels. She would take nothing that should justify any man in saying that she had been enriched by her acquaintance with the Duke of Omnium. It might be that she had been foolish, but she would be more foolish still were she to accept a reward for her folly. As it was there had been something of romance in it,—though the romance of friendship at the bedside of a sick and selfish old man had hardly been satisfactory.

Even in her close connection with the present Duchess there was something which was almost hollow. Had there not been a compact between them, never expressed, but not the less understood? Had not her dear friend, Lady Glen, agreed to bestow upon her support, fashion, and all kinds of worldly good things,—on condition that she never married the old Duke? She had liked Lady Glencora,—had enjoyed her friend's society, and been happy in her friend's company,—but she had always felt that Lady Glencora's attraction to herself had been simply on the score of the Duke. It was necessary that the Duke should be pampered and kept in good humour. An old man, let him be ever so old, can do what he likes with himself and his belongings. To keep the Duke out of harm's way Lady Glencora had opened her arms to Madame Goesler. Such, at least, was the interpretation which Madame Goesler chose to give to the history of the last three years. They had not, she thought, quite understood her. When once she had made up her mind not to marry the Duke, the Duke had been safe from her;—as his jewels and money should be safe now that he was dead.

Three years had passed by, and nothing had been done of that which she had intended to do. Three years had passed, which to her, with her desires, were so important. And yet she hardly knew what were her desires, and had never quite defined her intentions. She told herself on this very journey that the time had now gone by, and that in losing these three years she had lost everything. As yet,—so she declared to herself now,—the world had done but little for her. Two old men had loved her; one had become her husband, and the other had asked to become so;—and to both she had done her duty. To both she had been grateful, tender, and self-sacrificing. From the former she had, as his widow, taken wealth which she valued greatly; but the wealth alone had given her no happiness. From the latter, and from his family, she had accepted a certain position. Some persons, high in repute and fashion, had known her before, but everybody knew her now. And yet what had all this done for her? Dukes and duchesses, dinner-parties and drawing-rooms,—what did they all amount to? What was it that she wanted?

She was ashamed to tell herself that it was love. But she knew this,—that it was necessary for her happiness that she should devote herself to some one. All the elegancies and outward charms of life were delightful, if only they could be used as the means to some end. As an end themselves they were nothing. She had devoted herself to this old man who was now dead, and there had been moments in which she had thought that that sufficed. But it had not sufficed, and instead of being borne down by grief at the loss of her friend, she found herself almost rejoicing at relief from a vexatious burden. Had she been a hypocrite then? Was it her nature to be false? After that she reflected whether it might not be best for her to become a devotee,—it did not matter much in what branch of the Christian religion, so that she could assume some form of faith. The sour strictness of the confident Calvinist or the asceticism of St. Francis might suit her equally,—if she could only believe in Calvin or in St. Francis. She had tried to believe in the Duke of Omnium, but there she had failed. There had been a saint at whose shrine she thought she could have worshipped with a constant and happy devotion, but that saint had repulsed her from his altar.

Mr. Maule, Senior, not understanding much of all this, but still understanding something, thought that he might perhaps be the saint. He knew well that audacity in asking is a great merit in a middle-aged wooer. He was a good deal older than the lady, who, in spite of all her experiences, was hardly yet thirty. But then he was,—he felt sure,—very young for his age, whereas she was old. She was a widow; he was a widower. She had a house in town and an income. He had a place in the country and an estate. She knew all the dukes and duchesses, and he was a man of family. She could make him comfortably opulent. He could make her Mrs. Maule of Maule Abbey. She, no doubt, was good-looking. Mr. Maule, Senior, as he tied on his cravat, thought that even in that respect there was no great disparity between them. Considering his own age, Mr. Maule, Senior, thought there was not perhaps a better-looking man than himself about Pall Mall. He was a little stiff in the joints and moved rather slowly, but what was wanting in suppleness was certainly made up in dignity.

He watched his opportunity, and called in Park Lane on

the day after Madame Goesler's return. There was already between them an amount of acquaintance which justified his calling, and, perhaps, there had been on the lady's part something of that cordiality of manner which is wont to lead to intimate friendship. Mr. Maule had made himself agreeable, and Madame Goesler had seemed to be grateful. He was admitted, and on such an occasion it was impossible not to begin the conversation about the 'dear Duke.' Mr. Maule could afford to talk about the Duke, and to lay aside for a short time his own cause, as he had not suggested to himself the possibility of becoming pressingly tender on his own behalf on this particular occasion. Audacity in wooing is a great virtue, but a man must measure even his virtues. 'I heard that you had gone to Matching, as soon as the poor Duke was taken ill,' he said.

She was in mourning, and had never for a moment thought of denying the peculiarity of the position she had held in reference to the old man. She could not have been content to wear her ordinary coloured garments after sitting so long by the side of the dying man. A hired nurse may do so, but she had not been that. If there had been hypocrisy in her friendship the hypocrisy must be maintained to the end.

'Poor old man! I only came back yesterday.'

'I never had the pleasure of knowing his Grace,' said Mr. Maule. 'But I have always heard him named as a nobleman of whom England might well be proud.'

Madame Goesler was not at the moment inclined to tell lies on the matter, and did not think that England had much cause to be proud of the Duke of Omnium. 'He was a man who held a very peculiar position,' she said.

'Most peculiar;—a man of infinite wealth, and of that special dignity which I am sorry to say so many men of rank among us are throwing aside as a garment which is too much for them. We can all wear coats, but it is not every one that can carry a robe. The Duke carried his to the last.' Madame Goesler remembered how he looked with his nightcap on, when he had lost his temper because they would not let him have a glass of curaçoa. 'I don't know that we have any one left that can be said to be his equal,' continued Mr. Maule.

'No one like him, perhaps. He was never married, you know.'

'But was once willing to marry,' said Mr. Maule, 'if all that we hear be true.' Madame Goesler, without a smile and equally without a frown, looked as though the meaning of Mr. Maule's words had escaped her. 'A grand old gentleman! I don't know that anybody will ever say as much for his heir.'

'The men are very different.'

'Very different indeed. I dare say that Mr. Palliser, as Mr. Palliser, has been a useful man. But so is a coal-heaver a useful man. The grace and beauty of life will be clean gone when we all become useful men.'

'I don't think we are near that yet.'

'Upon my word, Madame Goesler, I am not so sure about it. Here are sons of noblemen going into trade on every side of us. We have earls dealing in butter, and marquises sending their peaches to market. There was nothing of that kind about the Duke. A great fortune had been entrusted to him, and he knew that it was his duty to spend it. He did spend it, and all the world looked up to him. It must have been a great pleasure to you to know him so well.'

Madame Goesler was saved the necessity of making any answer to this by the announcement of another visitor. The door was opened, and Phineas Finn entered the room. He had not seen Madame Goesler since they had been together at Harrington Hall, and had never before met Mr. Maule. When riding home with the lady after their unsuccessful attempt to jump out of the wood, Phineas had promised to call in Park Lane whenever he should learn that Madame Goesler was not at Matching. Since that the Duke had died, and the bond with Matching no longer existed. It seemed but the other day that they were talking about the Duke together, and now the Duke was gone. 'I see you are in mourning,' said Phineas, as he still held her hand. 'I must say one word to condole with you for your lost friend.'

'Mr. Maule and I were now speaking of him,' she said, as she introduced the two gentlemen. 'Mr. Finn and I had the pleasure of meeting your son at Harrington Hall a few weeks since, Mr. Maule.'

'I heard that he had been there. Did you know the Duke, Mr. Finn?'

'After the fashion in which such a one as I would know such a one as the Duke, I knew him. He probably had forgotten my existence.'

'He never forgot any one,' said Madame Goesler.

'I don't know that I was ever introduced to him,' continued Mr. Maule, 'and I shall always regret it. I was telling Madame Goesler how profound a reverence I had for the Duke's character.' Phineas bowed, and Madame Goesler, who was becoming tired of the Duke as a subject of conversation, asked some question as to what had been going on in the House. Mr. Maule, finding it to be improbable that he should be able to advance his cause on that occasion, took his leave. The moment he was gone Madame Goesler's manner changed altogether. She left her former seat and came near to Phineas, sitting on a sofa close to the chair he occupied; and as she did so she pushed her hair back from her face in a manner that he remembered well in former days.

'I am so glad to see you,' she said. 'Is it not odd that he should have gone so soon after what we were saying but the other day?'

'You thought then that he would not last long.'

'Long is comparative. I did not think he would be dead within six weeks, or I should not have been riding there. He was a burden to me, Mr. Finn.'

'I can understand that.'

'And yet I shall miss him sorely. He had given all the colour to my life which it possessed. It was not very bright, but still it was colour.'

'The house will be open to you just the same.'

'I shall not go there. I shall see Lady Glencora in town, of course; but I shall not go to Matching; and as to Gatherum

Castle, I would not spend another week there, if they would give it me. You haven't heard of his will?'

'No;—not a word. I hope he remembered you,—to mention your name. You hardly wanted more.'

'Just so. I wanted no more than that.'

'It was made, perhaps, before you knew him.'

'He was always making it, and always altering it. He left me money, and jewels of enormous value.'

'I am so glad to hear it.'

'But I have refused to take anything. Am I not right?'

'I don't know why you should refuse.'

'There are people who will say that—I was his mistress. If a woman be young, a man's age never prevents such scandal. I don't know that I can stop it, but I can perhaps make it seem to be less probable. And after all that has passed, I could not bear that the Pallisers should think that I clung to him for what I could get. I should be easier this way.'

'Whatever is best to be done, you will do it;—I know that.'

'Your praise goes beyond the mark, my friend. I can be both generous and discreet;—but the difficulty is to be true. I did take one thing,—a black diamond that he always wore. I would show it you, but the goldsmith has it to make it fit me. When does the great affair come off at the House?'

'The bill will be read again on Monday, the first.'

'What an unfortunate day!—You remember young Mr. Maule? Is he not like his father? And yet in manners they are as unlike as possible.'

'What is the father?' Phineas asked.

'A battered old beau about London, selfish and civil, pleasant and penniless, and I should think utterly without a principle. Come again soon. I am so anxious to hear that you are getting on. And you have got to tell me all about that shooting with the pistol.' Phineas as he walked away thought that Madame Goesler was handsomer even than she used to be.

CHAPTER XXXI

The Duke and Duchess in town

At the end of March the Duchess of Omnium, never more to be called Lady Glencora by the world at large, came up to London. The Duke, though he was now banished from the House of Commons, was nevertheless wanted in London; and what funereal ceremonies were left might be accomplished as well in town as at Matching Priory. No old Ministry could be turned out and no new Ministry formed without the assistance of the young Duchess. It was a question whether she should not be asked to be Mistress of the Robes, though those who asked it knew very well that she was the last woman in England to hamper herself by dependence on the Court. Up to London they came; and, though of course they went into no society, the house in Carlton Gardens was continually thronged with people who had some special reason for breaking the ordinary rules of etiquette in their desire to see how Lady Glencora carried herself as Duchess of Omnium. 'Do you think she's altered much?' said Aspasia Fitzgibbon, an elderly spinster, the daughter of Lord Claddagh, and sister of Laurence Fitzgibbon, member for one of the western Irish counties. 'I don't think she was quite so loud as she used to be.'

Mrs. Bonteen was of opinion that there was a change. 'She was always uncertain, you know, and would scratch like a cat if you offended her.'

'And won't she scratch now?' asked Miss Fitzgibbon.

'I'm afraid she'll scratch oftener. It was always a trick of hers to pretend to think nothing of rank;—but she values her place as highly as any woman in England.'

This was Mrs. Bonteen's opinion; but Lady Baldock, who was present, differed. This Lady Baldock was not the mother, but the sister-in-law of that Augusta Boreham who had lately become Sister Veronica John. 'I don't believe it,' said Lady Baldock. 'She always seems to me to be like a great schoolgirl who has been allowed too much of her own way. I think people give way to her too much, you know.' As Lady Baldock was herself the wife of a peer, she naturally did not stand so much in awe of a duchess as did Mrs. Bonteen, or Miss Fitzgibbon.

'Have you seen the young Duke?' asked Mr. Ratler of Barrington Erle.

'Yes; I have been with him this morning.'

'How does he like it?'

'He's bothered out of his life,—as a hen would be if you were to throw her into water. He's so shy, he hardly knows how to speak to you; and he broke down altogether when I said something about the Lords.'

'He'll not do much more.'

'I don't know about that,' said Erle. 'He'll get used to it, and go into harness again. He's a great deal too good to be lost.'

'He didn't give himself airs?'

'What!—Planty Pall! If I know anything of a man he's not the man to do that because he's a duke. He can hold his own against all comers, and always could. Quiet as he always seemed, he knew who he was, and who other people were. I don't think you'll find much difference in him when he has got over the annoyance.' Mr. Ratler, however, was of a different opinion. Mr. Ratler had known many docile members of the House of Commons who had become peers by the death of uncles and fathers, and who had lost all respect for him as soon as they were released from the crack of the whip. Mr. Ratler rather depised peers who had been members of the House of Commons, and who passed by inheritance from a scene of unparalleled use and influence to one of idle and luxurious dignity.

Soon after their arrival in London the Duchess wrote the following very characteristic letter:—

'Dear Lord Chiltern,

'Mr. Palliser——' Then having begun with a mistake, she scratched the word through with her pen. 'The Duke has asked me to write about Trumpeton Wood, as he knows nothing about it, and I know just as little. But if you say what you want, it shall be done. Shall we get foxes and put them there? Or ought there to be a special fox-keeper? You mustn't be angry because the poor old Duke was too feeble to take notice of the matter. Only speak, and it shall be done.

'Yours faithfully,

'Glencora O.

'Madame Goesler spoke to me about it; but at that time we were in trouble.'

The answer was as characteristic:—

'Dear Duchess of Omnium,

'Thanks. What is wanted, is that keepers should know that there are to be foxes. When keepers know that foxes are really expected, there always are foxes. The men latterly have known just the contrary. It is all a question of shooting. I don't mean to say a word against the late Duke. When he got old the thing became bad. No doubt it will be right now.

'Faithfully yours,
'CHILTERN.

'Our hounds have been poisoned in Trumpeton Wood. This would never have been done had not the keepers been against the hunting.'

Upon receipt of this she sent the letter to Mr. Fothergill, with a request that there might be no more shooting in Trumpeton Wood. 'I'll be shot if we'll stand that, you know,' said Mr. Fothergill to one of his underlings. 'There are two hundred and fifty acres in Trumpeton Wood, and we're never to kill another pheasant because Lord Chiltern is Master of the Brake Hounds. Property won't be worth having at that rate.'

The Duke by no means intended to abandon the world of politics, or even the narrower sphere of ministerial work, because he had been ousted from the House of Commons, and from the possibility of filling the office which he had best liked. This was proved to the world by the choice of his house for a meeting of the party on the 30th of March. As it happened, this was the very day on which he and the Duchess returned to London; but nevertheless the meeting was held there, and he was present at it. Mr. Gresham then repeated his reasons for opposing Mr. Daubeny's bill; and declared that even while doing so he would, with the approbation of his party, pledge himself to bring in a bill somewhat to the same effect, should he ever again find himself in power. And he declared that he would do this solely with the view of showing how strong was his opinion that such a measure should not be left in the hands of the Conservative party. It was doubted whether such a political proposition had ever before been made in England. It was a simple avowal that on this occasion men were to be regarded, and not measures. No doubt such is the case, and ever has been the case, with the majority of active politicians. The double pleasure of pulling down an opponent, and of raising oneself, is the charm of a politician's life. And by practice this becomes extended to so many branches, that the delights,—and also the disappointments,—are very widespread. Great satisfaction is felt by us because by some lucky conjunction of affairs our man, whom we never saw, is made Lord-Lieutenant of a county, instead of another man, of whom we know as little. It is a great thing to us that Sir Samuel Bobwig, an excellent Liberal, is seated high on the bench of justice, instead of that time-serving Conservative, Sir Alexander McSilk. Men and not measures are, no doubt, the very life of politics. But then it is not the fashion to say so in public places. Mr. Gresham was determined to introduce that fashion on the present occasion. He did not think very much of Mr. Daubeny's Bill. So he told his friends at the Duke's house. The Bill was full of faults,—went too far in one direction, and not far enough in another. It was not difficult to pick holes in the Bill. But the sin of sins consisted in this,—that it was to be passed, if passed at all, by the aid of men who would sin against their consciences by each vote they gave in its favour. What but treachery could be expected from an army in which every officer, and every private, was called upon to fight against his convictions? The meeting passed off with dissension, and it was agreed that the House of Commons should be called upon to reject the Church Bill simply because it was proposed from that side of the House on which the minority was sitting. As there were more than two hundred members present on the occasion, by none of whom were any objections raised, it seemed probable that Mr. Gresham might be successful. There was still, however, doubt in the minds of some men. 'It's all very well,' said Mr. Ratler, 'but Turnbull wasn't there, you know.'

But from what took place the next day but one in Park Lane it would almost seem that the Duchess had been there. She came at once to see Madame Goesler, having very firmly determined that the Duke's death should not have the appearance of interrupting her intimacy with her friend. 'Was it not very disagreeable,'—asked Madame Goesler,—'just the day you came to town?'

'We didn't think of that at all. One is not allowed to think of anything now. It was very improper, of course, because of the Duke's death;—but that had to be put on one side. And then it was quite contrary to etiquette that Peers and Commoners should be brought together. I think there was some idea of making sure of Plantagenet, and so they all came and wore out our carpets. There wasn't above a dozen peers; but they were enough to show that all the old landmarks have been upset. I don't think any one would have objected if I had opened the meeting myself, and called upon Mrs. Bonteen to second me.'

'Why Mrs. Bonteen?'

'Because next to myself she's the most talkative and political woman we have. She was at our house yesterday, and I'm not quite sure that she doesn't intend to cut me out.'

'We must put her down, Lady Glen.'

'Perhaps she'll put me down now that we're half shelved. The men did make such a racket, and yet no one seemed to speak for two minutes except Mr. Gresham, who stood upon my pet footstool, and kicked it almost to pieces.'

'Was Mr. Finn there?'

'Everybody was there, I suppose. What makes you ask particularly about Mr. Finn?'

'Because he's a friend.'

'That's come up again, has it? He's the handsome Irishman, isn't he, that came to Matching, the same day that brought you there?'

'He is an Irishman, and he was at Matching, that day.'

'He's certainly handsome. What a day that was, Marie! When one thinks of it all,—of all the perils and all the salvations, how strange it is! I wonder whether you would have liked it now if you were the Dowager Duchess.'

'I should have had some enjoyment, I suppose.'

'I don't know that it would have done us any harm, and yet

how keen I was about it. We can't give you the rank now, and you won't take the money.'

'Not the money, certainly.'

'Plantagenet says you'll have to take it;—but it seems to me he's always wrong. There are so many things that one must do that one doesn't do. He never perceives that everything gets changed every five years. So Mr. Finn is the favourite again?'

'He is a friend whom I like. I may be allowed to have a friend, I suppose.'

'A dozen, my dear;—and all of them good-looking. Good-bye, dear. Pray come to us. Don't stand off and make yourself disagreeable. We shan't be giving dinner parties, but you can come whenever you please. Tell me at once;—do you mean to be disagreeable?'

Then Madame Goesler was obliged to promise that she would not be more disagreeable than her nature had made her.

CHAPTER XXXII

The World becomes cold

A GREAT deal was said by very many persons in London as to the murderous attack which had been made by Mr. Kennedy on Phineas Finn in Judd Street, but the advice given by Mr. Slide in The People's Banner to the police was not taken. No public or official inquiry was made into the circumstance. Mr. Kennedy, under the care of his cousin, retreated to Scotland; and, as it seemed, there was to be an end of it. Throughout the month of March various smaller bolts were thrust both at Phineas and at the police by the editor of the above-named newspaper, but they seemed to fall without much effect. No one was put in prison; nor was any one ever examined. But, nevertheless, these missiles had their effect. Everybody knew that there had been a 'row' between Mr. Kennedy and Phineas Finn, and that the 'row' had been made about Mr. Kennedy's wife. Everybody knew that a pistol had been fired at Finn's head; and a great many people thought that there had been some cause for the assault. It was alleged at one club that the present member for Tankerville had spent the greater part of the last two years at Dresden, and at another that he had called on Mr. Kennedy twice, once down in Scotland, and once at the hotel in Judd Street, with a view of inducing that gentleman to concede to a divorce. There was also a very romantic story afloat as to an engagement which had existed between Lady Laura and Phineas Finn before the lady had been induced by her father to marry the richer suitor. Various details were given in corroboration of these stories. Was it not known that the Earl had purchased the submission of Phineas Finn by a seat for his borough of Loughton? Was it not known that Lord Chiltern, the brother of Lady Laura, had fought a duel with Phineas Finn? Was it not known that Mr. Kennedy himself had been as it were coerced into quiescence by the singular fact that he had been saved from garotters in the street by the opportune interference of Phineas Finn? It was even suggested that the scene with the garotters had been cunningly planned by Phineas Finn, that he might in this way be able to restrain the anger of the husband of the lady whom he loved. All these stories were very pretty; but as the reader, it is hoped, knows, they were all untrue. Phineas had made but one short visit to Dresden in his life. Lady Laura had been engaged to Mr. Kennedy before Phineas had ever spoken to her of his love. The duel with Lord Chiltern had been about another lady, and the seat at Loughton had been conferred upon Phineas chiefly on account of his prowess in extricating Mr. Kennedy from the garotters,—respecting which circumstance it may be said that as the meeting in the street was fortuitous, the reward was greater than the occasion seemed to require.

While all these things were being said Phineas became something of a hero. A man who is supposed to have caused a disturbance between two married people, in a certain rank of life, does generally receive a certain meed*of admiration. A man who was asked out to dinner twice a week before such rumours were afloat, would probably receive double that number of invitations afterwards. And then to have been shot at by a madman in a room, and to be the subject of the venom of a People's Banner, tends also to Fame. Other ladies besides Madame Goesler were anxious to have the story from the very lips of the hero, and in this way Phineas Finn became a conspicuous man. But Fame begets envy, and there were some who said that the member for Tankerville had injured his prospects with his party. It may be very well to give a dinner to a man who has caused the wife of a late Cabinet Minister to quarrel with her husband; but it can hardly be expected that he should be placed in office by the head of the party to which that late Cabinet Minister belonged. 'I never saw such a fellow as you are,' said Barrington Erle to him. 'You are always getting into a mess.'

'Nobody ought to know better than you how false all these calumnies are.' This he said because Erle and Lady Laura were cousins.

'Of course they are calumnies; but you had heard them before, and what made you go poking your head into the lion's mouth?'

Mr. Bonteen was very much harder upon him than was Barrington Erle. 'I never liked him from the first, and always knew he would not run straight. No Irishman ever does.' This was said to Viscount Fawn, a distinguished member of the Liberal party, who had but lately been married, and was known to have very strict notions as to the bonds of matrimony. He had been heard to say that any man who had interfered with the happiness of a married couple should be held to have committed a capital offence.

'I don't know whether the story about Lady Laura is true.'

'Of course it's true. All the world knows it to be true. He was always there; at Loughlinter, and at Saulsby, and in Portman Square after she had left her husband. The mischief he has done is incalculable. There's a Conservative sitting in poor Kennedy's seat for Dunross-shire.'

'That might have been the case anyway.'

'Nothing could have turned Kennedy out. Don't you remember how he behaved about the Irish Land Question? I hate such fellows.'

'If I thought it true about Lady Laura—'

Lord Fawn was again about to express his opinion in regard to matrimony, but Mr. Bonteen was too impetuous to listen to him. 'It's out of the question that he should come in again. At any rate if he does, I won't. I shall tell Gresham so very plainly. The women will do all that they can for him. They always do for a fellow of that kind.'

Phineas heard of it;—not exactly by any repetition of the words that were spoken, but by chance phrases, and from the looks of men. Lord Cantrip, who was his best friend among those who were certain to hold high office in a Liberal Government, did not talk to him cheerily,—did not speak as though he, Phineas, would as a matter of course have some place assigned to him. And he thought that Mr. Gresham was hardly as cordial to him as he might be when they met in the closer intercourse of the House. There was always a word or two spoken, and sometimes a shaking of hands. He had no right to complain. But yet he knew that something was wanting. We can generally read a man's purpose towards us in his manner, if his purposes are of much moment to us.

Phineas had written to Lady Laura, giving her an account of the occurrence in Judd Street on the 1st of March, and had received from her a short answer by return of post. It contained hardly more than a thanksgiving that his life had not been sacrificed, and in a day or two she had written again, letting him know that she had determined to consult her father. Then on the last day of the month he received the following letter:—

'Dresden, March 27th, 18—.

'My dear Friend,—

'At last we have resolved that we will go back to England, —almost at once. Things have gone so rapidly that I hardly know how to explain them all, but that is Papa's resolution. His lawyer, Mr. Forster, tells him that it will be best, and goes so far as to say that it is imperative on my behalf that some steps should be taken to put an end to the present state of things. I will not scruple to tell you that he is actuated chiefly by considerations as to money. It is astonishing to me that a man who has all his life been so liberal should now in his old age think so much about it. It is, however, in no degree for himself. It is all for me. He cannot bear to think that my fortune should be withheld from me by Mr. Kennedy while I have done nothing wrong. I was obliged to show him your letter, and what you said about the control of money took hold of his mind at once. He thinks that if my unfortunate husband be insane, there can be no difficulty in my obtaining a separation on terms which would oblige him or his friends to restore this horrid money.

'Of course I could stay if I chose. Papa would not refuse to find a home for me here. But I do agree with Mr. Forster that something should be done to stop the tongues of ill-conditioned people. The idea of having my name dragged through the newspapers is dreadful to me; but if this must be done one way or the other, it will be better that it should be done with truth. There is nothing that I need fear,—as you know so well.

'I cannot look forward to happiness anywhere. If the question of separation were once settled, I do not know whether I would not prefer returning here to remaining in London. Papa has got tired of the place, and wants, he says, to see Saulsby once again before he dies. What can I say in answer to this, but that I will go? We have sent to have the house in Portman Square got ready for us, and I suppose we shall be there about the 15th of next month. Papa has instructed Mr. Forster to tell Mr. Kennedy's lawyer that we are coming, and he is to find out, if he can, whether any interference in the management of the property has been as yet made by the family. Perhaps I ought to tell you that Mr. Forster has expressed surprise that you did not call on the police when the shot was fired. Of course I can understand it all. God bless you.

'Your affectionate friend,

'L. K.'

Phineas was obliged to console himself by reflecting that if she understood him of course that was everything. His first and great duty in the matter had been to her. If in performing that duty he had sacrificed himself, he must bear his undeserved punishment like a man. That he was to be punished he began to perceive too clearly. The conviction that Mr. Daubeny must recede from the Treasury Bench after the coming debate became every day stronger, and within the little inner circles of the Liberal party the usual discussions were made as to the Ministry which Mr. Gresham would, as a matter of course, be called upon to form. But in these discussions Phineas Finn did not find himself taking an assured and comfortable part. Laurence Fitzgibbon, his countryman, —who in the way of work had never been worth his salt,— was eager, happy, and without a doubt. Others of the old stagers, men who had been going in and out ever since they had been able to get seats in Parliament, stood about in clubs, and in lobbies, and chambers of the House, with all that busy, magpie air which is worn only by those who have high hopes of good things to come speedily. Lord Mount Thistle was more sublime and ponderous than ever, though they who best understood the party declared that he would never again be invited to undergo the cares of office. His lordship was one of those terrible political burdens, engendered originally by private friendship or family considerations, which one Minister leaves to another. Sir Gregory Grogram, the great Whig lawyer, showed plainly by his manner that he thought himself at last secure of reaching the reward for which he had been struggling all his life; for it was understood by all men who knew anything that Lord Weazeling was not to be asked again to sit on the Woolsack.* No better advocate or effective politician ever lived; but it was supposed that he lacked dignity for the office of first judge in the land. That most of the old lot would come back was a matter of course.

There would be the Duke,—the Duke of St. Bungay, who had for years past been 'the Duke' when Liberal administrations were discussed, and the second Duke, whom we know so well; and Sir Harry Coldfoot, and Legge Wilson, Lord Cantrip, Lord Thrift, and the rest of them. There would of course be Lord Fawn, Mr. Ratler, and Mr. Erle. The thing was so thoroughly settled that one was almost tempted to think that the Prime Minister himself would have no voice in the selections to be made. As to one office it was acknowledged on all sides that a doubt existed which would at last be found to be very injurious,—as some thought altogether crushing,—to the party. To whom would Mr. Gresham entrust the financial affairs of the country? Who would be the new Chancellor of the Exchequer? There were not a few who inferred that Mr. Bonteen would be promoted to that high office. During the last two years he had devoted himself to decimal coinage with a zeal only second to that displayed by Plantagenet Palliser, and was accustomed to say of himself that he had almost perished under his exertions. It was supposed that he would have the support of the present Duke of Omnium,—and that Mr. Gresham, who disliked the man, would be coerced by the fact that there was no other competitor. That Mr. Bonteen should go into the Cabinet would be gall and wormwood to many brother Liberals; but gall and wormwood such as this have to be swallowed. The rising in life of our familiar friends is, perhaps, the bitterest morsel of the bitter bread which we are called upon to eat in life. But we do eat it; and after a while it becomes food to us, —when we find ourselves able to use, on behalf, perhaps, of our children, the influence of those whom we had once hoped to leave behind in the race of life. When a man suddenly shoots up into power few suffer from it very acutely. The rise of a Pitt can have caused no heart-burning. But Mr. Bonteen had been a hack among the hacks, had filled the usual half-dozen places, had been a junior Lord, a Vice-President, a Deputy Controller, a Chief Commissioner, and a Joint Secretary. His hopes had been raised or abased among the places of £1,000, £1,200, or £1,500 a year. He had hitherto culminated at £2,000, and had been supposed with diligence to have worked himself up to the top of the ladder, as far as the ladder was accessible to him. And now he was spoken of in connection with one of the highest offices of the State! Of course this created much uneasiness, and gave rise to many prophecies of failure. But in the midst of it all no office was assigned to Phineas Finn; and there was a general feeling, not expressed, but understood, that his affair with Mr. Kennedy stood in his way.

Quintus Slide had undertaken to crush him! Could it be possible that so mean a man should be able to make good so monstrous a threat? The man was very mean, and the threat had been absurd as well as monstrous; and yet it seemed that it might be realised. Phineas was too proud to ask questions, even of Barrington Erle, but he felt that he was being 'left out in the cold,' because the editor of The People's Banner had said that no government could employ him; and at this moment, on the very morning of the day which was to usher in the great debate, which was to be so fatal to Mr. Daubeny and his Church Reform, another thunderbolt was hurled. The 'we' of The People's Banner had learned that the very painful matter, to which they had been compelled by a sense of duty to call the public attention in reference to the late member for Dunross-shire and the present member for Tankerville, would be brought before one of the tribunals of the country, in reference to the matrimonial differences between Mr. Kennedy and his wife. It would be in the remembrance of their readers that the unfortunate gentleman had been provoked to fire a pistol at the head of the member for Tankerville,—a circumstance which, though publicly known, had never been brought under the notice of the police. There was reason to hope that the mystery might now be cleared up, and that the ends of justice would demand that a certain document should be produced, which they,—the 'we,'—had been vexatiously restrained from giving to their readers, although it had been most carefully prepared for publication in the columns of The People's Banner. Then the thunderbolt went on to say that there was evidently a great move among the members of the so-called Liberal party, who seemed to think that it was only necessary that they should open their mouths wide enough in order that the sweets of office should fall into them. The 'we' were quite of a different opinion. The 'we' believed that no Minister for many a long day had been so firmly fixed on the Treasury Bench as was Mr. Daubeny at the present moment. But this at any rate might be inferred;—that should Mr. Gresham by any unhappy combination of circumstances be called upon to form a Ministry, it would be quite impossible for him to include within it the name of the member for Tankerville. This was the second great thunderbolt that fell,—and so did the work of crushing our poor friend proceed.

There was a great injustice in all this; at least so Phineas thought;—injustice, not only from the hands of Mr. Slide, who was unjust as a matter of course, but also from those who ought to have been his staunch friends. He had been enticed over to England almost with a promise of office, and he was sure that he had done nothing which deserved punishment, or even censure. He could not condescend to complain,—nor indeed as yet could he say that there was ground for complaint. Nothing had been done to him. Not a word had been spoken,—except those lying words in the newspapers which he was too proud to notice. On one matter, however, he was determined to be firm. When Barrington Erle had absolutely insisted that he should vote upon the Church Bill in opposition to all that he had said upon the subject at Tankerville, he had stipulated that he should have an opportunity in the great debate which would certainly take place of explaining his conduct,—or, in other words, that the privilege of making a speech should be accorded to him at a time in which very many members would no doubt attempt to speak and would attempt in vain. It may be imagined,—probably still is imagined by a great many,—that no such pledge as this could be given, that the right to speak depends simply on the Speaker's eye, and that energy at the moment in attracting attention would alone be of account to an eager orator. But

Phineas knew the House too well to trust to such a theory. That some preliminary assistance would be given to the travelling of the Speaker's eye, in so important a debate, he knew very well; and he knew also that a promise from Barrington Erle or from Mr. Ratler would be his best security. 'That will be all right, of course,' said Barrington Erle to him on the evening the day before the debate: 'We have quite counted on your speaking.' There had been a certain sullenness in the tone with which Phineas had asked his question as though he had been labouring under a grievance, and he felt himself rebuked by the cordiality of the reply. 'I suppose we had better fix it for Monday or Tuesday,' said the other. 'We hope to get it over by Tuesday, but there is no knowing. At any rate you shan't be thrown over.' It was almost on his tongue,—the entire story of his grievance, the expression of his feeling that he was not being treated as one of the chosen; but he restrained himself. He liked Barrington Erle well enough, but not so well as to justify him in asking for sympathy.

Nor had it been his wont in any of the troubles of his life to ask for sympathy from a man. He had always gone to some woman;—in old days to Lady Laura, or to Violet Effingham, or to Madame Goesler. By them he could endure to be petted, praised, or upon occasion even pitied. But pity or praise from any man had been distasteful to him. On the morning of the 1st of April he again went to Park Lane, not with any formed plan of telling the lady of his wrongs, but driven by a feeling that he wanted comfort, which might perhaps be found there. The lady received him very kindly, and at once inquired as to the great political tournament which was about to be commenced. 'Yes; we begin to-day,' said Phineas. 'Mr. Daubeny will speak, I should say, from half-past four till seven. I wonder you don't go and hear him.'

'What a pleasure! To hear a man speak for two hours and a half about the Church of England. One must be very hard driven for amusement! Will you tell me that you like it?'

'I like to hear a good speech.'

'But you have the excitement before you of making a good speech in answer. You are in the fight. A poor woman, shut up in a cage, feels there more acutely than anywhere else how insignificant a position she fills in the world.'

'You don't advocate the rights of women, Madame Goesler?

'Oh, no. Knowing our inferiority I submit without a grumble; but I am not sure that I care to go and listen to the squabbles of my masters. You may arrange it all among you, and I will accept what you do, whether it be good or bad,—as I must; but I cannot take so much interest in the proceeding as to spend my time in listening where I cannot speak, and in looking when I cannot be seen. You will speak?'

'Yes; I think so.'

'I shall read your speech, which is more than I shall do for most of the others. And when it is all over, will your turn come?'

'Not mine individually, Madame Goesler.'

'But it will be yours individually;—will it not?' she asked with energy. Then gradually, with half-pronounced sentences, he explained to her that even in the event of the formation of a Liberal Government, he did not expect that any place would be offered to him. 'And why not? We have been all speaking of it as a certainty.'

He longed to inquire who were the all of whom she spoke, but he could not do it without an egotism which would be distasteful to him. 'I can hardly tell;—but I don't think I shall be asked to join them.'

'You would wish it?'

'Yes;—talking to you I do not see why I should hesitate to say so.'

'Talking to me, why should you hesitate to say anything about yourself that is true? I can hold my tongue. I do not gossip about my friends. Whose doing is it?'

'I do not know that it is any man's doing.'

'But it must be. Everybody said that you were to be one of them if you could get the other people out. Is it Mr. Bonteen?'

'Likely enough. Not that I know anything of the kind; but

as I hate him from the bottom of my heart, it is natural to suppose that he has the same feeling in regard to me.'

'I agree with you there.'

'But I don't know that it comes from any feeling of that kind.'

'What does it come from?'

'You have heard all the calumny about Lady Laura Kennedy.'

'You do not mean to say that a story such as that has affected your position.'

'I fancy it has. But you must not suppose, Madame Goesler, that I mean to complain. A man must take these things as they come. No one has received more kindness from friends than I have, and few perhaps more favours from fortune. All this about Mr. Kennedy has been unlucky,—but it cannot be helped.'

'Do you mean to say that the morals of your party will be offended?' said Madame Goesler, almost laughing.

'Lord Fawn, you know, is very particular. In sober earnest one cannot tell how these things operate; but they do operate gradually. One's friends are sometimes very glad of an excuse for not befriending one.'

'Lady Laura is coming home?'

'Yes.'

'That will put an end to it.'

'There is nothing to put an end to except the foul-mouthed malice of a lying newspaper. Nobody believes anything against Lady Laura.'

'I'm not so sure of that. I believe nothing against her.'

'I'm sure you do not, Madame Goesler. Nor do I think that anybody does. It is too absurd for belief from beginning to end. Good-bye. Perhaps I shall see you when the debate is over.'

'Of course you will. Good-bye, and success to your oratory.' Then Madame Goesler resolved that she would say a few judicious words to her friend, the Duchess, respecting Phineas Finn.

CHAPTER XXXIII
The Two Gladiators

THE great debate was commenced with all the solemnities which are customary on such occasions, and which make men think for the day that no moment of greater excitement has ever blessed or cursed the country. Upon the present occasion London was full of clergymen. The specially clerical clubs,—the Oxford and Cambridge, the Old University, and the Athenaeum*,—were black with them. The bishops and deans, as usual, were pleasant in their manner and happy-looking, in spite of adverse circumstances. When one sees a bishop in the hours of the distress of the Church, one always thinks of the just and firm man who will stand fearless while the ruins of the world are falling about his ears. But the parsons from the country were a sorry sight to see. They were in earnest with all their hearts, and did believe,—not that the crack of doom was coming, which they could have borne with equanimity if convinced that their influence would last to the end,—but that the Evil One was to be made welcome upon the earth by Act of Parliament. It is out of nature that any man should think it good that his own order should be repressed, curtailed, and deprived of its power. If we go among cab-drivers or letter-carriers, among butlers or gamekeepers, among tailors or butchers, among farmers or graziers, among doctors or attorneys, we shall find in each set of men a conviction that the welfare of the community depends upon the firmness with which they,—especially they,—hold their own. This is so manifestly true with the Bar that no barrister in practice scruples to avow that barristers in practice are the salt of the earth. The personal confidence of a judge in his own position is beautiful, being salutary to the country, though not unfrequently damaging to the character of the man. But if this be so with men who are conscious of no higher influence than that exercised over the bodies and minds of their fellow creatures, how much stronger must be the feeling when the influence affects the soul! To the outsider, or layman, who simply uses a cab, or receives a letter, or goes to law, or has to be tried, these pretensions are ridiculous or annoying, according to the ascendancy of the pretender at the moment. But as the clerical pretensions are more exacting than all others, being put forward with an assertion that no answer is possible without breach of duty and sin, so are they more galling. The fight has been going on since the idea of a mitre first entered the heart of a priest,—since dominion in this world has found itself capable of sustentation by the exercise of fear as to the world to come. We do believe,—the majority among us does so,—that if we live and die in sin we shall after some fashion come to great punishment, and we believe also that by having pastors among us who shall be men of God, we may best aid ourselves and our children in avoiding this bitter end. But then the pastors and men of God can only be human,—cannot be altogether men of God; and so they have oppressed us, and burned us, and tortured us, and hence come to love palaces, and fine linen, and purple, and, alas, sometimes, mere luxury and idleness. The torturing and the burning, as also to speak truth the luxury and the idleness, have, among us, been already conquered, but the idea of ascendancy remains. What is a thoughtful man to do who acknowledges the danger of his soul, but cannot swallow his parson whole simply because he has been sent to him from some source in which he has no special confidence, perhaps by some distant lord, perhaps by a Lord Chancellor whose political friend has had a son with a tutor? What is he to do when, in spite of some fine linen and purple left among us, the provision for the man of God in his parish or district is so poor that no man of God fitted to teach him will come and take it? In no spirit of animosity to religion he begins to tell himself that Church and State together was a monkish combination, fit perhaps for monkish days, but no longer having fitness, and not much longer capable of existence in this country. But to the parson himself,—to the honest, hardworking, conscientious priest who does in his heart of hearts believe that no diminution in the general influence of his order can be made without ruin to the souls of men,—this opinion, when it becomes dominant, is as though the world were in truth breaking to pieces over his head. The world has been broken to pieces in the same way often;—but extreme Chaos does not come. The cabman and the letter-carrier always expect that Chaos will very nearly come when they are disturbed. The barristers are sure of Chaos when the sanctity of Benchers is in question. What utter Chaos would be promised to us could any one with impunity contemn the majesty of the House of Commons! But of all these Chaoses there can be no Chaos equal to that which in the mind of a zealous Oxford-bred constitutional country parson must attend that annihilation of his special condition which will be produced by the disestablishment of the Church. Of all good fellows he is the best good fellow. He is genial, hospitable, well-educated, and always has either a pretty wife or pretty daughters. But he has so extreme a belief in himself that he cannot endure to be told that absolute Chaos will not come at once if he be disturbed. And now disturbances,—ay, and utter dislocation and ruin were to come from the hands of a friend! Was it wonderful that parsons should be seen about Westminster in flocks with '*Et tu, Brute*' written on their faces as plainly as the law on the brows of a Pharisee?

The Speaker had been harassed for orders. The powers and prowess of every individual member had been put to the test. The galleries were crowded. Ladies' places had been balloted for with desperate enthusiasm, in spite of the sarcasm against the House which Madame Goesler had expressed. Two royal princes and a royal duke were accommodated within the House in an irregular manner. Peers swarmed in the passages, and were too happy to find standing room. Bishops jostled against lay barons with no other preference than that afforded to them by their broader shoulders. Men, and especially clergymen, came to the galleries loaded with sandwiches and flasks, prepared to hear all there was to be heard should the debate last from 4 P.M. to the same hour on the following morning. At two in the afternoon the entrances to the House were barred, and men of all ranks—deans, prebends, peers' sons, and baronets,—stood there patiently waiting till some powerful nobleman should let them through. The very ventilating chambers under the House were filled with courteous listeners, who had all pledged themselves that under no possible provocation would they even cough during the debate.

A few minutes after four, in a House from which hardly more than a dozen members were absent, Mr. Daubeny took his seat with that air of affected indifference to things around him which is peculiar to him. He entered slowly, amidst cheers from his side of the House, which no doubt were loud in proportion to the dismay of the cheerers as to the matter in hand. Gentlemen lacking substantial sympathy with their leader found it to be comfortable to deceive themselves, and raise their hearts at the same time by the easy enthusiasm of noise. Mr. Daubeny having sat down and covered his head just raised his hat from his brows, and then tried to look as though he were no more than any other gentleman present. But the peculiar consciousness of the man displayed itself even in his constrained absence of motion. You could see that he felt himself to be the beheld of all beholders, and that he enjoyed the position,—with some slight inward trepidation lest the effort to be made should not equal the greatness of the occasion. Immediately after him Mr. Gresham bustled up the centre of the House amidst a roar of good-humoured welcome. We have had many Ministers who have been personally dearer to their individual adherents in the House than the present leader of the Opposition and late Premier, but none, perhaps, who has been more generally respected by his party for earnestness and sincerity. On the present occasion there was a fierceness, almost a ferocity, in his very countenance, to the fire of which friends and enemies were equally anxious to add fuel,—the friends in order that so might these recreant Tories be more thoroughly annihilated, and the enemies, that their enemy's indiscretion might act back upon himself to his confusion. For, indeed, it never could be denied that as a Prime Minister Mr. Gresham could be very indiscreet.

A certain small amount of ordinary business was done, to the disgust of expectant strangers, which was as trivial as possible in its nature,—so arranged, apparently, that the importance of what was to follow might be enhanced by the force of contrast. And, to make the dismay of the novice stranger more thorough, questions were asked and answers were given in so low a voice, and Mr. Speaker uttered a word or two in so quick and shambling a fashion, that he, the novice stranger, began to fear that no word of the debate would reach him up there in his crowded back seat. All this, however, occupied but a few minutes, and at twenty minutes past four Mr. Daubeny was on his legs. Then the novice stranger found that, though he could not see Mr. Daubeny without the aid of an opera glass, he could hear every word that fell from his lips.

Mr. Daubeny began by regretting the hardness of his

position, in that he must, with what thoroughness he might be able to achieve, apply himself to two great subjects, whereas the right honourable gentleman opposite had already declared, with all the formality which could be made to attach itself to a combined meeting of peers and commoners, that he would confine himself strictly to one. The subject selected by the right honourable gentleman opposite on the present occasion was not the question of Church Reform. The right honourable gentleman had pledged himself with an almost sacred enthusiasm to ignore that subject altogether. No doubt it was the question before the House, and he, himself,—the present speaker,—must unfortunately discuss it at some length. The right honourable gentleman opposite would not, on this great occasion, trouble himself with anything of so little moment. And it might be presumed that the political followers of the right honourable gentleman would be equally reticent, as they were understood to have accepted his tactics without a dissentient voice. He, Mr. Daubeny, was the last man in England to deny the importance of the question which the right honourable gentleman would select for discussions in preference to that of the condition of the Church. That question was a very simple one, and might be put to the House in a very few words. Coming from the mouth of the right honourable gentleman, the proposition would probably be made in this form:—'That this House does think that I ought to be Prime Minister now, and as long as I may possess a seat in this House.' It was impossible to deny the importance of that question; but perhaps he, Mr. Daubeny, might be justified in demurring to the preference given to it over every other matter, let that matter be of what importance it might be to the material welfare of the country.

He made his point well; but he made it too often. And an attack of that kind, personal and savage in its nature, loses its effect when it is evident that the words have been prepared. A good deal may be done in dispute by calling a man an ass or a knave,—but the resolve to use the words should have been made only at the moment, and they should come hot from the heart. There was much neatness and some acuteness in Mr. Daubeny's satire, but there was no heat, and it was prolix. It had, however, the effect of irritating Mr. Gresham,—as was evident from the manner in which he moved his hat and shuffled his feet.

A man destined to sit conspicuously on our Treasury Bench, or on the seat opposite to it, should ask the gods for a thick skin as a first gift. The need of this in our national assembly is greater than elsewhere, because the differences between the men opposed to each other are smaller. When two foes meet together in the same Chamber, one of whom advocates the personal government of an individual ruler, and the other that form of State, which has come to be called a Red Republic, they deal, no doubt, weighty blows of oratory at each other, but blows which never hurt at the moment. They may cut each other's throats if they can find an opportunity; but they do not bite each other like dogs over a bone. But when opponents are almost in accord, as is always the case with our parliamentary gladiators, they are ever striving to give maddening little wounds through the joints of the harness. What is there with us to create the divergence necessary for debate but the pride of personal skill in the encounter? Who desires among us to put down the Queen, or to repudiate the National Debt, or to destroy religious worship, or even to disturb the ranks of society? When some small measure of reform has thoroughly recommended itself to the country,—so thoroughly that all men know that the country will have it,—then the question arises whether its details shall be arranged by the political party which calls itself Liberal,—or by that which is termed Conservative. The men are so near to each other in all their convictions and theories of life that nothing is left to them but personal competition for the doing of the thing that is to be done. It is the same in religion. The apostle of Christianity and the infidel can meet without a chance of a quarrel; but it is never safe to bring together two men who differ about a saint or a surplice.

Mr. Daubeny, having thus attacked and wounded his enemy, rushed boldly into the question of Church Reform, taking no little pride to himself and to his party that so great a blessing should be bestowed upon the country from so unexpected a source. 'See what we Conservatives can do. In fact we will conserve nothing when we find that you do not desire to have it conserved any longer. "Quod minime reris Graiâ pandetur ab urbe."' It was exactly the reverse of the complaint which Mr. Gresham was about to make. On the subject of the Church itself he was rather misty but very profound. He went into the question of very early Churches indeed, and spoke of the misappropriation of endowments in the time of Eli.* The establishment of the Levites had been no doubt complete; but changes had been effected as circumstances required. He was presumed to have alluded to the order of Melchisedek,* but he abstained from any mention of the name. He roamed very wide, and gave many of his hearers an idea that his erudition had carried him into regions in which it was impossible to follow him. The gist of his argument was to show that audacity in Reform was the very backbone of Conservatism. By a clearly pronounced disunion of Church and State the theocracy of Thomas à Becket* would be restored, and the people of England would soon again become the faithful flocks of faithful shepherds. By taking away the endowments from the parishes, and giving them back in some complicated way to the country, the parishes would be better able than ever to support their clergymen. Bishops would be bishops indeed, when they were no longer the creatures of a Minister's breath. As to the deans, not seeing a clear way to satisfy aspirants for future vacancies in the deaneries, he became more than usually vague, but seemed to imply that the Bill which was now with the leave of the House to be read a second time, contained no clause forbidding the appointment of deans, though the special stipend of the office must be matter of consideration with the new Church Synod.

The details of this part of his speech were felt to be dull by the strangers. As long as he would abuse Mr. Gresham, men could listen with pleasure; and could keep their attention fixed while he referred to the general Conservatism of the party which he had the honour of leading. There was a raciness in the promise of so much Church destruction from the chosen leader of the Church party, which was assisted by a conviction in the minds of most men that it was impossible for unfortunate Conservatives to refuse to follow this leader, let him lead where he might. There was a gratification in feeling that the country party was bound to follow, even should he take them into the very bowels of a mountain, as the pied piper did the children of Hamelin;—and this made listening pleasant. But when Mr. Daubeny stated the effect of his different clauses, explaining what was to be taken and what left,—with a fervent assurance that what was to be left would, under the altered circumstances, go much further than the whole had gone before,—then the audience became weary, and began to think that it was time that some other gentleman should be upon his legs. But at the end of the Minister's speech there was another touch of invective which went far to redeem him. He returned to that personal question to which his adversary had undertaken to confine himself, and expressed a holy horror at the political doctrine which was implied. He, during a prolonged Parliamentary experience, had encountered much factious opposition. He would even acknowledge that he had seen it exercised on both sides of the House, though he had always striven to keep himself free from its baneful influence. But never till now had he known a statesman proclaim his intention of depending upon faction, and upon faction alone, for the result which he desired to achieve. Let the right honourable gentleman raise a contest on either the principles or the details of the measure, and he would be quite content to abide the decision of the House; but he should regard such a raid as that threatened against him and his friends by the right honourable gentleman as unconstitutional, revolutionary, and tyrannical. He felt sure that an opposition so based, and so maintained, even if it be enabled by the heated feelings of the moment to obtain an unfortunate success in the House, would not be encouraged by the sympathy and support of the country at large. By these last words he was understood to signify that should he be beaten on the second reading, not in reference to the merits of the Bill, but simply on the issue as proposed by Mr. Gresham, he would again dissolve the House before he would resign. Now it was very well understood that there were Liberal members in the House who would prefer even the success of Mr. Daubeny to a speedy reappearance before their constituents.

Mr. Daubeny spoke till nearly eight, and it was surmised at the time that he had craftily arranged his oratory so as to embarrass his opponent. The House had met at four, and was to sit continuously till it was adjourned for the night. When this is the case, gentlemen who speak about eight o'clock are too frequently obliged to address themselves to empty benches. On the present occasion it was Mr. Gresham's intention to follow his opponent at once, instead of waiting, as is usual with a leader of his party, to the close of the debate. It was understood that Mr. Gresham would follow Mr. Daubeny, with the object of making a distinct charge against Ministers, so that the vote on this second reading of the Church Bill might in truth be a vote of want of confidence. But to commence his speech at eight o'clock when the House was hungry and uneasy, would be a trial. Had Mr. Daubeny closed an hour sooner there would, with a little stretching of the favoured hours, have been time enough. Members would not have objected to postpone their dinner till half-past eight, or perhaps nine, when their favourite orator was on his legs. But with Mr. Gresham beginning a great speech at eight, dinner would altogether become doubtful, and the disaster might be serious. It was not probable that Mr. Daubeny had even among his friends proclaimed any such strategy; but it was thought by the political speculators of the day that such an idea had been present to his mind.

But Mr. Gresham was not to be turned from his purpose. He waited for a few moments, and then rose and addressed the Speaker. A few members left the House;—gentlemen, doubtless, whose constitutions, weakened by previous service, could not endure prolonged fasting. Some who had nearly reached the door returned to their seats, mindful of Messrs. Roby and Ratler. But for the bulk of those assembled the interest of the moment was greater even than the love of dinner. Some of the peers departed, and it was observed that a bishop or two left the House; but among the strangers in the gallery, hardly a foot of space was gained. He who gave up his seat then, gave it up for the night.

Mr. Gresham began with a calmness of tone which seemed almost to be affected, but which arose from a struggle on his own part to repress that superabundant energy of which he was only too conscious. But the calmness soon gave place to warmth, which heated itself into violence before he had been a quarter of an hour upon his legs. He soon became even ferocious in his invective, and said things so bitter that he had himself no conception of their bitterness. There was this difference between the two men,—that whereas Mr. Daubeny hit always as hard as he knew how to hit, having premeditated each blow, and weighed its results beforehand, having calculated his power even to the effect of a blow repeated on a wound already given, Mr. Gresham struck right and left and straightforward with a readiness engendered by practice, and in his fury might have murdered his antagonist before he was aware that he had drawn blood. He began by refusing absolutely to discuss the merits of the bill. The right honourable gentleman had prided himself on his generosity as a Greek. He would remind the right honourable gentleman that presents from Greeks had ever been considered dangerous. 'It is their gifts, and only their gifts, that we fear,' he said. The political gifts of the right honourable gentleman, extracted by him from his unwilling colleagues and followers, had always been more bitter to the taste than Dead Sea apples.* That such gifts should not be bestowed on the country by unwilling hands, that reform should not come from those who themselves felt the necessity of no reform, he believed to be the wish not only of that House, but of the country at large. Would any gentleman on that bench, excepting the right honourable gentleman himself,—and he pointed to the crowded phalanx of the Government,—get up and declare that this measure of Church Reform, this severance of Church and State, was brought forward in consonance with his own long-cherished political conviction? He accused that party of being so bound to the chariot wheels of the right honourable gentleman, as to be unable to abide by their own convictions. And as to the right honourable gentleman himself, he would appeal to his followers opposite to say whether the right honourable gentleman was possessed of any one strong political conviction.

He had been accused of being unconstitutional, revolutionary, and tyrannical. If the House would allow him he would very shortly explain his idea of constitutional government as carried on in this country. It was based and built on majorities in that House, and supported solely by that power. There could be no constitutional government in this country that was not so maintained. Any other government must be both revolutionary and tyrannical. Any other government was a usurpation; and he would make bold to tell the right honourable gentleman that a Minister in this country who should recommend Her Majesty to trust herself to advisers not supported by a majority of the House of Commons, would plainly be guilty of usurping the powers of the State. He threw from him with disdain the charge which had been brought against himself of hankering after the sweets of office. He indulged and gloried in indulging the highest ambition of an English subject. But he gloried much more in the privileges and power of that House, within the walls of which was centred all that was salutary, all that was efficacious, all that was stable in the political constitution of his country. It had been his pride to have acted during nearly all his political life with that party which had commanded a majority, but he would defy his most bitter adversary, he would defy the right honourable gentleman himself, to point to any period of his career in which he had been unwilling to succumb to a majority when he himself had belonged to the minority.

He himself would regard the vote on this occasion as a vote of want of confidence. He took the line he was now taking because he desired to bring the House to a decision on that question. He himself had not that confidence in the right honourable gentleman which would justify him in accepting a measure on so important a subject as the union or severance of Church and State from his hands. Should the majority of the House differ from him and support the second reading of the Bill, he would at once so far succumb as to give his best attention to the clauses of the bill, and endeavour with the assistance of those gentlemen who acted with him to make it suitable to the wants of the country by omissions and additions as the clauses should pass through Committee. But before doing that he would ask the House to decide with all its solemnity and all its weight whether it was willing to accept from the hands of the right honourable gentleman any measure of reform on a matter so important as this now before them. It was nearly ten when he sat down; and then the stomach of the House could stand it no longer, and an adjournment at once took place.

On the next morning it was generally considered that Mr. Daubeny had been too long and Mr. Gresham too passionate. There were some who declared that Mr. Gresham had never been finer than when he described the privileges of the House of Commons; and others who thought that Mr. Daubeny's lucidity had been marvellous; but in this case, as in most others, the speeches of the day were generally thought to have been very inferior to the great efforts of the past.

CHAPTER XXXIV

The Universe

BEFORE the House met again, the quidnuncs about the clubs, on both sides of the question, had determined that Mr. Gresham's speech, whether good or not as an effort of oratory, would serve its intended purpose. He would be backed by a majority of votes, and it might have been very doubtful whether such would have been the case had he attempted to throw out the Bill on its merits. Mr. Ratler, by the time that prayers had been read, had become almost certain of success. There were very few Liberals in the House who were not anxious to declare by their votes that they had no confidence in Mr. Daubeny. Mr. Turnbull, the great Radical, and, perhaps, some two dozen with him, would support the second reading, declaring that they could not reconcile it with their consciences to record a vote in favour of a union of Church and State. On all such occasions as the present Mr. Turnbull was sure to make himself disagreeable to those who sat near to him in the House. He was a man who thought that so much was demanded of him in order that his independence might be doubted by none. It was nothing to him, he was wont to say, who called himself Prime Minister, or Secretary here, or President there. But then there would be quite as much of this independence on the Conservative as on the Liberal side of the House. Surely there would be more than two dozen gentlemen who would be true enough to the cherished principles of their whole lives to vote against such a Bill as this! It was the fact that there were so very few so true which added such a length to the faces of the country parsons. Six months ago not a country gentleman in England would have listened to such a proposition without loud protests as to its revolutionary wickedness. And now, under the sole pressure of one man's authority, the subject had become so common that men were assured that the thing would be done even though of all things that could be done it were the worst. 'It is no good any longer having any opinion upon anything,' one parson said to another, as they sat together at their club with their newspapers in their hands. 'Nothing frightens any one,—no infidelity, no wickedness, no revolution. All reverence is at an end, and the Holy of Holies is no more even to the worshipper than the threshold of the Temple.' Though it became known that the Bill would be lost, what comfort was there in that, when the battle was to be won, not by the chosen Israelites to whom the Church with all its appurtenances ought to be dear, but by a crew of Philistines who would certainly follow the lead of their opponents in destroying the holy structure?

On the Friday the debate was continued with much life on the Ministerial side of the House. It was very easy for them to cry Faction! Faction! and hardly necessary for them to do more. A few parrot words had been learned as to the expediency of fitting the great and increasing Church of England to the growing necessity of the age. That the CHURCH OF ENGLAND would still be the CHURCH OF ENGLAND was repeated till weary listeners were sick of the unmeaning words. But the zeal of the combatants was displayed on that other question. Faction was now the avowed weapon of the leaders of the so-called Liberal side of the House, and it was very easy to denounce the new doctrine. Every word that Mr. Gresham had spoken was picked in pieces, and the enormity of his theory was exhibited. He had boldly declared to them that they were to regard men and not measures, and they were to show by their votes whether they were prepared to accept such teaching. The speeches were, of course, made by alternate orators, but the firing from the Conservative benches was on this evening much the louder.

It would have seemed that with such an issue between them they might almost have consented to divide after the completion of the two great speeches. The course on which they were to run had been explained to them, and it was not probable that any member's intention as to his running would now be altered by anything that he might hear. Mr. Turnbull's two dozen defaulters were all known, and the two dozen and four true Conservatives were known also. But, nevertheless, a great many members were anxious to speak. It would be the great debate of the Session, and the subject to be handled,—that, namely, of the general merits and demerits of the two political parties,—was wide and very easy On that night it was past one o'clock when Mr. Turnbull adjourned the House.

'I'm afraid we must put you off till Tuesday,' Mr. Ratler said on the Sunday afternoon to Phineas Finn.

'I have no objection at all, so long as I get a fair place on that day.'

'There shan't be a doubt about that. Gresham particularly wants you to speak, because you are pledged to a measure of disestablishment. You can insist on his own views,—that even should such a measure be essentially necessary——'

'Which I think it is,' said Phineas.

'Still it should not be accepted from the old Church-and-State party.'

There was something pleasant in this to Phineas Finn,—something that made him feel for the moment that he had perhaps mistaken the bearing of his friend towards him. 'We are sure of a majority, I suppose,' he said.

'Absolutely sure,' said Ratler. 'I begin to think it will amount to half a hundred,—perhaps more.'

'What will Daubeny do?'

'Go out. He can't do anything else. His pluck is certainly wonderful, but even with his pluck he can't dissolve again. His Church Bill has given him a six months' run, and six months is something.'

'Is it true that Grogram is to be Chancellor?' Phineas asked the question, not from any particular solicitude as to the prospects of Sir Gregory Grogram, but because he was anxious to hear whether Mr. Ratler would speak to him with anything of the cordiality of fellowship respecting the new Government. But Mr. Ratler became at once discreet and close, and said that he did not think that anything as yet was known as to the Woolsack. Then Phineas retreated again within his shell, with a certainty that nothing would be done for him.

And yet to whom could this question of place be of such vital importance as it was to him? He had come back to his old haunts from Ireland, abandoning altogether the pleasant safety of an assured income, buoyed by the hope of office. He had, after a fashion, made his calculations. In the present disposition of the country it was, he thought, certain that the Liberal party must, for the next twenty years, have longer periods of power than their opponents; and he had thought also that were he in the House, some place would eventually be given to him. He had been in office before, and had been especially successful. He knew that it had been said of him that of the young debutants of latter years he had been the best. He had left his party by opposing them; but he had done so without creating any ill-will among the leaders of his party,—in a manner that had been regarded as highly honourable to him, and on departing had received expressions of deep regret from Mr. Gresham himself. When Barrington Erle had wanted him to return to his old work, his own chief doubt had been about the seat. But he had been bold and had adventured all, and had succeeded. There had been some little trouble about those pledges given at Tankerville, but he would be able to turn them even to the use of his party. It was quite true that nothing had been promised him; but Erle, when he had written, bidding him to come over from Ireland, must have intended him to understand that he would be again enrolled in the favoured regiment, should he be able to show himself as the possessor of a seat in the House. And yet,—yet he felt convinced that when the day should come it would be to him a day of disappointment, and that when the list should appear his name would not be on it. Madame Goesler had suggested to him that Mr. Bonteen might be his enemy, and he had replied by stating that he himself hated Mr. Bonteen. He now remembered that Mr. Bonteen had hardly spoken to him since his return to London, though there had not in fact been any quarrel between them. In this condition of mind he longed to speak openly to Barrington Erle, but he was restrained by a feeling of pride, and a still existing idea that no candidate for office, let his claim be what it might, should ask for a place. On that Sunday evening he saw Bonteen at the club. Men were going in and out with that feverish excitement which always prevails on the eve of a great parliamentary change. A large majority against the Government was considered to be certain; but there was an idea abroad that Mr. Daubeny had some scheme in his head by which to confute the immediate purport of his enemies. There was nothing to which the audacity of the man was not equal. Some said that he would dissolve the House,—which had hardly as yet been six months sitting. Others were of opinion that he would simply resolve not to vacate his place,—thus defying the majority of the House and all the ministerial traditions of the country. Words had fallen from him which made some men certain that such was his intention. That it should succeed ultimately was impossible. The whole country would rise against him. Supplies would be refused. In every detail of Government he would be impeded. But then,—such was the temper of the man,—it was thought that all these horrors would not deter him. There would be a blaze and a confusion, in which timid men would doubt whether the constitution would be burned to tinder or only illuminated; but that blaze and that confusion would be dear to Mr. Daubeny if he could stand as the centre figure,—the great pyrotechnist who did it all, red from head to foot with the glare of the squibs with which his own hands were filling all the spaces. The anticipation that some such display might take place made men busy and eager; so that on that Sunday evening they roamed about from one place of meeting to another, instead of sitting at home with their wives and daughters. There was at this time existing a small club,—so called though unlike other clubs,—which had entitled itself the Universe. The name was supposed to be a joke, as it was limited to ninety-nine members. It was domiciled in one simple and somewhat mean apartment. It was kept open only one hour before and one hour after midnight, and that only on two nights of the week, and that only when Parliament was sitting. Its attractions were not numerous, consisting chiefly of tobacco and tea. The conversation was generally listless and often desultory; and occasionally there would arise the great and terrible evil of a punster whom every one hated but no one had life enough to put down. But the thing had been a success, and men liked to be members of the Universe. Mr. Bonteen was a member, and so was Phineas Finn. On this Sunday evening the club was open, and Phineas, as he entered the room, perceived that his enemy was seated alone on a corner of a sofa. Mr. Bonteen was not a man who loved to be alone in public places, and was apt rather to make one of congregations, affecting popularity, and always at work increasing his influence. But on this occasion his own greatness had probably isolated him. If it were true that he was to be the new Chancellor of the Exchequer,—to ascend from demi-godhead to the perfect divinity of the Cabinet,—and to do so by a leap which would make him high even among first-class gods, it might be well for himself to look to himself and choose new congregations. Or, at least, it would be becoming that he should be chosen now instead of being a chooser. He was one who could weigh to the last ounce the importance of his position, and make most accurate calculations as to the effect of his intimacies. On that very morning Mr. Gresham had suggested to him that in the event of a Liberal Government being formed, he should hold the high office in question. This, perhaps, had not been done in the most flattering manner, as Mr. Gresham had deeply bewailed the loss of Mr. Palliser, and had almost demanded a pledge from Mr. Bonteen that he would walk exactly in Mr. Palliser's footsteps;—but the offer had been made, and could not be retracted; and Mr. Bonteen already felt the warmth of the halo of perfect divinity.

There are some men who seem to have been born to be Cabinet Ministers,—dukes mostly, or earls, or the younger sons of such,—who have been trained to it from their very cradles, and of whom we may imagine that they are subject to no special awe when they first enter into that august assembly, and feel but little personal elevation. But to the political aspirant not born in the purple of public life, this entrance upon the counsels of the higher deities must be accompanied by a feeling of supreme triumph, dashed by considerable misgivings. Perhaps Mr. Bonteen was revelling in his triumph;—perhaps he was anticipating his misgivings. Phineas, though disinclined to make any inquiries of a friend which might seem to refer to his own condition, felt no such reluctance in regard to one who certainly could not suspect him of asking a favour. He was presumed to be on terms of intimacy with the man, and he took his seat beside him, asking some question as to the debate. Now Mr. Bonteen had more than once expressed an opinion among his friends that Phineas Finn would throw his party over, and vote with the Government. The Ratlers and Erles and Fitzgibbons all knew that Phineas was safe, but Mr. Bonteen was still in doubt. It suited him to affect something more than doubt on the present occasion. 'I wonder that you should ask me,' said Mr. Bonteen.

'What do you mean by that?'

'I presume that you, as usual, will vote against us.'

'I never voted against my party but once,' said Phineas, 'and then I did it with the approbation of every man in it for whose good opinion I cared a straw.' There was insult in his tone as he said this, and something near akin to insult in his words.

'You must do it again now, or break every promise that you made at Tankerville.'

'Do you know what promise I made at Tankerville? I shall break no promise.'

'You must allow me to say, Mr. Finn, that the kind of independence which is practised by you and Mr. Monk, grand as it may be on the part of men who avowedly abstain from office, is a little dangerous when it is now and again adopted by men who have taken place. I like to be sure that the men who are in the same boat with me won't take it into their heads that their duty requires them to scuttle the ship.' Having so spoken, Mr. Bonteen, with nearly all the grace of a full-fledged Cabinet Minister, rose from his seat on the corner of the sofa and joined a small congregation.

Phineas felt that his ears were tingling and that his face was red. He looked round to ascertain from the countenances of others whether they had heard what had been said. Nobody had been close to them, and he thought that the conversation had been unnoticed. He knew now that he had been imprudent in addressing himself to Mr. Bonteen, though the question that he had first asked had been quite commonplace. As it was, the man, he thought, had been determined to affront him, and had made a charge against him which he could not allow to pass unnoticed. And then there was all the additional bitterness in it which arose from the conviction that Bonteen had spoken the opinion of other men as well as his own, and that he had plainly indicated that the gates of the official paradise were to be closed against the presumed offender. Phineas had before believed that it was to be so, but that belief had now become assurance. He got up in his misery to leave the room, but as he did so he met Laurence Fitzgibbon. 'You have heard the news about Bonteen?' said Laurence.

'What news?'

'He's to be pitchforked up to the Exchequer. They say it's quite settled. The higher a monkey climbs——;*you know the proverb.' So saying Laurence Fitzgibbon passed into the room, and Phineas Finn took his departure in solitude.

And so the man with whom he had managed to quarrel utterly was to be one in the Cabinet, a man whose voice would probably be potential in the selection of minor members of the Government. It seemed to him to be almost incredible that such a one as Mr. Bonteen should be chosen for such an office. He had despised almost as soon as he had known Mr. Bonteen, and had rarely heard the future manager of the finance of the country spoken of with either respect or regard. He had regarded Mr. Bonteen as a useful, dull, unscrupulous politician, well accustomed to Parliament, acquainted with the bye-paths and back doors of official life,—and therefore certain of employment when the Liberals were in power; but there was no one in the party he had thought less likely to be selected for high place. And yet this man was to be made Chancellor of the Exchequer, while he, Phineas Finn, very probably at this man's instance, was to be left out in the cold.

He knew himself to be superior to the man he hated, to have higher ideas of political life, and to be capable of greater political sacrifices. He himself had sat shoulder to shoulder with many men on the Treasury Bench whose political principles he had not greatly valued; but of none of them had he thought so little as he had done of Mr. Bonteen. And yet this Mr. Bonteen was to be the new Chancellor of the Exchequer!

He walked home to his lodgings in Marlborough Street, wretched because of his own failure;—doubly wretched because of the other man's success.

He laid awake half the night thinking of the words that had been spoken to him, and after breakfast on the following morning he wrote the following note to his enemy:—

'House of Commons, 5th April, 18—.

'DEAR MR. BONTEEN,

'It is matter of extreme regret to me that last night at the Universe I should have asked you some chance question about the coming division. Had I guessed to what it might have led, I should not have addressed you. But as it is I can hardly abstain from noticing what appeared to me to be a personal charge made against myself with a great want of the courtesy which is supposed to prevail among men who have acted together. Had we never done so my original question to you might perhaps have been deemed an impertinence.

'As it was, you accused me of having been dishonest to my party, and of having "scuttled the ship." On the occasion to which you alluded I acted with much consideration, greatly to the detriment of my own prospects,—and as I believed with the approbation of all who knew anything of the subject. If you will make inquiry of Mr. Gresham, or Lord Cantrip who was then my chief, I think that either will tell you that my conduct on that occasion was not such as to lay me open to reproach. If you will do this, I think that you cannot fail afterwards to express regret for what you said to me last night.

'Yours sincerely,
'PHINEAS FINN.

'Thos. Bonteen, Esq., M.P.'

He did not like the letter when he had written it, but he did not know how to improve it, and he sent it.

CHAPTER XXXV
Political Venom

On the Monday Mr. Turnbull opened the ball by declaring his reasons for going into the same lobby with Mr. Daubeny. This he did at great length. To him all the mighty pomp and all the little squabbles of office were, he said, as nothing. He would never allow himself to regard the person of the Prime Minister. The measure before the House ever had been and ever should be all in all to him. If the public weal were more regarded in that House, and the quarrels of men less considered, he thought that the service of the country would be better done. He was answered by Mr. Monk, who was sitting near him, and who intended to support Mr. Gresham. Mr. Monk was rather happy in pulling his old friend, Mr. Turnbull, to pieces, expressing his opinion that a difference in men meant a difference in measures. The characters of men whose principles were known were guarantees for the measures they would advocate. To him,—Mr. Monk,—it was matter of very great moment who was Prime Minister of England. He was always selfish enough to wish for a Minister with whom he himself could agree on the main questions of the day. As he certainly could not say that he had political confidence in the present Ministry, he should certainly vote against them on this occasion.

In the course of the evening Phineas found a letter addressed to himself from Mr. Bonteen. It was as follows:—

'*House of Commons, April 5th*, 18—.

'Dear Mr. Finn,

'I never accused you of dishonesty. You must have misheard or misunderstood me if you thought so. I did say that you had scuttled the ship;—and as you most undoubtedly did scuttle it,—you and Mr. Monk between you,—I cannot retract my words.

'I do not want to go to any one for testimony as to your

merits on the occasion. I accused you of having done nothing dishonourable or disgraceful. I think I said that there was danger in the practice of scuttling. I think so still, though I know that many fancy that those who scuttle do a fine thing. I don't deny that it's fine, and therefore you can have no cause of complaint against me.

'Yours truly,

'J. BONTEEN.'

He had brought a copy of his own letter in his pocket to the House, and he showed the correspondence to Mr. Monk. 'I would not have noticed it, had I been you,' said he.

'You can have no idea of the offensive nature of the remark when it was made.'

'It's as offensive to me as to you, but I should not think of moving in such a matter. When a man annoys you, keep out of his way. It is generally the best thing you can do.'

'If a man were to call you a liar?'

'But men don't call each other liars. Bonteen understands the world much too well to commit himself by using any word which common opinion would force him to retract. He says we scuttled the ship. Well;—we did. Of all the political acts of my life it is the one of which I am most proud. The manner in which you helped me has entitled you to my affectionate esteem. But we did scuttle the ship. Before you can quarrel with Bonteen you must be able to show that a metaphorical scuttling of a ship must necessarily be a disgraceful act. You see how he at once retreats behind the fact that it need not be so.'

'You wouldn't answer his letter.'

'I think not. You can do yourself no good by a correspondence in which you cannot get a hold of him. And if you did get a hold of him you would injure yourself much more than him. Just drop it.' This added much to our friend's misery, and made him feel that the weight of it was almost more than he could bear. His enemy had got the better of him at every turn. He had now rushed into a correspondence as to which he would have to own by his silence that he had been confuted. And yet he was sure that Mr. Bonteen had at the club insulted him most unjustifiably, and that if the actual truth were known, no man, certainly not Mr. Monk, would hesitate to say that reparation was due to him. And yet what could he do? He thought that he would consult Lord Cantrip, and endeavour to get from his late Chief some advice more palatable than that which had been tendered to him by Mr. Monk.

In the meantime animosities in the House were waxing very furious; and, as it happened, the debate took a turn that was peculiarly injurious to Phineas Finn in his present state of mind. The rumour as to the future promotion of Mr. Bonteen, which had been conveyed by Laurence Fitzgibbon to Phineas at the Universe, had, as was natural, spread far and wide, and had reached the ears of those who still sat on the Ministerial benches. Now it is quite understood among politicians in this country that no man should presume that he will have imposed upon him the task of forming a Ministry until he has been called upon by the Crown to undertake that great duty. Let the Gresham or the Daubeny of the day be ever so sure that the reins of the State chariot must come into his hands, he should not visibly prepare himself for the seat on the box till he has actually been summoned to place himself there. At this moment it was alleged that Mr. Gresham had departed from the reticence and modesty usual in such a position as his, by taking steps towards the formation of a Cabinet, while it was as yet quite possible that he might never be called upon to form any Cabinet. Late on this Monday night, when the House was quite full, one of Mr. Daubeny's leading lieutenants, a Secretary of State, Sir Orlando Drought by name,—a gentleman who if he had any heart in the matter must have hated this Church Bill from the very bottom of his heart, and who on that account was the more bitter against opponents who had not ceased to throw in his teeth his own political tergiversation,—fell foul of Mr. Gresham as to this rumoured appointment to the Chancellorship of the Exchequer. The reader will easily imagine the things that were said. Sir

Orlando had heard, and had been much surprised at hearing, that a certain honourable member of that House, who had long been known to them as a tenant of the Ministerial bench, had already been appointed to a high office. He, Sir Orlando, had not been aware that the office had been vacant, or that if vacant it would have been at the disposal of the right honourable gentleman; but he believed that there was no doubt that the place in question, with a seat in the Cabinet, had been tendered to, and accepted by, the honourable member to whom he alluded. Such was the rabid haste with which the right honourable gentleman opposite, and his colleagues, were attempting, he would not say to climb, but to rush into office, by opposing a great measure of Reform, the wisdom of which, as was notorious to all the world, they themselves did not dare to deny. Much more of the same kind was said, during which Mr. Gresham pulled about his hat, shuffled his feet, showed his annoyance to all the House, and at last jumped upon his legs.

'If,' said Sir Orlando Drought,—'if the right honourable gentleman wishes to deny the accuracy of any statements that I have made, I will give way to him for the moment, that he may do so.'

'I deny utterly, not only the accuracy, but every detail of the statement made by the right honourable gentleman opposite,' said Mr. Gresham, still standing and holding his hat in his hand as he completed his denial.

'Does the right honourable gentleman mean to assure me that he has not selected his future Chancellor of the Exchequer?'

'The right honourable gentleman is too acute not to be aware that we on this side of the House may have made such selection, and that yet every detail of the statement which he has been rash enough to make to the House may be——unfounded. The word, sir, is weak; but I would fain avoid the use of any words which, justifiable though they might be, would offend the feelings of the House. I will explain to the House exactly what has been done.'

Then there was a great hubbub—cries of 'Order,' 'Gresham,' 'Spoke,' 'Hear, hear,' and the like,—during which Sir Orlando Drought and Mr. Gresham both stood on their legs. So powerful was Mr. Gresham's voice that, through it all, every word that he said was audible to the reporters. His opponent hardly attempted to speak, but stood relying upon his right. Mr. Gresham said he understood that it was the desire of the House that he should explain the circumstances in reference to the charge that had been made against him, and it would certainly be for the convenience of the House that this should be done at the moment. The Speaker of course ruled that Sir Orlando was in possession of the floor, but suggested that it might be convenient that he should yield to the right honourable gentleman on the other side for a few minutes. Mr. Gresham, as a matter of course, succeeded. Rights and rules, which are bonds of iron to a little man, are packthread to a giant. No one in all that assembly knew the House better than did Mr. Gresham, was better able to take it by storm, or more obdurate in perseverance. He did make his speech, though clearly he had no right to do so. The House, he said, was aware, that by the most unfortunate demise of the late Duke of Omnium, a gentleman had been removed from this House to another place, whose absence from their counsels would long be felt as a very grievous loss. Then he pronounced a eulogy on Plantagenet Palliser, so graceful and well arranged, that even the bitterness of the existing opposition was unable to demur to it. The House was well aware of the nature of the labours which now for some years past had occupied the mind of the noble duke; and the paramount importance which the country attached to their conclusion. The noble duke no doubt was not absolutely debarred from a continuance of his work by the change which had fallen upon him; but it was essential that some gentleman, belonging to the same party with the noble duke, versed in office, and having a seat in that House, should endeavour to devote himself to the great measure which had occupied so much of the attention of the late Chancellor of the Exchequer. No doubt it must be fitting that the gentleman so selected should be at the Exchequer, in the event of their party coming into office. The honourable gentleman to whom allusion had been made had acted throughout with the present noble duke in arranging the details of the measure in question; and the probability of his being able to fill the shoes left vacant by the accession to the peerage of the noble duke had, indeed, been discussed;—but the discussion had been made in reference to the measure, and only incidentally in regard to the office. He, Mr. Gresham, held that he had done nothing that was indiscreet,—nothing that his duty did not demand. If right honourable gentlemen opposite were of a different opinion, he thought that that difference came from the fact that they were less intimately acquainted than he unfortunately had been with the burdens and responsibilities of legislation.

There was very little in the dispute which seemed to be worthy of the place in which it occurred, or of the vigour with which it was conducted; but it served to show the temper of the parties, and to express the bitterness of the political feelings of the day. It was said at the time, that never within the memory of living politicians had so violent an animosity displayed itself in the House as had been witnessed on this night. While Mr. Gresham was giving his explanation, Mr. Daubeny had arisen, and with a mock solemnity that was peculiar to him on occasions such as these, had appealed to the Speaker whether the right honourable gentleman opposite should not be called upon to resume his seat. Mr. Gresham had put him down with a wave of his hand. An affected stateliness cannot support itself but for a moment; and Mr. Daubeny had been forced to sit down when the Speaker did not at once support his appeal. But he did not forget that wave of the hand, nor did he forgive it. He was a man who in public life rarely forgot, and never forgave. They used to say of him that 'at home' he was kindly and forbearing, simple and unostentatious. It may be so. Who does not remember that horrible Turk, Jacob Asdrubal, the Old Bailey barrister, the terror of witnesses, the bane of judges,—who was gall and wormwood to all opponents. It was said of him that 'at home' his docile amiability was the marvel of his friends, and delight of his wife and daughters. 'At home,' perhaps, Mr. Daubeny might have been waved at, and have forgiven it; but men who saw the scene in the House of Commons knew that he would never forgive Mr. Gresham. As for Mr. Gresham himself, he triumphed at the moment, and exulted in his triumph.

Phineas Finn heard it all, and was disgusted to find that his enemy thus became the hero of the hour. It was, indeed, the opinion generally of the Liberal party that Mr. Gresham had not said much to flatter his new Chancellor of the Exchequer. In praise of Plantagenet Palliser he had been very loud, and he had no doubt said that which implied the capability of Mr. Bonteen, who, as it happened, was sitting next to him at the time; but he had implied also that the mantle which was to be transferred from Mr. Palliser to Mr. Bonteen would be carried by its new wearer with grace very inferior to that which had marked all the steps of his predecessor. Ratler, and Erle, and Fitzgibbon, and others had laughed in their sleeves at the expression, understood by them, of Mr. Gresham's doubt as to the qualifications of his new assistant, and Sir Orlando Drought, in continuing his speech, remarked that the warmth of the right honourable gentleman had been so completely expended in abusing his enemies that he had had none left for the defence of his friend. But to Phineas it seemed that this Bonteen, who had so grievously injured him, and whom he so thoroughly despised, was carrying off all the glories of the fight. A certain amount of consolation was, however, afforded to him. Between one and two o'clock he was told by Mr. Ratler that he might enjoy the privilege of adjourning the debate,—by which would accrue to him the right of commencing on the morrow,—and this he did at a few minutes before three.

CHAPTER XXXVI

Seventy-two

On the next morning Phineas, with his speech before him, was obliged for a while to forget, or at least to postpone, Mr. Bonteen and his injuries. He could not now go to Lord Cantrip, as the hours were too precious to him, and, as he felt, too short. Though he had been thinking what he would say ever since the debate had become imminent, and knew accurately the line which he would take, he had not as yet prepared a word of his speech. But he had resolved that he would not prepare a word otherwise than he might do by arranging certain phrases in his memory. There should be nothing written; he had tried that before in old days, and had broken down with the effort. He would load himself with no burden of words in itself so heavy that the carrying of it would incapacitate him for any other effort.

After a late breakfast he walked out far away, into the Regent's Park, and there, wandering among the uninteresting paths, he devised triumphs of oratory for himself. Let him resolve as he would to forget Mr. Bonteen, and that charge of having been untrue to his companions, he could not restrain himself from efforts to fit the matter after some fashion into his speech. Dim ideas of a definition of political honesty crossed his brain, bringing with him, however, a conviction that his thought must be much more clearly worked out than it could be on that day before he might venture to give it birth in the House of Commons. He knew that he had been honest two years ago in separating himself from his colleagues. He knew that he would be honest now in voting with them, apparently in opposition to the pledges he had given at Tankerville. But he knew also that it would behove him to abstain from speaking of himself unless he could do so in close reference to some point specially in dispute between the two parties. When he returned to eat a mutton chop at Great Marlborough Street at three o'clock he was painfully conscious that all his morning had been wasted. He had allowed his mind to run revel, instead of tying it down to the formation of sentences and construction of arguments.

He entered the House with the Speaker at four o'clock, and took his seat without uttering a word to any man. He seemed to be more than ever disjoined from his party. Hitherto, since he had been seated by the Judge's order, the former companions of his Parliamentary life,—the old men whom he had used to know,—had to a certain degree admitted him among them. Many of them sat on the front Opposition bench, whereas he, as a matter of course, had seated himself behind. But he had very frequently found himself next to some man who had held office and was living in the hope of holding it again, and had felt himself to be in some sort recognised as an aspirant. Now it seemed to him that it was otherwise. He did not doubt but that Bonteen had shown the correspondence to his friends, and that the Ratlers and Erles had conceded that he, Phineas, was put out of court by it. He sat doggedly still, at the end of a bench behind Mr. Gresham, and close to the gangway. When Mr. Gresham entered the House he was received with much cheering; but Phineas did not join in the cheer. He was studious to avoid any personal recognition of the future giver-away of places, though they two were close together; and he then fancied that Mr. Gresham had specially and most ungraciously abstained from any recognition of him. Mr. Monk, who sat near him, spoke a kind word to him. 'I shan't be very long,' said Phineas; 'not above twenty minutes, I should think.' He was able to assume an air of indifference, and yet at the moment he heartily wished himself back in Dublin. It was not now that he feared the task immediately before him, but that he was overcome by the feeling of general failure which had come upon him. Of what use was it to him or to any one else that he should be there in that assembly, with the privilege of making a speech that would influence no human being, unless his being there could be made a step to something beyond? While the usual preliminary work was being done, he looked round the House, and saw Lord Cantrip in the Peers' gallery. Alas! of what avail was that? He had always been able to bind to him individuals with whom he had been brought into close contact; but more than that was wanted in this most precarious of professions, in which now, for a second time, he was attempting to earn his bread.

At half-past four he was on his legs in the midst of a crowded House. The chance,—perhaps the hope,—of some such encounter as that of the former day, brought members into their seats, and filled the gallery with strangers. We may say, perhaps, that the highest duty imposed upon us as a nation is the management of India; and we may also say that in a great national assembly personal squabbling among its members is the least dignified work in which it can employ itself. But the prospect of an explanation,—or otherwise of a fight,—between two leading politicians will fill the House; and any allusion to our Eastern Empire will certainly empty it. An aptitude for such encounters is almost a necessary qualification for a popular leader in Parliament, as is a capacity for speaking for three hours to the reporters, and to the reporters only,—a necessary qualification for an Under-Secretary of State for India.

Phineas had the advantage of the temper of the moment in a House thoroughly crowded, and he enjoyed it. Let a man doubt ever so much his own capacity for some public exhibition which he has undertaken; yet he will always prefer to fail,—if fail he must,—before a large audience. But on this occasion there was no failure. That sense of awe from the surrounding circumstances of the moment, which had once been heavy on him, and which he still well remembered, had been overcome, and had never returned to him. He felt now that he should not lack words to pour out his own individual grievances were it not that he was prevented by a sense of the indiscretion of doing so. As it was, he did succeed in alluding to his own condition in a manner that brought upon him no reproach. He began by saying that he should not have added to the difficulty of the debate,—which was one simply of length,—were it not that he had been accused in advance of voting against a measure as to which he had pledged himself at the hustings to do all that he could to further it. No man was more anxious than he, an Irish Roman Catholic, to abolish that which he thought to be the anomaly of a State Church, and he did not in the least doubt that he should now be doing the best in his power with that object in voting against the second reading of the present bill. That such a measure should be carried by the gentlemen opposite, in their own teeth, at the bidding of the right honourable gentleman who led them, he thought to be impossible. Upon this he was hooted at from the other side with many gestures of indignant denial, and was, of course, equally cheered by those around him. Such interruptions are new breath to the nostrils of all orators, and Phineas enjoyed the noise. He repeated his assertion that it would be an evil thing for the country that the measure should be carried by men who in their hearts condemned it, and was vehemently called to order for this assertion about the hearts of gentlemen. But a speaker who can certainly be made amenable to authority for vilipending in debate the heart of any specified opponent, may with safety attribute all manner of ill to the agglomerated hearts of a party. To have told any individual Conservative,—Sir Orlando Drought for instance,—that he was abandoning all the convictions of his life, because he was a creature at the command of Mr. Daubeny, would have been an insult that would have moved even the Speaker from his serenity; but you can hardly be personal to a whole bench of Conservatives,—to bench above bench of Conservatives. The charge had been made and repeated over and over again, till all the Orlando Droughts were ready to cut some man's throat,—whether their own, or Mr. Daubeny's, or Mr. Gresham's, they hardly knew. It might probably have been Mr. Daubeny's for choice, had any real cutting of a throat been possible. It was now made again by Phineas Finn,—with the ostensible object of defending himself,—and he for the moment became the target for Conservative wrath. Some one asked him in fury by what right he took upon himself to judge of the motives of gentlemen on that side of the House of whom personally he knew nothing. Phineas replied that he did not at all doubt the motives of the honourable gentleman who asked the question, which he was sure were noble and patriotic. But unfortunately the whole country was convinced that the Conservative party as a body was supporting this measure, unwillingly, and at the bidding of one man;—and, for himself, he was bound to say that he agreed with the country. And so the row was renewed and prolonged, and the gentlemen assembled, members and strangers together, passed a pleasant evening.

Before he sat down, Phineas made one allusion to that former scuttling of the ship,—an accusation as to which had been made against him so injuriously by Mr. Bonteen. He himself, he said, had been called impractical, and perhaps he might allude to a vote which he had given in that House when last he had the honour of sitting there, and on giving which he resigned the office which he had then held. He had the gratification of knowing that he had been so far practical as to have then foreseen the necessity of a measure which had since been passed. And he did not doubt that he would hereafter be found to have been equally practical in the view that he had expressed on the hustings at Tankerville, for he was convinced that before long the anomaly of which he had spoken would cease to exist under the influence of a Government that would really believe in the work it was doing.

There was no doubt as to the success of his speech. The vehemence with which his insolence was abused by one after another of those who spoke later from the other side was ample evidence of its success. But nothing occurred then or at the conclusion of the debate to make him think that he had won his way back to Elysium. During the whole evening he exchanged not a syllable with Mr. Gresham,—who indeed was not much given to converse with those around him in the House. Erle said a few good-natured words to him, and Mr. Monk praised him highly. But in reading the general barometer of the party as regarded himself, he did not find that the mercury went up. He was wretchedly anxious, and angry with himself for his own anxiety. He scorned to say a word that should sound like an entreaty; and yet he had placed his whole heart on a thing which seemed to be slipping from him for the want of asking. In a day or two it would be known whether the present Ministry would or would not go out. That they must be out of office before a month was over seemed to him the opinion of everybody. His fate,—and what a fate it was!—would then be absolutely in the hands of Mr. Gresham. Yet he could not speak a word of his hopes and fears even to Mr. Gresham. He had given up everything in the world with the view of getting into office; and now that the opportunity had come,—an opportunity which if allowed to slip could hardly return again in time to be of service to him,—the prize was to elude his grasp!

But yet he did not say a word to any one on the subject that was so near his heart, although in the course of the night he spoke to Lord Cantrip in the gallery of the House. He told his friend that a correspondence had taken place between himself and Mr. Bonteen, in which he thought that he had been ill-used, and as to which he was quite anxious to ask His Lordship's advice. 'I heard that you and he had been tilting at each other,' said Lord Cantrip, smiling.

'Have you seen the letters?'

'No;—but I was told of them by Lord Fawn, who has seen them.'

'I knew he would show them to every newsmonger about the clubs,' said Phineas angrily.

'You can't quarrel with Bonteen for showing them to Fawn, if you intend to show them to me.'

'He may publish them at Charing Cross if he likes.'

'Exactly. I am sure that there will have been nothing in them prejudicial to you. What I mean is that if you think it necessary, with a view to your own character, to show them to me or to another friend, you cannot complain that he should do the same.'

An appointment was made at Lord Cantrip's house for the next morning, and Phineas could but acknowledge to himself that the man's manner to himself had been kind and constant. Nevertheless, the whole affair was going against him. Lord Cantrip had not said a word prejudicial to that wretch Bonteen; much less had he hinted at any future arrangements which would be comfortable to poor Phineas. They two, Lord Cantrip and Phineas, had at one period been on most intimate terms together;—had worked in the same office, and had thoroughly trusted each other. The elder of the two,—for Lord Cantrip was about ten years senior to Phineas,—had frequently expressed the most lively interest in the prospects of the other; and Phineas had felt that in any emergency he could tell his friend all his hopes and fears. But now he did not say a word of his position, nor did Lord Cantrip allude to it. They were to meet on the morrow in order that Lord Cantrip might read the correspondence;—but Phineas was sure that no word would be said about the Government.

At five o'clock in the morning the division took place, and the Government was beaten by a majority of 72. This was much higher than any man had expected. When the parties were marshalled in the opposite lobbies it was found that in the last moment the number of those Conservatives who dared to rebel against their Conservative leaders was swelled by the course which the debate had taken. There were certain men who could not endure to be twitted with having deserted the principles of their lives, when it was clear that nothing was to be gained by the party by such desertion.

CHAPTER XXXVII
The Conspiracy

On the morning following the great division Phineas was with his friend, Lord Cantrip, by eleven o'clock; and Lord Cantrip, when he had read the two letters in which were comprised the whole correspondence, made to our unhappy hero the following little speech. 'I do not think that you can do anything. Indeed, I am sure that Mr. Monk is quite right. I don't quite see what it is that you wish to do. Privately,—between our two selves,—I do not hesitate to say that Mr. Bonteen has intended to be ill-natured. I fancy that he is an ill-natured—or at any rate a jealous—man; and that he would be willing to run down a competitor in the race who had made his running after a fashion different from his own. Bonteen has been a useful man,—a very useful man; and the more so perhaps because he has not entertained any high political theory of his own. You have chosen to do so,—and undoubtedly when you and Monk left us, to our very great regret, you did scuttle the ship.'

'We had no intention of that kind.'

'Do not suppose that I blame you. That which was odious to the eyes of Mr. Bonteen was to my thinking high and honourable conduct. I have known the same thing done by members of a Government perhaps half-a-dozen times, and the men by whom it has been done have been the best and noblest of our modern statesmen. There has generally been a hard contest in the man's breast between loyalty to his party and strong personal convictions, the result of which has been an inability on the part of the struggler to give even a silent support to a measure which he has disapproved. That inability is no doubt troublesome at the time to the colleagues of the seceder, and constitutes an offence hardly to be pardoned by such gentlemen as Mr. Bonteen.'

'For Mr. Bonteen personally I care nothing.'

'But of course you must endure the ill-effects of his influence,—be they what they may. When you seceded from our Government you looked for certain adverse consequences. If you did not, where was your self-sacrifice? That such men as Mr. Bonteen should feel that you had scuttled the ship, and be unable to forgive you for doing so,—that is exactly the evil which you knew you must face. You have to face it now, and surely you can do so without showing your teeth. Hereafter, when men more thoughtful than Mr. Bonteen shall have come to acknowledge the high principle by which your conduct has been governed, you will receive your reward. I suppose Mr. Daubeny must resign now.'

'Everybody says so.'

'I am by no means sure that he will. Any other Minister since Lord North's time* would have done so, with such a majority against him on a vital measure; but he is a man who delights in striking out some wonderful course for himself.'

'A prime minister so beaten surely can't go on.'

'Not for long, one would think. And yet how are you to

turn him out? It depends very much on a man's power of endurance.'

'His colleagues will resign, I should think.'

'Probably;—and then he must go. I should say that that will be the way in which the matter will settle itself. Good morning, Finn;—and take my word for it, you had better not answer Mr. Bonteen's letter.'

Not a word had fallen from Lord Cantrip's friendly lips as to the probability of Phineas being invited to join the future Government. An attempt had been made to console him with the hazy promise of some future reward,—which however was to consist rather of the good opinion of good men than of anything tangible and useful. But even this would never come to him. What would good men know of him and of his self-sacrifice when he should have been driven out of the world by poverty, and forced probably to go to some New Zealand or back Canadian settlement to look for his bread? How easy, thought Phineas, must be the sacrifices of rich men, who can stay their time, and wait in perfect security for their rewards! But for such a one as he, truth to a principle was political annihilation. Two or three years ago he had done what he knew to be a noble thing;—and now, because he had done that noble thing, he was to be regarded as unfit for that very employment for which he was peculiarly fitted. But Bonteen and Co. had not been his only enemies. His luck had been against him throughout. Mr. Quintus Slide, with his People's Banner, and the story of that wretched affair in Judd Street, had been as strong against him probably as Mr. Bonteen's ill-word. Then he thought of Lady Laura, and her love for him. His gratitude to Lady Laura was boundless. There was nothing he would not do for Lady Laura,—were it in his power to do anything. But no circumstance in his career had been so unfortunate for him as this affection. A wretched charge had been made against him which, though wholly untrue, was as it were so strangely connected with the truth, that slanderers might not improbably be able almost to substantiate their calumnies. She would be in London soon, and he must devote himself to her service. But every act of friendship that he might do for her would be used as proof of the accusation that had been made against him. As he thought of all this he was walking towards Park Lane in order that he might call upon Madame Goesler according to his promise. As he went up to the drawing-room he met old Mr. Maule coming down, and the two bowed to each other on the stairs. In the drawing-room, sitting with Madame Goesler, he found Mrs. Bonteen. Now Mrs. Bonteen was almost as odious to him as was her husband.

'Did you ever know anything more shameful, Mr. Finn,' said Mrs. Bonteen, 'than the attack made upon Mr. Bonteen the night before last?' Phineas could see a smile on Madame Goesler's face as the question was asked;—for she knew, and he knew that she knew, how great was the antipathy between him and the Bonteens.

'The attack was upon Mr. Gresham, I thought,' said Phineas.

'Oh, yes; nominally. But of course everybody knows what was meant. Upon my word there is twice more jealousy among men than among women. Is there not, Madame Goesler?'

'I don't think any man could be more jealous than I am myself,' said Madame Goesler.

'Then you're fit to be a member of a Government, that's all. I don't suppose that there is a man in England has worked harder for his party than Mr. Bonteen.'

'I don't think there is,' said Phineas.

'Or made himself more useful in Parliament. As for work, only that his constitution is so strong, he would have killed himself.'

'He should take Thorley's mixture,—twice a day,' said Madame Goesler.

'Take!—he never has time to take anything. He breakfasts in his dressing-room, carries his lunch in his pocket, and dines with the division bell ringing him up between his fish and his mutton chop. Now he has got their decimal coinage in hand, and has not a moment to himself, even on Sundays!'

'He'll be sure to go to Heaven for it,—that's one comfort.'

'And because they are absolutely obliged to make him Chancellor of the Exchequer,—just as if he had not earned it, —everybody is so jealous that they are ready to tear him to pieces!'

'Who is everybody?' asked Phineas.

'Oh! I know. It wasn't only Sir Orlando Drought. Who told Sir Orlando? Never mind, Mr. Finn.'

'I don't in the least, Mrs. Bonteen.'

'I should have thought you would have been so triumphant,' said Madame Goesler.

'Not in the least, Madame Goesler. Why should I be triumphant? Of course the position is very high,—very high indeed. But it's no more than what I have always expected. If a man give up his life to a pursuit he ought to succeed. As for ambition, I have less of it than any woman. Only I do hate jealousy, Mr. Finn.' Then Mrs. Bonteen took her leave, kissing her dear friend, Madame Goesler, and simply bowing to Phineas.

'What a detestable woman!' said Phineas.

'I know of old that you don't love her.'

'I don't believe that you love her a bit better than I do, and yet you kiss her.'

'Hardly that, Mr. Finn. There has come up a fashion for ladies to pretend to be very loving, and so they put their faces together. Two hundred years ago ladies and gentlemen did the same thing with just as little regard for each other. Fashions change, you know.'

'That was a change for the worse, certainly, Madame Goesler.'

'It wasn't of my doing. So you've had a great victory.'

'Yes;—greater than we expected.'

'According to Mrs. Bonteen, the chief result to the country will be that the taxes will be so very safe in her husband's hands! I am sure she believes that all Parliament has been at work in order that he might be made a Cabinet Minister. I rather like her for it.'

'I don't like her, or her husband.'

'I do like a woman that can thoroughly enjoy her husband's success. When she is talking of his carrying about his food in his pocket she is completely happy. I don't think Lady Glencora ever cared in the least about her husband being Chancellor of the Exchequer.'

'Because it added nothing to her own standing.'

'That's very ill-natured, Mr. Finn; and I find that you are becoming generally ill-natured. You used to be the best-humoured of men.'

'I hadn't so much to try my temper as I have now, and then you must remember, Madame Goesler, that I regard these people as being especially my enemies.'

'Lady Glencora was never your enemy.'

'Nor my friend,—especially.'

'Then you wrong her. If I tell you something you must be discreet.'

'Am I not always discreet?'

'She does not love Mr. Bonteen. She has had too much of him at Matching. And as for his wife, she is quite as unwilling to be kissed by her as you can be. Her Grace is determined to fight your battle for you.'

'I want her to do nothing of the kind, Madame Goesler.'

'You will know nothing about it. We have put our heads to work, and Mr. Palliser,—that is, the new Duke,—is to be made to tell Mr. Gresham that you are to have a place. It is no good you being angry, for the thing is done. If you have enemies behind your back, you must have friends behind your back also. Lady Cantrip is to do the same thing.'

'For Heaven's sake, not.'

'It's all arranged. You'll be called the ladies' pet, but you mustn't mind that. Lady Laura will be here before it's arranged, and she will get hold of Mr. Erle.'

'You are laughing at me, I know.'

'Let them laugh that win. We thought of besieging Lord Fawn through Lady Chiltern, but we are not sure that anybody cares for Lord Fawn. The man we specially want now is the other Duke. We're afraid of attacking him through the Duchess because we think that he is inhumanly indifferent to anything that his wife says to him.'

'If that kind of thing is done I shall not accept place even if it is offered me.'

'Why not? Are you going to let a man like Mr. Bonteen bowl you over? Did you ever know Lady Glen fail in anything that she attempted? She is preparing a secret with the express object of making Mr. Ratler her confidant. Lord Mount Thistle is her slave, but then I fear Lord Mount Thistle is not of much use. She'll do anything and everything,—except flatter Mr. Bonteen.'

'Heaven forbid that anybody should do that for my sake.'

'The truth is that he made himself so disagreeable at Matching that Lady Glen is broken-hearted at finding that he is to seem to owe his promotion to her husband's favour. Now you know all about it.'

'You have been very wrong to tell me.'

'Perhaps I have, Mr. Finn. But I thought it better that you should know that you have friends at work for you. We believe,—or rather, the Duchess believes,—that falsehoods have been used which are as disparaging to Lady Laura Kennedy as they are injurious to you, and she is determined to put it right. Some one has told Mr. Gresham that you have been the means of breaking the hearts both of Lord Brentford and Mr. Kennedy,—two members of the late Cabinet,—and he must be made to understand that this is untrue. If only for Lady Laura's sake you must submit.'

'Lord Brentford and I are the best friends in the world.'

'And Mr. Kennedy is a madman,—absolutely in custody of his friends, as everybody knows; and yet the story has been made to work.'

'And you do not feel that all this is derogatory to me?'

Madame Goesler was silent for a moment, and then she answered boldly, 'Not a whit. Why should it be derogatory? It is not done with the object of obtaining an improper appointment on behalf of an unimportant man. When falsehoods of that kind are told you can't meet them in a straightforward way. I suppose I know with fair accuracy the sort of connection there has been between you and Lady Laura.' Phineas very much doubted whether she had any such knowledge; but he said nothing, though the lady paused a few moments for reply. 'You can't go and tell Mr. Gresham all that; nor can any friend do so on your behalf. It would be absurd.'

'Most absurd.'

'And yet it is essential to your interests that he should know it. When your enemies are undermining you, you must countermine or you'll be blown up.'

'I'd rather fight above ground.'

'That's all very well, but your enemies won't stay above ground. Is that newspaper man above ground? And for a little job of clever mining, believe me, that there is not a better engineer going than Lady Glen;—not but what I've known her to be very nearly "hoist with her own petard,"'—added Madame Goesler, as she remembered a certain circumstance in their joint lives.

All that Madame Goesler said was true. A conspiracy had been formed, in the first place at the instance of Madame Goesler, but altogether by the influence of the young Duchess, for forcing upon the future Premier the necessity of admitting Phineas Finn into his Government. On the Wednesday following the conclusion of the debate,—the day on the morning of which the division was to take place,—there was no House. On the Thursday, the last day on which the House was to sit before the Easter holidays, Mr. Daubeny announced his intention of postponing the declaration of his intentions till after the adjournment. The House would meet, he said, on that day week, and then he would make his official statement. This communication he made very curtly, and in a manner that was thought by some to be almost insolent to the House. It was known that he had been grievously disappointed by the result of the debate,—not probably having expected a majority since his adversary's strategy had been declared, but always hoping that the deserters from his own standard would be very few. The deserters had been very many, and Mr. Daubeny was majestic in his wrath.

Nothing, however, could be done till after Easter. The Ratlers of the Liberal party were very angry at the delay, declaring that it would have been much to the advantage of the country at large that the vacation week should have been used for constructing a Liberal Cabinet. This work of construction always takes time, and delays the business of the country. No one can have known better than did Mr. Daubeny how great was the injury of delay, and how advantageously the short holiday might have been used. With a majority of seventy-two against him, there could be no reason why he should not have at once resigned, and advised the Queen to send for Mr. Gresham. Nothing could be worse than his conduct. So said the Liberals, thirsting for office. Mr. Gresham himself did not open his mouth when the announcement was made;—nor did any man, marked for future office, rise to denounce the beaten statesman. But one or two independent Members expressed their great regret at the unnecessary delay which was to take place before they were informed who was to be the Minister of the Crown. But Mr. Daubeny, as soon as he had made his statement, stalked out of the House, and no reply whatever was made to the independent Members. Some few sublime and hot-headed gentlemen muttered the word 'impeachment.' Others, who were more practical and less dignified, suggested that the Prime Minister 'ought to have his head punched.'

It thus happened that all the world went out of town that week,—so that the Duchess of Omnium was down at Matching when Phineas called at the Duke's house in Carlton Terrace on Friday. With what object he had called he hardly knew himself; but he thought that he intended to assure the Duchess that he was not a candidate for office, and that he must deprecate her interference. Luckily,—or unluckily,—he did not see her, and he felt that it would be impossible to convey his wishes in a letter. The whole subject was one which would have defied him to find words sufficiently discreet for his object.

The Duke and Duchess of St. Bungay were at Matching for the Easter,—as also was Barrington Erle, and also that dreadful Mr. Bonteen, from whose presence the poor Duchess of Omnium could in these days never altogether deliver herself. 'Duke,' she said, 'you know Mr. Finn?'

'Certainly. It was not very long ago that I was talking to him.'

'He used to be in office, you remember.'

'Oh yes;—and a very good beginner he was. Is he a friend of Your Grace's?'

'A great friend. I'll tell you what I want you to do. You must have some place found for him.'

'My dear Duchess, I never interfere.'

'Why, Duke, you've made more Cabinets than any man living.'

'I fear, indeed, that I have been at the construction of more Governments than most men. It's forty years ago since Lord Melbourne*first did me the honour of consulting me. When asked for advice, my dear, I have very often given it. It has occasionally been my duty to say that I could not myself give my slender assistance to a Ministry unless I were supported by the presence of this or that political friend. But never in my life have I asked for an appointment as a personal favour; and I am sure you won't be angry with me if I say that I cannot begin to do so now.'

'But Mr. Finn ought to be there. He did so well before.'

'If so, let us presume that he will be there. I can only say, from what little I know of him, that I shall be happy to see him in any office to which the future Prime Minister may consider it to be his duty to appoint him.' 'To think,' said the Duchess of Omnium afterwards to her friend Madame Goesler,—'to think that I should have had that stupid old woman a week in the house, and all for nothing!'

'Upon my word, Duchess,' said Barrington Erle, 'I don't know why it is, but Gresham seems to have taken a dislike to him.'

'It's Bonteen's doing.'

'Very probably.'

'Surely you can get the better of that?'

'I look upon Phineas Finn, Duchess, almost as a child of my own. He has come back to Parliament altogether at my instigation.'

'Then you ought to help him.'

'And so I would if I could. Remember I am not the man I used to be when dear old Mr. Mildmay reigned. The truth is, I never interfere now unless I'm asked.'

'I believe that every one of you is afraid of Mr. Gresham.'

'Perhaps we are.'

'I'll tell you what. If he's passed over I'll make such a row that some of you shall hear it.'

'How fond all you women are of Phineas Finn.'

'I don't care that for him,' said the Duchess, snapping her fingers—'more than I do, that is, for any other mere acquaintance. The man is very well, as most men are.'

'Not all.'

'No, not all. Some are as little and jealous as a girl in her tenth season. He is a decently good fellow, and he is to be thrown over, because——'

'Because of what?'

'I don't choose to name any one. You ought to know all about it, and I do not doubt but you do. Lady Laura Kennedy is your own cousin.'

'There is not a spark of truth in all that.'

'Of course there is not; and yet he is to be punished. I know very well, Mr. Erle, that if you choose to put your shoulder to the wheel you can manage it; and I shall expect to have it managed.'

'Plantagenet,' she said the next day to her husband, 'I want you to do something for me.'

'To do something! What am I to do? It's very seldom you want anything in my line.'

'This isn't in your line at all, and yet I want you to do it.'

'Ten to one it's beyond my means.'

'No, it isn't. I know you can if you like. I suppose you are all sure to be in office within ten days or a fortnight?'

'I can't say, my dear. I have promised Mr. Gresham to be of use to him if I can.'

'Everybody knows all that. You're going to be Privy Seal, and to work just the same as ever at those horrible two farthings.'

'And what is it you want, Glencora?'

'I want you to say that you won't take any office unless you are allowed to bring in one or two friends with you.'

'Why should I do that? I shall not doubt any Cabinet chosen by Mr. Gresham.'

'I'm not speaking of the Cabinet; I allude to men in lower offices, lords, and Under-Secretaries, and Vice-people. You know what I mean.'

'I never interfere.'

'But you must. Other men do continually. It's quite a common thing for a man to insist that one or two others should come in with him.'

'Yes. If a man feels that he cannot sustain his own position without support, he declines to join the Government without it. But that isn't my case. The friends who are necessary to me in the Cabinet are the very men who will certainly be there. I would join no Government without the Duke; but——'

'Oh, the Duke—the Duke! I hate dukes—and duchesses too. I'm not talking about a duke. I want you to oblige me by making a point with Mr. Gresham that Mr. Finn shall have an office.'

'Mr. Finn!'

'Yes, Mr. Finn. I'll explain it all if you wish it.'

'My dear Glencora, I never interfere.'

'Who does interfere? Everybody says the same. Somebody interferes, I suppose. Mr. Gresham can't know everybody so well as to be able to fit all the pegs into all the holes without saying a word to anybody.'

'He would probably speak to Mr. Bonteen.'

'Then he would speak to a very disagreeable man, and one

I'm as sick of as I ever was of any man I ever knew. If you can't manage this for me, Plantagenet, I shall take it very ill. It's a little thing, and I'm sure you could have it done. I don't very often trouble you by asking for anything.'

The Duke in his quiet way was an affectionate man, and an indulgent husband. On the following morning he was closeted with Mr. Bonteen, two private secretaries, and a leading clerk from the Treasury for four hours, during which they were endeavouring to ascertain whether the commercial world of Great Britain would be ruined or enriched if twelve pennies were declared to contain fifty farthings. The discussion had been grievously burdensome to the minds of the Duke's assistants in it, but he himself had remembered his wife through it all. 'By the way,' he said, whispering into Mr. Bonteen's private ear as he led that gentleman away to lunch, 'if we do come in——'

'Oh, we must come in.'

'If we do, I suppose something will be done for that Mr. Finn. He spoke well the other night.'

Mr. Bonteen's face became very long. 'He helped to upset the coach when he was with us before.'

'I don't think that that is much against him.'

'Is he—a personal friend of Your Grace's?'

'No—not particularly. I never care about such things for myself; but Lady Glencora——'

'I think the Duchess can hardly know what has been his conduct to poor Kennedy. There was a most disreputable row at a public-house in London, and I am told that he behaved—very badly.'

'I never heard a word about it,' said the Duke.

'I'll tell you just the truth,' said Mr. Bonteen. 'I've been asked about him, and I've been obliged to say that he would weaken any Government that would give him office.'

'Oh, indeed!'

That evening the Duke told the Duchess nearly all that he had heard, and the Duchess swore that she wasn't going to be beaten by Mr. Bonteen.

CHAPTER XXXVIII

Once Again in Portman Square

On the Wednesday in Easter week Lord Brentford and Lady Laura Kennedy reached Portman Square from Dresden, and Phineas, who had remained in town, was summoned thither by a note written at Dover. 'We arrived here to-day, and shall be in town to-morrow afternoon, between four and five. Papa wants to see you especially. Can you manage to be with us in the Square at about eight? I know it will be inconvenient, but you will put up with inconvenience. I don't like to keep Papa up late; and if he is tired he won't speak to you as he would if you came early.—L. K.' Phineas was engaged to dine with Lord Cantrip; but he wrote to excuse himself,—telling the simple truth. He had been asked to see Lord Brentford on business, and must obey the summons.

He was shown into a sitting-room on the ground floor, which he had always known as the Earl's own room, and there he found Lord Brentford alone. The last time he had been there he had come to plead with the Earl on behalf of Lord Chiltern, and the Earl had then been a stern self-willed man, vigorous from a sense of power, and very able to maintain and to express his own feelings. Now he was a broken-down old man,—whose mind had been, as it were, unbooted and put into moral slippers for the remainder of its term of existence upon earth. He half shuffled up out of his chair as Phineas came up to him, and spoke as though every calamity in the world were oppressing him. 'Such a passage! Oh, very bad, indeed! I thought it would have been the death of me. Laura thought it better to come on.' The fact, however, had been that the Earl had so many objections to staying at Calais, that his daughter had felt herself obliged to yield to him.

'You must be glad at any rate to have got home,' said Phineas.

'Home! I don't know what you call home. I don't suppose I shall ever feel any place to be home again.'

'You'll go to Saulsby;—will you not?'

'How can I tell? If Chiltern would have kept the house up, of course I should have gone there. But he never would do anything like anybody else. Violet wants me to go to that place they've got there, but I shan't do that.'

'It's a comfortable house.'

'I hate horses and dogs, and I won't go.'

There was nothing more to be said on that point. 'I hope Lady Laura is well.'

'No, she's not. How should she be well? She's anything but well. She'll be in directly, but she thought I ought to see you first. I suppose this wretched man is really mad.'

'I am told so.'

'He never was anything else since I knew him. What are we to do now? Forster says it won't look well to ask for a separation only because he's insane. He tried to shoot you?'

'And very nearly succeeded.'

'Forster says that if we do anything, all that must come out.'

'There need not be the slightest hesitation as far as I am concerned, Lord Brentford.'

'You know he keeps all her money.'

'At present I suppose he couldn't give it up.'

'Why not? Why shouldn't he give it up? God bless my soul! Forty thousand pounds and all for nothing. When he married he declared that he didn't care about it! Money was nothing to him! So she lent it to Chiltern.'

'I remember.'

'But they hadn't been together a year before he asked for it. Now there it is;—and if she were to die to-morrow it would be lost to the family. Something must be done, you know. I can't let her money go in that way.'

'You'll do what Mr. Forster suggests, no doubt.'

'But he won't suggest anything. They never do. He doesn't

care what becomes of the money. It never ought to have been given up as it was.'

'It was settled, I suppose.'

'Yes;—if there were children. And it will come back to her if he dies first. But mad people never do die. That's a well-known fact. They've nothing to trouble them, and they live for ever. It'll all go to some cousin of his that nobody ever saw.'

'Not as long as Lady Laura lives.'

'But she does not get a penny of the income;—not a penny. There never was anything so cruel. He has published all manner of accusations against her.'

'Nobody believes a word of that, my lord.'

'And then when she is dragged forward by the necessity of vindicating her character, he goes mad and keeps all her money! There never was anything so cruel since the world began.'

This continued for half-an-hour, and then Lady Laura came in. Nothing had come, or could have come, from the consultation with the Earl. Had it gone on for another hour, he would simply have continued to grumble, and have persevered in insisting upon the hardships he endured. Lady Laura was in black, and looked sad, and old, and careworn; but she did not seem to be ill. Phineas could not but think at the moment how entirely her youth had passed away from her. She came and sat close by him, and began at once to speak of the late debate. 'Of course they'll go out,' she said.

'I presume they will.'

'And our party will come in.'

'Oh, yes;—Mr. Gresham, and the two dukes, and Lord Cantrip,—with Legge Wilson, Sir Harry Coldfoot, and the rest of them.'

'And you?'

Phineas smiled, and tried to smile pleasantly, as he answered, 'I don't know that they'll put themselves out by doing very much for me.'

'They'll do something.'

'I fancy not. Indeed, Lady Laura, to tell the truth at once, I know that they don't mean to offer me anything.'

'After making you give up your place in Ireland?'

'They didn't make me give it up. I should never dream of using such an argument to any one. Of course I had to judge for myself. There is nothing to be said about it;—only it is so.' As he told her this he strove to look light-hearted, and so to speak that she should not see the depth of his disappointment;—but he failed altogether. She knew him too well not to read his whole heart in the matter.

'Who has said it?' she asked.

'Nobody says things of that kind, and yet one knows.'

'And why is it?'

'How can I say? There are various reasons,—and, perhaps, very good reasons. What I did before makes men think that they can't depend on me. At any rate it is so.'

'Shall you not speak to Mr. Gresham?'

'Certainly not.'

'What do you say, Papa?'

'How can I understand it, my dear? There used to be a kind of honour in these things, but that's all old-fashioned now. Ministers used to think of their political friends; but in these days they only regard their political enemies. If you can make a Minister afraid of you, then it becomes worth his while to buy you up. Most of the young men rise now by making themselves thoroughly disagreeable. Abuse a Minister every night for half a session, and you may be sure to be in office the other half,—if you care about it.'

'May I speak to Barrington Erle?' asked Lady Laura.

'I had rather you did not. Of course I must take it as it comes.'

'But, my dear Mr. Finn, people do make efforts in such cases. I don't doubt but that at this moment there are a dozen men moving heaven and earth to secure something. No one has more friends than you have.'

Had not her father been present he would have told her what his friends were doing for him, and how unhappy such interferences made him; but he could not explain all this before the Earl. 'I would so much rather hear about yourself,' he said, again smiling.

'There is but little to say about us. I suppose Papa has told you?'

But the Earl had told him nothing, and indeed, there was nothing to tell. The lawyer had advised that Mr. Kennedy's friends should be informed that Lady Laura now intended to live in England, and that they should be invited to make to her some statement as to Mr. Kennedy's condition. If necessary he, on her behalf, would justify her departure from her husband's roof by a reference to the outrageous conduct of which Mr. Kennedy had since been guilty. In regard to Lady Laura's fortune, Mr. Forster said that she could no doubt apply for alimony, and that if the application were pressed at law she would probably obtain it;—but he could not recommend such a step at the present moment. As to the accusation which had been made against her character, and which had become public through the malice of the editor of The People's Banner, Mr. Forster thought that the best refutation would be found in her return to England. At any rate he would advise no further step at the present moment. Should any further libel appear in the columns of the newspaper, then the question might be again considered. Mr. Forster had already been in Portman Square, and this had been the result of the conference.

'There is not much comfort in it all,—is there?' said Lady Laura.

'There is no comfort in anything,' said the Earl.

When Phineas took his leave Lady Laura followed him out into the hall, and they went together into the large, gloomy dining-room,—gloomy and silent now, but which in former days he had known to be brilliant with many lights, and cheerful with eager voices. 'I must have one word with you,' she said, standing close to him against the table, and putting her hand upon his arm. 'Amidst all my sorrow, I have been so thankful that he did not——kill you.'

'I almost wish he had.'

'Oh, Phineas!—how can you say words so wicked! Would you have had him a murderer?'

'A madman is responsible for nothing.'

'Where should I have been? What should I have done? But of course you do not mean it. You have everything in life before you. Say some word to me more comfortable than that. You cannot think how I have looked forward to meeting you again. It has robbed the last month of half its sadness.' He put his arm round her waist and pressed her to his side, but he said nothing. 'It was so good of you to go to him as you did. How was he looking?'

'Twenty years older than when you saw him last.'

'But how in health?'

'He was thin and haggard.'

'Was he pale?'

'No; flushed and red. He had not shaved himself for days; nor, as I believe, had he been out of his room since he came up to London. I fancy that he will not live long.'

'Poor fellow;—unhappy man! I was very wrong to marry him, Phineas.'

'I have never said so;—nor, indeed, thought so.'

'But I have thought so; and I say it also,—to you. I owe him any reparation that I can make him; but I could not have lived with him. I had no idea, before, that the nature of two human beings could be so unlike. I so often remember what you told me of him,—here; in this house, when I first brought you together. Alas, how sad it has been!'

'Sad, indeed.'

'But can this be true that you tell me of yourself?

'It is quite true. I could not say so before your father, but it is Mr. Bonteen's doing. There is no remedy. I am sure of that. I am only afraid that people are interfering for me in a manner that will be as disagreeable to me as it will be useless.'

'What friends?' she asked.

He was still standing with his arm round her waist, and he did not like to mention the name of Madame Goesler.

'The Duchess of Omnium,—whom you remember as Lady Glencora Palliser.'

'Is she a friend of yours?'

'No;—not particularly. But she is an indiscreet woman, and hates Bonteen, and has taken it into her stupid head to interest herself in my concerns. It is no doing of mine, and yet I cannot help it.'

'She will succeed.'

'I don't want assistance from such a quarter; and I feel sure that she will not succeed.'

'What will you do, Phineas?'

'What shall I do? Carry on the battle as long as I can without getting into debt, and then—vanish.'

'You vanished once before,—did you not,—with a wife?'

'And now I shall vanish alone. My poor little wife! It seems all like a dream. She was so good, so pure, so pretty, so loving!'

'Loving! A man's love is so easily transferred;—as easily as a woman's hand;—is it not, Phineas? Say the word, for it is what you are thinking.'

'I was thinking of no such thing.'

'You must think it—You need not be afraid to reproach me. I could bear it from you. What could I not bear from you? Oh, Phineas;—if I had only known myself then, as I do now!'

'It is too late for regrets,' he said. There was something in the words which grated on her feelings, and induced her at length to withdraw herself from his arm. Too late for regrets! She had never told herself that it was not too late. She was the wife of another man, and therefore, surely it was too late. But still the word coming from his mouth was painful to her. It seemed to signify that for him at least the game was all over.

'Yes, indeed,' she said,—'if our regrets and remorse were at our own disposal! You might as well say that it is too late for unhappiness, too late for weariness, too late for all the misery that comes from a life's disappointment.'

'I should have said that indulgence in regrets is vain.'

'That is a scrap of philosophy which I have heard so often before! But we will not quarrel, will we, on the first day of my return?'

'I hope not.'

'And I may speak to Barrington?'

'No; certainly not.'

'But I shall. How can I help it? He will be here to-morrow, and will be full of the coming changes. How should I not mention your name? He knows—not all that has passed, but too much not to be aware of my anxiety. Of course your name will come up?'

'What I request,—what I demand is, that you ask no favour for me. Your father will miss you,—will he not? I had better go now.'

'Good night, Phineas.'

'Good night, dear friend.'

'Dearest, dearest friend,' she said. Then he left her, and without assistance, let himself out into the square. In her intercourse with him there was a passion the expression of which caused him sorrow and almost dismay. He did not say so even to himself, but he felt that a time might come in which she would resent the coldness of demeanour which it would be imperative upon him to adopt in his intercourse with her. He knew how imprudent he had been to stand there with his arm round her waist.

CHAPTER XXXIX
Cagliostro

It had been settled that Parliament should meet on the Thursday in Easter week, and it was known to the world at large that Cabinet Councils were held on the Friday previous, on the Monday, and on the Tuesday; but nobody knew what took place at those meetings. Cabinet Councils are, of course, very secret. What kind of oath the members take not to divulge any tittle of the proceedings at these awful conferences, the general public does not know; but it is presumed that oaths are taken very solemn, and it is known that they are very binding. Nevertheless, it is not an uncommon thing to hear openly at the clubs an account of what has been settled; and, as we all know, not a council is held as to which the editor of The People's Banner does not inform its readers next day exactly what took place. But as to these three Cabinet Councils there was an increased mystery abroad. Statements, indeed, were made, very definite and circumstantial, but then they were various,—and directly opposed one to another. According to The People's Banner, Mr. Daubeny had resolved, with that enduring courage which was his peculiar characteristic, that he would not be overcome by faction, but would continue to exercise all the functions of Prime Minister until he had had an opportunity of learning whether his great measure had been opposed by the sense of the country, or only by the tactics of an angry and greedy party. Other journals declared that the Ministry as a whole had decided on resigning. But the clubs were in a state of agonising doubt. At the great stronghold of conservative policy in Pall Mall men were silent, embarrassed, and unhappy. The party was at heart divorced from its leaders,—and a party without leaders is powerless. To these gentlemen there could be no triumph, whether Mr. Daubeny went out or remained in office. They had been betrayed;—but as a body were unable even to accuse the traitor. As regarded most of them they had accepted the treachery and bowed their heads beneath it, by means of their votes. And as to the few who had been staunch,—they also were cowed by a feeling that they had been instrumental in destroying their own power by endeavouring to protect a doomed institution. Many a thriving county member in those days expressed a wish among his friends that he had never meddled with the affairs of public life, and hinted at the Chiltern Hundreds.* On the other side, there was undoubtedly something of a rabid desire for immediate triumph, which almost deserved that epithet of greedy which was then commonly used by Conservatives in speaking of their opponents. With the Liberal leaders,—such men as Mr. Gresham and the two dukes,—the anxiety displayed was, no doubt, on behalf of the country. It is right, according to our constitution, that the Government should be entrusted to the hands of those whom the constituencies of the country have most trusted. And, on behalf of the country, it behoves the men in whom the country has placed its trust to do battle in season and out of season,—to carry on war internecine,—till the demands of the country are obeyed. A sound political instinct had induced Mr. Gresham on this occasion to attack his opponent simply on the ground of his being the leader only of a minority in the House of Commons. But from among Mr. Gresham's friends there had arisen a noise which sounded very like a clamour for place, and this noise of course became aggravated in the ears of those who were to be displaced. Now, during Easter week, the clamour became very loud. Could it be possible that the archfiend of a Minister would dare to remain in office till the end of a hurried Session, and then again dissolve Parliament? Men talked of rows in London,—even of revolution, and there were meetings in open places both by day and night. Petitions were to be prepared, and the country was to be made to express itself.

When, however, Thursday afternoon came, Mr. Daubeny 'threw up the sponge.' Up to the last moment the course which he intended to pursue was not known to the country at large. He entered the House very slowly,—almost with a languid air, as though indifferent to its performances, and took his seat at about half-past four. Every man there felt that there was insolence in his demeanour,—and yet there was nothing on which it was possible to fasten in the way of expressed complaint. There was a faint attempt at a cheer,—for good soldiers acknowledge the importance of supporting even an unpopular general. But Mr. Daubeny's soldiers on this occasion were not very good. When he had been seated about five minutes he rose, still very languidly, and began his statement. He and his colleagues, he said, in their attempt to legislate for the good of their country had been beaten in regard to a very great measure by a large majority, and in compliance with what he acknowledged to be the expressed opinion of the House, he had considered it to be his duty—as his colleagues had considered it to be theirs—to place their joint resignations in the hands of Her Majesty. This statement was received with considerable surprise, as it was not generally known that Mr. Daubeny had as yet even seen the Queen. But the feeling most predominant in the House was one almost of dismay at the man's quiescence. He and his colleagues had resigned, and he had recommended Her Majesty to send for Mr. Gresham. He spoke in so low a voice as to be hardly audible to the House at large, and then paused,—ceasing to speak, as though his work were done. He even made some gesture, as though stepping back to his seat;—deceived by which Mr. Gresham, at the other side of the table, rose to his legs. 'Perhaps,' said Mr. Daubeny,—'Perhaps the right honourable gentleman would pardon him, and the House would pardon him, if still, for a moment, he interposed between the House and the right honourable gentleman. He could well understand the impatience of the right honourable gentleman,—who no doubt was anxious to reassume that authority among them, the temporary loss of which he had not perhaps borne with all the equanimity which might have been expected from him. He would promise the House and the right honourable gentleman that he would not detain them long.' Mr. Gresham threw himself back into his seat, evidently not without annoyance, and his enemy stood for a moment looking at him. Unless they were angels these two men must at that moment have hated each other;—and it is supposed that they were no more than human. It was afterwards said that the little ruse of pretending to resume his seat had been deliberately planned by Mr. Daubeny with the view of seducing Mr. Gresham into an act of seeming impatience, and that these words about his opponent's failing equanimity had been carefully prepared.

Mr. Daubeny stood for a minute silent, and then began to

pour forth that which was really his speech on the occasion. Those flaccid half-pronounced syllables in which he had declared that he had resigned,—had been studiously careless, purposely flaccid. It was his duty to let the House know the fact, and he did his duty. But now he had a word to say in which he himself could take some little interest. Mr. Daubeny could be fiery or flaccid as it suited himself;—and now it suited him to be fiery. He had a prophecy to make, and prophets have ever been energetic men. Mr. Daubeny conceived it to be his duty to inform the House, and through the House the country, that now, at last, had the day of ruin come upon the British Empire, because it had bowed itself to the dominion of an unscrupulous and greedy faction. It cannot be said that the language which he used was unmeasured, because no word that he uttered would have warranted the Speaker in calling him to order; but, within the very wide bounds of parliamentary etiquette, there was no limit to the reproach and reprobation which he heaped on the House of Commons for its late vote. And his audacity equalled his insolence. In announcing his resignation, he had condescended to speak of himself and his colleagues; but now he dropped his colleagues as though they were unworthy of his notice, and spoke only of his own doings,—of his own efforts to save the country, which was indeed willing to be saved, but unable to select fitting instruments of salvation. 'He had been twitted,' he said, 'with inconsistency to his principles by men who were simply unable to understand the meaning of the word Conservatism. These gentlemen seemed to think that any man who did not set himself up as an apostle of constant change must therefore be bound always to stand still and see his country perish from stagnation. It might be that there were gentlemen in that House whose timid natures could not face the dangers of any movement; but for himself he would say that no word had ever fallen from his lips which justified either his friends or his adversaries in classing him among the number. If a man be anxious to keep his fire alight, does he refuse to touch the sacred coals as in the course of nature they are consumed?

Or does he move them with the salutary poker and add fresh fuel from the basket? They all knew that enemy to the comfort of the domestic hearth, who could not keep his hands for a moment from the fire-irons. Perhaps he might be justified if he said that they had been very much troubled of late in that House by gentlemen who could not keep their fingers from poker and tongs. But there had now fallen upon them a trouble of a nature much more serious in its effects than any that had come or could come from would-be reformers. A spirit of personal ambition, a wretched thirst for office, a hankering after the power and privileges of ruling, had not only actuated men,—as, alas, had been the case since first the need for men to govern others had arisen in the world,—but had been openly avowed and put forward as an adequate and sufficient reason for opposing a measure in disapprobation of which no single argument had been used! The right honourable gentleman's proposition to the House had been simply this;—"I shall oppose this measure, be it good or bad, because I desire, myself, to be Prime Minister, and I call upon those whom I lead in politics to assist me in doing so, in order that they may share the good things on which we may thus be enabled to lay our hands!"'

Then there arose a great row in the House, and there seemed to be a doubt whether the still existing Minister of the day would be allowed to continue his statement. Mr. Gresham rose to his feet, but sat down again instantly, without having spoken a word that was audible. Two or three voices were heard calling upon the Speaker for protection. It was, however, asserted afterwards that nothing had been said which demanded the Speaker's interference. But all moderate voices were soon lost in the enraged clamour of members on each side. The insolence showered upon those who generally supported Mr. Daubeny had equalled that with which he had exasperated those opposed to him; and as the words had fallen from his lips, there had been no purpose of cheering him from the conservative benches. But noise creates noise, and shouting is a ready and easy mode of contest. For a while it seemed as though the right side of the Speaker's chair was only beaten by the majority of lungs on the left side;—and in the midst of it all Mr. Daubeny still stood, firm on his feet, till gentlemen had shouted themselves silent,—and then he resumed his speech.

The remainder of what he said was profound, prophetic, and unintelligible. The gist of it, so far as it could be understood when the bran was bolted from it, consisted in an assurance that the country had now reached that period of its life in which rapid decay was inevitable, and that, as the mortal disease had already shown itself in its worst form, national decrepitude was imminent, and natural death could not long be postponed. They who attempted to read the prophecy with accuracy were of opinion that the prophet had intimated that had the nation, even in this its crisis, consented to take him, the prophet, as its sole physician and to obey his prescription with childlike docility, health might not only have been re-established, but a new juvenescence absolutely created. The nature of the medicine that should have been taken was even supposed to have been indicated in some very vague terms. Had he been allowed to operate he would have cut the tap-roots of the national cancer, have introduced fresh blood into the national veins, and resuscitated the national digestion, and he seemed to think that the nation, as a nation, was willing enough to undergo the operation, and be treated as he should choose to treat it;—but that the incubus of Mr. Gresham, backed by an unworthy House of Commons, had prevented, and was preventing, the nation from having its own way. Therefore the nation must be destroyed. Mr. Daubeny as soon as he had completed his speech took up his hat and stalked out of the House.

It was supposed at the time that the retiring Prime Minister had intended, when he rose to his legs, not only to denounce his opponents, but also to separate himself from his own unworthy associates. Men said that he had become disgusted with politics, disappointed, and altogether demoralized by defeat, and great curiosity existed as to the steps which might be taken at the time by the party of which he had hitherto been the leader. On that evening, at any rate, nothing was done. When Mr. Daubeny was gone, Mr. Gresham rose and said that in the present temper of the House he thought it best to postpone any statement from himself. He had received Her Majesty's commands only as he had entered that House, and in obedience to those commands, he should wait upon Her Majesty early to-morrow. He hoped to be able to inform the House at the afternoon sitting, what was the nature of the commands with which Her Majesty might honour him.

'What do you think of that?' Phineas asked Mr. Monk as they left the House together.

'I think that our Chatham*of to-day is but a very poor copy of him who misbehaved a century ago.'

'Does not the whole thing distress you?'

'Not particularly. I have always felt that there has been a mistake about Mr. Daubeny. By many he has been accounted as a statesman, whereas to me he has always been a political Cagliostro. Now a conjuror is I think a very pleasant fellow to have among us, if we know that he is a conjuror;—but a conjuror who is believed to do his tricks without sleight of hand is a dangerous man. It is essential that such a one should be found out and known to be a conjuror,—and I hope that such knowledge may have been communicated to some men this afternoon.'

'He was very great,' said Ratler to Bonteen. 'Did you not think so?'

'Yes, I did,—very powerful indeed. But the party is broken up to atoms.'

'Atoms soon come together again in politics,' said Ratler. 'They can't do without him. They haven't got anybody else. I wonder what he did when he got home.'

'Had some gruel and went to bed,' said Bonteen. 'They say these scenes in the House never disturb him at home.' From which conversations it may be inferred that Mr. Monk and Messrs. Ratler and Bonteen did not agree in their ideas respecting political conjurors.

CHAPTER XL

The Prime Minister is Hard Pressed

It can never be a very easy thing to form a Ministry. The one chosen chief is readily selected. Circumstances, indeed, have probably left no choice in the matter. Every man in the country who has at all turned his thoughts that way knows very well who will be the next Prime Minister when it comes to pass that a change is imminent. In these days the occupant of the throne can have no difficulty. Mr. Gresham recommends Her Majesty to send for Mr. Daubeny, or Mr. Daubeny for Mr. Gresham,—as some ten or a dozen years since Mr. Mildmay told her to send for Lord de Terrier, or Lord de Terrier for Mr. Mildmay. The Prime Minister is elected by the nation, but the nation, except in rare cases, cannot go below that in arranging details, and the man for whom the Queen sends is burdened with the necessity of selecting his colleagues. It may be,—probably must always be the case,—that this, that, and the other colleagues are clearly indicated to his mind, but then each of these colleagues may want his own inferior coadjutors, and so the difficulty begins, increases, and at length culminates. On the present occasion it was known at the end of a week that Mr. Gresham had not filled all his offices, and that there were difficulties. It was announced that the Duke of St. Bungay could not quite agree on certain points with Mr. Gresham, and that the Duke of Omnium would do nothing without the other Duke. The Duke of St. Bungay was very powerful, as there were three or four of the old adherents of Mr. Mildmay who would join no Government unless he was with them. Sir Harry Coldfoot and Lord Plinlimmon would not accept office without the Duke. The Duke was essential, and now, though the Duke's character was essentially that of a practical man who never raised unnecessary trouble, men said that the Duke was at the bottom of it all. The Duke did not approve of Mr. Bonteen.

Mr. Gresham, so it was said, insisted on Mr. Bonteen,—appealing to the other Duke. But that other Duke, our own special Duke, Planty Pall that was, instead of standing up for Mr. Bonteen, was cold and unsympathetic. He could not join the Ministry without his friend, the Duke of St. Bungay, and as to Mr. Bonteen, he thought that perhaps a better selection might be made.

Such were the club rumours which took place as to the difficulties of the day, and, as is generally the case, they were not far from the truth. Neither of the dukes had absolutely put a veto on poor Mr. Bonteen's elevation, but they had expressed themselves dissatisfied with the appointment, and the younger duke had found himself called upon to explain that although he had been thrown much into communication with Mr. Bonteen he had never himself suggested that that gentleman should follow him at the Exchequer. This was one of the many difficulties which beset the Prime Minister elect in the performance of his arduous duty.

Lady Glencora, as people would still persist in calling her, was at the bottom of it all. She had sworn an oath inimical to Mr. Bonteen, and did not leave a stone unturned in her endeavours to accomplish it. If Phineas Finn might find acceptance, then Mr. Bonteen might be allowed to enter Elysium. A second Juno, she would allow the Romulus she hated to sit in the seats of the blessed, to be fed with nectar,* and to have his name printed in the lists of unruffled Cabinet meetings,—but only on conditions. Phineas Finn must be allowed a seat also, and a little nectar,—though it were at the second table of the gods. For this she struggled, speaking her mind boldly to this and that member of her husband's party, but she struggled in vain. She could obtain no assurance on behalf of Phineas Finn. The Duke of St. Bungay would do nothing for her. Barrington Erle had declared himself powerless. Her husband had condescended to speak to Mr. Bonteen himself, and Mr. Bonteen's insolent answer had been reported to her. Then she went sedulously to work, and before a couple of days were over she did make her husband believe that Mr. Bonteen was not fit to be Chancellor of the Exchequer. This took place before Mr. Daubeny's statement, while the Duke and Duchess of St. Bungay were still at Matching,—while Mr. Bonteen, unconscious of what was being done, was still in the House. Before the two days were over, the Duke of St. Bungay had a very low opinion of Mr. Bonteen, but was quite ignorant of any connection between that low opinion and the fortunes of Phineas Finn.

'Plantagenet, of all your men that are coming up, your Mr. Bonteen is the worst. I often think that you are going down hill, both in character and intellect, but if you go as low as that I shall prefer to cross the water, and live in America.' This she said in the presence of the two dukes.

'What has Mr. Bonteen done?' asked the elder, laughing.

'He was boasting this morning openly of whom he intended to bring with him into the Cabinet.' Truth demands that the chronicler should say that this was a positive fib. Mr. Bonteen, no doubt, had talked largely and with indiscretion, but had made no such boast as that of which the Duchess accused him. 'Mr. Gresham will get astray if he doesn't allow some one to tell him the truth.'

She did not press the matter any further then, but what she had said was not thrown away. 'Your wife is almost right about that man,' the elder Duke said to the younger.

'It's Mr. Gresham's doing,—not mine,' said the younger.

'She is right about Gresham, too,' said the elder. 'With all his immense intellect and capacity for business no man wants more looking after.'

That evening Mr. Bonteen was singled out by the Duchess for her special attention, and in the presence of all who were there assembled he made himself an ass. He could not save himself from talking about himself when he was encouraged. On this occasion he offended all those feelings of official discretion and personal reticence which had been endeared to the old duke by the lessons which he had learned from former statesmen and by the experience of his own life. To be quiet, unassuming, almost affectedly modest in any mention of himself, low-voiced, reflecting always more than he resolved, and resolving always more than he said, had been his aim. Conscious of his high rank, and thinking, no doubt, much of the advantages in public life which his birth and position had given him, still he would never have ventured to speak of his own services as necessary to any Government. That he had really been indispensable to many he must have known, but not to his closest friend would he have said so in plain language. To such a man the arrogance of Mr. Bonteen was intolerable.

There is probably more of the flavour of political aristocracy to be found still remaining among our liberal leading statesmen than among their opponents. A conservative Cabinet is, doubtless, never deficient in dukes and lords, and the sons of such; but conservative dukes and lords are recruited here and there, and as recruits, are new to the business, whereas among the old Whigs a halo of statecraft has, for ages past, so strongly pervaded and enveloped certain great families, that the power in the world of politics thus produced still remains, and is even yet efficacious in creating a feeling of exclusiveness. They say that 'misfortune makes men acquainted with strange bedfellows'. The old hereditary Whig Cabinet ministers must, no doubt, by this time have learned to feel themselves at home with strange neighbours at their elbows. But still with them something of the feeling of high blood, of rank, and of living in a park with deer about it, remains. They still entertain a pride in their Cabinets, and have, at any rate, not as yet submitted themselves to a conjuror. The Charles James Fox*element of liberality still holds its own, and the fragrance of Cavendish*is essential. With no man was this feeling stronger than with the Duke of St. Bungay, though he well knew how to keep it in abeyance,—even to the extent of self-sacrifice. Bonteens must creep into the holy places. The faces which he loved to see,—born chiefly of other faces he had loved when young,—could not cluster around the sacred table without others which were much less welcome to him. He was wise enough to know that exclusiveness did not suit the nation, though human enough to feel that it must have been pleasant to himself. There must be Bonteens;—but when any Bonteen came up, who loomed before his eyes as specially disagreeable, it seemed to him to be a duty to close the door against such a one, if it could be closed without violence. A constant, gentle pressure against the door would tend to keep down the number of the Bonteens.

'I am not sure that you are not going a little too quick in regard to Mr. Bonteen,' said the elder duke to Mr. Gresham before he had finally assented to a proposition originated by himself,—that he should sit in the Cabinet without a portfolio.

'Palliser wishes it,' said Mr. Gresham, shortly.

'He and I think that there has been some mistake about that. You suggested the appointment to him, and he felt unwilling to raise an objection without giving the matter very mature consideration. You can understand that.'

'Upon my word I thought that the selection would be peculiarly agreeable to him.' Then the duke made a suggestion. Could not some special office at the Treasury be constructed for Mr. Bonteen's acceptance, having special reference to the question of decimal coinage?'

'But how about the salary?' asked Mr. Gresham. 'I couldn't propose a new office with a salary above £2,000.'

'Couldn't we make it permanent,' suggested the duke;—'with permission to hold a seat if he can get one?'

'I fear not,' said Mr. Gresham.

'He got into a very unpleasant scrape when he was Financial Secretary,' said the Duke.

But whither would'st thou, Muse? Unmeet
For jocund lyre are themes like these.
Shalt thou the talk of Gods repeat,
Debasing by thy strains effete
*Such lofty mysteries?**

The absolute words of a conversation so lofty shall no longer be attempted, but it may be said that Mr. Gresham was too wise to treat as of no account the objections of such a one as the Duke of St. Bungay. He saw Mr. Bonteen, and he saw the other duke, and difficulties arose. Mr. Bonteen made himself very disagreeable indeed. As Mr. Bonteen had never absolutely been as yet more than a demigod, our Muse, light as she is, may venture to report that he told Mr. Ratler that 'he'd be d—— if he'd stand it. If he were to be thrown over now, he'd make such a row, and would take such care that the fat should be in the fire, that his enemies, whoever they were, should wish that they had kept their fingers off him. He knew who was doing it.' If he did not know, his guess was right. In his heart he accused the young duchess, though he mentioned her name to no one. And it was the young duchess. Then there was made an insidious proposition to Mr. Gresham,—which reached him at last through Barrington Erle,—that matters would go quieter if Phineas Finn were placed in his old office at the Colonies instead of Lord Fawn, whose name had been suggested, and for whom,—as Barrington Erle declared,—no one cared a brass farthing. Mr. Gresham, when he heard this, thought that he began to smell a rat, and was determined to be on his guard. Why should the appointment of Mr. Phineas Finn make things go easier in regard to Mr. Bonteen? There must be some woman's fingers in the pie. Now Mr. Gresham was firmly resolved that no woman's fingers should have anything to do with his pie.

How the thing went from bad to worse, it would be bootless here to tell. Neither of the two dukes absolutely refused to join the Ministry; but they were persistent in their objection to Mr. Bonteen, and were joined in it by Lord Plinlimmon and Sir Harry Coldfoot. It was in vain that Mr. Gresham urged that he had no other man ready and fit to be Chancellor of the Exchequer. That excuse could not be accepted. There was Legge Wilson, who twelve years since had been at the Treasury, and would do very well. Now Mr. Gresham had always personally hated Legge Wilson,—and had, therefore, offered him the Board of Trade. Legge Wilson had disgusted him by accepting it, and the name had already been published in connection with the office. But in the lists which had appeared towards the end of the week, no name was connected with the office of Chancellor of the Exchequer, and no office was connected with the name of Mr. Bonteen. The editor of The People's Banner, however, expressed the gratification of that journal that even Mr. Gresham had not dared to propose Mr. Phineas Finn for any place under the Crown.

At last Mr. Bonteen was absolutely told that he could not be Chancellor of the Exchequer. If he would consent to give his very valuable services to the country with the view of carrying through Parliament the great measure of decimal coinage he should be President of the Board of Trade,—but without a seat in the Cabinet. He would thus become the Right Honourable Bonteen, which, no doubt, would be a great thing for him,—and, not busy in the Cabinet, must be able to devote his time exclusively to the great measure abovenamed. What was to become of 'Trade' generally, was not specially explained; but, as we all know, there would be a Vice-President to attend to details.

The proposition very nearly broke the man's heart. With a voice stopped by agitation, with anger flashing from his eyes, almost in a convulsion of mixed feelings, he reminded his chief of what had been said about his appointment in the House. Mr. Gresham had already absolutely defended it. After that did Mr. Gresham mean to withdraw a promise that had so formally been made? But Mr. Gresham was not to be caught in that way. He had made no promise;—had not even stated to the House that such appointment was to be made. A very improper question had been asked as to a rumour,—in answering which he had been forced to justify himself by explaining that discussions respecting the office had been necessary. 'Mr. Bonteen,' said Mr. Gresham, 'no one knows better than you the difficulties of a Minister. If you can act with us I shall be very grateful to you. If you cannot, I shall regret the loss of your services.' Mr. Bonteen took twenty-four hours to consider, and was then appointed

President of the Board of Trade without a seat in the Cabinet. Mr. Legge Wilson became Chancellor of the Exchequer. When the lists were completed, no office whatever was assigned to Phineas Finn. 'I haven't done with Mr. Bonteen yet,' said the young duchess to her friend Madame Goesler.

The secrets of the world are very marvellous, but they are not themselves half so wonderful as the way in which they become known to the world. There could be no doubt that Mr. Bonteen's high ambition had foundered, and that he had been degraded through the secret enmity of the Duchess of Omnium. It was equally certain that his secret enmity to Phineas Finn had brought this punishment on his head. But before the Ministry had been a week in office almost everybody knew that it was so. The rumours were full of falsehood, but yet they contained the truth. The duchess had done it. The duchess was the bosom friend of Lady Laura Kennedy, who was in love with Phineas Finn. She had gone on her knees to Mr. Gresham to get a place for her friend's favourite, and Mr. Gresham had refused. Consequently, at her bidding, half-a-dozen embryo Ministers—her husband among the number—had refused to be amenable to Mr. Gresham. Mr. Gresham had at last consented to sacrifice Mr. Bonteen, who had originally instigated him to reject the claims of Phineas Finn. That the degradation of the one man had been caused by the exclusion of the other all the world knew.

'It shuts the door to me for ever and ever,' said Phineas to Madame Goesler.

'I don't see that.'

'Of course it does. Such an affair places a mark against a man's name which will never be forgotten.'

'Is your heart set upon holding some trifling appointment under a Minister?'

'To tell you the truth, it is;—or rather it was. The prospect of office to me was more than perhaps to any other expectant. Even this man, Bonteen, has some fortune of his own, and can live if he be excluded. I have given up everything for the chance of something in this line.'

'Other lines are open.'

'Not to me, Madame Goesler. I do not mean to defend myself. I have been very foolish, very sanguine, and am now very unhappy.'

'What shall I say to you?'

'The truth.'

'In truth, then, I do not sympathise with you. The thing lost is too small, too mean to justify unhappiness.'

'But, Madame Goesler, you are a rich woman.'

'Well?'

'If you were to lose it all, would you not be unhappy? It has been my ambition to live here in London as one of a special set which dominates all other sets in our English world. To do so a man should have means of his own. I have none; and yet I have tried it,—thinking that I could earn my bread at it as men do at other professions. I acknowledge that I should not have thought so. No man should attempt what I have attempted without means, at any rate to live on if he fail; but I am not the less unhappy because I have been silly.'

'What will you do?'

'Ah,—what? Another friend asked me that the other day, and I told her that I should vanish.'

'Who was that friend?'

'Lady Laura.'

'She is in London again now?'

'Yes; she and her father are in Portman Square.'

'She has been an injurious friend to you.'

'No, by heaven,' exclaimed Phineas. 'But for her I should never have been here at all, never have had a seat in Parliament, never have been in office, never have known you.'

'And might have been the better without any of these things.'

'No man ever had a better friend than Lady Laura has been to me. Malice, wicked and false as the devil, has lately joined our names together to the incredible injury of both of us; but it has not been her fault.'

'You are energetic in defending her.'

'And so would she be in defending me. Circumstances threw us together and made us friends. Her father and her brother were my friends. I happened to be of service to her husband. We belonged to the same party. And therefore—because she has been unfortunate in her marriage—people tell lies of her.'

'It is a pity he should—not die, and leave her,' said Madame Goesler slowly.

'Why so?'

'Because then you might justify yourself in defending her by making her your wife.' She paused, but he made no answer to this. 'You are in love with her,' she said.

'It is untrue.'

'Mr. Finn!'

'Well, what would you have? I am not in love with her. To me she is no more than my sister. Were she as free as air I should not ask her to be my wife. Can a man and woman feel no friendship without being in love with each other?'

'I hope they may,' said Madame Goesler. Had he been lynx-eyed he might have seen that she blushed; but it required quick eyes to discover a blush on Madame Goesler's face. 'You and I are friends.'

'Indeed we are,' he said, grasping her hand as he took his leave.

PHINEAS REDUX

VOLUME II

CHAPTER XLI

'I hope I'm not Distrusted'

GERARD MAULE, as the reader has been informed, wrote three lines to his dearest Adelaide to inform her that his father would not assent to the suggestion respecting Maule Abbey which had been made by Lady Chiltern, and then took no further steps in the matter. In the fortnight next after the receipt of his letter nothing was heard of him at Harrington Hall, and Adelaide, though she made no complaint, was unhappy. Then came the letter from Mr. Spooner,—with all its rich offers, and Adelaide's mind was for a while occupied with wrath against her second suitor. But as the egregious folly of Mr. Spooner,—for to her thinking the aspirations of Mr. Spooner were egregiously foolish,—died out of her mind, her thoughts reverted to her engagement. Why did not the man come to her, or why did he not write?

She had received from Lady Chiltern an invitation to remain with them,—the Chilterns,—till her marriage. 'But, dear Lady Chiltern, who knows when it will be?' Adelaide had said. Lady Chiltern had good-naturedly replied that the longer it was put off the better for herself. 'But you'll be going to London or abroad before that day comes.' Lady Chiltern declared that she looked forward to no festivities which could under any circumstances remove her four-and-twenty hours travelling distance from the kennels. Probably she might go up to London for a couple of months as soon as the hunting was over, and the hounds had been drafted, and the horses had been coddled, and every covert had been visited. From the month of May till the middle of July she might, perhaps, be allowed to be in town, as communications by telegram could now be made day and night. After that, preparations for cub-hunting would be imminent, and, as a matter of course, it would be necessary that she should be at Harrington Hall at so important a period of the year. During those couple of months she would be very happy to have the companionship of her friend, and she hinted that Gerard Maule would certainly be in town. 'I begin to think it would have been better that I should never have seen Gerard Maule,' said Adelaide Palliser.

This happened about the middle of March, while hunting was still in force. Gerard's horses were standing in the neighbourhood, but Gerard himself was not there. Mr. Spooner, since that short, disheartening note had been sent to him by Lord Chiltern, had not been seen at Harrington. There was a Harrington Lawn Meet on one occasion, but he had not appeared till the hounds were at the neighbouring covert side. Nevertheless he had declared that he did not intend to give up the pursuit, and had even muttered something of the sort to Lord Chiltern. 'I am one of those fellows who stick to a thing, you know,' he said.

'I am afraid you had better give up sticking to her, because she's going to marry somebody else.'

'I've heard all about that, my lord. He's a very nice sort of young man, but I'm told he hasn't got his house ready yet for a family.' All which Lord Chiltern repeated to his wife. Neither of them spoke to Adelaide again about Mr. Spooner; but this did cause a feeling in Lady Chiltern's mind that perhaps this engagement with young Maule was a foolish thing, and that, if so, she was in a great measure responsible for the folly.

'Don't you think you'd better write to him?' she said, one morning.

'Why does he not write to me?'

'But he did,—when he wrote you that his father would not consent to give up the house. You did not answer him then.'

'It was two lines,—without a date. I don't even know where he lives.'

'You know his club?'

'Yes,—I know his club. I do feel, Lady Chiltern, that I have become engaged to marry a man as to whom I am altogether in the dark. I don't like writing to him at his club.'

'You have seen more of him here and in Italy than most girls see of their future husbands.'

'So I have,—but I have seen no one belonging to him. Don't you understand what I mean? I feel all at sea about him. I am sure he does not mean any harm.'

'Certainly he does not.'

'But then he hardly means any good.'

'I never saw a man more earnestly in love,' said Lady Chiltern.

'Oh yes,—he's quite enough in love. But——'

'But what?'

'He'll just remain up in London thinking about it, and never tell himself that there's anything to be done. And then, down here, what is my best hope? Not that he'll come to see me, but that he'll come to see his horse, and that so, perhaps, I may get a word with him.' Then Lady Chiltern suggested, with a laugh, that perhaps it might have been better that she should have accepted Mr. Spooner. There would have been no doubt as to Mr. Spooner's energy and purpose. 'Only that if there was not another man in the world I wouldn't marry him, and that I never saw any other man except Gerard Maule whom I even fancied I could marry.'

About a fortnight after this, when the hunting was all over, in the beginning of April, she did write to him as follows, and did direct her letter to his club. In the meantime Lord Chiltern had intimated to his wife that if Gerard Maule behaved badly he should consider himself to be standing in the place of Adelaide's father or brother. His wife pointed out to him that were he her father or her brother he could do nothing,—that in these days let a man behave ever so badly, no means of punishing was within reach of the lady's friends. But Lord Chiltern would not assent to this. He muttered something about a horsewhip, and seemed to suggest that one man could, if he were so minded, always have it out with another, if not in this way, then in that. Lady Chiltern protested, and declared that horsewhips could not under any circumstances be efficacious. 'He had better mind what he is about,' said Lord Chiltern. It was after this that Adelaide wrote her letter:—

'Harrington Hall, 5th April.

'DEAR GERARD,—

'I have been thinking that I should hear from you, and have been surprised,—I may say unhappy,—because I have not done so. Perhaps you thought I ought to have answered the three words which you wrote to me about your father; if so, I will apologise; only they did not seem to give me anything to say. I was very sorry that your father should have "cut up rough," as you call it, but you must remember that we both expected that he would refuse, and that we are only therefore where we thought we should be. I suppose we shall have to wait till Providence does something for us,—only, if so, it would be pleasanter to me to hear your own opinion about it.

'The Chilterns are surprised that you shouldn't have come back, and seen the end of the season. There were some very good runs just at last;—particularly one on last Monday. But on Wednesday Trumpeton Wood was again blank, and there was some row about wires. I can't explain it all; but you must come, and Lord Chiltern will tell you. I have gone down to see the horses ever so often;—but I don't care to go now as you never write to me. They are all three quite well, and Fan looks as silken and as soft as any lady need do.

'Lady Chiltern has been kinder than I can tell you. I go up to town with her in May, and shall remain with her while she is there. So far I have decided. After that my future home must, sir, depend on the resolution and determination, or perhaps on the vagaries and caprices, of him who is to be my future master. Joking apart, I must know to what I am to look forward before I can make up my mind whether I will or will not go back to Italy towards the end of the summer. If I do, I fear I must do so just in the hottest time of the year; but I shall not like to come down here again after leaving London. —unless something by that time has been settled.

'I shall send this to your club, and I hope that it will reach you. I suppose that you are in London.

'Good-bye, dearest Gerard.

'Yours most affectionately,

'ADELAIDE.

'If there is anything that troubles you, pray tell me. I ask you because I think it would be better for you that I should know. I sometimes think that you would have written if there had not been some misfortune. God bless you.'

Gerard was in London, and sent the following note by return of post:—

'————Club, Tuesday.

'DEAREST ADELAIDE,

'All right. If Chiltern can take me for a couple of nights, I'll come down next week, and settle about the horses, and will arrange everything.

'Ever your own, with all my heart,

'G. M.'

'He will settle about his horses, and arrange everything,' said Adelaide, as she showed the letter to Lady Chiltern. 'The horses first, and everything afterwards. The everything, of course, includes all my future happiness, the day of my marriage, whether to-morrow or in ten years' time, and the place where we shall live.'

'At any rate, he's coming.'

'Yes;—but when? He says next week, but he does not name any day. Did you ever hear or see anything so unsatisfactory.'

'I thought you would be glad to see him.'

'So I should be,—if there was any sense in him. I shall be glad, and shall kiss him.'

'I dare say you will.'

'And let him put his arm round my waist and be happy. He will be happy because he will think of nothing beyond. But what is to be the end of it?'

'He says that he will settle everything.'

'But he will have thought of nothing. What must I settle? That is the question. When he was told to go to his father, he went to his father. When he failed there the work was done, and the trouble was off his mind. I know him so well.'

'If you think so ill of him why did you consent to get into his boat?' said Lady Chiltern, seriously.

'I don't think ill of him. Why do you say that I think ill of him? I think better of him than of anybody else in the world;—but I know his fault, and, as it happens, it is a fault so very prejudicial to my happiness. You ask me why I got into his boat. Why does any girl get into a man's boat? Why did you get into Lord Chiltern's?'

'I promised to marry him when I was seven years old;—so he says.'

'But you wouldn't have done it, if you hadn't had a sort of feeling that you were born to be his wife. I haven't got into this man's boat yet; but I never can be happy unless I do, simply because——'

'You love him.'

'Yes;—just that. I have a feeling that I should like to be in his boat, and I shouldn't like to be anywhere else. After you have come to feel like that about a man I don't suppose it makes any difference whether you think him perfect or imperfect. He's just my own,—at least I hope so;—the one thing that I've got. If I wear a stuff* frock, I'm not going to despise it because it's not silk.'

'Mr. Spooner would be the stuff frock.'

'No;—Mr. Spooner is shoddy, and very bad shoddy, too.'

On the Saturday in the following week Gerard Maule did arrive at Harrington Hall,—and was welcomed as only accepted lovers are welcomed. Not a word of reproach was uttered as to his delinquencies. No doubt he got the kiss with which Adelaide had herself suggested that his coming would be rewarded. He was allowed to stand on the rug before the fire with his arm round her waist. Lady Chiltern smiled on him. His horses had been specially visited that morning, and a lively report as to their condition was made to him. Not a word was said on that occasion which could distress him. Even Lord Chiltern when he came in was gracious to him. 'Well, old fellow,' he said, 'you've missed your hunting.'

'Yes; indeed. Things kept me in town.'

'We had some uncommonly good runs.'

'Have the horses stood pretty well?' asked Gerard.

'I felt uncommonly tempted to borrow yours; and should have done so once or twice if I hadn't known that I should have been betrayed.'

'I wish you had, with all my heart,' said Gerard. And then they went to dress for dinner.

In the evening, when the ladies had gone to bed, Lord Chiltern took his friend off to the smoking-room. At Harrington Hall it was not unusual for the ladies and gentlemen to descend together into the very comfortable Pandemonium* which was so called, when,—as was the case at present,—the terms of intimacy between them were sufficient to warrant such a proceeding. But on this occasion Lady Chiltern went very discreetly upstairs, and Adelaide, with equal discretion, followed her. It had been arranged beforehand that Lord Chiltern should say a salutary word or two to the young man. Maule began about the hunting, asking questions about this and that, but his host stopped him at once. Lord Chiltern, when he had a task on hand, was always inclined to get through it at once,—perhaps with an energy that was too sudden in its effects. 'Maule,' he said, 'you ought to make up your mind what you mean to do about that girl.'

'Do about her! How?'

'You and she are engaged, I suppose?'

'Of course we are. There isn't any doubt about it.'

'Just so. But when things come to be like that, all delays are good fun to the man, but they're the very devil to the girl.'

'I thought it was always the other way up, and that girls wanted delay?'

'That's only a theoretical delicacy which never means much. When a girl is engaged she likes to have the day fixed. When there's a long interval the man can do pretty much as he pleases, while the girl can do nothing except think about him. Then it sometimes turns out that when he's wanted, he's not there.'

'I hope I'm not distrusted,' said Gerard, with an air that showed that he was almost disposed to be offended.

'Not in the least. The women here think you the finest paladin* in the world, and Miss Palliser would fly at my throat if she thought that I said a word against you. But she's in my house, you see; and I'm bound to do exactly as I should if she were my sister.'

'And if she were your sister?'

'I should tell you that I couldn't approve of the engagement unless you were prepared to fix the time of your marriage. And I should ask you where you intended to live.'

'Wherever she pleases. I can't go to Maule Abbey while my father lives, without his sanction.'

'And he may live for the next twenty years.'

'Or thirty.'

'Then you are bound to decide upon something else. It's no use saying that you leave it to her. You can't leave it to her. What I mean is this, that now you are here, I think you are bound to settle something with her. Good-night, old fellow.'

CHAPTER XLII

Boulogne

GERARD MAULE, as he sat upstairs half undressed in his bedroom that night didn't like it. He hardly knew what it was that he did not like,—but he felt that there was something wrong. He thought that Lord Chiltern had not been warranted in speaking to him with a tone of authority, and in talking of a brother's position,—and the rest of it. He had lacked the presence of mind for saying anything at the moment; but he must say something sooner or later. He wasn't going to be driven by Lord Chiltern. When he looked back at his own conduct he thought that it had been more than noble,—almost romantic. He had fallen in love with Miss Palliser, and spoken his love out freely, without any reference to money. He didn't know what more any fellow could have done. As to his marrying out of hand, the day after his engagement, as a man of fortune can do, everybody must have known that that was out of the question. Adelaide of course had known it. It had been suggested to him that he should consult his father as to living at Maule Abbey. Now if there was one thing he hated more than another, it was consulting his father; and yet he had done it. He had asked for a loan of the old house in perfect faith, and it was not his fault that it had been refused. He could not make a house to live in, nor could he coin a fortune. He had £800 a-year of his own, but of course he owed a little money. Men with such incomes always do owe a little money. It was almost impossible that he should marry quite at once. It was not his fault that Adelaide had no fortune of her own. When he fell in love with her he had been a great deal too generous to think of fortune, and that ought to be remembered now to his credit. Such was the sum of his thoughts, and his anger spread itself from Lord Chiltern even on to Adelaide herself. Chiltern would hardly have spoken in that way unless she had complained. She, no doubt, had been speaking to Lady Chiltern, and Lady

Chiltern had passed it on to her husband. He would have it out with Adelaide on the next morning,—quite decidedly. And he would make Lord Chiltern understand that he would not endure interference. He was quite ready to leave Harrington Hall at a moment's notice if he were ill-treated. This was the humour in which Gerard Maule put himself to bed that night.

On the following morning he was very late at breakfast,—so late that Lord Chiltern had gone over to the kennels. As he was dressing he had resolved that it would be fitting that he should speak again to his host before he said anything to Adelaide that might appear to impute blame to her. He would ask Chiltern whether anything was meant by what had been said over-night. But, as it happened, Adelaide had been left alone to pour out his tea for him, and,—as the reader will understand to have been certain on such an occasion,—they were left together for an hour in the breakfast parlour. It was impossible that such an hour should be passed without some reference to the grievance which was lying heavy on his heart. 'Late; I should think you are,' said Adelaide laughing. 'It is nearly eleven. Lord Chiltern has been out an hour. I suppose you never get up early except for hunting.'

'People always think it is so wonderfully virtuous to get up. What's the use of it?'

'Your breakfast is so cold.'

'I don't care about that. I suppose they can boil me an egg. I was very seedy when I went to bed.'

'You smoked too many cigars, sir.'

'No, I didn't; but Chiltern was saying things that I didn't like.' Adelaide's face at once became very serious. 'Yes, a good deal of sugar, please. I don't care about toast, and anything does for me. He has gone to the kennels, has he?'

'He said he should. What was he saying last night?'

'Nothing particular. He has a way of blowing up, you know; and he looks at one just as if he expected that everybody was to do just what he chooses.'

'You didn't quarrel.'

'Not at all; I went off to bed without saying a word. I hate jaws. I shall just put it right this morning; that's all.'

'Was it about me, Gerard?'

'It doesn't signify the least.'

'But it does signify. If you and he were to quarrel would it not signify to me very much? How could I stay here with them, or go up to London with them, if you and he had really quarrelled? You must tell me. I know that it was about me.' Then she came and sat close to him. 'Gerard,' she continued, 'I don't think you understand how much everything is to me that concerns you.'

When he began to reflect, he could not quite recollect what it was that Lord Chiltern had said to him. He did remember that something had been suggested about a brother and sister which had implied that Adelaide might want protection, but there was nothing unnatural or other than kind in the position which Lord Chiltern had declared that he would assume. 'He seemed to think that I wasn't treating you well,' said he, turning round from the breakfast-table to the fire, 'and that is a sort of thing I can't stand.'

'I have never said so, Gerard.'

'I don't know what it is that he expects, or why he should interfere at all. I can't bear to be interfered with. What does he know about it? He has had somebody to pay everything for him half-a-dozen times, but I have to look out for myself.'

'What does all this mean?'

'You would ask me, you know. I am bothered out of my life by ever so many things, and now he comes and adds his botheration.'

'What bothers you, Gerard? If anything bothers you, surely you will tell me. If there has been anything to trouble you since you saw your father why have you not written and told me? Is your trouble about me?'

'Well, of course it is, in a sort of way.'

'I will not be a trouble to you.'

'Now you are going to misunderstand me! Of course, you

are not a trouble to me. You know that I love you better than anything in the world.'

'I hope so.'

'Of course I do.' Then he put his arm round her waist and pressed her to his bosom. 'But what can a man do? When Lady Chiltern recommended that I should go to my father and tell him, I did it. I knew that no good could come of it. He wouldn't lift his hand to do anything for me.'

'How horrid that is!'

'He thinks it a shame that I should have my uncle's money, though he never had any more right to it than that man out there. He is always saying that I am better off than he is.'

'I suppose you are.'

'I am very badly off, I know that. People seem to think that £800 is ever so much, but I find it to be very little.'

'And it will be much less if you are married,' said Adelaide gravely.

'Of course, everything must be changed. I must sell my horses, and we must cut and run, and go and live at Boulogne, I suppose. But a man can't do that kind of thing all in a moment. Then Chiltern comes and talks as though he were Virtue personified. What business is it of his?'

Then Adelaide became still more grave. She had now removed herself from his embrace, and was standing a little apart from him on the rug. She did not answer him at first; and when she did so, she spoke very slowly. 'We have been rash, I fear; and have done what we have done without sufficient thought.'

'I don't say that at all.'

'But I do. It does seem now that we have been imprudent.' Then she smiled as she completed her speech. 'There had better be no engagement between us.'

'Why do you say that?'

'Because it is quite clear that it has been a trouble to you rather than a happiness.'

'I wouldn't give it up for all the world.'

'But it will be better. I had not thought about it as I should have done. I did not understand that the prospect of marrying would make you—so very poor. I see it now. You had better tell Lord Chiltern that it is—done with, and I will tell her the same. It will be better; and I will go back to Italy at once.'

'Certainly not. It is not done with, and it shall not be done with.'

'Do you think I will marry the man I love when he tells me that by—marrying—me, he will be—banished to—Bou—logne? You had better see Lord Chiltern; indeed you had.' And then she walked out of the room.

Then came upon him at once a feeling that he had behaved badly; and yet he had been so generous, so full of intentions to be devoted and true! He had never for a moment thought of breaking off the match, and would not think of it now. He loved her better than ever, and would live only with the intention of making her his wife. But he certainly should not have talked to her of his poverty, nor should he have mentioned Boulogne. And yet what should he have done? She would cross-question him about Lord Chiltern, and it was so essentially necessary that he should make her understand his real condition. It had all come from that man's unjustifiable interference,—as he would at once go and tell him. Of course he would marry Adelaide, but the marriage must be delayed. Everybody waits twelve months before they are married; and why should she not wait? He was miserable because he knew that he had made her unhappy;—but the fault had been with Lord Chiltern. He would speak his mind frankly to Chiltern, and then would explain with loving tenderness to his Adelaide that they would still be all in all to each other, but that a short year must elapse before he could put his house in order for her. After that he would sell his horses. That resolve was in itself so great that he did not think it necessary at the present moment to invent any more plans for the future. So he went out into the hall, took his hat, and marched off to the kennels.

At the kennels he found Lord Chiltern surrounded by the denizens of the hunt. His huntsman, with the kennelman and feeder, and two whips, and old Doggett were all there, and the Master of the Hounds was in the middle of his business. The dogs were divided by ages, as well as by sex, and were being brought out and examined. Old Doggett was giving advice,—differing almost always from Cox, the huntsman, as to the propriety of keeping this hound or of cashiering that. Nose, pace, strength, and docility were all questioned with an eagerness hardly known in any other business; and on each question Lord Chiltern listened to everybody, and then decided with a single word. When he had once resolved, nothing further urged by any man then could avail anything. Jove never was so autocratic, and certainly never so much in earnest. From the look of Lord Chiltern's brow it almost seemed as though this weight of empire must be too much for any mere man. Very little notice was taken of Gerard Maule when he joined the conclave, though it was felt in reference to him that he was sufficiently staunch a friend to the hunt to be trusted with the secrets of the kennel. Lord Chiltern merely muttered some words of greeting, and Cox lifted the old hunting-cap which he wore. For another hour the conference was held. Those who have attended such meetings know well that a morning on the flags is apt to be a long affair. Old Doggett, who had privileges, smoked a pipe, and Gerard Maule lit one cigar after another. But Lord Chiltern had become too thorough a man of business to smoke when so employed. At last the last order was given,—Doggett snarled his last snarl,—and Cox uttered his last 'My lord.' Then Gerard Maule and the Master left the hounds and walked home together.

The affair had been so long that Gerard had almost forgotten his grievance. But now as they got out together upon the park, he remembered the tone of Adelaide's voice as she left him, and remembered also that, as matters stood at present, it was essentially necessary that something should be said. 'I suppose I shall have to go and see that woman,' said Lord Chiltern.

'Do you mean Adelaide?' asked Maule, in a tone of infinite surprise.

'I mean this new Duchess, who I'm told is to manage everything herself. That man Fothergill is going on with just the old game at Trumpeton.'

'Is he, indeed? I was thinking of something else just at that moment. You remember what you were saying about Miss Palliser last night.'

'Yes.'

'Well;—I don't think, you know, you had a right to speak as you did.'

Lord Chiltern almost flew at his companion, as he replied, 'I said nothing. I do say that when a man becomes engaged to a girl, he should let her hear from him, so that they may know what each other is about.'

'You hinted something about being her brother.'

'Of course I did. If you mean well by her, as I hope you do, it can't fret you to think that she has got somebody to look after her till you come in and take possession. It is the commonest thing in the world when a girl is left all alone as she is.'

'You seemed to make out that I wasn't treating her well.'

'I said nothing of the kind, Maule; but if you ask me——'

'I don't ask you anything.'

'Yes, you do. You come and find fault with me for speaking last night in the most good-natured way in the world. And, therefore, I tell you now that you will be behaving very badly indeed, unless you make some arrangement at once as to what you mean to do.'

'That's your opinion,' said Gerard Maule.

'Yes, it is; and you'll find it to be the opinion of any man or woman that you may ask who knows anything about such things. And I'll tell you what, Master Maule, if you think you're going to face me down you'll find yourself mistaken. Stop a moment, and just listen to me. You haven't a much better friend than I am, and I'm sure she hasn't a better friend than my wife. All this has taken place under our roof, and I mean to speak my mind plainly. What do you propose to do about your marriage?'

'I don't propose to tell you what I mean to do.'

'Will you tell Miss Palliser,—or my wife?'

'That is just as I may think fit.'

'Then I must tell you that you cannot meet her at my house.'

'I'll leave it to-day.'

'You needn't do that either. You sleep on it, and then make up your mind. You can't suppose that I have any curiosity about it. The girl is fond of you, and I suppose that you are fond of her. Don't quarrel for nothing. If I have offended you, speak to Lady Chiltern about it.'

'Very well;—I will speak to Lady Chiltern.'

When they reached the house it was clear that something was wrong. Miss Palliser was not seen again before dinner, and Lady Chiltern was grave and very cold in her manner to Gerard Maule. He was left alone all the afternoon, which he passed with his horses and groom, smoking more cigars,—but thinking all the time of Adelaide Palliser's last words, of Lord Chiltern's frown, and of Lady Chiltern's manner to him. When he came into the drawing-room before dinner, Lady Chiltern and Adelaide were both there, and Adelaide immediately began to ask questions about the kennel and the huntsmen. But she studiously kept at a distance from him, and he himself felt that it would be impossible to resume at present the footing on which he stood with them both on the previous evening. Presently Lord Chiltern came in, and another man and his wife who had come to stay at Harrington. Nothing could be more dull than the whole evening. At least so Gerard found it. He did take Adelaide in to dinner, but he did not sit next to her at table, for which, however, there was an excuse, as, had he done so, the new-comer must have been placed by his wife. He was cross, and would not make an attempt to speak to his neighbour, and, though he tried once or twice to talk to Lady Chiltern—than whom, as a rule, no woman was ever more easy in conversation—he failed altogether. Now and again he strove to catch Adelaide's eye, but even in that he could not succeed. When the ladies left the room Chiltern and the new-comer—who was not a sporting man, and therefore did not understand the question—became lost in the mazes of Trumpeton Wood. But Gerard Maule did not put in a word; nor was a word addressed to him by Lord Chiltern. As he sat there sipping his wine, he made up his mind that he would leave Harrington Hall the next morning. When he was again in the drawing-room, things were conducted in just the same way. He spoke to Adelaide, and she answered him; but there was no word of encouragement—not a tone of comfort in her voice. He found himself driven to attempt conversation with the strange lady, and at last was made to play whist with Lady Chiltern and the two new-comers. Later on in the evening, when Adelaide had gone to her own chamber, he was invited by Lady Chiltern into her own sitting-room upstairs, and there the whole thing was explained to him. Miss Palliser had declared that the match should be broken off.

'Do you mean altogether, Lady Chiltern?'

'Certainly I do. Such a resolve cannot be a half-and-half arrangement.'

'But why?'

'I think you must know why, Mr. Maule.'

'I don't in the least. I won't have it broken off. I have as much right to have a voice in the matter as she has, and I don't in the least believe it's her doing.'

'Mr. Maule!'

'I do not care; I must speak out. Why does she not tell me so herself?'

'She did tell you so.'

'No, she didn't. She said something, but not that. I don't suppose a man was ever so used before; and it's all Lord Chiltern;—just because I told him that he had no right to interfere with me. And he has no right.'

'You and Oswald were away together when she told me that she had made up her mind. Oswald has hardly spoken to her since you have been in the house. He certainly has not spoken to her about you since you came to us.'

'What is the meaning of it, then?'

'You told her that your engagement had overwhelmed you with troubles.'

'Of course; there must be troubles.'

'And that——you would have to be banished to Boulogne when you were married.'

'I didn't mean her to take that literally.'

'It wasn't a nice way, Mr. Maule, to speak of your future life to the girl to whom you were engaged. Of course it was her hope to make your life happier, not less happy. And when you made her understand—as you did very plainly—that your married prospects filled you with dismay, of course she had no other alternative but to retreat from her engagement.'

'I wasn't dismayed.'

'It is not my doing, Mr. Maule.'

'I suppose she'll see me?'

'If you insist upon it she will; but she would rather not.'

Gerard, however, did insist, and Adelaide was brought to him there into that room before he went to bed. She was very gentle with him, and spoke to him in a tone very different from that which Lady Chiltern had used; but he found himself utterly powerless to change her. That unfortunate allusion to a miserable exile at Boulogne had completed the work which the former plaints had commenced, and had driven her to a resolution to separate herself from him altogether.

'Mr. Maule,' she said, 'when I perceived that our proposed marriage was looked upon by you as a misfortune, I could do nothing but put an end to our engagement.'

'But I didn't think it a misfortune.'

'You made me think that it would be unfortunate for you, and that is quite as strong a reason. I hope we shall part as friends.'

'I won't part at all,' he said, standing his ground with his back to the fire. 'I don't understand it, by heaven I don't. Because I said some stupid thing about Boulogne, all in joke——'

'It was not in joke when you said that troubles had come heavy on you since you were engaged.'

'A man may be allowed to know, himself, whether he was in joke or not. I suppose the truth is you don't care about me?'

'I hope, Mr. Maule, that in time it may come—not quite to that.'

'I think that you are—using me very badly. I think that you are—behaving—falsely to me. I think that I am—very—shamefully treated—among you. Of course I shall go. Of course I shall not stay in this house. A man can't make a girl keep her promise. No—I won't shake hands. I won't even say good-bye to you. Of course I shall go.' So saying he slammed the door behind him.

'If he cares for you he'll come back to you,' Lady Chiltern said to Adelaide that night, who at the moment was lying on her bed in a sad condition, frantic with headache.

'I don't want him to come back; I will never make him go to Boulogne.'

'Don't think of it, dear.'

'Not think of it! how can I help thinking of it? I shall always think of it. But I never want to see him again—never! How can I want to marry a man who tells me that I shall be a trouble to him? He shall never,—never have to go to Boulogne for me.'

CHAPTER XLIII
The Second Thunderbolt

THE quarrel between Phineas Finn and Mr. Bonteen had now become the talk of the town, and had taken many various phases. The political phase, though it was perhaps the best understood, was not the most engrossing. There was the personal phase,—which had reference to the direct altercation that had taken place between the two gentlemen, and to the correspondence between them which had followed, as to which phase it may be said that though there were many rumours abroad, very little was known. It was reported in some circles that the two aspirants for office had been within an ace of striking each other; in some, again, that a blow had passed,—and in others, further removed probably from the House of Commons and the Universe Club, that the Irishman had struck the Englishman, and that the Englishman had given the Irishman a thrashing. This was a phase that was very disagreeable to Phineas Finn. And there was a third,—which may perhaps be called the general social phase, and which unfortunately dealt with the name of Lady Laura Kennedy. They all, of course, worked into each other, and were enlivened and made interesting with the names of a great many big persons. Mr. Gresham, the Prime Minister, was supposed to be very much concerned in this matter. He, it was said, had found himself compelled to exclude Phineas Finn from the Government, because of the unfortunate alliance between him and the wife of one of his late colleagues, and had also thought it expedient to dismiss Mr. Bonteen from his Cabinet,—for it had amounted almost to dismissal,—because Mr. Bonteen had made indiscreet official allusion to that alliance. In consequence of this working in of the first and third phase, Mr. Gresham encountered hard usage from some friends and from many enemies. Then, of course, the scene at Macpherson's Hotel was commented on very generally. An idea prevailed that Mr. Kennedy, driven to madness by his wife's infidelity, which had become known to him through the quarrel between Phineas and Mr. Bonteen,—had endeavoured to murder his wife's lover, who had with the utmost effrontery invaded the injured husband's presence with a view of deterring him by threats from a publication of his wrongs. This murder had been nearly accomplished in the centre of the metropolis,—by daylight, as if that made it worse,—on a Sunday, which added infinitely to the delightful horror of the catastrophe; and yet no public notice had been taken of it! The would-be murderer had been a Cabinet Minister, and the lover who was so nearly murdered had been an Under-Secretary of State, and was even now a member of Parliament. And then it was positively known that the lady's father, who had always been held in the highest respect as a nobleman, favoured his daughter's lover, and not his daughter's husband. All which things together filled the public with dismay, and caused a delightful excitement, giving quite a feature of its own to the season.

No doubt general opinion was adverse to poor Phineas Finn, but he was not without his party in the matter. To oblige a friend by inflicting an injury on his enemy is often more easy than to confer a benefit on the friend himself. We have already seen how the young Duchess failed in her attempt to obtain an appointment for Phineas, and also how she succeeded in destroying the high hopes of Mr. Bonteen. Having done so much, of course she clung heartily to the side which she had adopted;—and, equally of course, Madame Goesler did the same. Between these two ladies there was a slight difference of opinion as to the nature of the alliance between Lady Laura and their hero. The Duchess was of opinion that young men are upon the whole averse to innocent alliances, and that, as Lady Laura and her husband certainly had long been separated, there was probably—something in it. 'Lord bless you, my dear,' the Duchess said, 'they were known to be lovers when they were at Loughlinter together before she married Mr. Kennedy. It has been the most romantic affair! She made her father give him a seat for his borough.'

'He saved Mr. Kennedy's life,' said Madame Goesler.

'That was one of the most singular things that ever happened. Laurence Fitzgibbon says that it was all planned,—that the garotters were hired, but unfortunately two policemen turned up at the moment, so the men were taken. I believe there is no doubt they were pardoned by Sir Henry Coldfoot, who was at the Home Office, and was Lord Brentford's great friend. I don't quite believe it all,—it would be too delicious; but a great many do.' Madame Goesler, however, was strong in her opinion that the report in reference to Lady Laura was scandalous. She did not believe a word of it, and was almost angry with the Duchess for her credulity.

It is probable that very many ladies shared the opinion of the Duchess; but not the less on that account did they take part with Phineas Finn. They could not understand why he should be shut out of office because a lady had been in love with him, and by no means seemed to approve the stern virtue of the Prime Minister. It was an interference with things which did not belong to him. And many asserted that Mr. Gresham was much given to such interference. Lady Cantrip, though her husband was Mr. Gresham's most intimate friend, was altogether of this party, as was also the Duchess of St. Bungay, who understood nothing at all about it, but who had once fancied herself to be rudely treated by Mrs. Bonteen. The young Duchess was a woman very strong in getting up a party; and the old Duchess, with many other matrons of high rank, was made to believe that it was incumbent on her to be a Phineas Finnite. One result of this was, that though Phineas was excluded from the Liberal Government, all Liberal drawing-rooms were open to him, and that he was a lion.*

Additional zest was given to all this by the very indiscreet conduct of Mr. Bonteen. He did accept the inferior office of President of the Board of Trade, an office inferior at least to that for which he had been designated, and agreed to fill it without a seat in the Cabinet. But having done so he could not bring himself to bear his disappointment quietly. He could not work and wait and make himself agreeable to those around him, holding his vexation within his own bosom. He was dark and sullen to his chief, and almost insolent to the Duke of Omnium. Our old friend Plantagenet Palliser was a man who hardly knew insolence when he met it. There was such an absence about him of all self-consciousness, he was so little given to think of his own personal demeanour and outward trappings,—that he never brought himself to question the manners of others to him. Contradiction he would take for simple argument. Strong difference of opinion even on the part of subordinates recommended itself to him. He could put up with apparent rudeness without seeing it, and always gave men credit for good intentions. And with it all he had an assurance in his own position,—a knowledge of the strength derived from his intellect, his industry, his rank, and his wealth,—which made him altogether fearless of others. When the little dog snarls, the big dog does not connect the snarl with himself, simply fancying that the little dog must be uncomfortable. Mr. Bonteen snarled a good deal, and the new Lord Privy Seal thought that the new President of the Board of Trade was not comfortable within himself. But at last the little dog took the big dog by the ear, and then the big dog put out his paw and knocked the little dog over. Mr. Bonteen was told that he had—forgotten himself; and there arose new rumours. It was soon reported that the Lord Privy Seal had refused to work out decimal coinage under the management, in the House of Commons, of the President of the Board of Trade.

Mr. Bonteen, in his troubled spirit, certainly did misbehave himself. Among his closer friends he declared very loudly that he didn't mean to stand it. He had not chosen to throw Mr. Gresham over at once, or to make difficulties at the moment;—but he would not continue to hold his present position or to support the Government without a seat in the Cabinet. Palliser had become quite useless,—so Mr. Bonteen said,—since his accession to the dukedom, and was quite unfit to deal with decimal coinage. It was a burden to kill any man, and he was not going to kill himself,—at any rate without the reward for which he had been working all his life, and to which he was fully entitled, namely, a seat in the Cabinet. Now there were Bonteenites in those days as well as Phineas Finnites. The latter tribe was for the most part feminine; but the former consisted of some half-dozen members of Parliament, who thought they saw their way in encouraging the forlorn hope of the unhappy financier.

A leader of a party is nothing without an organ, and an organ came forward to support Mr. Bonteen,—not very creditable to him as a Liberal, being a Conservative organ,—but not the less gratifying to his spirit, inasmuch as the organ not only supported him, but exerted its very loudest pipes in abusing the man whom of all men he hated the most. The People's Banner was the organ, and Mr. Quintus Slide was, of course, the organist. The following was one of the tunes he played, and was supposed by himself to be a second thunderbolt, and probably a conclusively crushing missile. This thunderbolt fell on Monday, the 3rd of May:—

'Early in last March we found it to be our duty to bring under public notice the conduct of the member for Tankerville in reference to a transaction which took place at a small hotel in Judd Street, and as to which we then ventured to call for the interference of the police. An attempt to murder the member for Tankerville had been made by a gentleman once well known in the political world, who,—as it is supposed,—had been driven to madness by wrongs inflicted on him in his dearest and nearest family relations. That the unfortunate gentleman is now insane we believe we may state as a fact. It had become our special duty to refer to this most discreditable transaction, from the fact that a paper, still in our hands, had been confided to us for publication by the wretched husband before his senses had become impaired,—which, however, we were debarred from giving to the public by an injunction served upon us in sudden haste by the Vice-Chancellor. We are far from imputing evil motives, or even indiscretion, to that functionary; but we are of opinion that the moral feeling of the country would have been served by the publication, and we are sure that undue steps were taken by the member for Tankerville to procure that injunction.

'No inquiries whatever were made by the police in reference to that attempt at murder, and we do expect that some member will ask a question on the subject in the House. Would such culpable quiescence have been allowed had not the unfortunate lady whose name we are unwilling to mention been the daughter of one of the colleagues of our present Prime Minister, the gentleman who fired the pistol another of them, and the presumed lover, who was fired at, also another? We think that we need hardly answer that question.

'One piece of advice which we ventured to give Mr. Gresham in our former article he has been wise enough to follow. We took upon ourselves to tell him that if, after what has occurred, he ventured to place the member for Tankerville again in office, the country would not stand it;—and he has abstained. The jaunty footsteps of Mr. Phineas Finn are not heard ascending the stairs of any office at about two in the afternoon, as used to be the case in one of those blessed Downing Street abodes about three years since. That scandal is, we think, over,—and for ever. The good-looking Irish member of Parliament who had been put in possession of a handsome salary by feminine influences, will not, we think, after what we have already said, again become a burden on the public purse. But we cannot say that we are as yet satisfied in this matter, or that we believe that the public has got to the bottom of it,—as it has a right to do in reference to all matters affecting the public service. We have never yet learned why it is that Mr. Bonteen, after having been nominated Chancellor of the Exchequer,—for the appointment to that office was declared in the House of Commons by the head of his party,—was afterwards excluded from the Cabinet, and placed in an office made peculiarly subordinate by the fact of that exclusion. We have never yet been told why this was done;—but we believe that we are justified in saying that it was managed through the influence of the member for Tankerville; and we are quite sure that the public service of the country has thereby been subjected to grievous injury.

'It is hardly our duty to praise any of that very awkward team of horses which Mr. Gresham drives with an audacity which may atone for his incapacity if no fearful accident should be the consequence; but if there be one among them whom we could trust for steady work up hill, it is Mr. Bonteen. We were astounded at Mr. Gresham's indiscretion in announcing the appointment of his new Chancellor of the Exchequer some weeks before he had succeeded in driving Mr. Daubeny from office;—but we were not the less glad to find that the finances of the country were to be entrusted to the hands of the most competent gentleman whom Mr. Gresham has induced to follow his fortunes. But Mr. Phineas Finn, with his female forces, has again interfered, and Mr. Bonteen has been relegated to the Board of Trade, without a seat in the Cabinet. We should not be at all surprised if, as the result of this disgraceful manœuvring, Mr. Bonteen found himself at the head of the Liberal party before the Session be over. If so, evil would have worked to good. But, be that as it may, we cannot but feel that it is a disgrace to the Government, a disgrace to Parliament, and a disgrace to the country that such results should come from the private scandals of two or three people among us by no means of the best class.'

CHAPTER XLIV

The Browborough Trial

THERE was another matter of public interest going on at this time which created a great excitement. And this, too, added to the importance of Phineas Finn, though Phineas was not the hero of the piece. Mr. Browborough, the late member for Tankerville, was tried for bribery. It will be remembered that when Phineas contested the borough in the autumn, this gentleman was returned. He was afterwards unseated, as the result of a petition before the judge, and Phineas was declared to be the true member. The judge who had so decided had reported to the Speaker that further inquiry before a commission into the practices of the late and former elections at Tankerville would be expedient, and such commission had sat in the months of January and February. Half the voters in Tankerville had been examined, and many who were not voters. The commissioners swept very clean, being new brooms, and in their report recommended that Mr. Browborough, whom they had themselves declined to examine, should be prosecuted. That report was made about the end of March, when Mr. Daubeny's great bill was impending. Then there arose a double feeling about Mr. Browborough, who had been regarded by many as a model member of Parliament, a man who never spoke, constant in his attendance, who wanted nothing, who had plenty of money, who gave dinners, to whom a seat in Parliament was the be-all and the end-all of life. It could not be the wish of any gentleman, who had been

accustomed to his slow step in the lobbies, and his burly form always quiescent on one of the upper seats just below the gangway on the Conservative side of the House, that such a man should really be punished. When the new laws regarding bribery came to take that shape the hearts of members revolted from the cruelty,—the hearts even of members on the other side of the House. As long as a seat was in question the battle should of course be fought to the nail. Every kind of accusation might then be lavished without restraint, and every evil practice imputed. It had been known to all the world,—

known as a thing that was a matter of course,—that at every election Mr. Browborough had bought his seat. How should a Browborough get a seat without buying it,—a man who could not say ten words, of no family, with no natural following in any constituency, distinguished by no zeal in politics, entertaining no special convictions of his own? How should such a one recommend himself to any borough unless he went there with money in his hand? Of course, he had gone to Tankerville with money in his hand, with plenty of money, and had spent it—like a gentleman. Collectively the House of Commons had determined to put down bribery with a very strong hand. Nobody had spoken against bribery with more fervour than Sir Gregory Grogram, who had himself, as Attorney-General, forged the chains for fettering future bribers. He was now again Attorney-General, much to his disgust, as Mr. Gresham had at the last moment found it wise to restore Lord Weazeling to the woolsack; and to his hands was to be entrusted the prosecution of Mr. Browborough. But it was observed by many that the job was not much to his taste. The House had been very hot against bribery,—and certain members of the existing Government, when the late Bill had been passed, had expressed themselves with almost burning indignation against the crime. But, through it all, there had been a slight undercurrent of ridicule attaching itself to the question of which only they who were behind the scenes were conscious. The House was bound to let the outside world know that all corrupt practices at elections were held to be abominable by the House; but Members of the House, as individuals, knew very well what had taken place at their own elections, and were aware of the cheques which they had drawn. Public-houses had been kept open as a matter of course, and nowhere perhaps had more beer been drunk than at Clovelly, the borough for which Sir Gregory Grogram sat. When it came to be a matter of individual prosecution against one whom they had all known, and who, as a member, had been inconspicuous and therefore inoffensive, against a heavy, rich, useful man who had been in nobody's way, many thought that it would amount to persecution. The idea of putting old Browborough into prison for conduct which habit had made second nature to a large proportion of the House was distressing to Members of Parliament generally. The recommendation for this prosecution was made to the House when Mr. Daubeny was in the first agonies of his great Bill, and he at once resolved to ignore the matter altogether, at any rate for the present. If he was to be driven out of power there could be no reason why his Attorney-General should prosecute his own ally and follower,—a poor, faithful creature, who had never in his life voted against his party, and who had always been willing to accept as his natural leader any one whom his party might select. But there were many who had felt that as Mr. Browborough must certainly now be prosecuted sooner or later,—for there could be no final neglecting of the Commissioners' report,—it would be better that he should be dealt with by natural friends than by natural enemies. The newspapers, therefore, had endeavoured to hurry the matter on, and it had been decided that the trial should take place at the Durham Spring Assizes, in the first week of May. Sir Gregory Grogram became Attorney-General in the middle of April, and he undertook the task upon compulsion. Mr. Browborough's own friends, and Mr. Browborough himself, declared very loudly that there would be the greatest possible cruelty in postponing the trial. His lawyers thought that his best chance lay in bustling the thing on, and were therefore able to show that the cruelty of delay would be extreme,—nay, that any postponement in such a matter would be unconstitutional, if not illegal. It would, of course, have been just as easy to show that hurry on the part of the prosecutor was cruel, and illegal, and unconstitutional, had it been considered that the best chance of acquittal lay in postponement.

And so the trial was forced forward, and Sir Gregory himself was to appear on behalf of the prosecuting House of Commons. There could be no doubt that the sympathies of the public generally were with Mr. Browborough, though there was as little doubt that he was guilty. When the evidence taken by the Commissioners had just appeared in the newspapers,—when first the facts of this and other elections at Tankerville were made public, and the world was shown how common it had been for Mr. Browborough to buy votes,—how clearly the knowledge of the corruption had been brought home to himself,—there had for a short week or so been a feeling against him. Two or three London papers had printed leading articles, giving in detail the salient points of the old sinner's criminality, and expressing a conviction that now, at least, would the real criminal be punished. But this had died away, and the anger against Mr. Browborough, even on the part of the most virtuous of the public press, had become no more than lukewarm. Some papers boldly defended him, ridiculed the Commissioners, and declared that the trial was altogether an absurdity. The People's Banner, setting at defiance with an admirable audacity all the facts as given in the Commissioners' report, declared that there was not one tittle of evidence against Mr. Browborough, and hinted that the trial had been got up by the malign influence of that doer of all evil, Phineas Finn. But men who knew better what was going on in the world than did Mr. Quintus Slide, were well aware that such assertions as these were both unavailing and unnecessary. Mr. Browborough was believed to be quite safe; but his safety lay in the indifference of his prosecutors,—certainly not in his innocence. Any one prominent in affairs can always see when a man may steal a horse and when a man may not look over a hedge. Mr. Browborough had stolen his horse, and had repeated the theft over and over again. The evidence of it all was forthcoming,—had, indeed, been already sifted. But Sir Gregory Grogram, who was prominent in affairs, knew that the theft might be condoned.

Nevertheless, the case came on at the Durham Assizes. Within the last two months Browborough had become quite a hero at Tankerville. The Church party had forgotten his broken pledges, and the Radicals remembered only his generosity. Could he have stood for the seat again on the day on which the judges entered Durham, he might have been returned without bribery. Throughout the whole county the prosecution was unpopular. During no portion of his Parliamentary career had Mr. Browborough's name been treated with so much respect in the grandly ecclesiastical city as now. He dined with the Dean on the day before the trial, and on the Sunday was shown by the head verger into the stall next to the Chancellor of the Diocese, with a reverence which seemed to imply that he was almost as graceful as a martyr. When he took his seat in the Court next to his attorney, everybody shook hands with him. When Sir Gregory got up to open his case, not one of the listeners then supposed that Mr. Browborough was about to suffer any punishment. He was arraigned before Mr. Baron Boultby, who had himself sat for a borough in his younger days, and who knew well how things were done. We are all aware how impassionately grand are the minds of judges, when men accused of crimes are brought before them for trial; but judges after all are men, and Mr. Baron Boultby, as he looked at Mr. Browborough, could not but have thought of the old days.

It was nevertheless necessary that the prosecution should be conducted in a properly formal manner, and that all the evidence should be given. There was a cloud of witnesses over from Tankerville,—miners, colliers, and the like,—having a very good turn of it at the expense of the poor borough. All these men must be examined, and their evidence would no doubt be the same now as when it was given with so damnable an effect before those clean-sweeping Commissioners. Sir Gregory's opening speech was quite worthy of Sir Gregory. It was essentially necessary, he said, that the atmosphere of our boroughs should be cleansed and purified from the taint of corruption. The voice of the country had spoken very plainly on the subject, and a verdict had gone forth that there should be no more bribery at elections. At the last election at Tankerville, and, as he feared, at some former elections, there had been manifest bribery. It would be for the jury to decide whether Mr. Browborough himself had been so connected with the acts of his agents as to be himself within the reach of the law. If it were found that he had brought himself within the reach of the law, the jury would no doubt say so, and in such case would do great service to the cause of purity; but if Mr. Browborough had not been personally cognisant of what his agents had done, then the jury would be bound to acquit him. A man was not necessarily guilty of bribery in the eye of the law because bribery had been committed, even though the bribery so committed had been sufficiently proved to deprive him of the seat which he would otherwise have enjoyed. Nothing could be clearer than the manner in which Sir Gregory explained it all to the jury; nothing more eloquent than his denunciations against bribery in general; nothing more mild than his allegations against Mr. Browborough individually.

In regard to the evidence Sir Gregory, with his two assistants, went through his work manfully. The evidence was given,—not to the same length as at Tankerville before the Commissioners,—but really to the same effect. But yet the record of the evidence as given in the newspapers seemed to be altogether different. At Tankerville there had been an indignant and sometimes an indiscreet zeal which had communicated itself to the whole proceedings. The general flavour of the trial at Durham was one of good-humoured raillery. Mr. Browborough's counsel in cross-examining the witnesses for the prosecution displayed none of that righteous wrath,—wrath righteous on behalf of injured innocence—which is so common with gentlemen employed in the defence of criminals; but bowed and simpered, and nodded at Sir Gregory in a manner that was quite pleasant to behold. Nobody scolded anybody. There was no roaring of barristers, no clenching of fists and kicking up of dust, no threats, no allusions to witnesses' oaths. A considerable amount of gentle fun was poked at the witnesses by the defending counsel, but not in a manner to give any pain. Gentlemen who acknowledged to have received seventeen shillings and sixpence for their votes at the last election were asked how they had invested their money.

Allusions were made to their wives, and a large amount of good-humoured sparring was allowed, in which the witnesses thought that they had the best of it. The men of Tankerville long remembered this trial, and hoped anxiously that there might soon be another. The only man treated with severity was poor Phineas Finn, and luckily for himself he was not present. His qualifications as member of Parliament for Tankerville were somewhat roughly treated. Each witness there, when he was asked what candidate would probably be returned for Tankerville at the next election, readily answered that Mr. Browborough would certainly carry the seat. Mr. Browborough sat in the Court throughout it all, and was the hero of the day.

The judge's summing up was very short, and seemed to have been given almost with indolence. The one point on which he insisted was the difference between such evidence of bribery as would deprive a man of his seat, and that which would make him subject to the criminal law. By the criminal law a man could not be punished for the acts of another. Punishment must follow a man's own act. If a man were to instigate another to murder he would be punished, not for the murder, but for the instigation. They were now administering the criminal law, and they were bound to give their verdict for an acquittal unless they were convinced that the man on his trial had himself,—wilfully and wittingly,—been guilty of the crime imputed. He went through the evidence, which was in itself clear against the old sinner, and which had been in no instance validly contradicted, and then left the matter to the jury. The men in the box put their heads together, and returned a verdict of acquittal without one moment's delay. Sir Gregory Grogram and his assistants collected their papers together. The judge addressed three or four words almost of compliment to Mr. Browborough, and the affair was over, to the manifest contentment of every one there present. Sir Gregory Grogram was by no means disappointed, and everybody, on his own side in Parliament and on the other, thought that he had done his duty very well. The clean-sweeping

Commissioners, who had been animated with wonderful zeal by the nature and novelty of their work, probably felt that they had been betrayed, but it may be doubted whether any one else was disconcerted by the result of the trial, unless it might be some poor innocents here and there about the country who had been induced to believe that bribery and corruption were in truth to be banished from the purlieus of Westminster.

Mr. Roby and Mr. Ratler, who filled the same office each for his own party, in and out, were both acquainted with each other, and apt to discuss parliamentary questions in the library and smoking-room of the House, where such discussions could be held on most matters. 'I was very glad that the case went as it did at Durham,' said Mr. Ratler.

'And so am I,' said Mr. Roby. 'Browborough was always a good fellow.'

'Not a doubt about it; and no good could have come from a conviction. I suppose there has been a little money spent at Tankerville.'

'And at other places one could mention,' said Mr. Roby.

'Of course there has;—and money will be spent again. Nobody dislikes bribery more than I do. The House, of course, dislikes it. But if a man loses his seat, surely that is punishment enough.'

'It's better to have to draw a cheque sometimes than to be out in the cold.'

'Nevertheless, members would prefer that their seats should not cost them so much,' continued Mr. Ratler. 'But the thing can't be done all at once. That idea of pouncing upon one man and making a victim of him is very disagreeable to me. I should have been sorry to have seen a verdict against Browborough. You must acknowledge that there was no bitterness in the way in which Grogram did it.'

'We all feel that,' said Mr. Roby,—who was, perhaps, by nature a little more candid than his rival,—'and when the time comes no doubt we shall return the compliment.'

The matter was discussed in quite a different spirit between two other politicians. 'So Sir Gregory has failed at Durham,' said Lord Cantrip to his friend, Mr. Gresham.

'I was sure he would.'

'And why?'

'Ah;—why? How am I to answer such a question? Did you think that Mr. Browborough would be convicted of bribery by a jury?'

'No, indeed,' answered Lord Cantrip.

'And can you tell me why?'

'Because there was no earnestness in the matter,—either with the Attorney-General or with any one else.'

'And yet,' said Mr. Gresham, 'Grogram is a very earnest man when he believes in his case. No member of Parliament will ever be punished for bribery as for a crime till members of Parliament generally look upon bribery as a crime. We are very far from that as yet. I should have thought a conviction to be a great misfortune.'

'Why so?'

'Because it would have created ill blood, and our own hands in this matter are not a bit cleaner than those of our adversaries. We can't afford to pull their houses to pieces before we have put our own in order. The thing will be done; but it must, I fear, be done slowly,—as is the case with all reforms from within.'

Phineas Finn, who was very sore and unhappy at this time, and who consequently was much in love with purity and anxious for severity, felt himself personally aggrieved by the acquittal. It was almost tantamount to a verdict against himself. And then he knew so well that bribery had been committed, and was so confident that such a one as Mr. Browborough could have been returned to Parliament by none other than corrupt means! In his present mood he would have been almost glad to see Mr. Browborough at the treadmill, and would have thought six months' solitary confinement quite inadequate to the offence. 'I never read anything in my life that disgusted me so much,' he said to his friend, Mr. Monk.

'I can't go along with you there.'

'If any man ever was guilty of bribery, he was guilty!'

'I don't doubt it for a moment.'

'And yet Grogram did not try to get a verdict.'

'Had he tried ever so much he would have failed. In a matter such as that,—political and not social in its nature,—a jury is sure to be guided by what it has, perhaps unconsciously, learned to be the feeling of the country. No disgrace is attached to their verdict, and yet everybody knows that Mr. Browborough had bribed, and all those who have looked into it know, too, that the evidence was conclusive.'

'Then are the jury all perjured,' said Phineas.

'I have nothing to say to that. No stain of perjury clings to them. They are better received in Durham to-day than they would have been had they found Mr. Browborough guilty. In business, as in private life, they will be held to be as trustworthy as before;—and they will be, for aught that we know, quite trustworthy. There are still circumstances in which a man, though on his oath, may be untrue with no more stain of falsehood than falls upon him when he denies himself at his front door though he happen to be at home.'

'What must we think of such a condition of things, Mr. Monk?'

'That it's capable of improvement. I do not know that we can think anything else. As for Sir Gregory Grogram and Baron Boultby and the jury, it would be waste of power to execrate them. In political matters it is very hard for a man in office to be purer than his neighbours,—and, when he is so, he becomes troublesome. I have found that out before to-day.'

With Lady Laura Kennedy, Phineas did find some sympathy;—but then she would have sympathised with him on any subject under the sun. If he would only come to her and sit with her she would fool him to the top of his bent. He had resolved that he would go to Portman Square as little as possible, and had been confirmed in that resolution by the scandal which had now spread everywhere about the town in reference to himself and herself. But still he went. He never left her till some promise of returning at some stated time had been extracted from him. He had even told her of his own scruples and of her danger,—and they had discussed together that last thunderbolt which had fallen from the Jove of The People's Banner. But she had laughed his caution to scorn. Did she not know herself and her own innocence? Was she not living in her father's house, and with her father? Should she quail beneath the stings and venom of such a reptile as Quintus Slide? 'Oh, Phineas,' she said, 'let us be braver than that.' He would much prefer to have stayed away,—but still he went to her. He was conscious of her dangerous love for him. He knew well that it was not returned. He was aware that it would be best for both that he should be apart. But yet he could not bring himself to wound her by his absence. 'I do not see why you should feel it so much,' she said, speaking of the trial at Durham.

'We were both on our trial,—he and I.'

'Everybody knows that he bribed and that you did not.'

'Yes;—and everybody despises me and pats him on the back. I am sick of the whole thing. There is no honesty in the life we lead.'

'You got your seat at any rate.'

'I wish with all my heart that I had never seen the dirty wretched place,' said he.

'Oh, Phineas, do not say that.'

'But I do say it. Of what use is the seat to me? If I could only feel that any one knew——'

'Knew what, Phineas?'

'It doesn't matter.'

'I understand. I know that you have meant to be honest, while this man has always meant to be dishonest. I know that you have intended to serve your country, and have wished to work for it. But you cannot expect that it should all be roses.'

'Roses! The nosegays which are worn down at Westminster are made of garlick and dandelions!'

CHAPTER XLV

Some Passages in the Life of Mr. Emilius

THE writer of this chronicle is not allowed to imagine that any of his readers have read the wonderful and vexatious adventures of Lady Eustace, a lady of good birth, of high rank, and of large fortune, who, but a year or two since, became almost a martyr to a diamond necklace which was stolen from her. With her history the present reader has but small concern, but it may be necessary that he should know that the lady in question, who had been a widow with many suitors, at last gave her hand and her fortune to a clergyman whose name was Joseph Emilius. Mr. Emilius, though not an Englishman by birth,—and, as was supposed, a Bohemian Jew in the earlier days of his career,—had obtained some reputation as a preacher in London, and had moved,—if not in fashionable circles,—at any rate in circles so near to fashion as to be brought within the reach of Lady Eustace's charms. They were married, and for some few months Mr. Emilius enjoyed a halcyon existence, the delights of which were, perhaps, not materially marred by the necessity which he felt of subjecting his young wife to marital authority. 'My dear,' he would say, 'you will know me better soon, and then things will be smooth.' In the meantime he drew more largely upon her money than was pleasing to her and to her friends, and appeared to have requirements for cash which were both secret and unlimited. At the end of twelve months Lady Eustace had run away from him, and Mr. Emilius had made overtures, by accepting which his wife would be enabled to purchase his absence at the cost of half her income. The arrangement was not regarded as being in every respect satisfactory, but Lady Eustace declared passionately that any possible sacrifice would be preferable to the company of Mr. Emilius. There had, however, been a rumour before her marriage that there was still living in his old country a Mrs. Emilius when he married Lady Eustace; and, though it had been supposed by those who were most nearly concerned with Lady Eustace that this report had been unfounded and malicious, nevertheless, when the man's claims became so exorbitant, reference was again made to the charge of bigamy. If it could be proved that Mr. Emilius had a wife living in Bohemia, a cheaper mode of escape would be found for the persecuted lady than that which he himself had suggested.

It had happened that, since her marriage with Mr. Emilius, Lady Eustace had become intimate with our Mr. Bonteen and his wife. She had been at one time engaged to marry Lord Fawn, one of Mr. Bonteen's colleagues, and during the various circumstances which had led to the disruption of that engagement, this friendship had been formed. It must be understood that Lady Eustace had a most desirable residence of her own in the country,—Portray Castle in Scotland,—and that it was thought expedient by many to cultivate her acquaintance. She was rich, beautiful, and clever; and, though her marriage with Mr. Emilius had never been looked upon as a success, still, in the estimation of some people, it added an interest to her career. The Bonteens had taken her up, and now both Mr. and Mrs. Bonteen were hot in pursuit of evidence which might prove Mr. Emilius to be a bigamist.

When the disruption of conjugal relations was commenced, Lady Eustace succeeded in obtaining refuge at Portray Castle without the presence of her husband. She fled from London during a visit he made to Brighton with the object of preaching to a congregation by which his eloquence was held in great esteem. He left London in one direction by the 5 P.M. express train on Saturday, and she in the other by the limited mail*at 8.45. A telegram, informing him of what had taken place, reached him the next morning at Brighton while he was at breakfast. He preached his sermon, charming the congregation by the graces of his extempore eloquence,—moving every woman there to tears,—and then was after his wife before the ladies had taken their first glass of sherry at luncheon. But her ladyship had twenty-four hours' start of him,—although he did his best; and when he reached Portray Castle the door was shut in his face. He endeavoured to obtain the aid of blacksmiths to open, as he said, his own hall door,—to obtain the aid of constables to compel the blacksmiths, of magistrates to compel the constables,—and even of a judge to compel the magistrates; but he was met on every side by a statement that the lady of the castle declared that she was not his wife, and that therefore he had no right whatever to demand that the door should be opened. Some other woman,—so he was informed that the lady said,—out in a strange country was really his wife. It was her intention to prove him to be a bigamist, and to have him locked up. In the meantime she chose to lock herself up in her own mansion. Such was the nature of the message that was delivered to him through the bars of the lady's castle.

How poor Lady Eustace was protected, and, at the same time, made miserable by the energy and unrestrained language of one of her own servants, Andrew Gowran by name, it hardly concerns us now to inquire. Mr. Emilius did not succeed in effecting an entrance; but he remained for some time in the neighbourhood, and had notices served on the tenants in regard to the rents, which puzzled the poor folk round Portray Castle very much. After a while Lady Eustace, finding that her peace and comfort imperatively demanded that she should prove the allegations which she had made, fled again from Portray Castle to London, and threw herself into the hands of the Bonteens. This took place just as Mr. Bonteen's hopes in regard to the Chancellorship of the Exchequer were beginning to soar high, and when his hands were very full of business. But with that energy for which he was so conspicuous, Mr. Bonteen had made a visit to Bohemia during his short Christmas holidays, and had there set people to work. When at Prague he had, he thought, very nearly unravelled the secret himself. He had found the woman whom he believed to be Mrs. Emilius, and who was now living somewhat merrily in Prague under another name. She acknowledged that in old days, when they were both young, she had been acquainted with a certain Yosef Mealyus, at a time in which he had been in the employment of a Jewish money-lender in the city; but,—as she declared,—she had never been married to him. Mr. Bonteen learned also that the gentleman now known as Mr. Joseph Emilius of the London Chapel had been known in his own country as Yosef Mealyus, the name which had been borne by the very respectable Jew who was his father. Then Mr. Bonteen had returned home, and, as we all know, had become engaged in matters of deeper import than even the deliverance of Lady Eustace from her thraldom.

Mr. Emilius made no attempt to obtain the person of his wife while she was under Mr. Bonteen's custody, but he did renew his offer to compromise. If the estate could not afford to give him the two thousand a year which he had first demanded, he would take fifteen hundred. He explained all this personally to Mr. Bonteen, who condescended to see him. He was very eager to make Mr. Bonteen understand how bad even then would be his condition. Mr. Bonteen was, of course, aware that he would have to pay very heavily for insuring his wife's life. He was piteous, argumentative, and at first gentle; but when Mr. Bonteen somewhat rashly told him that the evidence of a former marriage and of the present existence of the former wife would certainly be forthcoming, he defied Mr. Bonteen and his evidence,—and swore that if his claims were not satisfied, he would make use of the power which the English law gave him for the recovery of his wife's person. And as to her property,—it was his, not hers. From this time forward if she wanted to separate herself from him she must ask him for an allowance. Now, it certainly was the case that Lady Eustace had married the man without any sufficient precaution as to keeping her money in her own hands, and Mr. Emilius had insisted that the rents of the property which was hers for her life should be paid to him, and on his receipt only. The poor tenants had been noticed this way and noticed that till they had begun to doubt whether their safest course would not be to keep their rents in their own hands. But lately the lawyers of the Eustace family,—who were not, indeed, very fond of Lady Eustace personally,—came forward for the sake of the property, and guaranteed the tenants against all proceedings until the question of the legality of the marriage should be settled. So Mr. Emilius,—or the Reverend Mealyus, as everybody now called him,—went to law; and Lady Eustace went to law; and the Eustace family went to law;—but still, as yet, no evidence was forthcoming sufficient to enable Mr. Bonteen, as the lady's friend, to put the gentleman into prison.

It was said for a while that Mealyus had absconded. After his interview with Mr. Bonteen he certainly did leave England and made a journey to Prague. It was thought that he would not return, and that Lady Eustace would be obliged to carry on the trial, which was to liberate her and her property, in his absence. She was told that the very fact of his absence would go far with a jury, and she was glad to be freed from his presence in England. But he did return, declaring aloud that he would have his rights. His wife should be made to put herself into his hands, and he would obtain possession of the income which was his own. People then began to doubt. It was known that a very clever lawyer's clerk had been sent to Prague to complete the work there which Mr. Bonteen had commenced. But the clerk did not come back as soon as was expected, and news arrived that he had been taken ill. There was a rumour that he had been poisoned at his hotel; but, as the man was not said to be dead, people hardly believed the rumour. It became necessary, however, to send another lawyer's clerk, and the matter was gradually progressing to a very interesting complication.

Mr. Bonteen, to tell the truth, was becoming sick of it. When Emilius, or Mealyus, was supposed to have absconded, Lady Eustace left Mr. Bonteen's house, and located herself at one of the large London hotels; but when the man came back, bolder than ever, she again betook herself to the shelter of Mr. Bonteen's roof. She expressed the most lavish affection for Mrs. Bonteen, and professed to regard Mr. Bonteen as almost a political god, declaring her conviction that he, and he alone, as Prime Minister, could save the country, and became very loud in her wrath when he was robbed of his seat in the Cabinet. Lizzie Eustace, as her ladyship had always been called, was a clever, pretty, coaxing little woman, who knew how to make the most of her advantages. She had not been very wise in her life, having lost the friends who would have been truest to her, and confided in persons who had greatly injured her. She was neither true of heart or tongue, nor affectionate, nor even honest. But she was engaging; she could flatter; and could assume a reverential admiration which was very foreign to her real character. In these days she almost worshipped Mr. Bonteen, and could never be happy except in the presence of her dearest darling friend Mrs. Bonteen. Mr. Bonteen was tired of her, and Mrs. Bonteen was becoming almost sick of the constant kisses with which she was greeted; but Lizzie Eustace had got hold of them, and they could not turn her off.

'You saw The People's Banner, Mrs. Bonteen, on Monday?' Lady Eustace had been reading the paper in her friend's drawing-room. 'They seem to think that Mr. Bonteen must be Prime Minister before long.'

'I don't think he expects that, my dear.'

'Why not? Everybody says The People's Banner is the cleverest paper we have now. I always hated the very name of that Phineas Finn.'

'Did you know him?'

'Not exactly. He was gone before my time; but poor Lord Fawn used to talk of him. He was one of those conceited Irish upstarts that are never good for anything.'

'Very handsome, you know,' said Mrs. Bonteen.

'Was he? I have heard it said that a good many ladies admired him.'

'It was quite absurd; with Lady Laura Kennedy it was worse than absurd. And there was Lady Glencora, and Violet Effingham, who married Lady Laura's brother, and that Madame Goesler, whom I hate,—and ever so many others.'

'And is it true that it was he who got Mr. Bonteen so shamefully used?'

'It was his faction.'

'I do so hate that kind of thing,' said Lady Eustace, with righteous indignation; 'I used to hear a great deal about Government and all that when the affair was on between me and poor Lord Fawn, and that kind of dishonesty always disgusted me. I don't know that I think so much of Mr. Gresham after all.'

'He is a very weak man.'

'His conduct to Mr. Bonteen has been outrageous; and if he has done it just because that Duchess of Omnium has told him, I really do think that he is not fit to rule the nation. As for Mr. Phineas Finn, it is dreadful to think that a creature like that should be able to interfere with such a man as Mr. Bonteen.'

This was on Wednesday afternoon,—the day on which members of Parliament dine out,—and at that moment Mr. Bonteen entered the drawing-room, having left the House for his half-holiday at six o'clock. Lady Eustace got up, and gave him her hand, and smiled upon him as though he were indeed her god. 'You look so tired and so worried, Mr. Bonteen.'

'Worried;—I should think so.'

'Is there anything fresh?' asked his wife.

'That fellow Finn is spreading all manner of lies about me.'

'What lies, Mr. Bonteen?' asked Lady Eustace. 'Not new lies, I hope.'

'It all comes from Carlton Terrace.' The reader may perhaps remember that the young Duchess of Omnium lived in Carlton Terrace. 'I can trace it all there. I won't stand it if it goes on like this. A clique of stupid women to take up the cudgels for a coal-heaving sort of fellow like that, and sting one like a lot of hornets! Would you believe it?—the Duke almost refused to speak to me just now—a man for whom I have been working like a slave for the last twelve months!'

'I would not stand it,' said Lady Eustace.

'By the bye, Lady Eustace, we have had news from Prague.'

'What news?' said she, clasping her hands.

'That fellow Pratt we sent out is dead.'

'No!'

'Not a doubt but what he was poisoned; but they seem to think that nothing can be proved. Coulson is on his way out, and I shouldn't wonder if they served him the same.'

'And it might have been you!' said Lady Eustace, taking hold of her friend's arm with almost frantic affection.

Yes, indeed. It might have been the lot of Mr. Bonteen to have died at Prague—to have been poisoned by the machinations of the former Mrs. Mealyus, if such really had been the fortune of the unfortunate Mr. Pratt. For he had been quite as busy at Prague as his successor in the work. He had found out much, though not everything. It certainly had been believed that Yosef Mealyus was a married man, but he had brought the woman with him to Prague, and had certainly not married her in the city. She was believed to have come from Cracow, and Mr. Bonteen's zeal on behalf of his friend had not been sufficient to carry him so far East. But he had learned from various sources that the man and woman had been supposed to be married,—that she had borne the man's name, and that he had taken upon himself authority as her husband. There had been written communications with Cracow, and information was received that a man of the name of Yosef Mealyus had been married to a Jewess in that town. But this had been twenty years ago, and Mr. Emilius professed himself to be only thirty-five years old, and had in his possession a document from his synagogue professing to give a record of his birth, proving such to be his age. It was also ascertained that Mealyus was a name common at Cracow, and that there were very many of the family in Galicia.* Altogether the case was full of difficulty, but it was thought that Mr. Bonteen's evidence would be sufficient to save the property from the hands of the cormorant, at any rate till such time as better evidence of the first marriage could be obtained. It had been hoped that when the man went away he would not return; but he had returned, and it was now resolved that no terms should be kept with him and no payment offered to him. The house at Portray was kept barred, and the servants were ordered not to admit him. No money was to be paid to him, and he was to be left to take any proceedings at law which he might please,—while his adversaries were proceeding against him with all the weapons at their disposal. In the meantime his chapel was of course deserted, and the unfortunate man was left penniless in the world.

Various opinions prevailed as to Mr. Bonteen's conduct in the matter. Some people remembered that during the last autumn he and his wife had stayed three months at Portray Castle, and declared that the friendship between them and Lady Eustace had been very useful. Of these malicious people it seemed to be, moreover, the opinion that the connection might become even more useful if Mr. Emilius could be discharged. It was true that Mrs. Bonteen had borrowed a little money from Lady Eustace, but of this her husband knew nothing till the Jew in his wrath made the thing public. After all it had only been a poor £25, and the money had been repaid before Mr. Bonteen took his journey to Prague. Mr.

Bonteen was, however, unable to deny that the cost of that journey was defrayed by Lady Eustace, and it was thought mean in a man aspiring to be Chancellor of the Exchequer to have his travelling expenses paid for him by a lady. Many, however, were of opinion that Mr. Bonteen had been almost romantic in his friendship, and that the bright eyes of Lady Eustace had produced upon this dragon of business the wonderful effect that was noticed. Be that as it may, now, in the terrible distress of his mind at the political aspect of the times, he had become almost sick of Lady Eustace, and would gladly have sent her away from his house had he known how to do so without incurring censure.

CHAPTER XLVI

The Quarrel

On that Wednesday evening Phineas Finn was at The Universe. He dined at the house of Madame Goesler, and went from thence to the club in better spirits than he had known for some weeks past. The Duke and Duchess had been at Madame Goesler's, and Lord and Lady Chiltern, who were now up in town, with Barrington Erle, and,—as it had happened,—old Mr. Maule. The dinner had been very pleasant, and two or three words had been spoken which had tended to raise the heart of our hero. In the first place Barrington Erle had expressed a regret that Phineas was not at his old post at the Colonies, and the young Duke had re-echoed it. Phineas thought that the manner of his old friend Erle was more cordial to him than it had been lately, and even that comforted him. Then it was a delight to him to meet the Chilterns, who were always gracious to him. But perhaps his greatest pleasure came from the reception which was accorded by his hostess to Mr. Maule, which was of a nature not easy to describe. It had become evident to Phineas that Mr. Maule was constant in his attentions to Madame Goesler; and, though he had no purpose of his own in reference to the lady,

—though he was aware that former circumstances, circumstances of that previous life to which he was accustomed to look back as to another existence, made it impossible that he should have any such purpose,—still he viewed Mr. Maule with dislike. He had once ventured to ask her whether she really liked 'that old padded dandy.' She had answered that she did like the old dandy. Old dandies, she thought, were preferable to old men who did not care how they looked;—and as for the padding, that was his affair, not hers. She did not know why a man should not have a pad in his coat, as well as a woman one at the back of her head. But Phineas had known that this was her gentle raillery, and now he was delighted to find that she continued it, after a still more gentle fashion, before the man's face. Mr. Maule's manner was certainly peculiar. He was more than ordinarily polite,—and was afterwards declared by the Duchess to have made love like an old gander. But Madame Goesler, who knew exactly how to receive such attentions, turned a glance now and then upon Phineas Finn, which he could now read with absolute precision. 'You see how I can dispose of a padded old dandy directly he goes an inch too far.' No words could have said that to him more plainly than did these one or two glances;—and, as he had learned to dislike Mr. Maule, he was gratified.

Of course they all talked about Lady Eustace and Mr. Emilius. 'Do you remember how intensely interested the dear old Duke used to be when we none of us knew what had become of the diamonds?' said the Duchess.

'And how you took her part,' said Madame Goesler.

'So did you,—just as much as I; and why not? She was a most interesting young woman, and I sincerely hope we have not got to the end of her yet. The worst of it is that she has got into such—very bad hands. The Bonteens have taken her up altogether. Do you know her, Mr. Finn?'

'No, Duchess;—and am hardly likely to make her acquaintance while she remains where she is now.' The Duchess laughed and nodded her head. All the world knew by this time that she had declared herself to be the sworn enemy of the Bonteens.

And there had been some conversation on that terribly difficult question respecting the foxes in Trumpeton Wood.

'The fact is, Lord Chiltern,' said the Duke, 'I'm as ignorant as a child. I would do right if I knew how. What ought I to do? Shall I import some foxes?'

'I don't suppose, Duke, that in all England there is a spot in which foxes are more prone to breed.'

'Indeed. I'm very glad of that. But something goes wrong afterwards, I fear.'

'The nurseries are not well managed, perhaps,' said the Duchess.

'Gipsy kidnappers are allowed about the place,' said Madame Goesler.

'Gipsies!' exclaimed the Duke.

'Poachers!' said Lord Chiltern. 'But it isn't that we mind. We could deal with that ourselves if the woods were properly managed. A head of game and foxes can be reared together very well, if——'

'I don't care a straw for a head of game, Lord Chiltern. As far as my own tastes go, I would wish that there was neither a pheasant nor a partridge nor a hare on any property that I own. I think that sheep and barn-door fowls do better for everybody in the long run, and that men who cannot live without shooting should go beyond thickly-populated regions to find it. And, indeed, for myself, I must say the same about foxes. They do not interest me, and I fancy that they will gradually be exterminated.'

'God forbid!' exclaimed Lord Chiltern.

'But I do not find myself called upon to exterminate them myself,' continued the Duke. 'The number of men who amuse themselves by riding after one fox is too great for me to wish to interfere with them. And I know that my neighbours in the country conceive it to be my duty to have foxes for them. I will oblige them, Lord Chiltern, as far as I can without detriment to other duties.'

'You leave it to me,' said the Duchess to her neighbour, Lord Chiltern. 'I'll speak to Mr. Fothergill myself, and have it put right.' It unfortunately happened, however, that Lord Chiltern got a letter the very next morning from old Doggett telling him that a litter of young cubs had been destroyed that week in Trumpeton Wood.

Barrington Erle and Phineas went off to The Universe together, and as they went the old terms of intimacy seemed to be re-established between them. 'Nobody can be so sorry as I am,' said Barrington, 'at the manner in which things have gone. When I wrote to you, of course, I thought it certain that, if we came in, you would come with us.'

'Do not let that fret you.'

'But it does fret me,—very much. There are so many slips that of course no one can answer for anything.'

'Of course not. I know who has been my friend.'

'The joke of it is, that he himself is at present so utterly friendless. The Duke will hardly speak to him. I know that as a fact. And Gresham has begun to find something is wrong. We all hoped that he would refuse to come in without a seat in the Cabinet;—but that was too good to be true. They say he talks of resigning. I shall believe it when I see it. He'd better not play any tricks, for if he did resign, it would be accepted at once.' Phineas, when he heard this, could not help thinking how glorious it would be if Mr. Bonteen were to resign, and if the place so vacated, or some vacancy so occasioned, were to be filled by him!

They reached the club together, and as they went up the stairs, they heard the hum of many voices in the room. 'All the world and his wife are here to-night,' said Phineas. They overtook a couple of men at the door, so that there was something of the bustle of a crowd as they entered. There was a difficulty in finding places in which to put their coats and hats, —for the accommodation of The Universe is not great. There was a knot of men talking not far from them, and among the voices Phineas could clearly hear that of Mr. Bonteen. Ratler's he had heard before, and also Fitzgibbon's, though he had not distinguished any words from them. But those spoken by Mr. Bonteen he did distinguish very plainly. 'Mr. Phineas Finn, or some such fellow as that, would be after her at once,' said Mr. Bonteen. Then Phineas walked immediately among the knot of men and showed himself. As soon as he heard his name mentioned, he doubted for a moment what he would do. Mr. Bonteen when speaking had not known of his presence, and it might be his duty not to seem to have listened. But the speech had been made aloud, in the open room,—so that those who chose might listen;—and Phineas could not but have heard it. In that moment he resolved that he was bound to take notice of what he had heard. 'What is it, Mr. Bonteen, that Phineas Finn will do?' he asked.

Mr. Bonteen had been—dining. He was not a man by any

means habitually intemperate, and now any one saying that he was tipsy would have maligned him. But he was flushed with much wine, and he was a man whose arrogance in that condition was apt to become extreme. '*In vino veritas!*'*The sober devil can hide his cloven hoof; but when the devil drinks he loses his cunning and grows honest. Mr. Bonteen looked Phineas full in the face a second or two before he answered, and then said,—quite aloud—'You have crept upon us unawares, sir.'

'What do you mean by that, sir?' said Phineas. 'I have come in as any other man comes.'

'Listeners at any rate never hear any good of themselves.'

Then there were present among those assembled clear indications of disapproval of Bonteen's conduct. In these days, —when no palpable and immediate punishment is at hand for personal insolence from man to man,—personal insolence to one man in a company seems almost to constitute an insult to every one present. When men could fight readily, an arrogant word or two between two known to be hostile to each other was only an invitation to a duel, and the angry man was doing that for which it was known that he could be made to pay. There was, or it was often thought that there was, a real spirit in the angry man's conduct, and they who were his friends before became perhaps more his friends when he had thus shown that he had an enemy. But a different feeling prevails at present;—a feeling so different, that we may almost say that a man in general society cannot speak even roughly to any but his intimate comrades without giving offence to all around him. Men have learned to hate the nuisance of a row, and to feel that their comfort is endangered if a man prone to rows gets among them. Of all candidates at a club a known quarreller is more sure of blackballs*now than even in the times when such a one provoked duels. Of all bores he is the worst; and there is always an unexpressed feeling that such a one exacts more from his company than his share of attention. This is so strong, that too often the man quarrelled with, though he be as innocent as was Phineas on the present occasion, is made subject to the general aversion which is felt for men who misbehave themselves.

'I wish to hear no good of myself from you,' said Phineas, following him to his seat. 'Who is it that you said,—I should be after?' The room was full, and every one there, even they who had come in with Phineas, knew that Lady Eustace was the woman. Everybody at present was talking about Lady Eustace.

'Never mind,' said Barrington Erle, taking him by the arm. 'What's the use of a row?'

'No use at all;—but if you heard your name mentioned in such a manner you would find it impossible to pass it over. There is Mr. Monk;—ask him.'

Mr. Monk was sitting very quietly in a corner of the room with another gentleman of his own age by him,—one devoted to literary pursuits and a constant attendant at The Universe. As he said afterwards, he had never known any unpleasantness of that sort in the club before. There were many men of note in the room. There was a foreign minister, a member of the Cabinet, two ex-members of the Cabinet, a great poet, an exceedingly able editor, two earls, two members of the Royal Academy, the president of a learned society, a celebrated professor,—and it was expected that Royalty might come in at any minute, speak a few benign words, and blow a few clouds of smoke. It was abominable that the harmony of such a meeting should be interrupted by the vinous insolence of Mr. Bonteen, and the useless wrath of Phineas Finn. 'Really, Mr. Finn, if I were you I would let it drop,' said the gentleman devoted to literary pursuits.

Phineas did not much affect the literary gentleman, but in such a matter would prefer the advice of Mr. Monk to that of any man living. He again appealed to his friend. 'You heard what was said?'

'I heard Mr. Bonteen remark that you or somebody like you would in certain circumstances be after a certain lady. I thought it to be an ill-judged speech, and as your particular friend I heard it with great regret.'

'What a row about nothing!' said Mr. Bonteen, rising from his seat. 'We were speaking of a very pretty woman, and I was saying that some young fellow generally supposed to be fond of pretty women would soon be after her. If that offends your morals you must have become very strict of late.'

There was something in the explanation which, though very bad and vulgar, it was almost impossible not to accept. Such at least was the feeling of those who stood around Phineas Finn. He himself knew that Mr. Bonteen had intended to assert that he would be after the woman's money and not her beauty; but he had taste enough to perceive that he could not descend to any such detail as that. 'There are reasons, Mr. Bonteen,' he said, 'why I think you should abstain from mentioning my name in public. Your playful references should be made to your friends, and not to those who, to say the least of it, are not your friends.'

When the matter was discussed afterwards it was thought that Phineas Finn should have abstained from making the last speech. It was certainly evidence of great anger on his part. And he was very angry. He knew that he had been insulted,—and insulted by the man whom of all men he would feel most disposed to punish for any offence. He could not allow Mr. Bonteen to have the last word, especially as a certain amount of success had seemed to attend them. Fate at the moment was so far propitious to Phineas that outward circumstances saved him from any immediate reply, and thus left him in some degree triumphant. Expected Royalty arrived, and cast its salutary oil upon the troubled waters. The Prince, with some well-known popular attendant, entered the room, and for a moment every gentleman rose from his chair. It was but for a moment, and then the Prince became as any other gentleman, talking to his friends. One or two there present, who had perhaps peculiarly royal instincts, had crept up towards him so as to make him the centre of a little knot, but, otherwise, conversation went on much as it had done before the unfortunate arrival of Phineas. That quarrel, however, had been very distinctly trodden under foot by the Prince, for

Mr. Bonteen had found himself quite incapacitated from throwing back any missile in reply to the last that had been hurled at him.

Phineas took a vacant seat next to Mr. Monk,—who was deficient perhaps in royal instincts,—and asked him in a whisper his opinion of what had taken place. 'Do not think any more of it,' said Mr. Monk.

'That is so much more easily said than done. How am I not to think of it?'

'Of course I mean that you are to act as though you had forgotten it.'

'Did you ever know a more gratuitous insult? Of course he was talking of that Lady Eustace.'

'I had not been listening to him before, but no doubt he was. I need not tell you now what I think of Mr. Bonteen. He is not more gracious in my eyes than he is in yours. To-night I fancy he has been drinking, which has not improved him. You may be sure of this, Phineas,—that the less of resentful anger you show in such a wretched affair as took place just now, the more will be the blame attached to him and the less to you.'

'Why should any blame be attached to me?'

'I don't say that any will unless you allow yourself to become loud and resentful. The thing is not worth your anger.'

'I am angry.'

'Then go to bed at once, and sleep it off. Come with me, and we'll walk home together.'

'It isn't the proper thing, I fancy, to leave the room while the Prince is here.'

'Then I must do the improper thing,' said Mr. Monk. 'I haven't a key, and I musn't keep my servant up any longer. A quiet man like me can creep out without notice. Good night, Phineas, and take my advice about this. If you can't forget it, act and speak and look as though you had forgotten it.' Then Mr. Monk, without much creeping, left the room.

The club was very full, and there was a clatter of voices, and the clatter round the Prince was the noisiest and merriest.

Mr. Bonteen was there, of course, and Phineas as he sat alone could hear him as he edged his words in upon the royal ears. Every now and again there was a royal joke, and then Mr. Bonteen's laughter was conspicuous. As far as Phineas could distinguish the sounds no special amount of the royal attention was devoted to Mr. Bonteen. That very able editor, and one of the Academicians, and the poet, seemed to be the most honoured, and when the Prince went,—which he did when his cigar was finished,—Phineas observed with inward satisfaction that the royal hand, which was given to the poet, to the editor, and to the painter, was not extended to the President of the Board of Trade. And then, having taken delight in this, he accused himself of meanness in having even observed a matter so trivial. Soon after this a ruck of men left the club, and then Phineas rose to go. As he went down the stairs Barrington Erle followed him with Laurence Fitzgibbon, and the three stood for a moment at the door in the street talking to each other. Finn's way lay eastward from the club, whereas both Erle and Fitzgibbon would go westwards towards their homes. 'How well the Prince behaves at these sort of places!' said Erle.

'Princes ought to behave well,' said Phineas.

'Somebody else didn't behave very well,—eh, Finn, my boy?' said Laurence.

'Somebody else, as you call him,' replied Phineas, 'is very unlike a Prince, and never does behave well. To-night, however, he surpassed himself.'

'Don't bother your mind about it, old fellow,' said Barrington.

'I tell you what it is, Erle,' said Phineas. 'I don't think that I'm a vindictive man by nature, but with that man I mean to make it even some of these days. You know as well as I do what it is he has done to me, and you know also whether I have deserved it. Wretched reptile that he is! He has pretty nearly been able to ruin me,—and all from some petty feeling of jealousy.'

'Finn, me boy, don't talk like that,' said Laurence.

'You shouldn't show your hand,' said Barrington.

'I know what you mean, and it's all very well. After your different fashions you two have been true to me, and I don't care how much you see of my hand. That man's insolence angers me to such an extent that I cannot refrain from speaking out. He hasn't spirit enough to go out with me, or I would shoot him.'

'Blankenberg, eh!' said Laurence, alluding to the now notorious duel which had once been fought in that place between Phineas and Lord Chiltern.

'I would,' continued the angry man. 'There are times in which one is driven to regret that there has come an end to duelling, and there is left to one no immediate means of resenting an injury.'

As they were speaking Mr. Bonteen came out from the front door alone, and seeing the three men standing, passed on towards the left, eastwards. 'Good night, Erle,' he said. 'Good night, Fitzgibbon.' The two men answered him, and Phineas stood back in the gloom. It was about one o'clock and the night was very dark. 'By George, I do dislike that man,' said Phineas. Then, with a laugh, he took a life-preserver*out of his pocket, and made an action with it as though he were striking some enemy over the head. In those days there had been much garotting*in the streets, and writers in the Press had advised those who walked about at night to go armed with sticks. Phineas Finn had himself been once engaged with garotters,—as has been told in a former chronicle,—and had since armed himself, thinking more probably of the thing which he had happened to see than men do who had only heard of it. As soon as he had spoken, he followed Mr. Bonteen down the street, at the distance of perhaps a couple of hundred yards.

'They won't have a row,—will they?' said Erle.

'Oh, dear, no; Finn won't think of speaking to him; and you may be sure that Bonteen won't say a word to Finn. Between you and me, Barrington, I wish Master Phineas would give him a thorough good hiding.'

CHAPTER XLVII

What came of the quarrel

On the next morning at seven o'clock a superintendent of police called at the house of Mr. Gresham and informed the Prime Minister that Mr. Bonteen, the President of the Board of Trade, had been murdered during the night. There was no doubt of the fact. The body had been recognised, and information had been taken to the unfortunate widow at the house Mr. Bonteen had occupied in St. James's Place. The superintendent had already found out that Mr. Bonteen had been attacked as he was returning from his club late at night, —or rather, early in the morning, and expressed no doubt that he had been murdered close to the spot on which his body was found. There is a dark, uncanny-looking passage running from the end of Bolton Row, in May Fair, between the gardens of two great noblemen, coming out among the mews in

Berkeley Street, at the corner of Berkeley Square, just opposite to the bottom of Hay Hill. It was on the steps leading up from the passage to the level of the ground above that the body was found. The passage was almost as near a way as any from the club to Mr. Bonteen's house in St. James's Place; but the superintendent declared that gentlemen but seldom used the passage after dark, and he was disposed to think that the unfortunate man must have been forced down the steps by the ruffian who had attacked him from the level above. The murderer, so thought the superintendent, must have been cognizant of the way usually taken by Mr. Bonteen, and must have lain in wait for him in the darkness of the mouth of the passage. The superintendent had been at work on his inquiries since four in the morning, and had heard from Lady Eustace,—and from Mrs. Bonteen, as far as that poor distracted woman had been able to tell her story,—some account of the cause of quarrel between the respective husbands of those two ladies. The officer, who had not as yet heard a word of the late disturbance between Mr. Bonteen and Phineas Finn, was strongly of opinion that the Reverend Mr. Emilius had been the murderer. Mr. Gresham, of course, coincided in that opinion. What steps had been taken as to the arrest of Mr. Emilius? The superintendent was of opinion that Mr. Emilius was already in custody. He was known to be lodging close to the Marylebone Workhouse, in Northumberland Street, having removed to that somewhat obscure neighbourhood as soon as his house in Lowndes Square had been broken up by the running away of his wife and his consequent want of means. Such was the story as told to the Prime Minister at seven o'clock in the morning.

At eleven o'clock, at his private room at the Treasury Chambers, Mr. Gresham heard much more. At that time there were present with him two officers of the police force, his colleagues in the Cabinet, Lord Cantrip and the Duke of Omnium, three of his junior colleagues in the Government, Lord Fawn, Barrington Erle, and Laurence Fitzgibbon,—and Major Mackintosh, the chief of the London police. It was not exactly part of the duty of Mr. Gresham to investigate the circumstances of this murder; but there was so much in it that brought it closely home to him and his Government, that it became impossible for him not to concern himself in the business. There had been so much talk about Mr. Bonteen lately, his name had been so common in the newspapers, the ill-usage which he had been supposed by some to have suffered had been so freely discussed, and his quarrel, not only with Phineas Finn, but subsequently with the Duke of Omnium, had been so widely known,—that his sudden death created more momentary excitement than might probably have followed that of a greater man. And now, too, the facts of the past night, as they became known, seemed to make the crime more wonderful, more exciting, more momentous than it would have been had it been brought clearly home to such a wretch as the Bohemian Jew, Yosef Mealyus, who had contrived to cheat that wretched Lizzie Eustace into marrying him.

As regarded Yosef Mealyus the story now told respecting him was this. He was already in custody. He had been found in bed at his lodgings between seven and eight, and had, of course, given himself up without difficulty. He had seemed to be horror-struck when he heard of the man's death—,but had openly expressed his joy. 'He has endeavoured to ruin me, and has done me a world of harm. Why should I sorrow for him?'—he said to the policeman when rebuked for his inhumanity. But nothing had been found tending to implicate him in the crime. The servant declared that he had gone to bed before eleven o'clock, to her knowledge,—for she had seen him there,—and that he had not left the house afterwards. Was he in possession of a latch-key? It appeared that he did usually carry a latch-key, but that it was often borrowed from him by members of the family when it was known that he would not want it himself,—and that it had been so lent on this night. It was considered certain by those in the house that he had not gone out after he went to bed. Nobody in fact had left the house after ten; but in accordance with his usual custom Mr. Emilius had sent down the key as soon as he had found that he would not want it, and it had been all night in the custody of the mistress of the establishment. Nevertheless his clothes were examined minutely, but without affording any evidence against him. That Mr. Bonteen had been killed with some blunt weapon, such as a life-preserver, was assumed by the police, but no such weapon was in the possession of Mr. Emilius, nor had any such weapon yet been found. He was, however, in custody, with no evidence against him except that which was afforded by his known and acknowledged enmity to Mr. Bonteen.

So far, Major Mackintosh and the two officers had told their story. Then came the united story of the other gentlemen assembled,—from hearing which, however, the two police officers were debarred. The Duke and Barrington Erle had both dined in company with Phineas Finn at Madame Goesler's, and the Duke was undoubtedly aware that ill blood had existed between Finn and Mr. Bonteen. Both Erle and Fitzgibbon described the quarrel at the club, and described also the anger which Finn had expressed against the wretched man as he stood talking at the club door. His gesture of vengeance was remembered and repeated, though both the men who heard it expressed their strongest conviction that the murder had not been committed by him. As Erle remarked, the very expression of such a threat was almost proof that he had not at that moment any intention on his mind of doing such a deed as had been done. But they told also of the life-preserver which Finn had shown them, as he took it from the pocket of his outside coat, and they marvelled at the coincidences of the night. Then Lord Fawn gave further evidence, which seemed to tell very hardly upon Phineas Finn. He also had been at the club, and had left it just before Finn and the two other men had clustered at the door. He had walked very slowly, having turned down to Curzon Street and Bolton Row, from whence he made his way into Piccadilly by Clarges Street. He had seen nothing of Mr. Bonteen; but as he crossed over to Clarges Street he was passed at a very rapid pace by a man muffled in a top coat, who made his way straight along Bolton Row towards the passage which has been described. At the moment he had not connected the person of the man who passed him with any acquaintance of his own; but he now felt sure,—after what he had heard,—that the man was Mr. Finn. As he passed out of the club Finn was putting on his overcoat, and Lord Fawn had observed the peculiarity of the grey colour. It was exactly a similar coat, only with its collar raised, that had passed him in the street. The man, too, was of Mr. Finn's height and build. He had known Mr. Finn well, and the man stepped with Mr. Finn's step. Major Mackintosh thought that Lord Fawn's evidence was—'very unfortunate as regarded Mr. Finn.'

'I'm d—— if that idiot won't hang poor Phinny,' said Fitzgibbon afterwards to Erle. 'And yet I don't believe a word of it.'

'Fawn wouldn't lie for the sake of hanging Phineas Finn,' said Erle.

'No;—I don't suppose he's given to lying at all. He believes it all. But he's such a muddle-headed fellow that he can get himself to believe anything. He's one of those men who always unconsciously exaggerate what they have to say for the sake of the importance it gives them.' It might be possible that a jury would look at Lord Fawn's evidence in this light; otherwise it would bear very heavily, indeed, against Phineas Finn.

Then a question arose as to the road which Mr. Bonteen usually took from the club. All the members who were there present had walked home with him at various times,—and by various routes, but never by the way through the passage. It was supposed that on this occasion he must have gone by Berkeley Square, because he had certainly not turned down by the first street to the right, which he would have taken had he intended to avoid the square. He had been seen by Barrington Erle and Fitzgibbon to pass that turning. Otherwise they would have made no remark as to the possibility of a renewed quarrel between him and Phineas, should Phineas chance to overtake him;—for Phineas would certainly go by the square unless taken out of his way by some special purpose. The most direct way of all for Mr. Bonteen would have been that followed by Lord Fawn; but as he had not turned down this street, and had not been seen by Lord Fawn, who was known to walk very slowly, and had often been seen to go by Berkeley Square,—it was presumed that he had now taken that road. In this case he would certainly pass the end of the passage towards which Lord Fawn declared that he had seen the man hurrying whom he now supposed to have been Phineas Finn. Finn's direct road home would, as has been already said, have been through the square, cutting off the corner of the square, towards Bruton Street, and thence across Bond Street by Conduit Street to Regent Street, and so to Great Marlborough Street, where he lived. But it had been, no doubt, possible for him to have been on the spot on which Lord Fawn had seen the man; for, although in his natural course thither from the club he would have at once gone down the street to the right,—a course which both Erle and Fitzgibbon were able to say that he did not take, as they had seen him go beyond the turning,—nevertheless there had been ample time for him to have retraced his steps to it in time to have caught Lord Fawn, and thus to have deceived Fitzgibbon and Erle as to the route he had taken.

When they had got thus far Lord Cantrip was standing close to the window of the room at Mr. Gresham's elbow. 'Don't allow yourself to be hurried into believing it,' said Lord Cantrip.

'I do not know that we need believe it, or the reverse. It is a case for the police.'

'Of course it is;—but your belief and mine will have a weight. Nothing that I have heard makes me for a moment think it possible. I know the man.'

'He was very angry.'

'Had he struck him in the club I should not have been much surprised; but he never attacked his enemy with a bludgeon in a dark alley. I know him well.'

'What do you think of Fawn's story?'

'He was mistaken in his man. Remember;—it was a dark night.'

'I do not see that you and I can do anything,' said Mr. Gresham. 'I shall have to say something in the House as to the poor fellow's death, but I certainly shall not express a suspicion. Why should I?'

Up to this moment nothing had been done as to Phineas Finn. It was known that he would in his natural course of business be in his place in Parliament at four, and Major Mackintosh was of opinion that he certainly should be taken before a magistrate in time to prevent the necessity of arresting him in the House. It was decided that Lord Fawn, with Fitzgibbon and Erle, should accompany the police officer to Bow Street, and that a magistrate should be applied to for a warrant if he thought the evidence was sufficient. Major Mackintosh was of opinion that, although by no possibility could the two men suspected have been jointly guilty of the murder, still the circumstances were such as to justify the immediate arrest of both. Were Yosef Mealyus really guilty and to be allowed to slip from their hands, no doubt it might be very difficult to catch him. Facts did not at present seem to prevail against him; but, as the Major observed, facts are apt to alter considerably when they are minutely sifted. His character was half sufficient to condemn him;—and then with him there was an adequate motive, and what Lord Cantrip regarded as 'a possibility.' It was not to be conceived that from mere rage Phineas Finn would lay a plot for murdering a man in the street. 'It is on the cards, my lord,' said the Major, 'that he may have chosen to attack Mr. Bonteen without intending to murder him. The murder may afterwards have been an accident.'

It was impossible after this for even a Prime Minister and two Cabinet Ministers to go about their work calmly. The men concerned had been too well known to them to allow their minds to become clear of the subject. When Major Mackintosh went off to Bow Street with Erle and Laurence, it was certainly the opinion of the majority of those who had been present that the blow had been struck by the hand of Phineas Finn. And perhaps the worst aspect of it all was that there had been not simply a blow,—but blows. The constables had declared that the murdered man had been struck thrice about the head, and that the fatal stroke had been given on the side of his head after the man's hat had been knocked off. That Finn should have followed his enemy through the street, after such words as he had spoken, with the view of having the quarrel out in some shape, did not seem to be very improbable to any of them except Lord Cantrip;—and then had there been a scuffle, out in the open path, at the spot at which the angry man might have overtaken his adversary, it was not incredible to them that he should have drawn even such a weapon as a life-preserver from his pocket. But, in the case as it had occurred, a spot peculiarly traitorous had been selected, and the attack had too probably been made from behind. As yet there was no evidence that the murderer had himself encountered any ill-usage. And Finn, if he was the murderer, must, from the time he was standing at the club door, have contemplated a traitorous, dastardly attack. He must have counted his moments;—have returned slyly in the dark to the corner of the street which he had once passed;—have muffled his face in his coat;—and have then laid wait in a spot to which an honest man at night would hardly trust himself with honest purposes. 'I look upon it as quite out of the question,' said Lord Cantrip, when the three Ministers were left alone. Now Lord Cantrip had served for many months in the same office as Phineas Finn.

'You are simply putting your own opinion of the man against the facts,' said Mr. Gresham. 'But facts always convince, and another man's opinion rarely convinces.'

'I'm not sure that we know the facts yet,' said the Duke.

'Of course we are speaking of them as far as they have been told to us. As far as they go,—unless they can be upset and shown not to be facts,—I fear they would be conclusive to me on a jury.'

'Do you mean that you have heard enough to condemn him?' asked Lord Cantrip.

'Remember what we have heard. The murdered man had two enemies.'

'He may have had a third.'

'Or ten; but we have heard of but two.'

'He may have been attacked for his money,' said the Duke.

'But neither his money nor his watch were touched,' continued Mr. Gresham. 'Anger, or the desire of putting the man out of the way, has caused the murder. Of the two enemies one,—according to the facts as we now have them,—could not have been there. Nor is it probable that he could have known that his enemy would be on that spot. The other not only could have been there, but was certainly near the place at the moment,—so near that did he not do the deed himself, it is almost wonderful that it should not have been interrupted in its doing by his nearness. He certainly knew that the victim would be there. He was burning with anger against him at the moment. He had just threatened him. He had with him such an instrument as was afterwards used. A man believed to be him is seen hurrying to the spot by a witness whose credibility is beyond doubt. These are the facts such as we have them at present. Unless they can be upset, I fear they would convince a jury,—as they have already convinced those officers of the police.'

'Officers of the police always believe men to be guilty,' said Lord Cantrip.

'They don't believe the Jew clergyman to be guilty,' said Mr. Gresham.

'I fear that there will be enough to send Mr. Finn to a trial,' said the Duke.

'Not a doubt of it,' said Mr. Gresham.

'And yet I feel as convinced of his innocence as I do of my own,' said Lord Cantrip.

CHAPTER XLVIII

Mr. Maule's attempt

ABOUT three o'clock in the day the first tidings of what had taken place reached Madame Goesler in the following perturbed note from her friend the Duchess:—'Have you heard what took place last night? Good God! Mr. Bonteen was murdered as he came home from his club, and they say that it was done by Phineas Finn. Plantagenet has just come in from Downing Street, where everybody is talking about it. I can't get from him what he believes. One never can get anything from him. But I never will believe it;—nor will you, I'm sure. I vote we stick to him to the last. He is to be put in prison and tried. I can hardly believe that Mr. Bonteen has been murdered, though I don't know why he shouldn't as well as anybody else. Plantagenet talks about the great loss; I know which would be the greatest loss, and so do you. I'm going out now to try and find out something. Barrington Erle was there, and if I can find him he will tell me. I shall be home by half-past five. Do come, there's a dear woman; there is no one else I can talk to about it. If I'm not back, go in all the same, and tell them to bring you tea.

'Only think of Lady Laura,—with one mad and the other in Newgate! G. P.'

This letter gave Madame Goesler such a blow that for a few minutes it altogether knocked her down. After reading it once she hardly knew what it contained beyond a statement that Phineas Finn was in Newgate. She sat for a while with it in her hands, almost swooning; and then with an effort she recovered herself, and read the letter again. Mr. Bonteen murdered, and Phineas Finn,—who had dined with her only yesterday evening, with whom she had been talking of all the sins of the murdered man, who was her special friend, of whom she thought more than of any other human being, of whom she could not bring herself to cease to think,—accused of the murder! Believe it! The Duchess had declared with that sort of enthusiasm which was common to her, that she never would believe it. No, indeed! What judge of character would any one be who could believe that Phineas Finn could be guilty of a midnight murder? 'I vote we stick to him.' 'Stick to him!' Madame Goesler said, repeating the words to herself. 'What is the use of sticking to a man who does not want you?' How can a woman cling to a man who, having said that he did not want her, yet comes again within her influence, but does not unsay what he had said before? Nevertheless, if it should be that the man was in real distress,—in absolutely dire sorrow,—she would cling to him with a constancy which, as she thought, her friend the Duchess would hardly understand. Though they should hang him, she would bathe his body with her tears, and live as a woman should live who had loved a murderer to the last.

But she swore to herself that she would not believe it. Nay, she did not believe it. Believe it, indeed! It was simply impossible. That he might have killed the wretch in some struggle brought on by the man's own fault was possible. Had the man attacked Phineas Finn it was only too probable that there might have been such result. But murder, secret midnight murder, could not have been committed by the man she had chosen as her friend. And yet, through it all, there was a resolve that even though he should have committed murder she would be true to him. If it should come to the very worst, then would she declare the intensity of the affection with which she regarded the murderer. As to Mr. Bonteen, what the Duchess said was true enough; why should not he be killed as well as another? In her present frame of mind she felt very little pity for Mr. Bonteen. After a fashion a verdict of 'served him right' crossed her mind, as it had doubtless crossed that of the Duchess when she was writing her letter. The man had made himself so obnoxious that it was well that he should be out of the way. But not on that account would she believe that Phineas Finn had murdered him.

Could it be true that the man after all was dead? Marvellous reports, and reports marvellously false, do spread themselves about the world every day. But this report had come from the Duke, and he was not a man given to absurd rumours. He had heard the story in Downing Street, and if so it must be true. Of course she would go down to the Duchess at the hour fixed. It was now a little after three, and she ordered the carriage to be ready for her at a quarter past five. Then she told the servant, at first to admit no one who might call, and then to come up and let her know, if any one should come, without sending the visitor away. It might be that some one would come to her expressly from Phineas, or at least with tidings about this affair.

Then she read the letter again, and those few last words in it stuck to her thoughts like a burr. 'Think of Lady Laura, with one mad and the other in Newgate.' Was this man,—the only man whom she had ever loved,—more to Lady Laura Kennedy than to her; or rather, was Lady Laura more to him than was she herself? If so, why should she fret herself for his sake? She was ready enough to own that she could sacrifice everything for him, even though he should be standing as a murderer in the dock, if such sacrifice would be valued by him. He had himself told her that his feelings towards Lady Laura were simply those of an affectionate friend; but how could she believe that statement when all the world were saying the reverse? Lady Laura was a married woman,—a woman whose husband was still living,—and of course he was bound to make such an assertion when he and she were named together. And then it was certain,—Madame Goesler believed it to be certain,—that there had been a time in which Phineas had asked for the love of Lady Laura Standish. But he had never asked for her love. It had been tendered to him, and he had rejected it! And now the Duchess,—who, with all her inaccuracies, had that sharpness of vision which enables some men and women to see into facts,—spoke as though Lady Laura were to be pitied more than all others, because of the evil that had befallen Phineas Finn! Had not Lady Laura chosen her own husband; and was not the man, let him be ever so mad, still her husband? Madame Goesler was sore of heart, as well as broken down with sorrow, till at last, hiding her face on the pillow of the sofa, still holding the Duchess's letter in her hand, she burst into a fit of hysteric sobs.

Few of those who knew Madame Max Goesler well, as she lived in town and in country, would have believed that such could have been the effect upon her of the news which she had heard. Credit was given to her everywhere for good nature, discretion, affability, and a certain grace of demeanour which always made her charming. She was known to be generous, wise, and of high spirit. Something of her conduct to the old Duke had crept into general notice, and had been told, here and there, to her honour. She had conquered the good opinion of many, and was a popular woman. But there was not one among her friends who supposed her capable of becoming a victim to a strong passion, or would have suspected her of reckless weeping for any sorrow. The Duchess, who thought that she knew Madame Goesler well, would not have believed it to be true, even if she had seen it. 'You like people, but I don't think you ever love any one,' the Duchess had once said to her. Madame Goesler had smiled, and had seemed to assent. To enjoy the world,—and to know that the best enjoyment must come from witnessing the satisfaction of others, had apparently been her philosophy. But now she was prostrate because this man was in trouble, and because she had been told that his trouble was more than another woman could bear!

She was still sobbing and crushing the letter in her hand when the servant came up to tell her that Mr. Maule had called. He was below, waiting to know whether she would see him. She remembered at once that Mr. Maule had met Phineas at her table on the previous evening, and, thinking that he must have come with tidings respecting this great event, desired that he might be shown up to her. But, as it happened, Mr. Maule had not yet heard of the death of Mr. Bonteen. He had remained at home till nearly four, having a great object in view, which made him deem it expedient that he should go direct from his own rooms to Madame Goesler's house, and had not even looked in at his club. The reader will, perhaps, divine the great object. On this day he proposed to ask Madame Goesler to make him the happiest of men,—as he certainly would have thought himself for a time, had she consented to put him in possession of her large income. He had therefore padded himself with more than ordinary care,—reduced but not obliterated the greyness of his locks,—looked carefully to the fitting of his trousers, and spared himself those ordinary labours of the morning which might have robbed him of any remaining spark of his juvenility.

Madame Goesler met him more than half across the room as he entered it. 'What have you heard?' said she. Mr. Maule wore his sweetest smile, but he had heard nothing. He could only press her hand, and look blank,—understanding that there was something which he ought to have heard. She thought nothing of the pressure of her hand. Apt as she was to be conscious at an instant of all that was going on around her, she thought of nothing now but that man's peril, and of the truth or falsehood of the story that had been sent to her. 'You have heard nothing of Mr. Finn?'

'Not a word,' said Mr. Maule, withdrawing his hand. 'What has happened to Mr. Finn?' Had Mr. Finn broken his neck it would have been nothing to Mr. Maule. But the lady's solicitude was something to him.

'Mr. Bonteen has been——murdered!'

'Mr. Bonteen!'

'So I hear. I thought you had come to tell me of it.'

'Mr. Bonteen murdered! No;—I have heard nothing. I do not know the gentleman. I thought you said—Mr. Finn.'

'It is not known about London, then?'

'I cannot say, Madame Goesler. I have just come from home, and have not been out all the morning. Who has——murdered him?'

'Ah! I do not know. That is what I wanted you to tell me.'

'But what of Mr. Finn?'

'I also have not been out, Mr. Maule, and can give you no

information. I thought you had called because you knew that Mr. Finn had dined here.'

'Has Mr. Finn been murdered?'

'Mr. Bonteen! I said that the report was that Mr. Bonteen had been murdered.' Madame Goesler was now waxing angry,—most unreasonably. 'But I know nothing about it, and am just going out to make inquiry. The carriage is ordered.' Then she stood, expecting him to go; and he knew that he was expected to go. It was at any rate clear to him that he could not carry out his great design on the present occasion. 'This has so upset me that I can think of nothing else at present, and you must, if you please, excuse me. I would not have let you take the trouble of coming up, had not I thought that you were the bearer of some news.' Then she bowed, and Mr. Maule bowed; and as he left the room she forgot to ring the bell.

'What the deuce can she have meant about that fellow Finn?' he said to himself. 'They cannot both have been murdered.' He went to his club, and there he soon learned the truth. The information was given to him with clear and undoubting words. Phineas Finn and Mr. Bonteen had quarrelled at The Universe. Mr. Bonteen, as far as words went, had got the best of his adversary. This had taken place in the presence of the Prince, who had expressed himself as greatly annoyed by Mr. Finn's conduct. And afterwards Phineas Finn had waylaid Mr. Bonteen in the passage between Bolton Row and Berkeley Street, and had there—murdered him. As it happened, no one who had been at The Universe was at that moment present; but the whole affair was now quite well known, and was spoken of without a doubt.

'I hope he'll be hung, with all my heart,' said Mr. Maule, who thought that he could read the riddle which had been so unintelligible in Park Lane.

When Madame Goesler reached Carlton Terrace, which she did before the time named by the Duchess, her friend had not yet returned. But she went upstairs, as she had been desired, and they brought her tea. But the teapot remained untouched till past six o'clock, and then the Duchess returned. 'Oh, my dear, I am so sorry for being late. Why haven't you had tea?'

'What is the truth of it all?' said Madame Goesler, standing up with her fists clenched as they hung by her side.

'I don't seem to know nearly as much as I did when I wrote to you.'

'Has the man been—murdered?'

'Oh dear, yes. There's no doubt about that. I was quite sure of that when I sent the letter. I have had such a hunt. But at last I went up to the door of the House of Commons, and got Barrington Erle to come out to me.'

'Well?'

'Two men have been arrested.'

'Not Phineas Finn?'

'Yes; Mr. Finn is one of them. Is it not awful? So much more dreadful to me than the other poor man's death! One oughtn't to say so, of course.'

'And who is the other man? Of course he did it.'

'That horrid Jew preaching man that married Lizzie Eustace. Mr. Bonteen had been persecuting him, and making out that he had another wife at home in Hungary, or Bohemia, or somewhere.'

'Of course he did it.'

'That's what I say. Of course the Jew did it. But then all the evidence goes to show that he didn't do it. He was in bed at the time; and the door of the house was locked up so that he couldn't get out; and the man who did the murder hadn't got on his coat, but had got on Phineas Finn's coat.'

'Was there—blood?' asked Madame Goesler, shaking from head to foot.

'Not that I know. I don't suppose they've looked yet. But Lord Fawn saw the man, and swears to the coat.'

'Lord Fawn! How I have always hated that man! I wouldn't believe a word he would say.'

'Barrington doesn't think so much of the coat. But Phineas had a club in his pocket, and the man was killed by a club.

There hasn't been any other club found, but Phineas Finn took his home with him.'

'A murderer would not have done that.'

'Barrington says that the head policeman says that it is just what a very clever murderer would do.'

'Do you believe it, Duchess?'

'Certainly not;—not though Lord Fawn swore that he had seen it. I never will believe what I don't like to believe, and nothing shall ever make me.'

'He couldn't have done it.'

'Well;—for the matter of that, I suppose he could.'

'No, Duchess, he could not have done it.'

'He is strong enough,—and brave enough.'

'But not enough of a coward. There is nothing cowardly about him. If Phineas Finn could have struck an enemy with a club, in a dark passage, behind his back, I will never care to speak to any man again. Nothing shall make me believe it. If I did, I could never again believe in any one. If they told you that your husband had murdered a man, what would you say?'

'But he isn't your husband, Madame Max.'

'No;—certainly not. I cannot fly at them, when they say so, as you would do. But I can be just as sure. If twenty Lord Fawns swore that they had seen it, I would not believe them. Oh, God, what will they do with him!'

The Duchess behaved very well to her friend, saying not a single word to twit her with the love which she betrayed. She seemed to take it as a matter of course that Madame Goesler's interest in Phineas Finn should be as it was. The Duke, she said, could not come home to dinner, and Madame Goesler should stay with her. Both Houses were in such a ferment about the murder, that nobody liked to be away. Everybody had been struck with amazement, not simply,—not chiefly,— by the fact of the murder, but by the double destruction of the two men whose ill-will to each other had been of late so often the subject of conversation. So Madame Goesler remained at Carlton Terrace till late in the evening, and during the whole visit there was nothing mentioned but the murder of Mr. Bonteen and the peril of Phineas Finn. 'Some one will go and see him, I suppose,' said Madame Goesler.

'Lord Cantrip has been already,—and Mr. Monk.'

'Could not I go?'

'Well, it would be rather strong.'

'If we both went together?' suggested Madame Goesler. And before she left Carlton Terrace she had almost extracted a promise from the Duchess that they would together proceed to the prison and endeavour to see Phineas Finn.

CHAPTER XLIX

Showing what Mrs. Bunce said to the policeman

'WE HAVE left Adelaide Palliser down at the Hall. We are up here only for a couple of days to see Laura, and try to find out what had better be done about Kennedy.' This was said to Phineas Finn in his own room in Great Marlborough Street by Lord Chiltern, on the morning after the murder, between ten and eleven o'clock. Phineas had not as yet heard of the death of the man with whom he had quarrelled. Lord Chiltern had now come to him with some proposition which he as yet did not understand, and which Lord Chiltern certainly did not know how to explain. Looked at simply, the proposition was one for providing Phineas Finn with an income out of the wealth belonging, or that would belong, to the Standish family. Lady Laura's fortune would, it was thought, soon be at her own disposal. They who acted for her husband had assured the Earl that the yearly interest of the money should be at her ladyship's command as soon as the law would allow them so to plan it. Of Robert Kennedy's inability to act for himself there was no longer any doubt whatever, and there was, they said, no desire to embarrass the estate with so small a disputed matter as the income derived from £40,000. There was great pride of purse in the manner in which the information was conveyed;—but not the less on that account was it satisfactory to the Earl. Lady Laura's first thought about it referred to the imminent wants of Phineas Finn. How might it be possible for her to place a portion of her income at the command of the man she loved so that he should not feel disgraced by receiving it from her hand? She conceived some plan as to a loan to be made nominally by her brother,—a plan as to which it may at once be said that it could not be made to hold water for a minute. But she did succeed in inducing her brother to undertake the embassy, with the view of explaining to Phineas that there would be money for him when he wanted it. 'If I make it over to Papa, Papa can leave it him in his will; and if he wants it at once there can be no harm in your advancing to him what he must have at Papa's death.' Her brother had frowned angrily and had shaken his head. 'Think how he has been thrown over by all the party,' said Lady Laura. Lord Chiltern had disliked the whole affair,—had felt with dismay that his sister's name would become subject to reproach if it should be known that this young man was supported by her bounty. She, however, had persisted, and he had consented to see the young man, feeling sure that Phineas would refuse to bear the burden of the obligation.

But he had not touched the disagreeable subject when they were interrupted. A knocking of the door had been heard, and now Mrs. Bunce came upstairs, bringing Mr. Low with her. Mrs. Bunce had not heard of the tragedy, but she had at once perceived from the barrister's manner that there was some serious matter forward,—some matter that was probably not only serious, but also calamitous. The expression of her countenance announced as much to the two men, and the countenance of Mr. Low when he followed her into the room told the same story still more plainly. 'Is anything the matter?' said Phineas, jumping up.

'Indeed, yes,' said Mr. Low, who then looked at Lord Chiltern and was silent.

'Shall I go?' said Lord Chiltern. Mr. Low did not know him, and of course was still silent.

'This is my friend, Mr. Low. This is my friend, Lord Chiltern,' said Phineas, aware that each was well acquainted with the other's name. 'I do not know of any reason why you should go. What is it, Low?'

Lord Chiltern had come there about money, and it occurred to him that the impecunious young barrister might already be in some scrape on that head. In nineteen cases out of twenty, when a man is in a scrape, he simply wants money. 'Perhaps I can be of help,' he said.

'Have you heard, my Lord, what happened last night?' said Mr. Low, with his eyes fixed on Phineas Finn.

'I have heard nothing,' said Lord Chiltern.

'What has happened?' asked Phineas, looking aghast. He knew Mr. Low well enough to be sure that the thing referred to was of great and distressing moment.

'You, too, have heard nothing?'

'Not a word—that I know of.'

'You were at The Universe last night?'

'Certainly I was.'

'Did anything occur?'

'The Prince was there.'

'Nothing has happened to the Prince?' said Chiltern.

'His name has not been mentioned to me,' said Mr. Low. 'Was there not a quarrel?'

'Yes;'—said Phineas. 'I quarrelled with Mr. Bonteen.'

'What then?'

'He behaved like a brute;—as he always does. Thrashing a brute hardly answers nowadays, but if ever a man deserved a thrashing he does.'

'He has been murdered,' said Mr. Low.

. . . . .

The reader need hardly be told that, as regards this great offence, Phineas Finn was as white as snow. The maintenance of any doubt on that matter,—were it even desirable to maintain a doubt,—would be altogether beyond the power of the present writer. The reader has probably perceived, from the first moment of the discovery of the body on the steps at the end of the passage, that Mr. Bonteen had been killed by that ingenious gentleman, the Rev. Mr. Emilius, who found it to be worth his while to take the step with the view of suppressing his enemy's evidence as to his former marriage. But Mr. Low, when he entered the room, had been inclined to think that his friend had done the deed. Laurence Fitzgibbon, who had been one of the first to hear the story, and who had summoned Erle to go with him and Major Mackintosh to Downing Street, had, in the first place, gone to the house in Carey Street, in which Bunce was wont to work, and had sent him to Mr. Low. He, Fitzgibbon, had not thought it safe that he himself should warn his countryman, but he could not bear to think that the hare should be knocked over on its form, or that his friend should be taken by policemen without notice. So he had sent Bunce to Mr. Low, and Mr. Low had now come with his tidings.

'Murdered!' exclaimed Phineas.

'Who has murdered him?' said Lord Chiltern, looking first at Mr. Low and then at Phineas.

'That is what the police are now endeavouring to find out.' Then there was a pause, and Phineas stood up with his hand on his forehead, looking savagely from one to the other. A glimmer of an idea of the truth was beginning to cross his brain. Mr. Low was there with the object of asking him whether he had murdered the man! 'Mr. Fitzgibbon was with you last night,' continued Mr. Low.

'Of course he was.'

'It was he who has sent me to you.'

'What does it all mean?' asked Lord Chiltern. 'I suppose they do not intend to say that,—our friend, here,—murdered the man.'

'I begin to suppose that is what they intend to say,' rejoined Phineas, scornfully.

Mr. Low had entered the room, doubting indeed, but still inclined to believe,—as Bunce had very clearly believed,—that the hands of Phineas Finn were red with the blood of this man who had been killed. And, had he been questioned on such a matter, when no special case was before his mind, he would have declared of himself that a few tones from the voice, or a few glances from the eye, of a suspected man would certainly not suffice to eradicate suspicion. But now he was quite sure,—almost quite sure,—that Phineas was as innocent as himself. To Lord Chiltern, who had heard none of the details, the suspicion was so monstrous as to fill him with wrath. 'You don't mean to tell us, Mr. Low, that any one says that Finn killed the man?'

'I have come as his friend,' said Low, 'to put him on his guard. The accusation will be made against him.'

To Phineas, not clearly looking at it, not knowing very accurately what had happened, not being in truth quite sure that Mr. Bonteen was actually dead, this seemed to be a continuation of the persecution which he believed himself to have suffered from that man's hand. 'I can believe anything from that quarter,' he said.

'From what quarter?' asked Lord Chiltern. 'We had better let Mr. Low tell us what really has happened.'

Then Mr. Low told the story, as well as he knew it, describing the spot on which the body had been found. 'Often as I go to the club,' said Phineas, 'I never was through that passage in my life.' Mr. Low went on with his tale, telling how the man had been killed with some short bludgeon. 'I had that in my pocket,' said Finn, producing the life-preserver. 'I have almost always had something of the kind when I have been in London, since that affair of Kennedy's.' Mr. Low cast one glance at it,—to see whether it had been washed or scraped, or in any way cleansed. Phineas saw the glance, and was angry. 'There it is, as it is. You can make the most of it. I shall not touch it again till the policeman comes. Don't put your hand on it, Chiltern. Leave it there.' And the instrument was left lying on the table, untouched. Mr. Low went on with his story. He had heard nothing of Yosef Mealyus as connected with the murder, but some indistinct reference to Lord Fawn and the top-coat had been made to him. 'There is the coat, too,' said Phineas, taking it from the sofa on which he had flung it when he came home the previous night. It was a very light coat,—-fitted for May use,—lined with silk, and by no means suited for enveloping the face or person. But it had a collar which might be made to stand up. 'That at any rate was the coat I wore,' said Finn, in answer to some observation from the barrister. 'The man that Lord Fawn saw,' said Mr. Low, 'was, as I understand, enveloped in a heavy great coat.' 'So Fawn has got his finger in the pie!' said Lord Chiltern.

Mr. Low had been there an hour, Lord Chiltern remaining also in the room, when there came three men belonging to the police,—a superintendent and with him two constables. When the men were shown up into the room neither the bludgeon or the coat had been moved from the small table as Phineas had himself placed them there. Both Phineas and Chiltern had lit cigars, and they were all there sitting in silence. Phineas had entertained the idea that Mr. Low believed the charge, and that the barrister was therefore an enemy. Mr. Low had perceived this, but had not felt it to be his duty to declare his opinion of his friend's innocence. What he could do for his friend he would do; but, as he thought, he could serve him better now by silent observation than by protestation. Lord Chiltern, who had been implored by Phineas not to leave him, continued to pour forth unabating execrations on the monstrous malignity of the accusers. 'I do not know that there are any accusers,' said Mr. Low, 'except the circumstances which the police must, of course, investigate.' Then the men came, and the nature of their duty was soon explained. They must request Mr. Finn to go with them to Bow Street. They took possession of many articles besides the two which had been prepared for them,—the dress coat and shirt which Phineas had worn, and the boots. He had gone out to dinner with a Gibus hat,* and they took that. They took his umbrella and his latch-key. They asked, even, as to his purse and money;—but abstained from taking the purse when Mr. Low suggested that they could have no concern with that. As it happened, Phineas was at the moment wearing the shirt in which he had dined out on the previous day, and the men asked him whether he had any objection to change it in their presence,—as it might be necessary, after the examination, that it should be detained as evidence. He did so, in the presence of all the men assembled; but the humiliation of doing it almost broke his heart. Then they searched among his linen, clean and dirty, and asked questions of Mrs. Bunce in audible whispers behind the door. Whatever Mrs. Bunce could do to injure the cause of her favourite lodger by severity of manner, snubbing the policeman, and determination to give no information, she did do. 'Had a shirt washed? How do you suppose a gentleman's shirts are washed? You were brought up near enough to a washtub yourself to know more than I can tell you!' But the very respectable constable did not seem to be in the least annoyed by the landlady's amenities.

He was taken to Bow Street, going thither in a cab with the two policemen, and the superintendent followed them with Lord Chiltern and Mr. Low. 'You don't mean to say that you believe it?' said Lord Chiltern to the officer. 'We never believe and we never disbelieve anything, my Lord,' replied the man. Nevertheless, the superintendent did most firmly believe that Phineas Finn had murdered Mr. Bonteen.

At the police-office Phineas was met by Lord Cantrip and Barrington Erle, and soon became aware that both Lord Fawn and Fitzgibbon were present. It seemed that everything else was made to give way to this inquiry, as he was at once confronted by the magistrate. Everybody was personally very civil to him, and he was asked whether he would not wish to have professional advice while the charge was being made against him. But this he declined. He would tell the magistrate, he said, all he knew, but, at any rate for the present, he would have no need of advice. He was, at last, allowed to tell his own story,—after repeated cautions. There had been some words between him and Mr. Bonteen in the club; after which, standing at the door of the club with his friends, Mr. Erle and Mr. Fitzgibbon, who were now in court, he had seen Mr. Bonteen walk away towards Berkeley Square. He had soon followed, but had never overtaken Mr. Bonteen. When reaching the Square he had crossed over to the fountain standing there on the south side, and from thence had taken the shortest way up Bruton Street. He had seen Mr. Bonteen for the last time dimly, by the gaslight, at the corner of the Square. As far as he could remember, he himself had at the moment passed the fountain. He had not heard the sound of any struggle, or of words, round the corner towards Piccadilly. By the time that Mr. Bonteen would have reached the head of the steps leading into the passage, he would have been near Bruton Street, with his back completely turned to the scene of the murder. He had walked faster than Mr. Bonteen, having gradually drawn near to him; but he had determined in his own mind that he would not pass the man, or get so near him as to attract attention. Nor had he done so. He had certainly worn the grey coat which was now produced. The collar of it had not been turned up. The coat was nearly new, and to the best of his belief the collar had never been turned up. He had carried the life-preserver now produced with him because it had once before been necessary for him to attack garotters in the street. The life-preserver had never been used, and, as it happened, was quite new. It had been bought about a month since,—in consequence of some commotion about garotters which had just then taken place. But before the purchase of the life-preserver he had been accustomed to carry some stick or bludgeon at night. Undoubtedly he had quarelled with Mr. Bonteen before this occasion, and had bought this instrument since the commencement of the quarrel. He had not seen any one on his way from the Square to his own house with sufficient observation to enable him to describe such person. He could not remember that he had passed a policeman on his way home.

This took place after the hearing of such evidence as was then given. The statements made both by Erle and Fitzgibbon as to what had taken place in the club, and afterwards at the door, tallied exactly with that afterwards given by Phineas. An accurate measurement of the streets and ways concerned was already furnished. Taking the duration of time as surmised by Erle and Fitzgibbon to have passed after they had turned their back upon Phineas, a constable proved that the prisoner would have had time to hurry back to the corner of the street he had passed, and to be in the place where Lord Fawn saw the man,—supposing that Lord Fawn had walked at the rate of three miles an hour, and that Phineas had walked or run at twice that pace. Lord Fawn stated that he was walking very slow,—less he thought than three miles an hour, and that the man was hurrying very fast,—not absolutely running, but going as he thought at quite double his own pace. The two coats were shown to his lordship. Finn knew nothing of the other coat,—which had, in truth, been taken from the Rev. Mr. Emilius,—a rough, thick, brown coat, which had belonged to the preacher for the last two years. Finn's coat was grey in colour. Lord Fawn looked at the coats very attentively, and then said that the man he had seen had certainly not worn the brown coat. The night had been dark, but still he was sure that the coat had been grey. The collar had certainly been turned up. Then a tailor was produced who gave it as his opinion that Finn's coat had been lately worn with the collar raised.

It was considered that the evidence given was sufficient to make a remand imperative, and Phineas Finn was committed to Newgate. He was assured that every attention should be paid to his comfort, and was treated with great consideration. Lord Cantrip, who still believed in him, discussed the subject both with the magistrate and with Major Mackintosh. Of course the strictest search would be made for a second life-preserver, or any such weapon as might have been used. Search had already been made, and no such weapon had been as yet found. Emilius had never been seen with any such weapon. No one about Curzon Street or Mayfair could be found who had seen the man with the quick step and raised collar, who doubtless had been the murderer, except Lord Fawn,—so that no evidence was forthcoming tending to show that Phineas Finn could not have been that man. The evidence adduced to prove that Mr. Emilius,—or Mealyus, as he was henceforth called,—could not have been on the spot was so very strong, that the magistrate told the constables that that man must be released on the next examination unless something could be adduced against him.

The magistrate, with the profoundest regret, was unable to agree with Lord Cantrip in his opinion that the evidence adduced was not sufficient to demand the temporary committal of Mr. Finn.

CHAPTER L

What the Lords and Commons said about the murder

WHEN the House met on that Thursday at four o'clock everybody was talking about the murder, and certainly four-fifths of the members had made up their minds that Phineas Finn was the murderer. To have known a murdered man is something, but to have been intimate with a murderer is certainly much more. There were many there who were really sorry for poor Bonteen,—of whom without a doubt the end had come in a very horrible manner; and there were more there who were personally fond of Phineas Finn,—to whom the future of the young member was very sad, and the fact that he should have become a murderer very awful. But, nevertheless, the occasion was not without its consolations. The business of the House is not always exciting, or even interesting. On this afternoon there was not a member who did not feel that something had occurred which added an interest to Parliamentary life.

Very soon after prayers Mr. Gresham entered the House, and men who had hitherto been behaving themselves after a most unparliamentary fashion, standing about in knots, talking by no means in whispers, moving in and out of the House rapidly, all crowded into their places. Whatever pretence of business had been going on was stopped in a moment, and Mr. Gresham rose to make his statement. 'It was with the deepest regret,—nay, with the most profound sorrow,—that he was called upon to inform the House that his right honourable friend and colleague, Mr. Bonteen, had been basely and cruelly murdered during the past night.' It was odd then to see how the name of the man, who, while he was alive and a member of that House, could not have been pronounced in that assembly without disorder, struck the members almost with dismay. 'Yes, his friend Mr. Bonteen, who had so lately filled the office of President of the Board of Trade, and whose loss the country and that House could so ill bear, had been beaten to death in one of the streets of the metropolis by the arm of a dastardly ruffian during the silent watches of the night.' Then Mr. Gresham paused, and every one expected that some further statement would be made. 'He did not know that he had any further communication to make on the subject. Some little time must elapse before he could fill the office. As for adequately supplying the loss, that would be impossible. Mr. Bonteen's services to the country, especially in reference to decimal coinage, were too well known to the House to allow of his holding out any such hope.' Then he sat down without having as yet made an allusion to Phineas Finn.

But the allusion was soon made. Mr. Daubeny rose, and with much graceful and mysterious circumlocution asked the Prime Minister whether it was true that a member of the House had been arrested, and was now in confinement on the charge of having been concerned in the murder of the late much-lamented President of the Board of Trade. He—Mr. Daubeny—had been given to understand that such a charge had been made against an honourable member of that House, who had once been a colleague of Mr. Bonteen's, and who had always supported the right honourable gentleman opposite. Then Mr. Gresham rose again. 'He regretted to say that the honourable member for Tankerville was in custody on that charge. The House would of course understand that he only made that statement as a fact, and that he was offering no opinion as to who was the perpetrator of the murder. The case seemed to be shrouded in great mystery. The two gentlemen had unfortunately differed, but he did not at all think that the House would on that account be disposed to attribute guilt so black and damning to a gentleman they had all known so well as the honourable member for Tankerville.' So much and no more was spoken publicly, to the reporters; but members continued to talk about the affair the whole evening.

There was nothing, perhaps, more astonishing than the absence of rancour or abhorrence with which the name of Phineas was mentioned, even by those who felt most certain of his guilt. All those who had been present at the club acknowledged that Bonteen had been the sinner in reference to the transaction there; and it was acknowledged to have been almost a public misfortune that such a man as Bonteen should have been able to prevail against such a one as Phineas Finn in regard to the presence of the latter in the Government. Stories which were exaggerated, accounts worse even than the truth, were bandied about as to the perseverance with which the murdered man had destroyed the prospects of the supposed murderer, and robbed the country of the services of a good workman. Mr. Gresham, in the official statement which he had made, had, as a matter of course, said many fine things about Mr. Bonteen. A man can always have fine things said about him for a few hours after his death. But in the small private conferences which were held the fine things said all referred to Phineas Finn. Mr. Gresham had spoken of a 'dastardly ruffian in the silent watches,' but one would have almost thought from overhearing what was said by various gentlemen in different parts of the House that upon the whole Phineas Finn was thought to have done rather a good thing in putting poor Mr. Bonteen out of the way.

And another pleasant feature of excitement was added by the prevalent idea that the Prince had seen and heard the row. Those who had been at the club at the time of course knew that this was not the case; but the presence of the Prince at The Universe between the row and the murder had really been a fact, and therefore it was only natural that men should allow themselves the delight of mixing the Prince with the whole concern. In remote circles the Prince was undoubtedly supposed to have had a great deal to do with the matter, though whether as abettor of the murdered or of the murderer was never plainly declared. A great deal was said about the Prince that evening in the House, so that many members were able to enjoy themselves thoroughly.

'What a godsend for Gresham,' said one gentleman to Mr. Ratler very shortly after the strong eulogium which had been uttered on poor Mr. Bonteen by the Prime Minister.

'Well,—yes; I was afraid that the poor fellow would never have got on with us.'

'Got on! He'd have been a thorn in Gresham's side as long as he held office. If Finn should be acquitted, you ought to do something handsome for him.' Whereupon Mr. Ratler laughed heartily.

'It will pretty nearly break them up,' said Sir Orlando Drought, one of Mr. Daubeny's late Secretaries of State to Mr. Roby, Mr. Daubeny's late patronage secretary.

'I don't quite see that. They'll be able to drop their decimal coinage with a good excuse, and that will be a great comfort. They are talking of getting Monk to go back to the Board of Trade.'

'Will that strengthen them?'

'Bonteen would have weakened them. The man had got beyond himself, and lost his head. They are better without him.'

'I suppose Finn did it?' asked Sir Orlando.

'Not a doubt about it, I'm told. The queer thing is that he should have declared his purpose beforehand to Erle. Gresham says that all that must have been part of his plan,—so as to make men think afterwards that he couldn't have done it. Grogram's idea is that he had planned the murder before he went to the club.'

'Will the Prince have to give evidence?'

'No, no,' said Mr. Roby. 'That's all wrong. The Prince

had left the club before the row commenced. Confucius Putt says that the Prince didn't hear a word of it. He was talking to the Prince all the time.' Confucius Putt was the distinguished artist with whom the Prince had shaken hands on leaving the club.

Lord Drummond was in the Peers' Gallery, and Mr. Boffin was talking to him over the railings. It may be remembered that those two gentlemen had conscientiously left Mr. Daubeny's Cabinet because they had been unable to support him in his views about the Church. After such sacrifice on their parts their minds were of course intent on Church matters. 'There doesn't seem to be a doubt about it,' said Mr. Boffin.

'Cantrip won't believe it,' said the peer.

'He was at the Colonies with Cantrip, and Cantrip found him very agreeable. Everybody says that he was one of the pleasantest fellows going. This makes it out of the question that they should bring in any Church bill this Session.'

'Do you think so?'

'Oh yes;—certainly. There will be nothing else thought of now till the trial.'

'So much the better,' said his Lordship. 'It's an ill wind that blows no one any good. Will they have evidence for a conviction?'

'Oh dear yes; not a doubt about it. Fawn can swear to him,' said Mr. Boffin.

Barrington Erle was telling his story for the tenth time when he was summoned out of the Library to the Duchess of Omnium, who had made her way up into the lobby. 'Oh, Mr. Erle, do tell me what you really think,' said the Duchess.

'That is just what I can't do.'

'Why not?'

'Because I don't know what to think.'

'He can't have done it, Mr. Erle.'

'That's just what I say to myself, Duchess.'

'But they do say that the evidence is so very strong against him.'

'Very strong.'

'I wish we could get that Lord Fawn out of the way.'

'Ah;—but we can't.'

'And will they—hang him?'

'If they convict him, they will.'

'A man we all knew so well! And just when we had made up our minds to do everything for him. Do you know I'm not a bit surprised. I've felt before now as though I should like to have done it myself.'

'He could be very nasty, Duchess!'

'I did so hate that man. But I'd give,—oh, I don't know what I'd give to bring him to life again this minute. What will Lady Laura do?' In answer to this, Barrington Erle only shrugged his shoulders. Lady Laura was his cousin. 'We mustn't give him up, you know, Mr. Erle.'

'What can we do?'

'Surely we can do something. Can't we get it in the papers that he must be innocent,—so that everybody should be made to think so? And if we could get hold of the lawyers, and make them not want to—to destroy him! There's nothing I wouldn't do. There's no getting hold of a judge, I know.'

'No, Duchess. The judges are stone.'

'Not that they are a bit better than anybody else,—only they like to be safe.'

'They do like to be safe.'

'I'm sure we could do it if we put our shoulders to the wheel. I don't believe, you know, for a moment that he murdered him. It was done by Lizzie Eustace's Jew.'

'It will be sifted, of course.'

'But what's the use of sifting if Mr. Finn is to be hung while it's being done? I don't think anything of the police. Do you remember how they bungled about that woman's necklace?* I don't mean to give him up, Mr. Erle; and I expect you to help me.' Then the Duchess returned home, and, as we know, found Madame Goesler at her house.

Nothing whatever was done that night, either in the Lords or Commons. A 'statement' about Mr. Bonteen was made in the Upper as well as in the Lower House, and after that statement any real work was out of the question. Had Mr.

Bonteen absolutely been Chancellor of the Exchequer, and in the Cabinet when he was murdered, and had Phineas Finn been once more an Under-Secretary of State, the commotion and excitement could hardly have been greater. Even the Duke of St. Bungay had visited the spot,—well known to him, as there the urban domains meet of two great Whig peers, with whom and whose predecessors he had long been familiar. He also had known Phineas Finn, and not long since had said civil words to him and of him. He, too, had, of late days, especially disliked Mr. Bonteen, and had almost insisted that the man now murdered should not be admitted into the Cabinet. He had heard what was the nature of the evidence;—had heard of the quarrel, the life-preserver, and the grey coat. 'I suppose he must have done it,' said the Duke of St. Bungay to himself as he walked away up Hay Hill.

CHAPTER LI

'You think it shameful'

THE tidings of what had taken place first reached Lady Laura Kennedy from her brother on his return to Portman Square after the scene in the police court. The object of his visit to Finn's lodgings has been explained, but the nature of Lady Laura's vehemence in urging upon her brother the performance of a very disagreeable task has not been sufficiently described. No brother would willingly go on such a mission from a married sister to a man who had been publicly named as that sister's lover;—and no brother could be less likely to do so than Lord Chiltern. But Lady Laura had been very stout in her arguments, and very strong-willed in her purpose. The income arising from this money,—which had been absolutely her own,—would again be exclusively her own should the claim to it on behalf of her husband's estate be abandoned. Surely she might do what she liked with her own. If her brother would not assist her in making this arrangement, it must be done by other means. She was quite willing that it should appear to come to Mr. Finn from her father and not from herself. Did her brother think any ill of her? Did he believe in the calumnies of the newspapers? Did he or his wife for a moment conceive that she had a lover? When he looked at her, worn out, withered, an old woman before her time, was it possible that he should so believe? She herself asked him these questions. Lord Chiltern of course declared that he had no suspicion of the kind, 'No;—indeed,' said Lady Laura. 'I defy any one to suspect me who knows me. And if so, why am not I as much entitled to help a friend as you might be? You need not even mention my name.' He endeavoured to make her understand that her name would be mentioned, and others would believe and would say evil things. 'They cannot say worse than they have said,' she continued. 'And yet what harm have they done to me,—or you?' Then he demanded why she desired to go so far out of her way with the view of spending her money upon one who was in no way connected with her. 'Because I like him better than any one else,' she answered, boldly. 'There is very little left for which I care at all;—but I do care for his prosperity. He was once in love with me and told me so,—but I had chosen to give my hand to Mr. Kennedy. He is not in love with me now,—nor I with him; but I choose to regard him as my friend.' He assured her over and over again that Phineas Finn would certainly refuse to touch her money;—but this she declined to believe. At any rate the trial might be made. He would not refuse money left to him by will, and why should he not now enjoy that which was intended for him? Then she explained how certain it was that he must speedily vanish out of the world altogether, unless some assurance of an income were made to him. So Lord Chiltern went on his mission, hardly meaning to make the offer, and confident that it would be refused if made. We know the nature of the new trouble in which he found Phineas Finn enveloped. It was such that Lord Chiltern did not open his mouth about money, and now, having witnessed the scene at the police-office, he had come back to tell his tale to his sister. She was sitting with his wife when he entered the room.

'Have you heard anything?' he asked at once.

'Heard what?' said his wife.

'Then you have not heard it. A man has been murdered.'

'What man?' said Lady Laura, jumping suddenly from her seat. 'Not Robert!' Lord Chiltern shook his head. 'You do not mean that Mr. Finn has been——killed!' Again he shook his head; and then she sat down as though the asking of the two questions had exhausted her.

'Speak, Oswald,' said his wife. 'Why do you not tell us? Is it one whom we knew?'

'I think that Laura used to know him. Mr. Bonteen was murdered last night in the streets.'

'Mr. Bonteen! The man who was Mr. Finn's enemy,' said Lady Chiltern.

'Mr. Bonteen!' said Lady Laura, as though the murder of twenty Mr. Bonteens were nothing to her.

'Yes;—the man whom you talk of as Finn's enemy. It would be better if there were no such talk.'

'And who killed him?' said Lady Laura, again getting up and coming close to her brother.

'Who was it, Oswald?' asked his wife; and she also was now too deeply interested to keep her seat.

'They have arrested two men,' said Lord Chiltern;—'that Jew who married Lady Eustace, and——' But there he paused. He had determined beforehand that he would tell his sister the double arrest that the doubt this implied might lessen the weight of the blow; but now he found it almost impossible to mention the name.

'Who is the other, Oswald?' said his wife.

'Not Phineas,' screamed Lady Laura.

'Yes, indeed; they have arrested him, and I have just come from the court.' He had no time to go on, for his sister was crouching prostrate on the floor before him. She had not fainted. Women do not faint under such shocks. But in her agony she had crouched down rather than fallen, as though it were vain to attempt to stand upright with so crushing a weight of sorrow on her back. She uttered one loud shriek, and then covering her face with her hands burst out into a wail of sobs. Lady Chiltern and her brother both tried to raise her, but she would not be lifted. 'Why will you not hear me through, Laura?' said he.

'You do not think he did it?' said his wife.

'I'm sure he did not,' replied Lord Chiltern.

The poor woman, half-lying, half-seated, on the floor, still hiding her face with her hands, still bursting with half suppressed sobs, heard and understood both the question and the answer. But the fact was not altered to her,—nor the condition of the man she loved. She had not yet begun to think whether it were possible that he should have been guilty of such a crime. She had heard none of the circumstances, and knew nothing of the manner of the man's death. It might be that

Phineas had killed the man, bringing himself within the reach of the law, and that yet he should have done nothing to merit her reproaches;—hardly even her reprobation! Hitherto she felt only the sorrow, the annihilation of the blow;—but not the shame with which it would overwhelm the man for whom she so much coveted the good opinion of the world.

'You hear what he says, Laura.'

'They are determined to destroy him,' she sobbed out, through her tears.

'They are not determined to destroy him at all,' said Lord Chiltern. 'It will have to go by evidence. You had better sit up and let me tell you all. I will tell you nothing till you are seated again. You disgrace yourself by sprawling there.'

'Do not be hard to her, Oswald.'

'I am disgraced,' said Lady Laura, slowly rising and placing herself again on the sofa. 'If there is anything more to tell, you can tell it. I do not care what happens to me now, or who knows it. They cannot make my life worse than it is.'

Then he told all the story,—of the quarrel, and the position of the streets, of the coat, and the bludgeon, and the three blows, each on the head, by which the man had been killed. And he told them also how the Jew was said never to have been out of his bed, and how the Jew's coat was not the coat Lord Fawn had seen, and how no stain of blood had been found about the raiment of either of the men. 'It was the Jew who did it, Oswald, surely,' said Lady Chiltern.

'It was not Phineas Finn who did it,' he replied.

'And they will let him go again?'

'They will let him go when they find out the truth, I suppose. But those fellows blunder so, I would never trust them. He will get some sharp lawyer to look into it; and then perhaps everything will come out. I shall go and see him to-morrow. But there is nothing further to be done.'

'And I must see him,' said Lady Laura slowly.

Lady Chiltern looked at her husband, and his face became redder than usual with an angry flush. When his sister had pressed him to take her message about the money, he had assured her that he suspected her of no evil. Nor had he ever thought evil of her. Since her marriage with Mr. Kennedy, he had seen but little of her or of her ways of life. When she had separated herself from her husband he had approved of the separation, and had even offered to assist her should she be in difficulty. While she had been living a sad lonely life at Dresden, he had simply pitied her, declaring to himself and his wife that her lot in life had been very hard. When these calumnies about her and Phineas Finn had reached his ears,—or his eyes,—as such calumnies always will reach the ears and eyes of those whom they are most capable of hurting, he had simply felt a desire to crush some Quintus Slide, or the like, into powder for the offence. He had received Phineas in his own house with all his old friendship. He had even this morning been with the accused man as almost his closest friend. But, nevertheless, there was creeping into his heart a sense of the shame with which he would be afflicted, should the world really be taught to believe that the man had been his sister's lover. Lady Laura's distress on the present occasion was such as a wife might show, or a girl weeping for her lover, or a mother for her son, or a sister for a brother; but was extravagant and exaggerated in regard to such friendship as might be presumed to exist between the wife of Mr. Robert Kennedy and the member for Tankerville. He could see that his wife felt this as he did, and he thought it necessary to say something at once, that might force his sister to moderate at any rate her language, if not her feelings. Two expressions of face were natural to him; one eloquent of good humour, in which the reader of countenances would find some promise of coming frolic;—and the other, replete with anger, sometimes to the extent almost of savagery. All those who were dependent on him were wont to watch his face with care and sometimes with fear. When he was angry it would almost seem that he was about to use personal violence on the object of his wrath. At the present moment he was rather grieved than enraged; but there came over his face that look of wrath with which all who knew him were so well acquainted. 'You cannot see him,' he said.

'Why not I, as well as you?'

'If you do not understand, I cannot tell you. But you must not see him;—and you shall not.'

'Who will hinder me?'

'If you put me to it, I will see that you are hindered. What is the man to you that you should run the risk of evil tongues, for the sake of visiting him in gaol? You cannot save his life,—though it may be that you might endanger it.'

'Oswald,' she said very slowly, 'I do not know that I am in any way under your charge, or bound to submit to your orders.'

'You are my sister.'

'And I have loved you as a sister. How should it be possible that my seeing him should endanger his life?'

'It will make people think that the things are true which have been said.'

'And will they hang him because I love him? I do love him. Violet knows how well I have always loved him.' Lord Chiltern turned his angry face upon his wife. Lady Chiltern put her arm round her sister-in-law's waist, and whispered some words into her ear. 'What is that to me?' continued the half-frantic woman. 'I do love him. I have always loved him. I shall love him to the end. He is all my life to me.'

'Shame should prevent your telling it,' said Lord Chiltern.

'I feel no shame. There is no disgrace in love. I did disgrace myself when I gave the hand for which he asked to another man, because,—because——' But she was too noble to tell her brother even then that at the moment of her life to which she was alluding she had married the rich man, rejecting the poor man's hand, because she had given up all her fortune to the payment of her brother's debts. And he, though he had well known what he had owed to her, and had never been easy till he had paid the debt, remembered nothing of all this now. No lending and paying back of money could alter the nature either of his feelings or his duty in such an emergency as this. 'And, mind you,' she continued, turning to her sister-in-law, 'there is no place for the shame of which he is thinking,' and she pointed her finger out at her brother. 'I love him,—as a mother might love her child, I fancy; but he has no love for me; none;—none. When I am with him, I am only a trouble to him. He comes to me, because he is good; but he would sooner be with you. He did love me once;—but then I could not afford to be so loved.'

'You can do no good by seeing him,' said her brother.

'But I will see him. You need not scowl at me as though you wished to strike me. I have gone through that which makes me different from other women, and I care not what they say of me. Violet understands it all;—but you understand nothing.'

'Be calm, Laura,' said her sister-in-law, 'and Oswald will do all that can be done.'

'But they will hang him.'

'Nonsense!' said her brother. 'He has not been as yet committed for his trial. Heaven knows how much has to be done. It is as likely as not that in three days' time he will be out at large, and all the world will be running after him just because he has been in Newgate.'

'But who will look after him?'

'He has plenty of friends. I will see that he is not left without everything that he wants.'

'But he will want money.'

'He has plenty of money for that. Do you take it quietly, and not make a fool of yourself. If the worst comes to the worst——'

'Oh, heavens!'

'Listen to me, if you can listen. Should the worst come to the worst, which I believe to be altogether impossible,—mind, I think it next to impossible, for I have never for a moment believed him to be guilty,—we will,—visit him,—together. Good-bye now. I am going to see that friend of his, Mr. Low.' So saying Lord Chiltern went, leaving the two women together.

'Why should he be so savage with me?' said Lady Laura.

'He does not mean to be savage.'

'Does he speak to you like that? What right has he to tell me of shame? Has my life been so bad, and his so good? Do you think it shameful that I should love this man?' She sat looking into her friend's face, but her friend for a while hesitated to answer. 'You shall tell me, Violet. We have known each other so well that I can bear to be told by you. Do not you love him?'

'I, love him!—certainly not.'

'But you did.'

'Not as you mean. Who can define love, and say what it is? There are so many kinds of love. We say that we love the Queen.'

'Psha!'

'And we are to love all our neighbours. But as men and women talk of love, I never at any moment of my life loved any man but my husband. Mr. Finn was a great favourite with me,—always.'

'Indeed he was.'

'As any other man might be,—or any woman. He is so still, and with all my heart I hope that this may be untrue.'

'It is false as the Devil. It must be false. Can you think of the man,—his sweetness, the gentle nature of him, his open, free speech, and courage, and believe that he would go behind his enemy and knock his brains out in the dark? I can conceive it of myself, that I should do it, much easier than of him.'

'Oswald says it is false.'

'But he says it as partly believing that it is true. If it be true I will hang myself. There will be nothing left among men or women fit to live for. You think it shameful that I should love him.'

'I have not said so.'

'But you do.'

'I think there is cause for shame in your confessing it.'

'I do confess it.'

'You ask me, and press me, and because we have loved one

another so well I must answer you. If a woman, a married woman,—be oppressed by such a feeling, she should lay it down at the bottom of her heart, out of sight, never mentioning it, even to herself.'

'You talk of the heart as though we could control it.'

'The heart will follow the thoughts, and they may be controlled. I am not passionate, perhaps, as you are, and I think I can control my heart. But my fortune has been kind to me, and I have never been tempted. Laura, do not think I am preaching to you.'

'Oh no;—but your husband; think of him, and think of mine! You have babies.'

'May God make me thankful. I have every good thing on earth that God can give.'

'And what have I? To see that man prosper in life, who they tell me is a murderer; that man who is now in a felon's gaol,—whom they will hang for ought we know,—to see him go forward and justify my thoughts of him! that yesterday was all I had. To-day I have nothing,—except the shame with which you and Oswald say that I have covered myself.'

'Laura, I have never said so.'

'I saw it in your eye when he accused me. And I know that it is shameful. I do know that I am covered with shame. But I can bear my own disgrace better than his danger.' After a long pause,—a silence of probably some fifteen minutes,—she spoke again. 'If Robert should die,—what would happen then?'

'It would be—a release, I suppose,' said Lady Chiltern in a voice so low, that it was almost a whisper.

'A release indeed;—and I would become that man's wife the next day, at the foot of the gallows;—if he would have me. But he would not have me.'

CHAPTER LII

Mr. Kennedy's will

MR. KENNEDY had fired a pistol at Phineas Finn in Macpherson's Hotel with the manifest intention of blowing out the brains of his presumed enemy, and no public notice had been taken of the occurrence. Phineas himself had been only too willing to pass the thing by as a trifling accident, if he might be allowed to do so, and the Macphersons had been by far too true to their great friend to think of giving him in charge to the police. The affair had been talked about, and had come to the knowledge of reporters and editors. Most of the newspapers had contained paragraphs giving various accounts of the matter; and one or two had followed the example of The People's Banner in demanding that the police should investigate the matter. But the matter had not been investigated. The police were supposed to know nothing about it,—as how should they, no one having seen or heard the shot but they who were determined to be silent? Mr. Quintus Slide had been indignant all in vain, so far as Mr. Kennedy and his offence had been concerned. As soon as the pistol had been fired and Phineas had escaped from the room, the unfortunate man had sunk back in his chair, conscious of what he had done, knowing that he had made himself subject to the law, and expecting every minute that constables would enter the room to seize him. He had seen his enemy's hat lying on the floor, and, when nobody would come to fetch it, had thrown it down the stairs. After that he had sat waiting for the police, with the pistol, still loaded in every barrel but one, lying by his side,—hardly repenting the attempt, but trembling for the result,—till Macpherson, the landlord, who had been brought home from chapel, knocked at his door. There was very little said between them; and no positive allusion was made to the shot that had been fired; but Macpherson succeeded in getting the pistol into his possession,— as to which the unfortunate man put no impediment in his way, and he managed to have it understood that Mr. Kennedy's cousin should be summoned on the following morning. 'Is anybody else coming?' Robert Kennedy asked, when the landlord was about to leave the room. 'Naebody as I ken o', yet, laird,' said Macpherson, 'but likes they will.' Nobody, however, did come, and the 'laird' had spent the evening by himself in very wretched solitude.

On the following day the cousin had come, and to him the whole story was told. After that, no difficulty was found in taking the miserable man back to Loughlinter, and there he had been for the last two months in the custody of his more wretched mother and of his cousin. No legal steps had been taken to deprive him of the management either of himself or of his property,—so that he was in truth his own master. And he exercised his mastery in acts of petty tyranny about his domain, becoming more and more close-fisted in regard to money, and desirous, as it appeared, of starving all living things about the place,—cattle, sheep, and horses, so that the value of their food might be saved. But every member of the establishment knew that the laird was 'nae just himself', and consequently his orders were not obeyed. And the laird knew the same of himself, and, though he would give the orders not only resolutely, but with imperious threats of penalties to follow disobedience, still he did not seem to expect compliance. While he was in this state, letters addressed to him came for a while into his own hands, and thus more than one reached him from Lord Brentford's lawyer, demanding that restitution should be made of the interest arising from Lady Laura's fortune. Then he would fly out into bitter wrath, calling his wife foul names, and swearing that she should never have a farthing of his money to spend upon her paramour. Of course it was his money, and his only. All the world knew that. Had she not left his roof, breaking her marriage vows, throwing aside every duty, and bringing him down to his present state of abject misery? Her own fortune! If she wanted the interest of her wretched money, let her come to Loughlinter and receive it there. In spite of all her wickedness, her cruelty, her misconduct, which had brought him,—as he now said,—to the verge of the grave, he would still give her shelter and room for repentance. He recognised his vows, though she did not. She should still be his wife, though she had utterly disgraced both herself and him. She should still be his wife, though she had so lived as to make it impossible that there should be any happiness in their household.

It was thus he spoke when first one and then another letter came from the Earl's lawyer, pointing out to him the injustice to which Lady Laura was subjected by the loss of her fortune. No doubt these letters would not have been written in the line assumed had not Mr. Kennedy proved himself to be unfit to have the custody of his wife by attempting to shoot the man whom he accused of being his wife's lover. An act had been done, said the lawyer, which made it quite out of the question that Lady Laura should return to her husband. To this, when speaking of the matter to those around him,—which he did with an energy which seemed to be foreign to his character,—Mr. Kennedy made no direct allusion; but he swore most positively that not a shilling should be given up. The fear of policemen coming down to Loughlinter to take account of that angry shot had passed away; and, though he knew, with an uncertain knowledge, that he was not in all respects obeyed as he used to be,—that his orders were disobeyed by stewards and servants, in spite of his threats of dismissal,—he still felt that he was sufficiently his own master to defy the Earl's attorney and to maintain his claim upon his wife's person. Let her return to him first of all!

But after a while the cousin interfered still further; and Robert Kennedy, who so short a time since had been a member of the Government, graced by permission to sit in the Cabinet, was not allowed to open his own post-bag. He had written a letter to one person, and then again to another, which had induced those who received them to return answers to the cousin. To Lord Brentford's lawyer he had used a few very strong words. Mr. Forster had replied to the cousin, stating how grieved Lord Brentford would be, how much grieved would be Lady Laura, to find themselves driven to take steps in reference to what they conceived to be the unfortunate condition of Mr. Robert Kennedy; but that such steps must be taken unless some arrangement could be made which should be at any rate reasonable. Then Mr. Kennedy's post-bag was taken from him; the letters which he wrote were not sent;—and he took to his bed. It was during this condition of affairs that the cousin took upon himself to intimate to Mr. Forster that the managers of Mr. Kennedy's estate were by no means anxious of embarrassing their charge by so trumpery an additional matter as the income derived from Lady Laura's forty thousand pounds.

But things were in a terrible confusion at Loughlinter, Rents were paid as heretofore on receipts given by Robert Kennedy's agent; but the agent could only pay the money to Robert Kennedy's credit at his bank. Robert Kennedy's cheques would, no doubt, have drawn the money out again;—but it was almost impossible to induce Robert Kennedy to sign a cheque. Even in bed he inquired daily about his money, and knew accurately the sum lying at his banker's; but he could be persuaded to disgorge nothing. He postponed from day to day the signing of certain cheques that were brought to him, and alleged very freely that an attempt was being made to rob him. During all his life he had been very generous in subscribing to public charities; but now he stopped all his subscriptions. The cousin had to provide even for the payment of wages, and things went very badly at Loughlinter. Then there arose the question whether legal steps should be taken for placing the management of the estate in other hands, on the ground of the owner's insanity. But the wretched old mother begged that this might not be done;—and Dr. Macnuthrie, from Callender, was of opinion that no steps should be taken at present. Mr. Kennedy was very ill,—very ill indeed; would take no nourishment, and seemed to be sinking under the pressure of his misfortunes. Any steps such as those suggested would probably send their friend out of the world at once.

In fact Robert Kennedy was dying;—and in the first week of May, when the beauty of the spring was beginning to show itself on the braes of Loughlinter, he did die. The old woman, his mother, was seated by his bedside, and into her ears he murmured his last wailing complaint. 'If she had the fear of God before her eyes, she would come back to me.' 'Let us pray that He may soften her heart,' said the old lady. 'Eh, mother;—nothing can soften the heart Satan has hardened, till it be hard as the nether millstone.' And in that faith he died believing, as he had ever believed, that the spirit of evil was stronger than the spirit of good.

For some time past there had been perturbation in the mind of that cousin, and of all other Kennedys of that ilk, as to the nature of the will of the head of the family. It was feared lest he should have been generous to the wife who was believed by them all to have been so wicked and treacherous to her husband;—and so it was found to be when the will was read. During the last few months no one near him had dared to speak to him of his will, for it had been known that his condition of mind rendered him unfit to alter it; nor had he ever alluded to it himself. As a matter of course there had been a settlement, and it was supposed that Lady Laura's own money would revert to her; but when it was found that in addition to this the Loughlinter estate became hers for life, in the event of Mr. Kennedy dying without a child, there was great consternation among the Kennedys generally. There were but two or three of them concerned, and for those there was money enough; but it seemed to them now that the bad wife, who had utterly refused to acclimatise herself to the soil to which she had been transplanted, was to be rewarded for her wicked stubbornness. Lady Laura would become mistress of her own fortune and of all Loughlinter, and would be once more a free woman, with all the power that wealth and fashion can give. Alas, alas! it was too late now for the taking of any steps to sever her from her rich inheritance!

'And the false harlot will come and play havoc here, in my son's mansion,' said the old woman with extremest bitterness.

The tidings were conveyed to Lady Laura through her lawyer, but did not reach her in full till some eight or ten days after the news of her husband's death. The telegram announcing that event had come to her at her father's house in Portman Square, on the day after that on which Phineas had been arrested, and the Earl had of course known that his great longing for the recovery of his wife's fortune had been now realised. To him there was no sorrow in the news. He had only known Robert Kennedy as one who had been thoroughly disagreeable to himself, and who had persecuted his daughter throughout their married life. There had come no happiness, —not even prosperity,—through the marriage. His daughter had been forced to leave the man's house,—and had been forced also to leave her money behind her. Then she had been driven abroad, fearing persecution, and had only dared to return when the man's madness became so notorious as to annul his power of annoying her. Now by his death, a portion of the injury which he had inflicted on the great family of Standish would be remedied. The money would come back,—together with the stipulated jointure,—and there could no longer be any question of return. The news delighted the old Lord,—and he was almost angry with his daughter because she also would not confess her delight.

'Oh, Papa, he was my husband.'

'Yes, yes, no doubt. I was always against it, you will remember.'

'Pray do not talk in that way now, Papa. I know that I was not to him what I should have been.'

'You used to say it was all his fault.'

'We will not talk of it now, Papa. He is gone, and I remember his past goodness to me.'

She clothed herself in the deepest of mourning, and made herself a thing of sorrow by the sacrificial uncouthness of her garments. And she tried to think of him;—to think of him, and not to think of Phineas Finn. She remembered with real sorrow the words she had spoken to her sister-in-law, in which she had declared, while still the wife of another man, that she would willingly marry Phineas at the foot even of the gallows if she were free. She was free now; but she did not repeat her assertion. It was impossible not to think of Phineas in his present strait, but she abstained from speaking of him as far as she could, and for the present never alluded to her former purpose of visiting him in his prison.

From day to day, for the first few days of her widowhood, she heard what was going on. The evidence against him became stronger and stronger, whereas the other man, Yosef Mealyus, had been already liberated. There were still many who felt sure that Mealyus had been the murderer, among whom were all those who had been ranked among the staunch friends of our hero. The Chilterns so believed, and Lady Laura; the Duchess so believed, and Madame Goesler. Mr. Low felt sure of it, and Mr. Monk and Lord Cantrip; and nobody was more sure than Mrs. Bunce. There were many who professed that they doubted; men such as Barrington Erle, Laurence Fitzgibbon, the two Dukes,—though the younger Duke never expressed such doubt at home,—and Mr. Gresham himself. Indeed, the feeling of Parliament in general was one of great doubt. Mr. Daubeny never expressed an opinion one way or the other, feeling that the fate of two second-class Liberals could not be matter of concern to him; —but Sir Orlando Drought, and Mr. Roby, and Mr. Boffin, were as eager as though they had not been Conservatives, and were full of doubt. Surely, if Phineas Finn were not the murderer, he had been more ill-used by Fate than had been any man since Fate first began to be unjust. But there was also a very strong party by whom no doubt whatever was entertained as to his guilt,—at the head of which, as in duty bound, was the poor widow, Mrs. Bonteen. She had no doubt as to the hand by which her husband had fallen, and clamoured loudly for the vengeance of the law. All the world, she said, knew how bitter against her husband had been this wretch, whose villainy had been exposed by her dear, gracious lord; and now the evidence against him was, to her thinking, complete. She was supported strongly by Lady Eustace, who, much as she wished not to be the wife of the Bohemian Jew, thought even that preferable to being known as the widow of a murderer who had been hung. Mr. Ratler, with one or two others in the House, was certain of Finn's guilt. The People's Banner, though it prefaced each one of its daily paragraphs on the subject with a statement as to the manifest duty of an influential newspaper to abstain from the expression of any opinion on such a subject till the question had been decided by a jury, nevertheless from day to day recapitulated the evidence against the Member for Tankerville, and showed how strong were the motives which had existed for such a deed. But, among those who were sure of Finn's guilt, there was no one more sure than Lord Fawn, who had seen the coat and the height of the man,—and the step. He declared among his intimate friends that of course he could not swear to the person. He could not venture, when upon his oath, to give an opinion. But the man who had passed him at so quick a pace had been half a foot higher than Mealyus;—of that there could be no doubt. Nor could there be any doubt as to the grey coat. Of course there might be other men with grey coats besides Mr. Phineas Finn,—and other men half a foot taller than Yosef Mealyus. And there might be other men with that peculiarly energetic step. And the man who hurried by him might not have been the man who murdered Mr. Bonteen. Of all that Lord Fawn could say nothing. But what he did say,—of that he was sure. And all those who knew him were well aware that in his own mind he was convinced of the guilt of Phineas Finn. And there was another man equally convinced. Mr. Maule, Senior, remembered well the manner in which Madame Goesler spoke of Phineas Finn in reference to the murder, and was quite sure that Phineas was the murderer.

For a couple of days Lord Chiltern was constantly with the

poor prisoner, but after that he was obliged to return to Harrington Hall. This he did a day after the news arrived of the death of his brother-in-law. Both he and Lady Chiltern had promised to return home, having left Adelaide Palliser alone in the house, and already they had overstayed their time. 'Of course I will remain with you,' Lady Chiltern had said to her sister-in-law; but the widow had preferred to be left alone. For these first few days,—when she must make pretence of sorrow because her husband had died; and had such real cause for sorrow in the miserable condition of the man she loved,—she preferred to be alone. Who could sympathise with her now, or with whom could she speak of her grief? Her father was talking to her always of her money;—but from him she could endure it. She was used to him, and could remember when he spoke to her of her forty thousand pounds, and of her twelve hundred a year of jointure, that it had not always been with him like that. As yet nothing had been heard of the will, and the Earl did not in the least anticipate any further accession of wealth from the estate of the man whom they had all hated. But his daughter would now be a rich woman; and was yet young, and there might still be splendour. 'I suppose you won't care to buy land,' he said.

'Oh, Papa, do not talk of buying anything yet.'

'But, my dear Laura, you must put your money into something. You can get very nearly 5 per cent. from Indian Stock.'

'Not yet, Papa,' she said. But he proceeded to explain to her how very important an affair money is, and that persons who have got money cannot be excused for not considering what they had better do with it. No doubt she could get 4 per cent. on her money by buying up certain existing mortgages on the Saulsby property,—which would no doubt be very convenient if, hereafter, the money should go to her brother's child. 'Not yet, Papa,' she said again, having, however, already made up her mind that her money should have a different destination.

She could not interest her father at all in the fate of Phineas Finn. When the story of the murder had first been told to him, he had been amazed,—and, no doubt, somewhat gratified, as we all are, at tragic occurrences which do not concern ourselves. But he could not be made to tremble for the fate of Phineas Finn. And yet he had known the man during the last few years most intimately, and had had much in common with him. He had trusted Phineas in respect to his son, and had trusted him also in respect to his daughter. Phineas had been his guest at Dresden; and, on his return to London, had been the first friend he had seen, with the exception of his lawyer. And yet he could hardly be induced to express the slightest interest as to the fate of this friend who was to be tried for murder. 'Oh;—he's committed, is he? I think I remember that Protheroe once told me that, in thirty-nine cases out of forty, men committed for serious offences have been guilty of them.' The Protheroe here spoken of as an authority in criminal matters was at present Lord Weazeling, the Lord Chancellor.

'But Mr. Finn has not been guilty, Papa.'

'There is always the one chance out of forty. But, as I was saying, if you like to take up the Saulsby mortgages, Mr. Forster can't be told too soon.'

'Papa, I shall do nothing of the kind,' said Lady Laura. And then she rose and walked out of the room.

At the end of ten days from the death of Mr. Kennedy, there came the tidings of the will. Lady Laura had written to Mrs. Kennedy a letter which had taken her much time in composition, expressing her deep sorrow, and condoling with the old woman. And the old woman had answered. 'Madam, I am too old now to express either grief or anger. My dear son's death, caused by domestic wrong, has robbed me of any remaining comfort which the undeserved sorrows of his latter years had not already dispelled. Your obedient servant, Sarah Kennedy.' From which it may be inferred that she had also taken considerable trouble in the composition of her letter. Other communications between Loughlinter and Portman Square there were none, but there came through the lawyers a statement of Mr. Kennedy's will, as far as the interests of Lady Laura were concerned. This reached Mr. Forster first, and he brought it personally to Portman Square. He asked for Lady Laura, and saw her alone. 'He has bequeathed to you the use of Loughlinter for your life, Lady Laura.'

'To me!'

'Yes, Lady Laura. The will is dated in the first year of his marriage, and has not been altered since.'

'What can I do with Loughlinter? I will give it back to them.' Then Mr. Forster explained that the legacy referred not only to the house and immediate grounds,—but to the whole estate known as the domain of Loughlinter. There could be no reason why she should give it up, but very many why she should not do so. Circumstanced as Mr. Kennedy had been, with no one nearer to him than a first cousin, with a property purchased with money saved by his father,—a property to which no cousin could by inheritance have any claim,—he could not have done better with it than to leave it to his widow in fault of any issue of his own. Then the lawyer explained that were she to give it up, the world would of course say that she had done so from a feeling of her own unworthiness. 'Why should I feel myself to be unworthy?' she asked. The lawyer smiled, and told her that of course she would retain Loughlinter.

Then, at her request, he was taken to the Earl's room and there repeated the good news. Lady Laura preferred not to hear her father's first exultations. But while this was being done she also exulted. Might it not still be possible that there should be before her a happy evening to her days; and that she might stand once more beside the falls of Linter, contented, hopeful, nay, almost glorious, with her hand in his to whom she had once refused her own on that very spot?

CHAPTER LIII

None but the brave deserve the fair

THOUGH Mr. Robert Kennedy was lying dead at Loughlinter, and though Phineas Finn, a member of Parliament, was in prison, accused of murdering another member of Parliament, still the world went on with its old ways, down in the neighbourhood of Harrington Hall and Spoon Hall as at other places. The hunting with the Brake hounds was now over for the season,—had indeed been brought to an auspicious end three weeks since,—and such gentlemen as Thomas Spooner had time on their hands to look about their other concerns. When a man hunts five days a week, regardless of distances, and devotes a due proportion of his energies to the necessary circumstances of hunting, the preservation of foxes, the maintenance of good humour with the farmers, the proper compensation for poultry really killed by four-legged favourites, the growth and arrangement of coverts, the lying-in of vixens, and the subsequent guardianship of nurseries, the persecution of enemies, and the warm protection of friends, —when he follows the sport, accomplishing all the concomitant duties of a true sportsman, he has not much time left for anything. Such a one as Mr. Spooner of Spoon Hall finds that his off day is occupied from breakfast to dinner with grooms, keepers, old women with turkeys' heads, and gentlemen in velveteens with information about wires and unknown earths. His letters fall naturally to the Sunday afternoon, and are hardly written before sleep overpowers him. Many a large fortune has been made with less of true devotion to the work than is given to hunting by so genuine a sportsman as Mr. Spooner.

Our friend had some inkling of this himself, and felt that many of the less important affairs of his life were neglected because he was so true to the one great object of his existence. He had wisely endeavoured to prevent wrack and ruin among the affairs of Spoon Hall,—and had thoroughly succeeded by joining his cousin Ned with himself in the administration of his estate,—but there were things which Ned with all his zeal and all his cleverness could not do for him. He was conscious that had he been as remiss in the matter of hunting, as that hard-riding but otherwise idle young scamp, Gerard Maule, he might have succeeded much better than he had hitherto done with Adelaide Palliser. 'Hanging about and philandering, that's what they want,' he said to his cousin Ned.

'I suppose it is,' said Ned. 'I was fond of a girl once myself, and I hung about a good deal. But we hadn't sixpence between us.'

'That was Polly Maxwell. I remember. You behaved very badly then.'

'Very badly, Tom; about as bad as a man could behave,—and she was as bad. I loved her with all my heart, and I told her so. And she told me the same. There never was anything worse. We had just nothing between us, and nobody to give us anything.'

'It doesn't pay; does it, Ned, that kind of thing?'

'It doesn't pay at all. I wouldn't give her up,—nor she me. She was about as pretty a girl as I remember to have seen.'

'I suppose you were a decent-looking fellow in those days yourself. They say so, but I never quite believed it.'

'There wasn't much in that,' said Ned. 'Girls don't want a man to be good-looking, but that he should speak up and not be afraid of them. There were lots of fellows came after her. You remember Blinks, of the Carabineers.* He was full of money, and he asked her three times. She is an old maid to this day, and is living as companion to some crusty crochetty countess.'

'I think you did behave badly, Ned. Why didn't you set her free?'

'Of course, I behaved badly. And why didn't she set me free, if you come to that? I might have found a female Blinks of my own,—only for her. I wonder whether it will come against us when we die, and whether we shall be brought up together to receive punishment.'

'Not if you repent, I suppose,' said Tom Spooner, very seriously.

'I sometimes ask myself whether she has repented. I made her swear that she'd never give me up. She might have broken her word a score of times, and I wish she had.'

'I think she was a fool, Ned.'

'Of course she was a fool. She knows that now, I dare say. And perhaps she has repented. Do you mean to try it again with that girl at Harrington Hall?'

Mr. Thomas Spooner did mean to try it again with the girl at Harrington Hall. He had never quite trusted the note which he had got from his friend Chiltern, and had made up his mind that, to say the least of it, there had been very little friendship shown in the letter. Had Chiltern meant to have stood to him 'like a brick,' as he ought to have stood by his right hand man in the Brake country, at any rate a fair chance might have been given him. 'Where the devil would he be in such a country as this without me,'—Tom had said to his cousin,—'not knowing a soul, and with all the shooting men against him? I might have had the hounds myself,—and might have 'em now if I cared to take them. It's not standing by a fellow as he ought to do. He writes to me, by George, just as he might do to some fellow who never had a fox about his place.'

'I suppose he didn't put the two things together,' said Ned Spooner.

'I hate a fellow that can't put two things together. If I stand to you you've a right to stand to me. That's what you mean by putting two things together. I mean to have another shy at her. She has quarrelled with that fellow Maule altogether. I've learned that from the gardener's girl at Harrington.'

Yes,—he would make another attempt. All history, all romance, all poetry and all prose, taught him that perseverance in love was generally crowned with success,—that true love rarely was crowned with success except by perseverance. Such a simple little tale of boy's passion as that told him by his cousin had no attraction for him. A wife would hardly be worth having, and worth keeping, so won. And all proverbs were on his side. 'None but the brave deserve the fair,' said his cousin. 'I shall stick to it,' said Tom Spooner. 'Labor omnia vincit,'*said his cousin. But what should be his next step? Gerard Maule had been sent away with a flea in his ear, —so, at least, Mr. Spooner asserted, and expressed an undoubting opinion that this imperative dismissal had come from the fact that Gerard Maule, when 'put through his facings'* about income was not able to 'show the money.' 'She's not one of your Polly Maxwells, Ned.' Ned said that he supposed she was not one of that sort. 'Heaven knows I couldn't show the money,' said Ned, 'but that didn't make her any wiser.' Then Tom gave it as his opinion that Miss Palliser was one of those young women who won't go anywhere without having everything about them. 'She could have her own carriage with me, and her own horses, and her own maid, and everything.'

'Her own way into the bargain,' said Ned. Whereupon Tom Spooner winked, and suggested that that might be as things turned out after the marriage. He was quite willing to run his chance for that.

But how was he to get at her to prosecute his suit? As to writing to her direct,—he didn't much believe in that. 'It looks as though one were afraid of her, you know;—which I ain't the least. I stood up to her before, and I wasn't a bit more nervous than I am at this moment. Were you nervous in that affair with Miss Maxwell?'

'Ah;—it's a long time ago. There wasn't much nervousness there.'

'A sort of milkmaid affair?'

'Just that.'

'That is different, you know. I'll tell you what I'll do. I'll just drive slap over to Harrington and chance it. I'll take the two bays in the phaeton.* Who's afraid?'

'There's nothing to be afraid of,' said Ned.

'Old Chiltern is such a d—— cantankerous fellow, and perhaps Lady C. may say that I oughtn't to have taken advantage of her absence. But, what's the odds? If she takes me there'll be an end of it. If she don't, they can't eat me.'

'The only thing is whether they'll let you in.'

'I'll try at any rate,' said Tom, 'and you shall go over with me. You won't mind trotting about the grounds while I'm carrying on the war inside? I'll take the two bays, and Dick Farren behind, and I don't think there's a prettier got-up trap in the county. We'll go to-morrow.'

And on the morrow they did start, having heard on that very morning of the arrest of Phineas Finn. 'By George, don't it feel odd,' said Tom just as they started,—'a fellow that we used to know down here, having him out hunting and all that, and now he's—a murderer! Isn't it a coincidence?'

'It startles one,' said Ned.

'That's what I mean. It's such a strange thing that it should be the man we know ourselves. These things always are happening to me. Do you remember when poor Fred Fellows got his bad fall and died the next year? You weren't here then.'

'I've heard you speak of it.'

'I was in the very same field, and should have been the man to pick him up, only the hounds had just turned to the left. It's very odd that these coincidences always are happening to some men and never do happen to others. It makes one feel that he's marked out, you know.'

'I hope you'll be marked out by victory to-day.'

'Well;—yes. That's more important just now than Mr. Bonteen's murder. Do you know, I wish you'd drive. These horses are pulling, and I don't want to be all in a flurry when I get to Harrington.' Now it was a fact very well known to all concerned with Spoon Hall, that there was nothing as to which the Squire was so jealous as the driving of his own horses. He would never trust the reins to a friend, and even Ned had hardly ever been allowed the honour of the whip when sitting with his cousin. 'I'm apt to get red in the face when I'm overheated,' said Tom as he made himself comfortable and easy in the left hand seat.

There were not many more words spoken during the journey. The lover was probably justified in feeling some trepidation. He had been quite correct in suggesting that the matter between him and Miss Palliser bore no resemblance at all to that old affair between his cousin Ned and Polly Maxwell. There had been as little trepidation as money in that case,—simply love and kisses, parting, despair, and a broken heart. Here things were more august. There was plenty of money, and, let affairs go as they might, there would be no broken heart. But that perseverance in love of which Mr. Spooner intended to make himself so bright an example does require some courage. The Adelaide Pallisers of the world have a way of making themselves uncommonly unpleasant to a man when they refuse him for the third or fourth time. They allow themselves sometimes to express a contempt which is almost akin to disgust, and to speak to a lover as though he were no better than a footman. And then the lover is bound to bear it all, and when he has borne it, finds it so very difficult to get out of the room. Mr. Spooner had some idea of all this as his cousin drove him up to the door, at what he then thought a very fast pace. 'D—— it all,' he said, 'you needn't have brought them up so confoundedly hot.' But it was not of the horses that he was really thinking, but of the colour of his own nose. There was something working within him which had flurried him, in spite of the tranquillity of his idle seat.

Not the less did he spring out of the phaeton with a quite youthful jump. It was well that every one about Harrington Hall should know how alert he was on his legs; a little weather-beaten about the face he might be; but he could get in and out of his saddle as quickly as Gerard Maule even yet; and for a short distance would run Gerard Maule for a ten-pound note. He dashed briskly up to the door, and rang the bell as though he feared neither Adelaide nor Lord Chiltern any more than he did his own servants at Spoon Hall. 'Was Miss Palliser at home?' The maid-servant who opened the door told him that Miss Palliser was at home, with a celerity which he certainly had not expected. The male members of the establishment were probably disporting themselves in the absence of their master and mistress, and Adelaide Palliser was thus left to the insufficient guardianship of young women who were altogether without discretion. 'Yes, sir; Miss Palliser is at home.' So said the indiscreet female, and Mr. Spooner was for the moment confounded by his own success. He had hardly told himself what reception he had expected, or whether, in the event of the servant informing him at the front door that the young lady was not at home he would make any further immediate effort to prolong the siege so as to force an entry; but now, when he had carried the very fortress by surprise, his heart almost misgave him. He certainly had not thought, when he descended from his chariot like a young Bacchus in quest of his Ariadne, that he should so soon be enabled to repeat the tale of his love. But there he was, confronted with Ariadne before he had had a moment to shake his godlike locks or arrange the divinity of his thoughts. 'Mr. Spooner,' said the maid, opening the door.

'Oh dear!' exclaimed Ariadne, feeling the vainness of her wish to fly from the god. 'You know, Mary, that Lady Chiltern is up in London.'

'But he didn't ask for Lady Chiltern, Miss.' Then there was a pause, during which the maid-servant managed to shut the door and to escape.

'Lord Chiltern is up in London,' said Miss Palliser, rising from her chair, 'and Lady Chiltern is with him. They will be at home, I think, to-morrow, but I am not quite sure.' She looked at him rather as Diana might have looked at poor Orion than as any Ariadne at any Bacchus;*and for a moment Mr. Spooner felt that the pale chillness of the moon was entering in upon his very heart and freezing the blood in his veins.

'Miss Palliser——' he began.

But Adelaide was for the moment an unmitigated Diana. 'Mr. Spooner,' she said, 'I cannot for an instant suppose that you wish to say anything to me.'

'But I do,' said he, laying his hand upon his heart.

'Then I must declare that—that—that you ought not to. And I hope you won't. Lady Chiltern is not in the house, and I think that—that you ought to go away. I do, indeed.'

But Mr. Spooner, though the interview had been commenced with unexpected and almost painful suddenness, was too much a man to be driven off by the first angry word. He remembered that this Diana was but mortal; and he remembered, too, that though he had entered in upon her privacy he had done so in a manner recognised by the world as lawful. There was no reason why he should allow himself to be congealed,—or even banished out of the grotto of the nymph,—without speaking a word on his own behalf. Were he to fly now, he must fly for ever; whereas, if he fought now,—fought well, even though not successfully at the moment,—he might fight again. While Miss Palliser was scowling at him he resolved upon fighting. 'Miss Palliser,' he said, 'I did not come to see Lady Chiltern; I came to see you. And now that I have been happy enough to find you I hope you will listen to me for a minute. I shan't do you any harm.'

'I'm not afraid of any harm, but I cannot think that you have anything to say that can do anybody any good.' She sat down, however, and so far yielded. 'Of course I cannot make you go away, Mr. Spooner; but I should have thought, when I asked you——'

Mr. Spooner also seated himself, and uttered a sigh. Making love to a sweet, soft, blushing, willing, though silent girl is a pleasant employment; but the task of declaring love to a stony-hearted, obdurate, ill-conditioned Diana is very disagreeable for any gentleman. And it is the more so when the gentleman really loves,—or thinks that he loves,—his Diana. Mr. Spooner did believe himself to be verily in love. Having sighed, he began: 'Miss Palliser, this opportunity of declaring to you the state of my heart is too valuable to allow me to give it up without—without using it.'

It can't be of any use.'

'Oh, Miss Palliser,—if you knew my feelings!'

'But I know my own.'

'They may change, Miss Palliser.'

'No, they can't.'

'Don't say that, Miss Palliser.'

'But I do say it. I say it over and over again. I don't know what any gentleman can gain by persecuting a lady. You oughtn't to have been shown up here at all.'

Mr. Spooner knew well that women have been won even at the tenth time of asking, and this with him was only the third. 'I think if you knew my heart——' he commenced.

'I don't want to know your heart.'

'You might listen to a man, at any rate.'

'I don't want to listen. It can't do any good. I only want you to leave me alone, and go away.'

'I don't know what you take me for,' said Mr. Spooner, beginning to wax angry.

'I haven't taken you for anything at all. This is very disagreeable and very foolish. A lady has a right to know her own mind, and she has a right not to be persecuted.' She would have referred to Lord Chiltern's letter had not all the hopes of her heart been so terribly crushed since that letter had been written. In it he had openly declared that she was already engaged to be married to Mr. Maule, thinking that he would thus put an end to Mr. Spooner's little adventure. But since the writing of Lord Chiltern's letter that unfortunate reference had been made to Boulogne, and every particle of her happiness had been destroyed. She was a miserable, blighted young woman, who had quarrelled irretrievably with her lover, feeling greatly angry with herself because she had made the quarrel, and yet conscious that her own self-respect had demanded the quarrel. She was full of regret, declaring to herself from morning to night that, in spite of all his manifest wickedness in having talked of Boulogne, she never could care at all for any other man. And now there was this aggravation to her misery,—this horrid suitor, who disgraced her by making those around her suppose it to be possible that she should ever accept him; who had probably heard of her quarrel, and had been mean enough to suppose that therefore there might be a chance for himself! She did despise him, and wanted him to understand that she despised him.

'I believe I am in a condition to offer my hand and fortune to any young lady without impropriety,' said Mr. Spooner.

'I don't know anything about your condition.'

'But I will tell you everything.'

'I don't want to know anything about it.'

'I have an estate of——'

'I don't want to know about your estate. I won't hear about your estate. It can be nothing to me.'

'It is generally considered to be a matter of some importance.'

'It is of no importance to me, at all, Mr. Spooner; and I won't hear anything about it. If all the parish belonged to you, it would not make any difference.'

'All the parish does belong to me, and nearly all the next,' replied Mr. Spooner, with great dignity.

'Then you'd better find some lady who would like to have two parishes. They haven't any weight with me at all.' At that moment she told herself how much she would prefer even Bou——logne, to Mr. Spooner's two parishes.

'What is it that you find so wrong about me?' asked the unhappy suitor.

Adelaide looked at him, and longed to tell him that his nose was red. And, though she would not quite do that, she could not bring herself to spare him. What right had he to come to her,—a nasty, red-nosed old man, who knew nothing about anything but foxes and horses,—to her, who had never given him the encouragement of a single smile? She could not allude to his nose, but in regard to his other defects she would not spare him. 'Our tastes are not the same, Mr. Spooner.'

'You are very fond of hunting.'

'And our ages are not the same.'

'I always thought that there should be a difference of age,' said Mr. Spooner, becoming very red.

'And,—and,—and,—it's altogether quite preposterous. I don't believe that you can really think it yourself.'

'But I do.'

'Then you must unthink it. And, indeed, Mr. Spooner, since you drive me to say so,—I consider it to be very unmanly of you, after what Lord Chiltern told you in his letter.'

'But I believe that is all over.'

Then her anger flashed up very high. 'And if you do believe it, what a mean man you must be to come to me when you must know how miserable I am, and to think that I should be driven to accept you after losing him! You never could have been anything to me. If you wanted to get married at all, you should have done it before I was born.' This was hard upon the man, as at that time he could not have been much more than twenty. 'But you don't know anything of the difference in people if you think that any girl would look at you, after having been——loved by Mr. Maule. Now, as you do not seem inclined to go away, I shall leave you.' So saying, she walked off with stately step, out of the room, leaving the door open behind her to facilitate her escape.

She had certainly been very rude to him, and had treated him very badly. Of that he was sure. He had conferred upon her what is commonly called the highest compliment which a gentleman can pay to a lady, and she had insulted him;—had doubly insulted him. She had referred to his age, greatly exaggerating his misfortune in that respect; and she had compared him to that poor beggar Maule in language most offensive. When she left him, he put his hand beneath his waistcoat, and turned with an air almost majestic towards the window. But in an instant he remembered that there was nobody there to see how he bore his punishment, and he sank down into human nature. 'Damnation!' he said, as he put his hands into his trousers pockets.

Slowly he made his way down into the hall, and slowly he opened for himself the front door, and escaped from the house on to the gravel drive. There he found his cousin Ned still seated in the phaeton, and slowly driving round the circle in front of the hall door. The squire succeeded in gaining such command over his own gait and countenance that his cousin divined nothing of the truth as he clambered up into his seat. But he soon showed his temper. 'What the devil have you got the reins in this way for?'

'The reins are all right,' said Ned.

'No they ain't;—they're all wrong.' And then he drove down the avenue to Spoon Hall as quickly as he could make the horses trot.

'Did you see her?' said Ned, as soon as they were beyond the gates.

'See your grandmother.'

'Do you mean to say that I'm not to ask?'

'There's nothing I hate so much as a fellow that's always asking questions,' said Tom Spooner. 'There are some men so d——d thick-headed that they never know when they ought to hold their tongue.'

For a minute or two Ned bore the reproof in silence, and then he spoke. 'If you are unhappy, Tom, I can bear a good deal; but don't overdo it,—unless you want me to leave you.'

'She's the d——t vixen that ever had a tongue in her head,' said Tom Spooner, lifting his whip and striking the poor off-horse in his agony. Then Ned forgave him.

CHAPTER LIV

The Duchess takes counsel

PHINEAS FINN, when he had been thrice remanded before the Bow Street magistrate, and four times examined, was at last committed to be tried for the murder of Mr. Bonteen. This took place on Wednesday, May 19th, a fortnight after the murder. But during those fourteen days little was learned, or even surmised, by the police, in addition to the circumstances which had transpired at once. Indeed the delay, slight as it was, had arisen from a desire to find evidence that might affect Mr. Emilius, rather than with a view to strengthen that which did affect Phineas Finn. But no circumstance could be found tending in any way to add to the suspicion to which the converted Jew was made subject by his own character, and by the supposition that he would have been glad to get rid of Mr. Bonteen. He did not even attempt to run away,—for which attempt certain pseudo-facilities were put in his way by police ingenuity. But Mr. Emilius stood his ground and courted inquiry. Mr. Bonteen had been to him, he said, a very bitter, unjust, and cruel enemy. Mr. Bonteen had endeavoured to rob him of his dearest wife;—had charged him with bigamy;—had got up false evidence in the hope of ruining him. He had undoubtedly hated Mr. Bonteen, and might probably have said so. But, as it happened, through God's mercy, he was enabled to prove that he could not possibly have been at the scene of the murder when the murder was committed. During that hour of the night he had been in his own bed; and, had he been out, could not have re-entered the house without calling up the inmates. But, independently of his alibi, Mealyus was able to rely on the absolute absence of any evidence against him. No grey coat could be traced to his hands, even for an hour. His height was very much less than that attributed by Lord Fawn to the man whom he had seen hurrying to the spot. No weapon was found in his possession by which the deed could have been done. Inquiry was made as to the purchase of life-preservers, and the reverend gentleman was taken to half-a-dozen shops at which such instruments had lately been sold. But there had been a run upon life preservers, in consequence of recommendations as to their use given by certain newspapers;—and it was found as impossible to trace one particular purchase as it would be that of a loaf of bread. At none of the half-dozen shops to which he was taken was Mr. Emilius remembered; and then all further inquiry in that direction was abandoned, and Mr. Emilius was set at liberty. 'I forgive my persecutors from the bottom of my heart,' he said,—'but God will requite it to them.'

In the meantime Phineas was taken to Newgate, and was there confined, almost with the glory and attendance of a State prisoner. This was no common murder, and no common murderer. Nor were they who interested themselves in the matter the ordinary rag, tag, and bobtail of the people,—the mere wives and children, or perhaps fathers and mothers, or brothers and sisters of the slayer or the slain. Dukes and Earls, Duchesses and Countesses, Members of the Cabinet, great statesmen, Judges, Bishops, and Queen's Counsellors, beautiful women, and women of highest fashion, seemed for a while to think of but little else than the fate of Mr. Bonteen and the fate of Phineas Finn. People became intimately acquainted with each other through similar sympathies in this matter, who had never before spoken to or seen each other. On the day after the full committal of the man, Mr. Low received a most courteous letter from the Duchess of Omnium, begging him to call in Carlton Terrace if his engagements would permit him to do so. The Duchess had heard that Mr. Low was devoting all his energies to the protection of Phineas Finn; and, as a certain friend of hers,—a lady,—was doing the same, she was anxious to bring them together. Indeed, she herself was equally prepared to devote her energies for the present to the same object. She had declared to all her friends, —especially to her husband and to the Duke of St. Bungay, —her absolute conviction of the innocence of the accused man, and had called upon them to defend him. 'My dear,' said the elder Duke, 'I do not think that in my time any innocent man has ever lost his life upon the scaffold.'

'Is that a reason why our friend should be the first instance?' said the Duchess.

'He must be tried according to the laws of his country,' said the younger Duke.

'Plantagenet, you always speak as if everything were perfect, whereas you know very well that everything is imperfect. If that man is—is hung, I——'

'Glencora,' said her husband, 'do not connect yourself with the fate of a stranger from any misdirected enthusiasm.'

'I do connect myself. If that man be hung—I shall go into mourning for him. You had better look to it.'

Mr. Low obeyed the summons, and called on the Duchess. But, in truth, the invitation had been planned by Madame

Goesler, who was present when the lawyer, about five o'clock in the afternoon, was shown into the presence of the Duchess. Tea was immediately ordered, and Mr. Low was almost embraced. He was introduced to Madame Goesler, of whom he did not before remember that he had heard the name, and was at once given to understand that the fate of Phineas was now in question. 'We know so well,' said the Duchess, 'how true you are to him.'

'He is an old friend of mine,' said the lawyer, 'and I cannot believe him to have been guilty of a murder.'

'Guilty!—he is no more guilty than I am. We are as sure of that as we are of the sun. We know that he is innocent;—do we not, Madame Goesler? And we, too, are very dear friends of his;—that is, I am.'

'And so am I,' said Madame Goesler, in a voice very low and sweet, but yet so energetic as to make Mr. Low almost rivet his attention upon her.

'You must understand, Mr. Low, that Mr. Finn is a man horribly hated by certain enemies. That wretched Mr. Bonteen hated his very name. But there are other people who think very differently of him. He must be saved.'

'Indeed I hope he may,' said Mr. Low.

'We wanted to see you for ever so many reasons. Of course you understand that,—that any sum of money can be spent that the case may want.'

'Nothing will be spared on that account certainly,' said the lawyer.

'But money will do a great many things. We would send all round the world if we could get evidence against that other man,—Lady Eustace's husband, you know.'

'Can any good be done by sending all round the world?'

'He went back to his own home not long ago,—in Poland, I think,' said Madame Goesler. 'Perhaps he got the instrument there, and brought it with him.' Mr. Low shook his head. 'Of course we are very ignorant;—but it would be a pity that everything should not be tried.'

'He might have got in and out of the window, you know,'

said the Duchess. Still Mr. Low shook his head. 'I believe things can always be found out, if only you take trouble enough. And trouble means money;—does it not? We wouldn't mind how many thousand pounds it cost; would we, Marie?'

'I fear that the spending of thousands can do no good,' said Mr. Low.

'But something must be done. You don't mean to say that Mr. Finn is to be hung because Lord Fawn says that he saw a man running along the street in a grey coat.'

'Certainly not.'

'There is nothing else against him;—nobody else saw him.'

'If there be nothing else against him he will be acquitted.'

'You think then,' said Madame Goesler, 'that there will be no use in tracing what the man Mealyus did when he was out of England. He might have bought a grey coat then, and have hidden it till this night, and then have thrown it away.' Mr. Low listened to her with close attention, but again shook his head. 'If it could be shown that the man had a grey coat at that time it would certainly weaken the effect of Mr. Finn's grey coat.'

'And if he bought a bludgeon there, it would weaken the effect of Mr. Finn's bludgeon. And if he bought rope to make a ladder it would show that he had got out. It was a dark night, you know, and nobody would have seen it. We have been talking it all over, Mr. Low, and we really think you ought to send somebody.'

'I will mention what you say to the gentlemen who are employed on Mr. Finn's defence.'

'But will not you be employed?' Then Mr. Low explained that the gentlemen to whom he referred were the attorneys who would get up the case on their friend's behalf, and that as he himself practised in the Courts of Equity only, he could not defend Mr. Finn on his trial.

'He must have the very best men,' said the Duchess.

'He must have good men, certainly.'

'And a great many. Couldn't we get Sir Gregory Grogram?'

Mr. Low shook his head. 'I know very well that if you get men who are really,—really swells, for that is what it is, Mr. Low,—and pay them well enough, and so make it really an important thing, they can browbeat any judge and hoodwink any jury. I daresay it is very dreadful to say so, Mr. Low; but, nevertheless, I believe it, and as this man is certainly innocent it ought to be done. I daresay it's very shocking, but I do think that twenty thousand pounds spent among the lawyers would get him off.'

'I hope we can get him off without expending twenty thousand pounds, Duchess.'

'But you can have the money and welcome;—cannot he, Madame Goesler?'

'He could have double that, if double were necessary.'

'I would fill the court with lawyers for him,' continued the Duchess. 'I would cross-examine the witnesses off their legs. I would rake up every wicked thing that horrid Jew has done since he was born. I would make witnesses speak. I would give a carriage and pair of horses to every one of the jurors' wives, if that would do any good. You may shake your head, Mr. Low; but I would. And I'd carry Lord Fawn off to the Antipodes, too;—and I shouldn't care if you left him there. I know that this man is innocent, and I'd do anything to save him. A woman, I know, can't do much;—but she has this privilege, that she can speak out what men only think. I'd give them two carriages and two pairs of horses a-piece if I could do it that way.'

Mr. Low did his best to explain to the Duchess that the desired object could hardly be effected after the fashion she proposed, and he endeavoured to persuade her that justice was sure to be done in an English court of law. 'Then why are people so very anxious to get this lawyer or that to bamboozle the witnesses?' said the Duchess. Mr. Low declared it to be his opinion that the poorest man in England was not more likely to be hung for a murder he had not committed than the richest. 'Then why would you, if you were accused, have ever so many lawyers to defend you?' Mr. Low went on to explain. 'The more money you spend,' said the Duchess, 'the more fuss you make. And the longer a trial is about and the greater the interest, the more chance a man has to escape. If a man is tried for three days you always think he'll get off, but if it lasts ten minutes he is sure to be convicted and hung. I'd have Mr. Finn's trial made so long that they never could convict him. I'd tire out all the judges and juries in London. If you get lawyers enough they may speak for ever.' Mr. Low endeavoured to explain that this might prejudice the prisoner. 'And I'd examine every member of the House of Commons, and all the Cabinet, and all their wives. I'd ask them all what Mr. Bonteen had been saying. I'd do it in such a way as a trial was never done before;—and I'd take care that they should know what was coming.'

'And if he were convicted afterwards?'

'I'd buy up the Home Secretary. It's very horrid to say so, of course, Mr. Low; and I dare say there is nothing wrong ever done in Chancery. But I know what Cabinet Ministers are. If they could get a majority by granting a pardon they'd do it quick enough.'

'You are speaking of a liberal Government, of course, Duchess.'

'There isn't twopence to choose between them in that respect. Just at this moment I believe Mr. Finn is the most popular member of the House of Commons; and I'd bring all that to bear. You can't but know that if everything of that kind is done it will have an effect. I believe you could make him so popular that the people would pull down the prison rather than have him hung;—so that a jury would not dare to say he was guilty.'

'Would that be justice, ladies?' asked the just man.

'It would be success, Mr. Low,—which is a great deal the better thing of the two.'

'If Mr. Finn were found guilty, I could not in my heart believe that that would be justice,' said Madame Goesler.

Mr. Low did his best to make them understand that the plan of pulling down Newgate by the instrumentality of

Phineas Finn's popularity, or of buying up the Home Secretary by threats of Parliamentary defection, would hardly answer their purpose. He would, he assured them, suggest to the attorneys employed the idea of searching for evidence against the man Mealyus in his own country, and would certainly take care that nothing was omitted from want of means. 'You had better let us put a cheque in your hands,' said the Duchess. But to this he would not assent. He did admit that it would be well to leave no stone unturned, and that the turning of such stones must cost money;—but the money, he said, would be forthcoming. 'He's not a rich man himself,' said the Duchess. Mr. Low assured her that if money were really wanting he would ask for it. 'And now,' said the Duchess, 'there is one other thing that we want. Can we see him?'

'You, yourself?'

'Yes;—I myself, and Madame Goesler. You look as if it would be very wicked.' Mr. Low thought that it would be wicked;—that the Duke would not like it; and that such a visit would occasion ill-natured remarks. 'People do visit him, I suppose. He's not locked up like a criminal.'

'I visit him,' said Mr. Low, 'and one or two other friends have done so. Lord Chiltern has been with him, and Mr. Erle.'

'Has no lady seen him?' asked the Duchess.

'Not to my knowledge.'

'Then it's time some lady should do so. I suppose we could be admitted. If we were his sisters they'd let us in.'

'You must excuse me, Duchess, but——'

'Of course I will excuse you. But what?'

'You are not his sisters.'

'If I were engaged to him, to be his wife?—' said Madame Goesler, standing up. 'I am not so. There is nothing of that kind. You must not misunderstand me. But if I were?'

'On that plea I presume you could be admitted.'

'Why not as a friend? Lord Chiltern is admitted as his friend.'

'Because of the prudery of a prison,' said the Duchess. 'All things are wrong to the lookers after wickedness, my dear. If it would comfort him to see us, why should he not have that comfort?'

'Would you have gone to him in his own lodgings?' asked Mr. Low.

'I would,—if he'd been ill,' said Madame Goesler.

'Madam,' said Mr. Low, speaking with a gravity which for a moment had its effect even upon the Duchess of Omnium, 'I think, at any rate, that if you visit Mr. Finn in prison, you should do so through the instrumentality of his Grace, your husband.'

'Of course you suspect me of all manner of evil.'

'I suspect nothing;—but I am sure that it should be so.'

'It shall be so,' said the Duchess. 'Thank you, sir. We are much obliged to you for your wise counsel.'

'I am obliged to you,' said Madame Goesler, 'because I know that you have his safety at heart.'

'And so am I,' said the Duchess, relenting, and giving him her hand. 'We are really ever so much obliged to you. You don't quite understand about the Duke; and how should you? I never do anything without telling him, but he hasn't time to attend to things.'

'I hope I have not offended you.'

'Oh dear, no. You can't offend me unless you mean it. Good-bye,—and remember to have a great many lawyers, and all with new wigs; and let them all get in a great rage that anybody should suppose it possible that Mr. Finn is a murderer. I'm sure I am. Good-bye, Mr. Low.'

'You'll never be able to get to him,' said the Duchess, as soon as they were alone.

'I suppose not.'

'And what good could you do? Of course I'd go with you if we could get in;—but what would be the use?'

'To let him know that people do not think him guilty.'

'Mr. Low will tell him that. I suppose, too, we can write to him. Would you mind writing?'

'I would rather go.'

'You might as well tell the truth when you are about it. You are breaking your heart for him.'

'If he were to be condemned, and——executed, I should break my heart. I could never appear bright before the world again.'

'That is just what I told Plantagenet. I said I would go into mourning.'

'And I should really mourn. And yet were he free tomorrow he would be no more to me than any other friend.'

'Do you mean you would not marry him?'

'No;—I would not. Nor would he ask me. I will tell you what will be his lot in life,—if he escapes from the present danger.'

'Of course he will escape. They don't really hang innocent men.'

'Then he will become the husband of Lady Laura Kennedy.'

'Poor fellow! If I believed that, I should think it cruel to help him escape from Newgate.'

CHAPTER LV

Phineas in Prison

PHINEAS FINN himself, during the fortnight in which he was carried backwards and forwards between his prison and the Bow Street Police-office, was able to maintain some outward show of manly dignity,—as though he felt that the terrible accusation and great material inconvenience to which he was subjected were only, and could only be, temporary in their nature, and that the truth would soon prevail. During this period he had friends constantly with him,—either Mr. Low, or Lord Chiltern, or Barrington Erle, or his landlord, Mr. Bunce, who, in these days, was very true to him. And he was very frequently visited by the attorney, Mr. Wickerby, who had been expressly recommended to him for this occasion. If anybody could be counted upon to see him through his difficulty it was Wickerby. But the company of Mr.

Wickerby was not pleasant to him, because, as far as he could judge, Mr. Wickerby did not believe in his innocence. Mr. Wickerby was willing to do his best for him; was, so to speak, moving heaven and earth on his behalf; was fully conscious that this case was a great affair, and in no respect similar to those which were constantly placed in his hands; but there never fell from him a sympathetic expression of assurance of his client's absolute freedom from all taint of guilt in the matter. From day to day, and ten times a day, Phineas would express his indignant surprise that any one should think it possible that he had done this deed, but to all these expressions Mr. Wickerby would make no answer whatever. At last Phineas asked him the direct question. 'I never suspect anybody of anything,' said Mr. Wickerby. 'Do you believe in my innocence?' demanded Phineas. 'Everybody is entitled to be believed innocent till he has been proved to be guilty,' said Mr. Wickerby. Then Phineas appealed to his friend Mr. Low, asking whether he might not be allowed to employ some lawyer whose feelings would be more in unison with his own. But Mr. Low adjured him to make no change. Mr. Wickerby understood the work and was a most zealous man. His client was entitled to his services, but to nothing more than his services. And so Mr. Wickerby carried on the work, fully believing that Phineas Finn had in truth murdered Mr. Bonteen.

But the prisoner was not without sympathy and confidence. Mr. Low, Lord Chiltern, and Lady Chiltern, who, on one occasion, came to visit him with her husband, entertained no doubts prejudicial to his honour. They told him perhaps almost more than was quite true of the feelings of the world in his favour. He heard of the friendship and faith of the Duchess of Omnium, of Madame Goesler, and of Lady Laura Kennedy,—hearing also that Lady Laura was now a widow. And then at length his two sisters came over to him from Ireland, and wept and sobbed, and fell into hysterics in his presence. They were sure that he was innocent, as was every one, they said, throughout the length and breadth of Ireland. And Mrs.

Bunce, who came to see Phineas in his prison, swore that she would tear the judge from his bench if he did not at once pronounce a verdict in favour of her darling without waiting for any nonsense of a jury. And Bunce, her husband, having

convinced himself that his lodger had not committed the murder, was zealous in another way, taking delight in the case, and proving that no jury could find a verdict of guilty.

During that week Phineas, buoyed up by the sympathy of his friends, and in some measure supported by the excitement of the occasion, carried himself well, and bore bravely the terrible misfortune to which he had been subjected by untoward circumstances. But when the magistrate fully committed him, giving the first public decision on the matter from the bench, declaring to the world at large that on the evidence as given, prima facie, he, Phineas Finn, must be regarded as the murderer of Mr. Bonteen, our hero's courage almost gave way. If such was now the judicial opinion of the magistrate, how could he expect a different verdict from a jury in two months' time, when he would be tried before a final court? As far as he could understand, nothing more could be learned on the matter. All the facts were known that could be known,—as far as he, or rather his friends on his behalf, were able to search for facts. It seemed to him that there was no tittle whatever of evidence against him. He had walked straight home from his club with the life-preserver in his pocket, and had never turned to the right or to the left. Till he found himself committed, he would not believe that any serious and prolonged impediment could be thrown in the way of his liberty. He would not believe that a man altogether innocent could be in danger of the gallows on a false accusation. It had seemed to him that the police had kept their hold on him with a rabid ferocity, straining every point with the view of showing that it was possible that he should have been the murderer. Every policeman who had been near him, carrying him backward and forward from his prison, or giving evidence as to the circumstances of the locality and of his walk home on that fatal night, had seemed to him to be an enemy. But he had looked for impartiality from the magistrate,—and now the magistrate had failed him. He had seen in court the faces of men well known to him,—men known in the world,—with whom he had been on pleasant terms in Parliament, who had sat upon the bench while he was standing as a culprit between two constables; and they who had been his familiar friends had appeared at once to have been removed from him by some unmeasurable distance. But all that he had, as it were, discounted, believing that a few hours,—at the very longest a few days,—would remove the distance; but now he was sent back to his prison, there to await his trial for the murder.

And it seemed to him that his committal startled no one but himself. Could it be that even his dearest friends thought it possible that he had been guilty? When that day came, and he was taken back to Newgate on his last journey there from Bow Street, Lord Chiltern had returned for a while to Harrington Hall, having promised that he would be back in

London as soon as his business would permit; but Mr. Low came to him almost immediately to his prison room. 'This is a pleasant state of things,' said Phineas, with a forced laugh. But as he laughed he also sobbed, with a low, irrepressible, convulsive movement in his throat.

'Phineas, the time has come in which you must show yourself to be a man.'

'A man! Oh, yes, I can be a man. A murderer you mean. I shall have to be——hung, I suppose.'

'May God, in His mercy, forbid.'

'No;—not in His mercy; in His justice. There can be no need for mercy here,—not even from Heaven. When they take my life may He forgive my sins through the merits of my Saviour. But for this there can be no mercy. Why do you not speak? Do you mean to say that I am guilty?'

'I am sure that you are innocent.'

'And yet, look here. What more can be done to prove it than has been done? That blundering fool will swear my life away.' Then he threw himself on his bed, and gave way to his sobs.

That evening he was alone,—as, indeed, most of his evenings had been spent, and the minutes were minutes of agony to him. The external circumstances of his position were as comfortable as circumstances would allow. He had a room to himself looking out through heavy iron bars into one of the courts of the prison. The chamber was carpeted, and was furnished with bed and chairs and two tables. Books were allowed him as he pleased, and pen and ink. It was May, and no fire was necessary. At certain periods of the day he could walk alone in the court below,—the restriction on such liberty being that at other certain hours the place was wanted for other prisoners. As far as he knew no friend who called was denied to him, though he was by no means certain that his privilege in that respect would not be curtailed now that he had been committed for trial. His food had been plentiful and well cooked, and even luxuries, such as fish and wine and fruit, had been supplied to him. That the fruit had come from the hot-houses of the Duchess of Omnium, and the wine from Mr. Low's cellar, and the fish and lamb and spring vegetables, the cream and coffee and fresh butter from the unrestricted orders of another friend, that Lord Chiltern had sent him champagne and cigars, and that Lady Chiltern had given directions about the books and stationery, he did not know. But as far as he could be consoled by such comforts, there had been the consolation. If lamb and salad could make him happy he might have enjoyed his sojourn in Newgate. Now, this evening, he was past all enjoyment. It was impossible that he should read. How could a man fix his attention on any book, with a charge of murder against himself affirmed by the deliberate decision of a judge? And he knew himself to be as innocent as the magistrate himself. Every now and then he would rise from his bed, and almost rush across the room as though he would dash his head against the wall. Murder! They really believed that he had deliberately murdered the man;—he, Phineas Finn, who had served his country with repute, who had sat in Parliament, who had prided himself on living with the best of his fellow-creatures, who had been the friend of Mr. Monk and of Lord Cantrip, the trusted intimate of such women as Lady Laura and Lady Chiltern, who had never put his hand to a mean action, or allowed his tongue to speak a mean word! He laughed in his wrath, and then almost howled in his agony. He thought of the young loving wife who had lived with him little more than for one fleeting year, and wondered whether she was looking down upon him from Heaven, and how her spirit would bear this accusation against the man upon whose bosom she had slept, and in whose arms she had gone to her long rest. 'They can't believe it,' he said aloud. 'It is impossible. Why should I have murdered him?' And then he remembered an example in Latin from some rule of grammar, and repeated it to himself over and over again.— 'No one at an instant,—of a sudden,—becomes most base.'* It seemed to him that there was such a want of knowledge of human nature in the supposition that it was possible that he should have committed such a crime. And yet——there he was, committed to take his trial for the murder of Mr. Bonteen.

The days were long, and it was daylight till nearly nine. Indeed the twilight lingered, even through those iron bars, till after nine. He had once asked for candles, but had been told that they could not be allowed him without an attendant in the room,—and he had dispensed with them. He had been treated doubtless with great respect, but nevertheless he had been treated as a prisoner. They hardly denied him anything that he asked, but when he asked for that which they did not choose to grant they would annex conditions which induced him to withdraw his request. He understood their ways now, and did not rebel against them.

On a sudden he heard the key in the door, and the man who attended him entered the room with a candle in his hand. A lady had come to call, and the governor had given permission for her entrance. He would return for the light,—and for the lady, in half an hour. He had said all this before Phineas could see who the lady was. And when he did see the form of her who followed the gaoler, and who stood with hesitating steps behind him in the doorway, he knew her by her sombre solemn raiment, and not by her countenance. She was dressed from head to foot in the deepest weeds of widowhood, and a heavy veil fell from her bonnet over her face. 'Lady Laura, is it you?' said Phineas, putting out his hand. Of course it was Lady Laura. While the Duchess of Omnium and Madame Goesler were talking about such a visit, allowing themselves to be deterred by the wisdom of Mr. Low, she had made her way through bolts and bars, and was now with him in his prison.

'Oh, Phineas!' She slowly raised her veil, and stood gazing at him. 'Of all my troubles this,—to see you here,—is the heaviest.'

'And of all my consolations to see you here is the greatest.' He should not have so spoken. Could he have thought of things as they were, and have restrained himself, he should not have uttered words to her which were pleasant but not true. There came a gleam of sunshine across her face as she listened to him, and then she threw herself into his arms, and wept upon his shoulder. 'I did not expect that you would have found me,' he said.

She took the chair opposite to that on which he usually sat, and then began her tale. Her cousin, Barrington Erle, had brought her there, and was below, waiting for her in the Governor's house. He had procured an order for her admission that evening, direct from Sir Harry Coldfoot, the Home Secretary,—which, however, as she admitted, had been given under the idea that she and Erle were to see him together. 'But I would not let him come with me,' she said. 'I could not have spoken to you, had he been here;—could I?'

'It would not have been the same, Lady Laura.' He had thought much of his mode of addressing her on occasions before this, at Dresden and at Portman Square, and had determined that he would always give her her title. Once or twice he had lacked the courage to be so hard to her. Now as she heard the name the gleam of sunshine passed from her altogether. 'We hardly expected that we should ever meet in such a place as this?' he said.

'I cannot understand it. They cannot really think you killed him.' He smiled, and shook his head. Then she spoke of her own condition. 'You have heard what has happened? You know that I am—a widow?'

'Yes;—I had heard,' And then he smiled again. 'You will have understood why I could not come to you,—as I should have done but for this little accident.'

'He died on the day that they arrested you. Was it not strange that such a double blow should fall together? Oswald, no doubt, told you all.'

'He told me of your husband's death.'

'But not of his will? Perhaps he has not seen you since he heard it.' Lord Chiltern had heard of the will before his last visit to Phineas in Newgate, but had not chosen then to speak of his sister's wealth.

'I have heard nothing of Mr. Kennedy's will.'

'It was made immediately after our marriage,—and he never changed it, though he had so much cause of anger against me.'

'He has not injured you, then,—as regards money.'

'Injured me! No, indeed. I am a rich woman,—very rich. All Loughlinter is my own,—for life. But of what use can it be to me?' He in his present state could tell her of no uses for such a property. 'I suppose, Phineas, it cannot be that you are really in danger?'

'In the greatest danger, I fancy.'

'Do you mean that they will say—you are guilty?'

'The magistrates have said so already.'

'But surely that is nothing. If I thought so, I should die. If I believed it, they should never take me out of the prison while you are here. Barrington says that it cannot be. Oswald and Violet are sure that such a thing can never happen. It was that Jew who did it.'

'I cannot say who did it. I did not.'

'You! Oh, Phineas! The world must be mad when any can believe it!'

'But they do believe it?' This, he said, meaning to ask a question as to that outside world.

'We do not. Barrington says——'

'What does Barrington say?'

'That there are some who do;—just a few, who were Mr. Bonteen's special friends.'

'The police believe it. That is what I cannot understand;—men who ought to be keen-eyed and quick-witted. That magistrate believes it. I saw men in the Court who used to know me well, and I could see that they believed it. Mr. Monk was here yesterday.'

'Does he believe it?'

'I asked him, and he told me—no. But I did not quite trust him as he told me. There are two or three who believe me innocent.'

'Who are they?'

'Low, and Chiltern, and his wife;—and that man Bunce,

and his wife. If I escape from this,—if they do not hang me,—I will remember them. And there are two other women who know me well enough not to think me a murderer.'

'Who are they, Phineas?'

'Madame Goesler, and the Duchess of Omnium.'

'Have they been here?' she asked, with jealous eagerness.

'Oh, no. But I hear that it is so,—and I know it. One learns to feel even from hearsay what is in the minds of people.'

'And what do I believe, Phineas? Can you read my thoughts?'

'I know them of old, without reading them now.' Then he put forth his hand and took hers. 'Had I murdered him in real truth, you would not have believed it.'

'Because I love you, Phineas.'

Then the key was again heard in the door, and Barrington Erle appeared with the gaolers. The time was up, he said, and he had come to redeem his promise. He spoke cordially to his old friend, and grasped the prisoner's hand cordially,—but not the less did he believe that there was blood on it, and Phineas knew that such was his belief. It appeared on his arrival that Lady Laura had not at all accomplished the chief object of her visit. She had brought with her various cheques, all drawn by Barrington Erle on his banker,—amounting altogether to many hundreds of pounds,—which it was intended that Phineas should use from time to time for the necessities of his trial. Barrington Erle explained that the money was in fact to be a loan from Lady Laura's father, and was simply passed through his banker's account. But Phineas knew that the loan must come from Lady Laura, and he positively refused to touch it. His friend, Mr. Low, was managing all that for him, and he would not embarrass the matter by a fresh account. He was very obstinate, and at last the cheques were taken away in Barrington Erle's pocket.

'Good-night, old fellow,' said Erle, affectionately. 'I'll see you again before long. May God send you through it all.'

'Good-night, Barrington. It was kind of you to come to me.' Then Lady Laura, watching to see whether her cousin would leave her alone for a moment with the object of her idolatry, paused before she gave him her hand. 'Good-night, Lady Laura,' he said.

'Good-night!' Barrington Erle was now just outside the door.

'I shall not forget your coming here to me.'

'How should we, either of us, forget it?'

'Come, Laura,' said Barrington Erle, 'we had better make an end of it.'

'But if I should never see him again!'

'Of course you will see him again.'

'When! and where! Oh, God,—if they should murder him!' Then she threw herself into his arms, and covered him with kisses, though her cousin had returned into the room and stood over her as she embraced him.

'Laura,' said he, 'you are doing him an injury. How should he support himself if you behave like this! Come away.'

'Oh, my God, if they should kill him!' she exclaimed. But she allowed her cousin to take her in his arms, and Phineas Finn was left alone without having spoken another word to either of them.

CHAPTER LVI

The Meager family

On the day after the committal a lady, who had got out of a cab at the corner of Northumberland Street, in the Marylebone Road, walked up that very uninviting street, and knocked at a door just opposite to the deadest part of the dead wall of the Marylebone Workhouse. Here lived Mrs. and Miss Meager,—and also on occasions Mr. Meager, who, however, was simply a trouble and annoyance in the world, going about to race-courses, and occasionally, perhaps, to worse places, and being of no slightest use to the two poor hard-worked women,—mother and daughter,—who endeavoured to get their living by letting lodgings. The task was difficult, for it is not everybody who likes to look out upon the dead wall of a workhouse, and they who do are disposed to think that their willingness that way should be considered in the rent. But Mr. Emilius, when the cruelty of his wife's friends deprived him of the short-lived luxury of his mansion in Lowndes Square, had found in Northumberland Street a congenial retreat, and had for a while trusted to Mrs. and Miss Meager for all his domestic comforts. Mr. Emilius was always a favourite with new friends, and had not as yet had his Northumberland Street gloss rubbed altogether off him when Mr. Bonteen was murdered. As it happened, on that night, or rather early in the day, for Meager had returned to the bosom of his family after a somewhat prolonged absence in the provinces, and therefore the date had become specially remarkable in the Meager family from the double event,—Mr. Meager had declared that unless his wife could supply him with a five-pound note he must cut his throat instantly. His wife and daughter had regretted the necessity, but had declared the alternative to be out of the question. Whereupon Mr. Meager had endeavoured to force the lock of an old bureau with a carving-knife, and there had been some slight personal encounter,—after which he had had some gin and had gone to bed. Mrs. Meager remembered the day very well indeed, and Miss Meager, when the police came the next morning, had accounted for her black eye by a tragical account of a fall she had had against the bed-post in the dark. Up to that period Mr. Emilius had been everything that was sweet and good,—an excellent, eloquent clergyman, who was being ill-treated by his wife's wealthy relations, who was soft in his manners and civil in his words, and never gave more trouble than was necessary. The period, too, would have been one of comparative prosperity to the Meager ladies, —but for that inopportune return of the head of the family,— as two other lodgers had been inclined to look out upon the dead wall, or else into the cheerful back-yard; which circumstance came to have some bearing upon our story, as Mrs. Meager had been driven by the press of her increased household to let that good-natured Mr. Emilius know that if 'he didn't mind it' the latch-key might be an accommodation on occasions. To give him his due, indeed, he had, when first taking the rooms, offered to give up the key when not intending to be out at night.

After the murder Mr. Emilius had been arrested, and had been kept in durance for a week. Miss Meager had been sure that he was innocent; Mrs. Meager had trusted the policemen, who evidently thought that the clergyman was guilty. Of the policemen who were concerned on the occasion, it may be said in a general way that they believed that both the gentlemen had committed the murder,—so anxious were they not to be foiled in the attempts at discovery which their duty called upon them to make. Mr. Meager had left the house on the morning of the arrest, having arranged that little matter of the five-pound note by a compromise. When the policeman came for Mr. Emilius, Mr. Meager was gone. For a day or two the lodger's rooms were kept vacant for the clergyman till Mrs. Meager became quite convinced that he had committed the murder, and then all his things were packed up and placed in the passage. When he was liberated he returned to the house, and expressed unbounded anger at what had been done. He took his two boxes away in a cab, and was seen no more by the ladies of Northumberland Street.

But a further gleam of prosperity fell upon them in consequence of the tragedy which had been so interesting to them. Hitherto the inquiries made at their house had had reference solely to the habits and doings of their lodger during the last few days; but now there came to them a visitor who made a more extended investigation; and this was one of their own sex. It was Madame Goesler who got out of the cab at the workhouse corner, and walked from thence to Mrs. Meager's house. This was her third appearance in Northumberland Street, and at each coming she had spoken kind words, and had left behind her liberal recompense for the trouble which she gave. She had no scruples as to paying for the evidence which she desired to obtain,—no fear of any questions which might afterwards be asked in cross-examination. She dealt out sovereigns—womanfully, and had had Mrs. and Miss Meager at her feet. Before the second visit was completed they were both certain that the Bohemian converted Jew had murdered Mr. Bonteen, and were quite willing to assist in hanging him.

'Yes, Ma'am,' said Mrs. Meager, 'he did take the key with him. Amelia remembers we were a key short at the time he was away.' The absence here alluded to was that occasioned by the journey which Mr. Emilius took to Prague, when he heard that evidence of his former marriage was being sought against him in his own country.

'That he did,' said Amelia, 'because we were put out ever so. And he had no business, for he was not paying for the room.'

'You have only one key.'

'There is three, Ma'am. The front attic has one regular because he's on a daily paper, and of course he doesn't get to bed till morning. Meager always takes another, and we can't get it from him ever so.'

'And Mr. Emilius took the other away with him?' asked Madame Goesler.

'That he did, Ma'am. When he came back he said it had been in a drawer,—but it wasn't in the drawer. We always knows what's in the drawers.'

'The drawer wasn't left locked, then?'

'Yes, it was, Ma'am, and he took that key—unbeknownst to us,' said Mrs. Meager. 'But there is other keys that open the drawers. We are obliged in our line to know about the lodgers, Ma'am.'

This was certainly no time for Madame Goesler to express disapprobation of the practices which were thus divulged. She smiled, and nodded her head, and was quite sympathetic with Mrs. Meager. She had learned that Mr. Emilius had taken the latch-key with him to Bohemia, and was convinced that a dozen other latch-keys might have been made after the pattern without any apparent detection by the London police. 'And now about the coat, Mrs. Meager.'

'Well, Ma'am?'

'Mr. Meager has not been here since?'

'No, Ma'am. Mr. Meager, Ma'am, isn't what he ought to be. I never do own it up, only when I'm driven. He hasn't been home.'

'I suppose he still has the coat.'

'Well, Ma'am, no. We sent a young man after him, as you said, and the young man found him at the Newmarket Spring.'

'Some water cure?' asked Madame Goesler.

'No, Ma'am. It ain't a water cure, but the races. He hadn't got the coat. He does always manage a tidy great coat when November is coming on, because it covers everything, and is respectable, but he mostly parts with it in April. He gets short, and then he—just pawns it.'

'But he had it the night of the murder?'

'Yes, Ma'am, he had. Amelia and I remembered it especial.

When we went to bed, which we did soon after ten, it was left in this room, lying there on the sofa.' They were now sitting in the little back parlour, in which Mrs. and Miss Meager were accustomed to live.

'And it was there in the morning?'

'Father had it on when he went out,' said Amelia.

'If we paid him he would get it out of the pawnshop, and bring it to us, would he not?' asked the lady.

To this Mrs. Meager suggested that it was quite on the cards that Mr. Meager might have been able to do better with his coat by selling it, and if so, it certainly would have been sold, as no prudent idea of redeeming his garment for the next winter's wear would ever enter his mind. And Mrs. Meager seemed to think that such a sale would not have taken place between her husband and any old friend. 'He wouldn't know where he sold it,' said Mrs. Meager.

'Anyways he'd tell us so,' said Amelia.

'But if we paid him to be more accurate?' said Madame Goesler.

'They is so afraid of being took up themselves,' said Mrs. Meager. There was, however, ample evidence that Mr. Meager had possessed a grey great coat, which during the night of the murder had been left in the little sitting-room, and which they had supposed to have lain there all night. To this coat Mr. Emilius might have had easy access. 'But then it was a big man that was seen, and Emilius isn't no ways a big man. Meager's coat would be too long for him, ever so much.'

'Nevertheless we must try and get the coat,' said Madame Goesler. 'I'll speak to a friend about it. I suppose we can find your husband when we want him?'

'I don't know, Ma'am. We never can find him; but then we never do want him,—not now. The police know him at the races, no doubt. You won't go and get him into trouble, Ma'am, worse than he is? He's always been in trouble, but I wouldn't like to be means of making it worse on him than it is.'

Madame Goesler, as she again paid the woman for her services, assured her that she would do no injury to Mr. Meager. All that she wanted of Mr. Meager was his grey coat, and that not with any view that could be detrimental either to his honour or to his safety, and she was willing to pay any reasonable price,—or almost any unreasonable price, —for the coat. But the coat must be made to be forthcoming if it were still in existence, and had not been as yet torn to pieces by the shoddy makers.

'It ain't near come to that yet,' said Amelia. 'I don't know that I ever see father more respectable,—that is, in the way of a great coat.'

CHAPTER LVII

The beginning of the search for the key and the coat

WHEN Madame Goesler revealed her plans and ideas to Mr. Wickerby, the attorney, who had been employed to bring Phineas Finn through his troubles, that gentleman evidently did not think much of the unprofessional assistance which the lady proposed to give him. 'I'm afraid it is far-fetched, Ma'am,—if you understand what I mean,' said Mr. Wickerby. Madame Goesler declared that she understood very well what Mr. Wickerby meant, but that she could hardly agree with him. 'According to that the gentleman must have plotted the murder more than a month before he committed it,' said Mr. Wickerby.

'And why not?'

'Murder plots are generally the work of a few hours at the longest, Madame Goesler. Anger, combined with an indifference to self-sacrifice, does not endure the wear of many days. And the object here was insufficient. I don't think we can ask to have the trial put off in order to find out whether a false key may have been made in Prague.'

'And you will not look for the coat?'

'We can look for it, and probably get it, if the woman has

not lied to you; but I don't think it will do us any good. The woman probably is lying. You have been paying her very liberally, so that she has been making an excellent livelihood out of the murder. No jury would believe her. And a grey coat is a very common thing. After all, it would prove nothing. It would only let the jury know that Mr. Meager had a grey coat as well as Mr. Finn. That Mr. Finn wore a grey coat on that night is a fact which we can't upset. If you got hold of Meager's coat you wouldn't be a bit nearer to proof that Emilius had worn it.'

'There would be the fact that he might have worn it.'

'Madame Goesler, indeed it would not help our client. You see what are the difficulties in our way. Mr. Finn was on the spot at the moment, or so near it as to make it certainly possible that he might have been there. There is no such evidence as to Emilius, even if he could be shown to have had a latch-key. The man was killed by such an instrument as Mr. Finn had about him. There is no evidence that Mr. Emilius had such an instrument in his hand. A tall man in a grey coat was seen hurrying to the spot at the exact hour. Mr. Finn is a tall man and wore a grey coat at the time. Emilius is not a tall man, and, even though Meager had a grey coat, there is no evidence to show that Emilius ever wore it. Mr. Finn had quarelled violently with Mr. Bonteen within the hour. It does not appear that Emilius ever quarelled with Mr. Bonteen, though Mr. Bonteen had exerted himself in opposition to Emilius.'

'Is there to be no defence, then?'

'Certainly there will be a defence, and such a defence as I think will prevent any jury from being unanimous in convicting my client. Though there is a great deal of evidence against him, it is all—what we call circumstantial.'

'I understand, Mr. Wickerby.'

'Nobody saw him commit the murder.'

'Indeed no,' said Madame Goesler.

'Although there is personal similarity, there is no personal identity. There is no positive proof of anything illegal on his part, or of anything that would have been suspicious had no murder been committed,—such as the purchase of poison, or carrying of a revolver. The life-preserver, had no such instrument been unfortunately used, might have been regarded as a thing of custom.'

'But I am sure that that Bohemian did murder Mr. Bonteen, said Madame Goesler, with enthusiasm.

'Madame,' said Mr. Wickerby, holding up both his hands, 'I can only wish that you could be upon the jury.'

'And you won't try to show that the other man might have done it?'

'I think not. Next to an alibi that breaks down;—you know what an alibi is, Madame Goesler?'

'Yes, Mr. Wickerby; I know what an alibi is.'

'Next to an alibi that breaks down, an unsuccessful attempt to affix the fault on another party is the most fatal blow which a prisoner's counsel can inflict upon him. It is always taken by the jury as so much evidence against him. We must depend altogether on a different line of defence.'

'What line, Mr. Wickerby?'

'Juries are always unwilling to hang,'—Madame Goesler shuddered as the horrid word was broadly pronounced,—'and are apt to think that simply circumstantial evidence cannot be suffered to demand so disagreeable a duty. They are peculiarly averse to hanging a gentleman, and will hardly be induced to hang a member of Parliament. Then Mr. Finn is very good-looking, and has been popular,—which is all in his favour. And we shall have such evidence on the score of character as was never before brought into one of our courts. We shall have half the Cabinet. There will be two dukes.' Madame Goesler, as she listened to the admiring enthusiasm of the attorney while he went on with his list, acknowledged to herself that her dear friend, the Duchess, had not been idle. 'There will be three Secretaries of State. The Secretary of State for the Home Department himself will be examined. I am not quite sure that we mayn't get the Lord Chancellor. There will be Mr. Monk,—about the most popular man in

England,—who will speak of the prisoner as his particular friend. I don't think any jury would hang a particular friend of Mr. Monk's. And there will be ever so many ladies. That has never been done before, but we mean to try it.' Madame Goesler had heard all this, and had herself assisted in the work. 'I rather think we shall get four or five leading members of the Opposition, for they all disliked Mr. Bonteen. If we could manage Mr. Daubeny and Mr. Gresham, I think we might reckon ourselves quite safe. I forgot to say that the Bishop of Barchester has promised.'

'All that won't prove his innocence, Mr. Wickerby.' Mr. Wickerby shrugged his shoulders. 'If he be acquitted after that fashion men then will say—that he was guilty.'

'We must think of his life first, Madame Goesler,' said the attorney.

Madame Goesler when she left the attorney's room was very ill-satisfied with him. She desired some adherent to her cause who would with affectionate zeal resolve upon washing Phineas Finn white as snow in reference to the charge now made against him. But no man would so resolve who did not believe in his innocence,—as Madame Goesler believed herself. She herself knew that her own belief was romantic and unpractical. Nevertheless, the conviction of the guilt of that other man, towards which she still thought that much could be done if that coat were found and the making of a secret key were proved, was so strong upon her that she would not allow herself to drop it. It would not be sufficient for her that Phineas Finn should be acquitted. She desired that the real murderer should be hung for the murder, so that all the world might be sure,—as she was sure,—that her hero had been wrongfully accused.

'Do you mean that you are going to start yourself?' the Duchess said to her that same afternoon.

'Yes, I am.'

'Then you must be very far gone in love, indeed.'

'You would do as much, Duchess, if you were free as I am. It isn't a matter of love at all. It's womanly enthusiasm for the cause one has taken up.'

'I'm quite as enthusiastic,—only I shouldn't like to go to Prague in June.'

'I'd go to Siberia in January if I could find out that that horrid man really committed the murder.'

'Who are going with you?'

'We shall be quite a company. We have got a detective policeman, and an interpreter who understands Czech and German to go about with the policeman, and a lawyer's clerk, and there will be my own maid.'

'Everybody will know all about it before you get there.'

'We are not to go quite together. The policeman and the interpreter are to form one party, and I and my maid another. The poor clerk is to be alone. If they get the coat, of course you'll telegraph to me.'

'Who is to have the coat?'

'I suppose they'll take it to Mr. Wickerby. He says he doesn't want it,—that it would do no good. But I think that if we could show that the man might very easily have been out of the house,—that he had certainly provided himself with means of getting out of the house secretly,—the coat would be of service. I am going at any rate; and shall be in Paris to-morrow morning.'

'I think it very grand of you, my dear; and for your sake I hope he may live to be Prime Minister. Perhaps, after all, he may give Plantagenet his "Garter." '

When the old Duke died, a Garter became vacant, and had of course fallen to the gift of Mr. Gresham. The Duchess had expected that it would be continued in the family, as had been the Lieutenancy of Barsetshire, which also had been held by the old Duke. But the Garter had been given to Lord Cantrip, and the Duchess was sore. With all her Radical propensities and inclination to laugh at dukes and marquises, she thought very much of Garters and Lieutenancies;—but her husband would not think of them at all, and hence there were words between them. The Duchess had declared that the Duke should insist on having the Garter. 'These are things that men do not ask for,' the Duke had said.

'Don't tell me, Plantagenet, about not asking. Everybody asks for everything nowadays.'

'Your everybody is not correct, Glencora. I never yet asked for anything,—and never shall. No honour has any value in my eyes unless it comes unasked.' Thereupon it was that the Duchess now suggested that Phineas Finn, when Prime Minister, might perhaps bestow a Garter upon her husband.

And so Madame Goesler started for Prague with the determination of being back, if possible, before the trial began. It was to be commenced at the Old Bailey towards the end of June, and people already began to foretell that it would extend over a very long period. The circumstances seemed to be simple; but they who understood such matters declared that the duration of a trial depended a great deal more on the public interest felt in the matter than upon its own nature. Now it was already perceived that no trial of modern days had ever been so interesting as would be this trial. It was already known that the Attorney-General, Sir Gregory Grogram, was to lead the case for the prosecution, and that the Solicitor-General, Sir Simon Slope, was to act with him. It had been thought to be due to the memory and character of Mr. Bonteen, who when he was murdered had held the office of President of the Board of Trade, and who had very nearly been Chancellor of the Exchequer, that so unusual a task should be imposed on these two high legal officers of the Government. No doubt there would be a crowd of juniors with them, but it was understood that Sir Gregory Grogram would himself take the burden of the task upon his own shoulders. It was declared everywhere that Sir Gregory did believe Phineas Finn to be guilty, but it was also declared that Sir Simon Slope was convinced he was innocent. The defence was to be entrusted to the well-practised but now aged hands of that most experienced practitioner Mr. Chaffanbrass, than whom no barrister living or dead ever rescued more culprits from the fangs of the law. With Mr. Chaffanbrass, who quite late in life had consented to take a silk gown, was to be associated Mr. Serjeant Birdbolt,—who was said to be employed in order that the case might be in safe hands should the strength of Mr. Chaffanbrass fail him at the last moment; and Mr. Snow, who was supposed to handle a witness more judiciously than any of the rising men, and that subtle, courageous, eloquent, and painstaking youth, Mr. Golightly, who now, with no more than ten or fifteen years' practice, was already known to be earning his bread and supporting a wife and family.

But the glory of this trial would not depend chiefly on the array of counsel, nor on the fact that the Lord Chief Justice himself would be the judge, so much as on the social position of the murdered man and of the murderer. Noble lords and great statesmen would throng the bench of the court to see Phineas Finn tried, and all the world who could find an entrance would do the same to see the great statesmen and the noble lords. The importance of such an affair increases like a snowball as it is rolled on. Many people talk much, and then very many people talk very much more. The under-sheriffs of the City, praiseworthy gentlemen not hitherto widely known to fame, became suddenly conspicuous and popular, as being the dispensers of admissions to seats in the court. It had been already admitted by judges and counsel that sundry other cases must be postponed, because it was known that the Bonteen murder would occupy at least a week. It was supposed that Mr. Chaffanbrass would consume a whole day at the beginning of the trial in getting a jury to his mind,—a matter on which he was known to be very particular,—and another whole day at the end of the trial in submitting to the jury the particulars of all the great cases on record in which circumstantial evidence was known to have led to improper verdicts. It was therefore understood that the last week in June would be devoted to the trial, to the exclusion of all other matters of interest. When Mr. Gresham, hard pressed by Mr. Turnbull for a convenient day, offered that gentleman Thursday, the 24th of June, for suggesting to the House a little proposition of his own with reference to the English Church establishment, Mr. Turnbull openly repudiated the offer, because on that day the trial of Phineas Finn would be commenced. 'I hope,' said Mr. Gresham, 'that the work of the country will not be impeded by that unfortunate affair.' 'I am afraid,' said Mr. Turnbull, 'that the right honourable gentleman will find that the member for Tankerville will on that day monopolise the attention of this House.' The remark was thought to have been made in very bad taste, but nobody doubted its truth. Perhaps the interest was enhanced among politicians by the existence very generally of an opinion that though Phineas Finn had murdered Mr. Bonteen, he would certainly be acquitted. Nothing could then prevent the acquitted murderer from resuming his seat in the House, and gentlemen were already beginning to ask themselves after what fashion it would become them to treat him. Would the Speaker catch his eye when he rose to speak? Would he still be 'Phineas' to the very large number of men with whom his general popularity had made him intimate? Would he be cold-shouldered at the clubs, and treated as one whose hands were red with blood? or would he become more popular than ever, and receive an ovation after his acquittal?

In the meantime Madame Goesler started on her journey for Prague.

CHAPTER LVIII

The two Dukes

It was necessary that the country should be governed, even though Mr. Bonteen had been murdered;—and in order that it should be duly governed it was necessary that Mr. Bonteen's late place at the Board of Trade should be filled. There was some hesitation as to the filling it, and when the arrangement was completed people were very much surprised indeed. Mr. Bonteen had been appointed chiefly because it was thought that he might in that office act as a quasi House of Commons deputy to the Duke of Omnium in carrying out his great scheme of a five-farthinged penny and a ten-pennied shilling. The Duke, in spite of his wealth and rank and honour, was determined to go on with his great task. Life would be nothing to him now unless he could at least hope to arrange the five farthings. When his wife had bullied him about the Garter he had declared to her, and with perfect truth, that he had never asked for anything. He had gone on

to say that he never would ask for anything; and he certainly did not think that he was betraying himself with reference to that assurance when he suggested to Mr. Gresham that he would himself take the place left vacant by Mr. Bonteen —of course retaining his seat in the Cabinet.

'I should hardly have ventured to suggest such an arrangement to your Grace,' said the Prime Minister.

'Feeling that it might be so, I thought that I would venture to ask,' said the Duke. 'I am sure you know that I am the last man to interfere as to place or the disposition of power.'

'Quite the last man,' said Mr. Gresham.

'But it has always been held that the Board of Trade is not incompatible with the Peerage.'

'Oh dear, yes.'

'And I can feel myself nearer to this affair of mine there than I can elsewhere.'

Mr. Gresham of course had no objection to urge. This great nobleman, who was now asking for Mr. Bonteen's shoes, had been Chancellor of the Exchequer, and would have remained Chancellor of the Exchequer had not the mantle of his nobility fallen upon him. At the present moment he held an office in which peers are often temporarily shelved, or put away, perhaps, out of harm's way for the time, so that they may be brought down and used when wanted, without having received crack or detriment from that independent action into which a politician is likely to fall when his party is 'in' but he is still 'out'. He was Lord Privy Seal,—a Lordship of State which does carry with it a status and a seat in the Cabinet, but does not necessarily entail any work. But the present Lord, who cared nothing for status, and who was much more intent on his work than he was even on his seat in the Cabinet, was possessed by what many of his brother politicians regarded as a morbid dislike to pretences. He had not been happy during his few weeks of the Privy Seal, and had almost envied Mr. Bonteen the realities of the Board of Trade. 'I think upon the whole it will be best to make the change,' he said to Mr. Gresham. And Mr. Gresham was delighted.

But there were one or two men of mark,—one or two who were older than Mr. Gresham probably, and less perfect in their Liberal sympathies,—who thought that the Duke of Omnium was derogating from his proper position in the step which he was now taking. Chief among these was his friend the Duke of St. Bungay, who alone perhaps could venture to argue the matter with him. 'I almost wish that you had spoken to me first,' said the elder Duke.

'I feared that I should find you so strongly opposed to my resolution.'

'If it was a resolution.'

'I think it was,' said the younger. 'It was a great misfortune to me that I should have been obliged to leave the House of Commons.'

'You should not feel it so.'

'My whole life was there,' said he who, as Plantagenet Palliser, had been so good a commoner.

'But your whole life should certainly not be there now,—nor your whole heart. On you the circumstances of your birth have imposed duties quite as high, and I will say quite as useful, as any which a career in the House of Commons can put within the reach of a man.'

'Do you think so, Duke?'

'Certainly I do. I do think that the England which we know could not be the England that she is but for the maintenance of a high-minded, proud, and self-denying nobility. And though with us there is no line dividing our very broad aristocracy into two parts, a higher and a lower, or a greater and a smaller, or a richer and a poorer, nevertheless we all feel that the success of our order depends chiefly on the conduct of those whose rank is the highest and whose means are the greatest. To some few, among whom you are conspicuously one, wealth has been given so great and rank so high that much of the welfare of your country depends on the manner in which you bear yourself as the Duke of Omnium.'

'I would not wish to think so.'

'Your uncle so thought. And, though he was a man very different from you, not inured to work in his early life, with fewer attainments, probably a slower intellect, and whose general conduct was inferior to your own,—I speak freely because the subject is important,—he was a man who understood his position and the requirements of his order very thoroughly. A retinue almost Royal, together with an expenditure which Royalty could not rival, secured for him the respect of the nation.'

'Your life has not been as was his, and you have won a higher respect.'

'I think not. The greater part of my life was spent in the House of Commons, and my fortune was never much more than the tenth of his. But I wish to make no such comparison.'

'I must make it, if I am to judge which I would follow.'

'Pray understand me, my friend,' said the old man, energetically. 'I am not advising you to abandon public life in order that you may live in repose as a great nobleman. It would not be in your nature to do so, nor could the country afford to lose your services. But you need not therefore take your place in the arena of politics as though you were still Plantagenet Palliser, with no other duties than those of a politician,—as you might so well have done had your uncle's titles and wealth descended to a son.'

'I wish they had,' said the regretful Duke.

'It cannot be so. Your brother perhaps wishes that he were a Duke, but it has been arranged otherwise. It is vain to repine. Your wife is unhappy because your uncle's Garter was not at once given to you.'

'Glencora is like other women,—of course.'

'I share her feelings. Had Mr. Gresham consulted me, I should not have scrupled to tell him that it would have been for the welfare of his party that the Duke of Omnium should be graced with any and every honour in his power to bestow. Lord Cantrip is my friend, almost as warmly as are you; but the country would not have missed the ribbon from the breast of Lord Cantrip. Had you been more the Duke, and less the slave of your country, it would have been sent to you. Do I make you angry by speaking so?'

'Not in the least. I have but one ambition.'

'And that is——?'

'To be the serviceable slave of my country.'

'A master is more serviceable than a slave,' said the old man.

'No; no; I deny it. I can admit much from you, but I cannot admit that. The politician who becomes the master of his country sinks from the statesman to the tyrant.'

'We misunderstand each other, my friend. Pitt, and Peel, and Palmerston, were not tyrants, though each assumed and held for himself to the last the mastery of which I speak. Smaller men who have been slaves, have been as patriotic as they, but less useful. I regret that you should follow Mr. Bonteen in his office.'

'Because he was Mr. Bonteen.'

'All the circumstances of the transfer of office occasioned by your uncle's death seem to me to make it undesirable. I would not have you make yourself too common. This very murder adds to the feeling. Because Mr. Bonteen has been lost to us, the Minister has recourse to you.'

'It was my own suggestion.'

'But who knows that it was so? You, and I, and Mr. Gresham—and perhaps one or two others.'

'It is too late now, Duke; and, to tell the truth of myself, not even you can make me other than I am. My uncle's life to me was always a problem which I could not understand. Were I to attempt to walk in his ways I should fail utterly, and become absurd. I do not feel the disgrace of following Mr. Bonteen.'

'I trust you may at least be less unfortunate.'

'Well;—yes. I need not expect to be murdered in the streets because I am going to the Board of Trade. I shall have made no enemy by my political success.'

'You think that—Mr. Finn—did do that deed?' asked the elder Duke.

'I hardly know what I think. My wife is sure that he is innocent.'

'The Duchess is enthusiastic always.'

'Many others think the same. Lord and Lady Chiltern are sure of that.'

'They were always his best friends.'

'I am told that many of the lawyers are sure that it will be impossible to convict him. If he be acquitted I shall strive to think him innocent. He will come back to the House, of course.'

'I should think he would apply for the Hundreds,' said the Duke of St. Bungay.

'I do not see why he should. I would not in his place. If he be innocent, why should he admit himself unfit for a seat in

Parliament? I tell you what he might do;—resign, and then throw himself again upon his constituency.' The other Duke shook his head, thereby declaring his opinion that Phineas Finn was in truth the man who had murdered Mr. Bonteen.

When it was publicly known that the Duke of Omnium had stepped into Mr. Bonteen's shoes, the general opinion certainly coincided with that given by the Duke of St. Bungay. It was not only that the late Chancellor of the Exchequer should not have consented to fill so low an office, or that the Duke of Omnium should have better known his own place, or that he should not have succeeded a man so insignificant as Mr. Bonteen. These things, no doubt, were said,—but more was said also. It was thought that he should not have gone to an office which had been rendered vacant by the murder of a man who had been placed there merely to assist himself. If the present arrangement was good, why should it not have been made independently of Mr. Bonteen? Questions were asked about it in both Houses, and the transfer no doubt did have the effect of lowering the man in the estimation of the political world. He himself felt that he did not stand so high with his colleagues as when he was Chancellor of the Exchequer; not even so high as when he held the Privy Seal. In the printed lists of those who attended the Cabinets his name generally was placed last, and an opponent on one occasion thought, or pretended to think, that he was no more than Postmaster-General. He determined to bear all this without wincing,—but he did wince. He would not own to himself that he had been wrong, but he was sore,—as a man is sore who doubts about his own conduct; and he was not the less so because he strove to bear his wife's sarcasms without showing that they pained him.

'They say that poor Lord Fawn is losing his mind,' she said to him.

'Lord Fawn! I haven't heard anything about it.'

'He was engaged to Lady Eustace once, you remember. They say that he'll be made to declare why he didn't marry her if this bigamy case goes on. And then it's so unfortunate that he should have seen the man in the grey coat; I hope he won't have to resign.'

'I hope not, indeed.'

'Because, of course, you'd have to take his place as Under-Secretary.' This was very awkward;—but the husband only smiled, and expressed a hope that if he did so he might himself be equal to his new duties. 'By the bye, Plantagenet, what do you mean to do about the jewels?'

'I haven't thought about them. Madame Goesler had better take them.'

'But she won't.'

'I suppose they had better be sold.'

'By auction?'

'That would be the proper way.'

'I shouldn't like that at all. Couldn't we buy them ourselves, and let the money stand till she choose to take it? It's an affair of trade, I suppose, and you're at the head of all that now.' Then again she asked him some question about the Home Secretary, with reference to Phineas Finn; and when he told her that it would be highly improper for him to speak to that officer on such a subject, she pretended to suppose that the impropriety would consist in the interference of a man holding so low a position as he was. 'Of course it is not the same now,' she said, 'as it used to be when you were at the Exchequer.' All which he took without uttering a word of anger, or showing a sign of annoyance. 'You only get two thousand a year, do you, at the Board of Trade, Plantagenet?'

'Upon my word, I forget. I think it's two thousand five hundred.'

'How nice! It was five at the Exchequer, wasn't it?'

'Yes; five thousand at the Exchequer.'

'When you're a Lord of the Treasury it will only be one;—will it?'

'What a goose you are, Glencora. If it suited me to be a Lord of the Treasury, what difference would the salary make?'

'Not the least;—nor yet the rank, or the influence, or the prestige, or the general fitness of things. You are above all such sublunary ideas. You would clean Mr. Gresham's shoes for him, if—the service of your country required it.' These last words she added in a tone of voice very similar to that which her husband himself used on occasions.

'I would even allow you to clean them,—if the service of the country required it,' said the Duke.

But, though he was magnanimous, he was not happy, and perhaps the intense anxiety which his wife displayed as to the fate of Phineas Finn added to his discomfort. The Duchess, as the Duke of St. Bungay had said, was enthusiastic, and he never for a moment dreamed of teaching her to change her nature; but it would have been as well if her enthusiasm at the present moment could have been brought to display itself on some other subject. He had been brought to feel that Phineas Finn had been treated badly when the good things of Government were being given away, and that this had been caused by the jealous prejudices of the man who had been since murdered. But an expectant Under-Secretary of State, let him have been ever so cruelly left out in the cold, should not murder the man by whom he has been ill-treated. Looking at all the evidence as best he could, and listening to the opinions of others, the Duke did think that Phineas had been guilty. The murder had clearly been committed by a personal enemy, not by a robber. Two men were known to have entertained feelings of enmity against Mr. Bonteen; as to one of whom he was assured that it was impossible that he should have been on the spot. As to the other it seemed equally manifest that he must have been there. If it were so, it would have been much better that his wife should not display her interest publicly in the murderer's favour. But the Duchess, wherever she went, spoke of the trial as a persecution; and seemed to think that the prisoner should already be treated as a hero and a martyr. 'Glencora,' he said to her, 'I wish that you could drop the subject of this trial till it be over.'

'But I can't.'

'Surely you can avoid speaking of it.'

'No more than you can avoid your decimals. Out of the full heart the mouth speaks, and my heart is very full. What harm do I do?'

'You set people talking of you.'

'They have been doing that ever since we were married;—but I do not know that they have made out much against me. We must go after our nature, Plantagenet. Your nature is decimals. I run after units.' He did not deem it wise to say anything further,—knowing that to this evil also of Phineas Finn the gods would at last vouchsafe an ending.*

CHAPTER LIX
Mrs. Bonteen

At the time of the murder, Lady Eustace, whom we must regard as the wife of Mr. Emilius till it be proved that he had another wife when he married her, was living as the guest of Mr. Bonteen. Mr. Bonteen had pledged himself to prove the bigamy, and Mrs. Bonteen had opened her house and her heart to the injured lady. Lizzie Eustace, as she had always been called, was clever, rich, and pretty, and knew well how to ingratiate herself with the friend of the hour. She was a greedy, grasping little woman, but, when she had before her a sufficient object, she could appear to pour all that she had into her friend's lap with all the prodigality of a child. Perhaps Mrs. Bonteen had liked to have things poured into her lap. Perhaps Mr. Bonteen had enjoyed the confidential tears of a pretty woman. It may be that the wrongs of a woman doomed to live with Mr. Emilius as his wife had touched their hearts. Be that as it might, they had become the acknowledged friends and supporters of Lady Eustace, and she was living with them in their little house in St. James's Place on that fatal night.

Lizzie behaved herself very well when the terrible tidings were brought home. Mr. Bonteen was so often late at the

House or at his club that his wife rarely sat up for him; and when the servants were disturbed between six and seven o'clock in the morning, no surprise had as yet been felt at his absence. The sergeant of police who had brought the news sent for the maid of the unfortunate lady, and the maid, in her panic, told her story to Lady Eustace before daring to communicate it to her mistress. Lizzie Eustace, who in former days had known something of policemen, saw the man, and learned from him all that there was to learn. Then, while the sergeant remained on the landing place, outside, to support her, if necessary, with the maid by her side to help her, kneeling by the bed, she told the wretched woman what had happened. We need not witness the paroxysms of the widow's misery, but we may understand that Lizzie Eustace was from that moment more strongly fixed than ever in her friendship with Mrs. Bonteen.

When the first three or four days of agony and despair had passed by, and the mind of the bereaved woman was able to turn itself from the loss to the cause of the loss, Mrs. Bonteen became fixed in her certainty that Phineas Finn had murdered her husband, and seemed to think that it was the first and paramount duty of the present Government to have the murderer hung,—almost without a trial. When she found that, at the best, the execution of the man she so vehemently hated could not take place for two months after the doing of the deed, even if then, she became almost frantic in her anger. Surely they would not let him escape! What more proof could be needed? Had not the miscreant quarrelled with her husband, and behaved abominably to him but a few minutes before the murder? Had he not been on the spot with the murderous instrument in his pocket? Had he not been seen by Lord Fawn hastening on the steps of her dear and doomed husband? Mrs. Bonteen, as she sat enveloped in her new weeds, thirsting for blood, could not understand that further evidence should be needed, or that a rational doubt should remain in the mind of any one who knew the circumstances. It was to her as though she had seen the dastard blow struck, and with such conviction as this on her mind did she insist on talking of the coming trial to her inmate, Lady Eustace. But Lizzie had her own opinion, though she was forced to leave it unexpressed in the presence of Mrs. Bonteen. She knew the man who claimed her as his wife, and did not think that Phineas Finn was guilty of the murder. Her Emilius,—her Yosef Mealyus, as she had delighted to call him, since she had separated herself from him,—was, as she thought, the very man to commit a murder. He was by no means degraded in her opinion by the feeling. To commit great crimes is the line of life that comes naturally to some men, and was, as she thought, a line less objectionable than that which confines itself to small crimes. She almost felt that the audacity of her husband in doing such a deed redeemed her from some of the ignominy to which she had subjected herself by her marriage with a runaway who had another wife living. There was a dash of adventure about it which was almost gratifying. But these feelings she was obliged, at any rate for the present, to keep to herself. Not only must she acknowledge the undoubted guilt of Phineas Finn for the sake of her friend, Mrs. Bonteen; but she must consider carefully whether she would gain or lose more by having a murderer for her husband. She did not relish the idea of being made a widow by the gallows. She was still urgent as to the charge of bigamy, and should she succeed in proving that the man had never been her husband, then she did not care how soon they might hang him. But for the present it was better for all reasons that she should cling to the Phineas Finn theory,—feeling certain that it was the bold hand of her own Emilius who had struck the blow.

She was by no means free from the solicitations of her husband, who knew well where she was, and who still adhered to his purpose of reclaiming his wife and his wife's property. When he was released by the magistrate's order, and had recovered his goods from Mr. Meager's house, and was once more established in lodgings, humbler, indeed, than those in Northumberland Street, he wrote the following letter to her who had been for one blessed year the partner of his joys, and his bosom's mistress:—

'3, Jellybag Street, Edgware Road,
'May 26, 18——

'Dearest Wife,—

'You will have heard to what additional sorrow and disgrace I have been subjected through the malice of my enemies. But all in vain! Though princes and potentates have been arrayed against me,'—the princes and potentates had no doubt been Lord Chiltern and Mr. Low,—'innocence has prevailed, and I have come out from the ordeal white as bleached linen or unsullied snow. The murderer is in the hands of justice, and though he be the friend of kings and princes,'—Mr. Emilius had probably heard that the Prince had been at the club with Phineas,—'yet shall justice be done upon him, and the truth of the Lord shall be made to prevail. Mr. Bonteen has been very hostile to me, believing evil things of me, and instigating you, my beloved, to believe evil of me. Nevertheless, I grieve for his death. I lament bitterly that he should have been cut off in his sins, and hurried before the judgment seat of the great Judge without an hour given to him for repentance. Let us pray that the mercy of the Lord may be extended even to him. I beg that you will express my deepest commiseration to his widow, and assure her that she has my prayers.

'And now, my dearest wife, let me approach my own affairs. As I have come out unscorched from the last fiery furnace which has been heated for me by my enemies seven times hot, so shall I escape from that other fire with which the poor man who has gone from us endeavoured to envelop me. If they have made you believe that I have any wife but yourself they have made you believe a falsehood. You, and you only, have my hand. You, and you only, have my heart. I know well what attempts are being made to suborn false evidence in my old country, and how the follies of my youth are being pressed against me,—how anxious are proud

Englishmen that the poor Bohemian should be robbed of the beauty and wit and wealth which he had won for himself. But the Lord fights on my side, and I shall certainly prevail.

'If you will come back to me all shall be forgiven. My heart is as it ever was. Come, and let us leave this cold and ungenial country and go to the sunny south; to the islands of the blest,'—Mr. Emilius during his married life had not quite fathomed the depths of his wife's character, though, no doubt, he had caught some points of it with sufficient accuracy,—'where we may forget these blood-stained sorrows, and mutually forgive each other. What happiness, what joys can you expect in your present mode of life? Even your income,—which in truth is my income,—you cannot obtain, because the tenants will not dare to pay it in opposition to my legal claims. But of what use is gold? What can purple do for us, and fine linen, and rich jewels, without love and a contented heart? Come, dearest, once more to your own one, who will never remember aught of the sad rupture which enemies have made, and we will hurry to the setting sun, and recline on mossy banks, and give up our souls to Elysium.' As Lizzie read this she uttered an exclamation of disgust. Did the man after all know so little of her as to suppose that she, with all her experiences, did not know how to keep her own life and her own pocket separate from her romance? She despised him for this, almost as much as she respected him for the murder.

'If you will only say that you will see me, I will be at your feet in a moment. Till the solemnity with which the late tragical event must have filled you shall have left you leisure to think of all this, I will not force myself into your presence, or seek to secure by law rights which will be much dearer to me if they are accorded by your own sweet goodwill. And in the meantime, I will agree that the income shall be drawn, provided that it be equally divided between us. I have been sorely straitened in my circumstances by these last events. My congregation is of course dispersed. Though my innocence has been triumphantly displayed, my name has been tarnished. It is with difficulty that I find a spot where to lay my weary head. I am ahungered and athirst;—and my very garments are parting from me in my need. Can it be that you willingly doom me to such misery because of my love for you? Had I been less true to you, it might have been otherwise.

'Let me have an answer at once, and I will instantly take steps about the money if you will agree.

'Your truly most loving husband,

'JOSEPH EMILIUS.

'To Lady Eustace, wife of the Rev. Joseph Emilius.'

When Lizzie had read the letter twice through she resolved that she would show it to her friend. 'I know it will reopen the floodgates of your grief,' she said; 'but unless you see it, how can I ask from you the advice which is so necessary to me?' But Mrs. Bonteen was a woman sincere at any rate in this,—that the loss of her husband had been to her so crushing a calamity that there could be no reopening of the floodgates. The grief that cannot bear allusion to its causes has generally something of affectation in its composition. The floodgates with this widowed one had never yet been for a moment closed. It was not that her tears were ever flowing, but that her heart had never yet for a moment ceased to feel that its misery was incapable of alleviation. No utterances concerning her husband could make her more wretched than she was. She took the letter and read it through. 'I daresay he is a bad man,' said Mrs. Bonteen.

'Indeed he is,' said the bad man's wife.

'But he was not guilty of this crime.'

'Oh, no;—I am sure of that,' said Lady Eustace, feeling certain at the same time that Mr. Bonteen had fallen by her husband's hands.

'And therefore I am glad they have given him up. There can be no doubt now about it.'

'Everybody knows who did it now,' said Lady Eustace.

'Infamous ruffian! My poor dear lost one always knew what he was. Oh that such a creature should have been allowed to come among us.'

'Of course he'll be hung, Mrs. Bonteen.'

'Hung! I should think so! What other end would be fit for him? Oh, yes; they must hang him. But it makes one think that the world is too hard a place to live in, when such a one as he can cause so great a ruin.'

'It has been very terrible.'

'Think what the country has lost! They tell me that the Duke of Omnium is to take my husband's place; but the Duke cannot do what he did. Every one knows that for real work there was no one like him. Nothing was more certain than that he would have been Prime Minister,—oh, very soon. They ought to pinch him to death with red-hot tweezers.'

But Lady Eustace was anxious at the present moment to talk about her own troubles. 'Of course, Mr. Emilius did not commit the murder.'

'Phineas Finn committed it,' said the half-maddened woman, rising from her chair. 'And Phineas Finn shall hang by his neck till he is dead.'

'But Emilius has certainly got another wife in Prague.'

'I suppose you know. He said it was so, and he was always right.'

'I am sure of it,—just as you are sure of this horrid Mr. Finn.'

'The two things can't be named together, Lady Eustace.'

'Certainly not. I wouldn't think of being so unfeeling. But he has written me this letter, and what must I do? It is very dreadful about the money, you know.'

'He cannot touch your money. My dear one always said that he could not touch it.'

'But he prevents me from touching it. What they give me only comes by a sort of favour from the lawyer. I almost wish that I had compromised.'

'You would not be rid of him that way.'

'No;—not quite rid of him. You see I never had to take

that horrid name because of the title. I suppose I'd better send the letter to the lawyer.'

'Send it to the lawyer, of course. That is what he would have done. They tell me that the trial is to be on the 24th of June. Why should they postpone it so long? They know all about it. They always postpone everything. If he had lived, there would be an end of that before long.'

Lady Eustace was tired of the virtues of her friend's martyred lord, and was very anxious to talk of her own affairs. She was still holding her husband's letter open in her hand, and was thinking how she could force her friend's dead lion to give place for a while to her own live dog, when a servant announced that Mr. Camperdown, the attorney, was below. In former days there had been an old Mr. Camperdown, who was vehemently hostile to poor Lizzie Eustace; but now, in her new troubles, the firm that had ever been true to her first husband had taken up her case for the sake of the family and her property—and for the sake of the heir, Lizzie Eustace's little boy; and Mr. Camperdown's firm had, next to Mr. Bonteen, been the depository of her trust. He had sent clerks out to Prague,—one who had returned ill,—as some had said poisoned, though the poison had probably been nothing more than the diet natural to Bohemians. And then another had been sent. This, of course, had all been previous to Madame Goesler's self-imposed mission,—which, though it was occasioned altogether by the suspected wickednesses of Mr. Emilius, had no special reference to his matrimonial escapades. And now Mr. Camperdown was down stairs. 'Shall I go down to him, dear Mrs. Bonteen?'

'He may come here if you please.'

'Perhaps I had better go down. He will disturb you.'

'My darling lost one always thought that there should be two present to hear such matters. He said it was safer.' Mr. Camperdown, junior, was therefore shown upstairs to Mrs. Bonteen's drawing-room.

'We have found it all out, Lady Eustace,' said Mr. Camperdown.

'Found out what?'

'We've got Madame Mealyus over here.'

'No!' said Mrs. Bonteen, with her hands raised. Lady Eustace sat silent, with her mouth open.

'Yes, indeed;—and photographs of the registry of the marriage from the books of the synagogue at Cracow. His signature was Yosef Mealyus, and his handwriting isn't a bit altered. I think we could have proved it without the lady; but of course it was better to bring her if possible.'

'Where is she?' asked Lizzie, thinking that she would like to see her own predecessor.

'We have her safe, Lady Eustace. She's not in custody; but as she can't speak a word of English or French, she finds it more comfortable to be kept in private. We're afraid it will cost a little money.'

'Will she swear that she is his wife?' asked Mrs. Bonteen.

'Oh, yes; there'll be no difficulty about that. But her swearing alone mightn't be enough.'

'Surely that settles it all,' said Lady Eustace.

'For the money that we shall have to pay,' said Mr. Camperdown, 'we might probably have got a dozen Bohemian ladies to come and swear that they were married to Yosef Mealyus at Cracow. The difficulty has been to bring over documentary evidence which will satisfy a jury that this is the woman she says she is. But I think we've got it.'

'And I shall be free!' said Lady Eustace, clasping her hands together.

'It will cost a good deal, I fear,' said Mr. Camperdown.

'But I shall be free! Oh, Mr. Camperdown, there is not a woman in all the world who cares so little for money as I do. But I shall be free from the power of that horrid man who has entangled me in the meshes of his sinful life.' Mr. Camperdown told her that he thought that she would be free, and went on to say that Yosef Mealyus had already been arrested, and was again in prison. The unfortunate man had not therefore long enjoyed that humbler apartment which he had found for himself in Jellybag Street.

When Mr. Camperdown went, Mrs. Bonteen followed him out to the top of the stairs. 'You have heard about the trial, Mr. Camperdown?' He said that he knew that it was to take place at the Central Criminal Court in June. 'Yes; I don't know why they have put it off so long. People know that he did it—eh?' Mr. Camperdown, with funereal sadness, declared that he had never looked into the matter. 'I cannot understand that everybody should not know it,' said Mrs. Bonteen.

CHAPTER LX

Two days before the Trial

THERE was a scene in the private room of Mr. Wickerby, the attorney in Hatton Garden, which was very distressing indeed to the feelings of Lord Fawn, and which induced his lordship to think that he was being treated without that respect which was due to him as a peer and a member of the Government. There were present at this scene Mr. Chaffanbrass, the old barrister, Mr. Wickerby himself, Mr. Wickerby's confidential clerk, Lord Fawn, Lord Fawn's solicitor,—that same Mr. Camperdown whom we saw in the last chapter calling upon Lady Eustace,—and a policeman. Lord Fawn had been invited to attend, with many protestations of regret as to the trouble thus imposed upon him, because the very important nature of the evidence about to be given by him at the forthcoming trial seemed to render it expedient that some questions should be asked. This was on Tuesday, the 22nd June, and the trial was to be commenced on the following Thursday. And there was present in the room, very conspicuously, an old heavy grey great coat, as to which Mr. Wickerby had instructed Mr. Chaffanbrass that evidence was forthcoming, if needed, to prove that that coat was lying on the night of the murder in a downstairs room in the house in which Yosef Mealyus was then lodging. The reader will remember the history of the coat. Instigated by Madame Goesler, who was still absent from England, Mr. Wickerby had traced the coat, and had purchased the coat, and was in a position to prove that this very coat was the coat which Mr. Meager had brought home with him to Northumberland Street on that day. But Mr. Wickerby was of opinion that the coat had better not be used. 'It does not go far enough,' said Mr. Wickerby. 'It don't go very far, certainly,' said Mr. Chaffanbrass. 'And if you try to show that another man has done it, and he hasn't,' said Mr. Wickerby, 'it always tells against you with a jury.' To this Mr. Chaffanbrass made no reply, preferring to form his own opinion, and to keep it to himself when formed. But in obedience to his instructions, Lord Fawn was asked to attend at Mr. Wickerby's chambers, in the cause of truth, and the coat was brought out on the occasion. 'Was that the sort of coat the man wore, my lord?' said Mr. Chaffanbrass as Mr. Wickerby held up the coat to view. Lord Fawn walked round and round the coat, and looked at it very carefully before he would vouchsafe a reply. 'You see it is a grey coat,' said Mr. Chaffanbrass, not speaking at all in the tone which Mr. Wickerby's note had induced Lord Fawn to expect.

'It is grey,' said Lord Fawn.

'Perhaps it's not the same shade of grey, Lord Fawn. You see, my lord, we are most anxious not to impute guilt where guilt doesn't lie. You are a witness for the Crown, and, of course, you will tell the Crown lawyers all that passes here. Were it possible, we would make this little preliminary inquiry in their presence;—but we can hardly do that. Mr. Finn's coat was a very much smaller coat.'

'I should think it was,' said his lordship, who did not like being questioned about coats.

'You don't think the coat the man wore when you saw him was a big coat like that? You think he wore a little coat?'

'He wore a grey coat,' said Lord Fawn.

'This is grey;—a coat shouldn't be greyer than that.'

'I don't think Lord Fawn should be asked any more questions on the matter till he gives his evidence in court,' said Mr. Camperdown.

'A man's life depends on it, Mr. Camperdown,' said the barrister. 'It isn't a matter of cross-examination. If I bring that coat into court I must make a charge against another man by the very act of doing so. And I will not do so unless I believe that other man to be guilty. It's an inquiry I can't postpone till we are before the jury. It isn't that I want to trump up a case against another man for the sake of extricating my client on a false issue. Lord Fawn doesn't want to hang Mr. Finn if Mr. Finn be not guilty.'

'God forbid!' said his lordship.

'Mr. Finn couldn't have worn that coat, or a coat at all like it.'

'What is it you do want to learn, Mr. Chaffanbrass?' asked Mr. Camperdown.

'Just put on the coat, Mr. Scruby.' Then at the order of the barrister, Mr. Scruby, the attorney's clerk, did put on Mr. Meager's old great coat, and walked about the room in it. 'Walk quick,' said Mr. Chaffanbrass;—and the clerk did 'walk quick.' He was a stout, thick-set little man, nearly half a foot shorter than Phineas Finn. 'Is that at all like the figure?' asked Mr. Chaffanbrass.

'I think it is like the figure,' said Lord Fawn.

'And like the coat?'

'It's the same colour as the coat.'

'You wouldn't swear it was not the coat?'

'I am not on my oath at all, Mr. Chaffanbrass.'

'No, my lord;—but to me your word is as good as your oath. If you think it possible that was the coat——'

'I don't think anything about it at all. When Mr. Scruby hurries down the room in that way he looks as the man looked when he was hurrying under the lamp-post. I am not disposed to say any more at present.'

'It's a matter of regret to me that Lord Fawn should have come here at all,' said Mr. Camperdown, who had been summoned to meet his client at the chambers, but had come with him.

'I suppose his lordship wishes us to know all that he knew,

seeing that it's a question of hanging the right man or the wrong one. I never heard such trash in my life. Take it off, Mr. Scruby, and let the policeman keep it. I understand Lord Fawn to say that the man's figure was about the same as yours. My client, I believe, stands about twelve inches taller. Thank you, my lord;—we shall get at the truth at last, I don't doubt.' It was afterwards said that Mr. Chaffanbrass's conduct had been very improper in enticing Lord Fawn to Mr. Wickerby's chambers; but Mr. Chaffanbrass never cared what any one said. 'I don't know that we can make much of it,' he said, when he and Mr. Wickerby were alone, 'but it may be as well to bring it into court. It would prove nothing against the Jew even if that fellow,'—he meant Lord Fawn,—'could be made to swear that the coat worn was exactly similar to this. I am thinking now about the height.'

'I don't doubt but you'll get him off.'

'Well;—I may do so. They ought not to hang any man on such evidence as there is against him, even though there were no moral doubt of his guilt. There is nothing really to connect Mr. Phineas Finn with the murder,—nothing tangible. But there is no saying nowadays what a jury will do. Juries depend a great deal more on the judge than they used to do. If I were on trial for my life, I don't think I'd have counsel at all.'

'No one could defend you as well as yourself, Mr. Chaffanbrass.'

'I didn't mean that. No;—I shouldn't defend myself. I should say to the judge, "My lord, I don't doubt the jury will do just as you tell them, and you'll form your own opinion quite independent of the arguments".'

'You'd be hung, Mr. Chaffanbrass.'

'No; I don't know that I should,' said Mr. Chaffanbrass, slowly. 'I don't think I could affront a judge of the present day into hanging me. They've too much of what I call thick-skinned honesty for that. It's the temper of the time to resent nothing,—to be mealy-mouthed and mealy-hearted. Jurymen are afraid of having their own opinion, and almost always shirk a verdict when they can.'

'But we do get verdicts.'

'Yes; the judge gives them. And they are mealy-mouthed verdicts, tending to equalise crime and innocence, and to make men think that after all it may be a question whether fraud is violence, which, after all, is manly, and to feel that we cannot afford to hate dishonesty. It was a bad day for the commercial world, Mr. Wickerby, when forgery ceased to be capital.'

'It was a horrid thing to hang a man for writing another man's name to a receipt for thirty shillings.'

'We didn't do it, but the fact that the law held certain frauds to be hanging matters operated on the minds of men in regard to all fraud. What with the joint-stock working of companies, and the confusion between directors who know nothing and managers who know everything, and the dislike of juries to tread upon people's corns, you can't punish dishonest trading. Caveat emptor* is the only motto going, and the worst proverb that ever came from dishonest stony-hearted Rome. With such a motto as that to guide us no man dare trust his brother. Caveat lex*;—and let the man who cheats cheat at his peril.'

'You'd give the law a great deal to do.'

'Much less than at present. What does your Caveat emptor come to? That every seller tries to pick the eyes out of the head of the purchaser. Sooner or later the law must interfere, and Caveat emptor falls to the ground. I bought a horse the other day; my daughter wanted something to look pretty, and like an old ass as I am I gave a hundred and fifty pounds for the brute. When he came home he wasn't worth a feed of corn.'

'You had a warranty, I suppose?'

'No, indeed! Did you ever hear of such an old fool?'

'I should have thought any dealer would have taken him back for the sake of his character.'

'Any dealer would; but—I bought him of a gentleman.'

'Mr. Chaffanbrass!'

'I ought to have known better, oughtn't I? Caveat emptor.'

'It was just giving away your money, you know.'

'A great deal worse than that. I could have given the—gentleman—a hundred and fifty pounds, and not have minded it much. I ought to have had the horse killed, and gone to a dealer for another. Instead of that,—I went to an attorney.'

'Oh, Mr. Chaffanbrass;—the idea of your going to an attorney.'

'I did then. I never had so much honest truth told me in my life.'

'By an attorney!'

'He said that he did think I'd been born long enough to have known better than that! I pleaded on my own behalf that the gentleman said the horse was all right. "Gentleman!" exclaimed my friend. "You go to a gentleman for a horse; you buy a horse from a gentleman without a warranty; and then you come to me! Didn't you ever hear of Caveat emptor, Mr. Chaffanbrass? What can I do for you?" That's what my friend, the attorney, said to me.'

'And what came of it, Mr. Chaffanbrass? Arbitration, I should say?'

'Just that;—with the horse eating his head off every meal at ever so much per week,—till at last I fairly gave in from sheer vexation. So the—gentleman—got my money, and I added something to my stock of experience. Of course, that's only my story, and it may be that the gentleman could tell it another way. But I say that if my story be right the doctrine of Caveat emptor does not encourage trade. I don't know how we got to all this from Mr. Finn. I'm to see him to-morrow.'

'Yes;—he is very anxious to speak to you.'

'What's the use of it, Wickerby? I hate seeing a client.—What comes of it?'

'Of course he wants to tell his own story.'

'But I don't want to hear his own story. What good will his own story do me? He'll tell me either one of two things. He'll swear he didn't murder the man——'

'That's what he'll say.'

'Which can have no effect upon me one way or the other;

or else he'll say that he did,—which would cripple me altogether.'

'He won't say that, Mr. Chaffanbrass.'

'There's no knowing what they'll say. A man will go on swearing by his God that he is innocent, till at last, in a moment of emotion, he breaks down, and out comes the truth. In such a case as this I do not in the least want to know the truth about the murder.'

'That is what the public wants to know.'

'Because the public is ignorant. The public should not wish to know anything of the kind. What we should all wish to get at is the truth of the evidence about the murder. The man is to be hung not because he committed the murder,—as to which no positive knowledge is attainable; but because he has been proved to have committed the murder,—as to which proof, though it be enough for hanging, there must always be attached some shadow of doubt. We were delighted to hang Palmer,—but we don't know that he killed Cook.* A learned man who knew more about it than we can know seemed to think that he didn't. Now the last man to give us any useful insight into the evidence is the prisoner himself. In nineteen cases out of twenty a man tried for murder in this country committed the murder for which he is tried.'

'There really seems to be a doubt in this case.'

'I dare say. If there be only nineteen guilty out of twenty, there must be one innocent; and why not Mr. Phineas Finn? But, if it be so, he, burning with the sense of injustice, thinks that everybody should see it as he sees it. He is to be tried, because, on investigation, everybody sees it just in a different light. In such case he is unfortunate, but he can't assist me in liberating him from his misfortune. He sees what is patent and clear to him,—that he walked home on that night without meddling with any one. But I can't see that, or make others see it, because he sees it.'

'His manner of telling you may do something.'

'If it do, Mr. Wickerby, it is because I am unfit for my business. If he have the gift of protesting well, I am to think him innocent; and, therefore, to think him guilty, if he be unprovided with such eloquence! I will neither believe or disbelieve anything that a client says to me,—unless he confess his guilt, in which case my services can be but of little avail. Of course I shall see him, as he asks it. We had better meet there,—say at half-past ten.' Whereupon Mr. Wickerby wrote to the governor of the prison begging that Phineas Finn might be informed of the visit.

Phineas had now been in gaol between six and seven weeks, and the very fact of his incarceration had nearly broken his spirits. Two of his sisters, who had come from Ireland to be near him, saw him every day, and his two friends, Mr. Low and Lord Chiltern, were very frequently with him; Lady Laura Kennedy had not come to him again; but he heard from her frequently through Barrington Erle. Lord Chiltern rarely spoke of his sister,—alluding to her merely in connection with her father and her late husband. Presents still came to him from various quarters,—as to which he hardly knew whence they came. But the Duchess and Lady Chiltern and Lady Laura all catered for him,—while Mrs. Bunce looked after his wardrobe, and saw that he was not cut down to prison allowance of clean shirts and socks. But the only friend whom he recognised as such was the friend who would freely declare a conviction of his innocence. They allowed him books and pens and paper, and even cards, if he chose to play at Patience with them or build castles. The paper and pens he could use because he could write about himself. From day to day he composed a diary in which he was never tired of expatiating on the terrible injustice of his position. But he could not read. He found it to be impossible to fix his attention on matters outside himself. He assured himself from hour to hour that it was not death he feared,—not even death from the hangman's hand. It was the condemnation of those who had known him that was so terrible to him; the feeling that they with whom he had aspired to work and live, the leading men and women of his day, Ministers of the Government and their wives, statesmen and their daughters, peers and members of the House in which he himself had sat;—that these should think that, after all, he had been a base adventurer unworthy of their society! That was the sorrow that broke him down, and drew him to confess that his whole life had been a failure.

Mr. Low had advised him not to see Mr. Chaffanbrass;—but he had persisted in declaring that there were instructions which no one but himself could give to the counsellor whose duty it would be to defend him at the trial. Mr. Chaffanbrass came at the hour fixed, and with him came Mr. Wickerby. The old barrister bowed courteously as he entered the prison room, and the attorney introduced the two gentlemen with more than all the courtesy of the outer world. 'I am sorry to see you here, Mr. Finn,' said the barrister.

'It's a bad lodging, Mr. Chaffanbrass, but the term will soon be over. I am thinking a good deal more of my next abode.'

'It has to be thought of, certainly,' said the barrister. 'Let us hope that it may be all that you would wish it to be. My services shall not be wanting to make it so.'

'We are doing all we can, Mr. Finn,' said Mr. Wickerby.

'Mr. Chaffanbrass,' said Phineas, 'there is one special thing that I want you to do.' The old man, having his own idea as to what was coming, laid one of his hands over the other, bowed his head, and looked meek. 'I want you to make men believe that I am innocent of this crime.'

This was better than Mr. Chaffanbrass expected. 'I trust that we may succeed in making twelve men believe it,' said he.

'Comparatively I do not care a straw for the twelve men. It is not to them especially that I am anxious that you should address yourself——'

'But that will be my bounden duty, Mr. Finn.'

'I can well believe, sir, that though I have myself been bred a lawyer, I may not altogether understand the nature of an advocate's duty to his client. But I would wish something more to be done than what you intimate.'

'The duty of an advocate defending a prisoner is to get a verdict of acquittal if he can, and to use his own discretion in making the attempt.'

'But I want something more to be attempted, even if in the struggle something less be achieved. I have known men to be so acquitted that every man in court believed them to be guilty.'

'No doubt;—and such men have probably owed much to their advocates.'

'It is not such a debt that I wish to owe. I know my own innocence.'

'Mr. Chaffanbrass takes that for granted,' said Mr. Wickerby.

'To me it is a matter of astonishment that any human being should believe me to have committed this murder. I am lost in surprise when I remember that I am here simply because I walked home from my club with a loaded stick in my pocket. The magistrate, I suppose, thought me guilty.'

'He did not think about it, Mr. Finn. He went by the evidence;—the quarrel, your position in the streets at the time, the colour of the coat you wore and that of the coat worn by the man whom Lord Fawn saw in the street; the doctor's evidence as to the blows by which the man was killed; and the nature of the weapon which you carried. He put these things together, and they were enough to entitle the public to demand that a jury should decide. He didn't say you were guilty. He only said that the circumstances were sufficient to justify a trial.'

'If he thought me innocent he would not have sent me here.'

'Yes, he would;—if the evidence required that he should do so.'

'We will not argue about that, Mr. Chaffanbrass.'

'Certainly not, Mr. Finn.'

'Here I am, and to-morrow I shall be tried for my life. My life will be nothing to me unless it can be made clear to all the world that I am innocent. I would be sooner hung for this,—with the certainty at my heart that all England on the next day would ring with the assurance of my innocence, than be acquitted and afterwards be looked upon as a murderer.' Phineas, when he was thus speaking, had stepped out into the middle of the room, and stood with his head thrown back, and his right hand forward. Mr. Chaffanbrass, who was himself an ugly, dirty old man, who had always piqued himself on being indifferent to appearance, found himself struck by the beauty and grace of the man whom he now saw for the first time. And he was struck, too, by his client's eloquence, though he had expressly declared to the attorney that it was his duty to be superior to any such influence. 'Oh, Mr. Chaffanbrass, for the love of Heaven, let there be no quibbling.'

'We never quibble, I hope, Mr. Finn.'

'No subterfuges, no escaping by a side wind, no advantage taken of little forms, no objection taken to this and that as though delay would avail us anything.'

'Character will go a great way, we hope.'

'It should go for nothing. Though no one would speak a word for me, still am I innocent. Of course the truth will be known some day.'

'I'm not so sure of that, Mr. Finn.'

'It will certainly be known some day. That it should not be known as yet is my misfortune. But in defending me I would have you hurl defiance at my accusers. I had the stick in my pocket,—having heretofore been concerned with ruffians in the street. I did quarrel with the man, having been insulted by him at the club. The coat which I wore was such as they say. But does that make a murderer of me?'

'Somebody did the deed, and that somebody could probably say all that you say.'

'No, sir;—he, when he is known, will be found to have been skulking in the streets; he will have thrown away his weapon; he will have been secret in his movements; he will have hidden his face, and have been a murderer in more than the deed. When they came to me in the morning did it seem to them that I was a murderer? Has my life been like that? They who have really known me cannot believe that I have been guilty. They who have not known me, and do believe, will live to learn their error.'

He then sat down and listened patiently while the old

lawyer described to him the nature of the case,—wherein lay his danger, and wherein what hope there was of safety. There was no evidence against him other than circumstantial evidence, and both judges and jury were wont to be unwilling to accept such, when uncorroborated, as sufficient in cases of life and death. Unfortunately, in this case the circumstantial evidence was very strong against him. But, on the other hand, his character, as to which men of great mark would speak with enthusiasm, would be made to stand very high. 'I would not have it made to stand higher than it is,' said Phineas. As to the opinion of the world afterwards, Mr. Chaffanbrass went on to say, of that he must take his chance. But surely he himself might fight better for it living than any friend could do for him after his death. 'You must believe me in this, Mr. Finn, that a verdict of acquittal from the jury is the one object that we must have before us.'

'The one object that I shall have before me is the verdict of the public,' said Phineas. 'I am treated with so much injustice in being thought a murderer that they can hardly add anything to it by hanging me.'

When Mr. Chaffanbrass left the prison he walked back with Mr. Wickerby to the attorney's chambers in Hatton Garden, and he lingered for awhile on the Viaduct expressing his opinion of his client. 'He's not a bad fellow, Wickerby.'

'A very good sort of fellow, Mr. Chaffanbrass.'

'I never did,—and I never will,—express an opinion of my own as to the guilt or innocence of a client till after the trial is over. But I have sometimes felt as though I would give the blood out of my veins to save a man. I never felt in that way more strongly than I do now.'

'It'll make me very unhappy, I know, if it goes against him,' said Mr. Wickerby.

'People think that the special branch of the profession into which I have chanced to fall is a very low one,—and I do not know whether, if the world were before me again, I would allow myself to drift into an exclusive practice in criminal courts.'

'Yours has been a very useful life, Mr. Chaffanbrass.'

'But I often feel,' continued the barrister, paying no attention to the attorney's last remark, 'that my work touches the heart more nearly than does that of gentlemen who have to deal with matters of property and of high social claims. People think I am savage,—savage to witnesses.'

'You can frighten a witness, Mr. Chaffanbrass.'

'It's just the trick of the trade that you learn, as a girl learns the notes of her piano. There's nothing in it. You forget it all the next hour. But when a man has been hung whom you have striven to save, you do remember that. Good-morning, Mr. Wickerby. I'll be there a little before ten. Perhaps you may have to speak to me.'

CHAPTER LXI

The beginning of the Trial

THE task of seeing an important trial at the Old Bailey is by no means a pleasant business, unless you be what the denizens of the Court would call 'one of the swells,'—so as to enjoy the privilege of being a benchfellow with the judge on the seat of judgment. And even in that case the pleasure is not unalloyed. You have, indeed, the gratification of seeing the man whom all the world has been talking about for the last nine days, face to face, and of being seen in a position which causes you to be acknowledged as a man of mark; but the intolerable stenches of the Court and its horrid heat come up to you there, no doubt, as powerfully as they fall on those below. And then the tedium of a prolonged trial, in which the points of interest are apt to be few and far between, grows upon you till you begin to feel that though the Prime Minister who is out should murder the Prime Minister who is in, and all the members of the two Cabinets were to be called in evidence, you would not attend the trial, though the seat of honour next to the judge were accorded to you. Those bewigged ones, who are the performers, are so insufferably long in their parts, so arrogant in their bearing,—so it strikes you, though doubtless the fashion of working has been found to be efficient for the purposes they have in hand,—and so uninteresting in their repetition, that you first admire, and then

question, and at last execrate the imperturbable patience of the judge, who might, as you think, force the thing through in a quarter of the time without any injury to justice. And it will probably strike you that the length of the trial is proportioned not to the complicity but to the importance, or rather to the public interest, of the case,—so that the trial which has been suggested of a disappointed and bloody-minded ex-Prime Minister would certainly take at least a fortnight, even though the Speaker of the House of Commons and the Lord Chancellor had seen the blow struck, whereas a collier may knock his wife's brains out in the dark and be sent to the gallows with a trial that shall not last three hours. And yet the collier has to be hung,—if found guilty,—and no one thinks that his life is improperly endangered by reckless haste. Whether lives may not be improperly saved by the more lengthened process is another question.

But the honours of such benchfellowship can be accorded but to few, and the task becomes very tiresome when the spectator has to enter the Court as an ordinary mortal. There are two modes open to him, either of which is subject to grievous penalties. If he be the possessor of a decent coat and hat, and can scrape any acquaintance with any one concerned, he may get introduced to that overworked and greatly perplexed official, the under-sheriff, who will stave him off if possible,—knowing that even an under-sheriff cannot make space elastic,—but, if the introduction has been acknowledged as good, will probably find a seat for him if he persevere to the end. But the seat when obtained must be kept in possession from morning to evening, and the fight must be renewed from day to day. And the benches are hard, and the space is narrow, and you feel that the under-sheriff would prod you with his sword if you ventured to sneeze, or to put to your lips the flask which you have in your pocket. And then, when all the benchfellows go out to lunch at half-past one, and you are left to eat your dry sandwich without room for your elbows, a feeling of unsatisfied ambition will pervade you. It is all very well to be the friend of an under-sheriff, but if you could but have known the judge, or have been a cousin of the real sheriff, how different it might have been with you!

But you may be altogether independent, and, as a matter of right, walk into an open English court of law as one of the British public. You will have to stand of course,—and to commence standing very early in the morning if you intend to succeed in witnessing any portion of the performance. And when you have made once good your entrance as one of the British public, you are apt to be a good deal knocked about, not only by your public brethren, but also by those who have to keep the avenues free for witnesses, and who will regard you from first to last as a disagreeable excrescence on the officialities of the work on hand. Upon the whole it may be better for you, perhaps, to stay at home and read the record of the affair as given in the next day's *Times*. Impartial reporters, judicious readers, and able editors between them will preserve for you all the kernel, and will save you from the necessity of having to deal with the shell.

At this trial there were among the crowd who succeeded in entering the Court three persons of our acquaintance who had resolved to overcome the various difficulties. Mr. Monk, who had formerly been a Cabinet Minister, was seated on the bench,—subject, indeed, to the heat and stenches, but priviledged to eat the lunch. Mr. Quintus Slide, of The People's Banner,—who knew the Court well, for in former days he had worked many an hour in it as a reporter,—had obtained the good graces of the under-sheriff. And Mr. Bunce, with all the energy of the British public, had forced his way in among the crowd, and had managed to wedge himself near to the dock, so that he might be able by a hoist of the neck to see his lodger as he stood at the bar. Of these three men, Bunce was assured that the prisoner was innocent,—led to such assurance partly by belief in the man, and partly by an innate spirit of opposition to all exercise of restrictive power. Mr. Quintus Slide was certain of the prisoner's guilt, and gave himself considerable credit for having assisted in running down the criminal. It seemed to be natural to Mr. Quintus Slide that a man who had openly quarrelled with the Editor of The People's Banner should come to the gallows. Mr. Monk, as Phineas himself well knew, had doubted. He had received the suspected murderer into his warmest friendship, and was made miserable even by his doubts. Since the circumstances of the case had come to his knowledge, they had weighed upon his mind so as to sadden his whole life. But he was a man who could not make his reason subordinate to his feelings. If the evidence against his friend was strong enough to send his friend for trial, how should he dare to discredit the evidence because the man was his friend? He had visited Phineas in prison, and Phineas had accused him of doubting. 'You need not answer me,' the unhappy man had said, 'but do not come unless you are able to tell me from your heart that you are sure of my innocence. There is no person living who could comfort me by such assurance as you could do.' Mr. Monk had thought about it very much, but he had not repeated his visit.

At a quarter past ten the Chief Justice was on the bench, with a second judge to help him, and with lords and distinguished commoners and great City magnates crowding the long seat between him and the doorway; the Court was full, so that you would say that another head could not be made to appear; and Phineas Finn, the member for Tankerville, was in the dock. Barrington Erle, who was there to see,—as one of the great ones, of course,—told the Duchess of Omnium that night that Phineas was thin and pale, and in many respects an altered man,—but handsomer than ever.

'He bore himself well?' asked the Duchess.

'Very well,—very well indeed. We were there for six hours, and he maintained the same demeanour throughout. He never spoke but once, and that was when Chaffanbrass began his fight about the jury.'

'What did he say?'

'He addressed the judge, interrupting Slope, who was arguing that some man would make a very good juryman, and declared that it was not by his wish that any objection was raised against any gentleman.'

'What did the judge say?'

'Told him to abide by his counsel. The Chief Justice was very civil to him,—indeed better than civil.'

'We'll have him down to Matching, and make ever so much of him,' said the Duchess.

'Don't go too fast, Duchess, for he may have to hang poor Phineas yet.'

'Oh dear; I wish you wouldn't use that word. But what did he say?'

'He told Finn that as he had thought fit to employ counsel for his defence,—in doing which he had undoubtedly acted wisely,—he must leave the case to the discretion of his counsel.'

'And then poor Phineas was silenced?'

'He spoke another word. "My lord," said he, "I for my part wish that the first twelve men on the list might be taken." But old Chaffanbrass went on just the same. It took them two hours and a half before they could swear a jury.'

'But, Mr. Erle,—taking it altogether,—which way is it going?'

'Nobody can even guess as yet. There was ever so much delay besides that about the jury. It seemed that somebody had called him Phinees instead of Phineas, and that took half an hour. They begin with the quarrel at the club, and are to call the first witness to-morrow morning. They are to examine Ratler about the quarrel, and Fitzgibbon, and Monk, and, I believe, old Bouncer, the man who writes, you know. They all heard what took place.'

'So did you?'

'I have managed to escape that. They can't very well examine all the club. But I shall be called afterwards as to what took place at the door. They will begin with Ratler.'

'Everybody knows there was a quarrel, and that Mr. Bonteen had been drinking, and that he behaved as badly as a man could behave.'

'It must all be proved, Duchess.'

'I'll tell you what, Mr. Erle. If,—if,—if this ends badly for Mr. Finn I'll wear mourning to the day of my death. I'll go to the Drawing Room in mourning, to show what I think of it.'

Lord Chiltern, who was also on the bench, took his account of the trial home to his wife and sister in Portman Square. At this time Miss Palliser was staying with them, and the three ladies were together when the account was brought to them. In that house it was taken as doctrine that Phineas Finn was innocent. In the presence of her brother, and before her sister-in-law's visitor, Lady Laura had learned to be silent on the subject, and she now contented herself with listening, knowing that she could relieve herself by speech when alone with Lady Chiltern. 'I never knew anything so tedious in my life,' said the Master of the Brake hounds. 'They have not done anything yet.'

'I suppose they have made their speeches?' said his wife.

'Sir Gregory Grogram opened the case, as they call it; and a very strong case he made of it. I never believe anything that a lawyer says when he has a wig on his head and a fee in his hand. I prepare myself beforehand to regard it all as mere words, supplied at so much the thousand. I know he'll say whatever he thinks most likely to forward his own views. But upon my word he put it very strongly. He brought it all within so very short a space of time! Bonteen and Finn left the club within a minute of each other. Bonteen must have been at the top of the passage five minutes afterwards, and Phineas at that moment could not have been above two hundred yards from him. There can be no doubt of that.'

'Oswald, you don't mean to say that it's going against him!' exclaimed Lady Chiltern.

'It's not going any way at present. The witnesses have not been examined. But so far, I suppose, the Attorney-General was right. He has got to prove it all, but so much no doubt he can prove. He can prove that the man was killed with some blunt weapon, such as Finn had. And he can prove that exactly at the same time a man was running to the spot very like to Finn, and that by a route which would not have been his route, but by using which he could have placed himself at that moment where the man was seen.'

'How very dreadful!' said Miss Palliser.

'And yet I feel that I know it was that other man,' said Lady Chiltern. Lady Laura sat silent through it all, listening with her eyes intent on her brother's face, with her elbow on the table and her brow on her hand. She did not speak a word till she found herself alone with her sister-in-law, and then it was hardly more than a word. 'Violet, they will murder him!' Lady Chiltern endeavoured to comfort her, telling her that as yet they had heard but one side of the case; but the wretched woman only shook her head. 'I know they will murder him,' she said, 'and then when it is too late they will find out what they have done!'

On the following day the crowd in Court was if possible greater, so that the benchfellows were very much squeezed indeed. But it was impossible to exclude from the high seat such men as Mr. Ratler and Lord Fawn when they were required in the Court as witnesses;—and not a man who had obtained a seat on the first day was willing to be excluded on the second. And even then the witnesses were not called at once. Sir Gregory Grogram began the work of the day by saying that he had heard that morning for the first time that one of his witnesses had been,—'tampered with' was the word that he unfortunately used,—by his learned friend on the other side. He alluded, of course, to Lord Fawn, and poor Lord Fawn, sitting up there on the seat of honour, visible to all the world, became very hot and very uncomfortable. Then there arose a vehement dispute between Sir Gregory, assisted by Sir Simon, and old Mr. Chaffanbrass, who rejected with disdain any assistance from the gentler men who were with him. 'Tampered with! That word should be recalled by the honourable gentleman who was at the head of the bar, or—or ——'. Had Mr. Chaffanbrass declared that as an alternative he would pull the Court about their ears, it would have been no more than he meant. Lord Fawn had been invited,—not summoned to attend; and why? In order that no suspicion of guilt might be thrown on another man, unless the knowledge that was in Lord Fawn's bosom, and there alone, would justify such a line of defence. Lord Fawn had been attended by his own solicitor, and might have brought the Attorney-General with him had he so pleased. There was a great deal said on both sides, and something said also by the judge. At last Sir Gregory withdrew the objectionable word, and substituted in lieu of it an assertion that his witness had been 'indiscreetly questioned.' Mr. Chaffanbrass would not for a moment admit the indiscretion, but bounced about in his place, tearing his wig almost off his head, and defying every one in the Court. The judge submitted to Mr. Chaffanbrass that he had been indiscreet.—

'I never contradicted the Bench yet, my lord,' said Mr. Chaffanbrass,—at which there was a general titter throughout the bar,—'but I must claim the privilege of conducting my own practice according to my own views. In this Court I am subject to the Bench. In my own chamber I am subject only to the law of the land.' The judge looking over his spectacles said a mild word about the profession at large. Mr. Chaffanbrass, twisting his wig quite on one side, so that it nearly fell on Mr. Serjeant Birdbott's face, muttered something as to having seen more work done in that Court than any other living lawyer, let his rank be what it might. When the little affair was over, everybody felt that Sir Gregory had been vanquished.

Mr. Ratler, and Laurence Fitzgibbon, and Mr. Monk, and Mr. Bouncer were examined about the quarrel at the club, and proved that the quarrel had been a very bitter quarrel. They all agreed that Mr. Bonteen had been wrong, and that the prisoner had had cause for anger. Of the three distinguished legislators and statesmen above named Mr. Chaffanbrass refused to take the slightest notice. 'I have no question to put to you,' he said to Mr. Ratler. 'Of course there was a quarrel. We all know that.' But he did ask a question or two of Mr. Bouncer. 'You write books, I think, Mr. Bouncer?'

'I do,' said Mr. Bouncer, with dignity. Now there was no peculiarity in a witness to which Mr. Chaffanbrass was so much opposed as an assumption of dignity.

'What sort of books, Mr. Bouncer?'

'I write novels,' said Mr. Bouncer, feeling that Mr. Chaffanbrass must have been ignorant indeed of the polite literature of the day to make such a question necessary.

'You mean fiction.'

'Well, yes; fiction,—if you like that word better.'

'I don't like either, particularly. You have to find plots, haven't you?'

Mr. Bouncer paused a moment. 'Yes; yes,' he said. 'In writing a novel it is necessary to construct a plot.'

'Where do you get 'em from?'

'Where do I get 'em from?'

'Yes,—where do you find them? You take them from the French mostly;—don't you?' Mr. Bouncer became very red. 'Isn't that the way our English writers get their plots?'

'Sometimes,—perhaps.'

'Your's ain't French then?'

'Well;—no;—that is——I won't undertake to say that—that——'

'You won't undertake to say that they're not French.'

'Is this relevant to the case before us, Mr. Chaffanbrass?' asked the judge.

'Quite so, my lud. We have a highly-distinguished novelist before us, my lud, who, as I have reason to believe, is intimately acquainted with the French system of the construction of plots. It is a business which the French carry to perfection. The plot of a novel should, I imagine, be constructed in accordance with human nature?'

'Certainly,' said Mr. Bouncer.

'You have murders in novels?'

'Sometimes,' said Mr. Bouncer, who had himself done many murders in his time.

'Did you ever know a French novelist have a premeditated murder committed by a man who could not possibly have conceived the murder ten minutes before he committed it;—with whom the cause of the murder anteceded the murder no more than ten minutes?' Mr. Bouncer stood thinking for a while. 'We will give you your time, because an answer to the question from you will be important testimony.'

'I don't think I do,' said Mr. Bouncer, who in his confusion had been quite unable to think of the plot of a single novel.

'And if there were such a French plot that would not be the plot that you would borrow?'

'Certainly not,' said Mr. Bouncer.

'Did you ever read poetry, Mr. Bouncer?'

'Oh yes;—I read a great deal of poetry.'

'Shakespeare, perhaps?' Mr. Bouncer did not condescend to do more than nod his head. 'There is a murder described in

Hamlet. Was that supposed by the poet to have been devised suddenly?'

'I should say not.'

'So should I, Mr. Bouncer. Do you remember the arrangements for the murder in *Macbeth*? That took a little time in concocting;—didn't it?'

'No doubt it did.'

'And when Othello murdered Desdemona, creeping up to her in her sleep, he had been thinking of it for some time?'

'I suppose he had.'

'Do you ever read English novels as well as French, Mr. Bouncer?' The unfortunate author again nodded his head. 'When Amy Robsart* was lured to her death, there was some time given to the preparation,—eh?'

'Of course there was.'

'Of course there was. And Eugene Aram,* when he murdered a man in Bulwer's novel, turned the matter over in his mind before he did it?'

'He was thinking a long time about it, I believe.'

'Thinking about it a long time! I rather think he was. Those great masters of human nature, those men who knew the human heart, did not venture to describe a secret murder as coming from a man's brain without premeditation?'

'Not that I can remember.'

'Such also is my impression. But now, I bethink me of a murder that was almost as sudden as this is supposed to have been. Didn't a Dutch smuggler murder a Scotch lawyer, all in a moment as it were?'

'Dirk Hatteraick did murder Glossop in *The Antiquary** very suddenly;—but he did it from passion.'

'Just so, Mr. Bouncer. There was no plot there, was there? No arrangement; no secret creeping up to his victim; no escape even?'

'He was chained.'

'So he was; chained like a dog;—and like a dog he flew at his enemy. If I understand you, then, Mr. Bouncer, you would not dare so to violate probability in a novel, as to produce a murderer to the public who should contrive a secret hidden murder,—contrive it and execute it, all within a quarter of an hour?'

Mr. Bouncer, after another minute's consideration, said that he thought he would not do so. 'Mr. Bouncer,' said Mr. Chaffanbrass, 'I am uncommonly obliged to our excellent friend, Sir Gregory, for having given us the advantage of your evidence.'

CHAPTER LXII

Lord Fawn's evidence

A CROWD of witnesses were heard on the second day after Mr. Chaffanbrass had done with Mr. Bouncer, but none of them were of much interest to the public. The three doctors were examined as to the state of the dead man's head when he was picked up, and as to the nature of the instrument with which he had probably been killed; and the fact of Phineas Finn's life-preserver was proved,—in the middle of which he begged that the Court would save itself some little trouble, as he was quite ready to acknowledge that he had walked home with the short bludgeon, which was then produced, in his pocket. 'We would acknowledge a great deal if they would let us,' said Mr. Chaffanbrass. 'We acknowledge the quarrel, we acknowledge the walk home at night, we acknowledge the bludgeon, and we acknowledge a grey coat.' But that happened towards the close of the second day, and they had not then reached the grey coat. The question of the grey coat was commenced on the third morning,—on the Saturday,—which day, as was well known, would be opened with the examination of Lord Fawn. The anxiety to hear Lord Fawn undergo his penance was intense, and had been greatly increased by the conviction that Mr. Chaffanbrass would resent upon him the charge made by the Attorney-General as to tampering with a witness. 'I'll tamper with him by-and-bye,' Mr. Chaffanbrass had whispered to Mr. Wickerby, and the whispered threat had been spread abroad. On the table before Mr. Chaffanbrass, when he took his place in the Court on the Saturday, was laid a heavy grey coat, and on the opposite side of the table, just before the Solicitor-General, was laid another grey coat, of much lighter material. When Lord Fawn saw the two coats as he took his seat on the bench his heart failed him.

He was hardly allowed to seat himself before he was called upon to be sworn. Sir Simon Slope, who was to examine him, took it for granted that his lordship could give his evidence from his place on the bench, but to this Mr. Chaffanbrass objected. He was very well aware, he said, that such a practice was usual. He did not doubt but that in his time he had examined some hundreds of witnesses from the bench. In nineteen cases out of twenty there could be no objection to such a practice. But in this case the noble lord would have to give evidence not only as to what he had seen, but as to what he then saw. It would be expedient that he should see colours as nearly as possible in the same light as the jury, which he would do if he stood in the witness-box. And there might arise questions of identity, in speaking of which it would be well that the noble lord should be as near as possible to the thing or person to be identified. He was afraid that he must trouble the noble lord to come down from the Elysium of the bench. Whereupon Lord Fawn descended, and was sworn in at the witness-box.

His treatment from Sir Simon Slope was all that was due from a Solicitor-General to a distinguished peer who was a member of the same Government as himself. Sir Simon put his questions so as almost to reassure the witness; and very quickly,—only too quickly,—obtained from him all the information that was needed on the side of the prosecution. Lord Fawn, when he had left the club, had seen both Mr. Bonteen and Mr. Finn preparing to follow him, but he had gone alone, and had never seen Mr. Bonteen since. He walked very slowly down into Curzon Street and Bolton Row, and when there, as he was about to cross the road at the top of

Clarges Street,—as he believed, just as he was crossing the street,—he saw a man come at a very fast pace out of the mews which runs into Bolton Row, opposite to Clarges Street, and from thence hurry very quickly towards the passage

which separates the gardens of Devonshire and Lansdowne Houses. It had already been proved that had Phineas Finn retraced his steps after Erle and Fitzgibbon had turned their backs upon him, his shortest and certainly most private way to the spot on which Lord Fawn had seen the man would have been by the mews in question. Lord Fawn went on to say that the man wore a grey coat,—as far as he could judge it was such a coat as Sir Simon now showed him; he could not at all identify the prisoner; he could not say whether the man he had seen was as tall as the prisoner; he thought that as far as he could judge, there was not much difference in the height.

He had not thought of Mr. Finn when he saw the man hurrying along, nor had he troubled his mind about the man. That was the end of Lord Fawn's evidence-in-chief, which he would gladly have prolonged to the close of the day could he thereby have postponed the coming horrors of his cross-examination. But there he was,—in the clutches of the odious, dirty, little man, hating the little man, despising him because he was dirty, and nothing better than an Old Bailey barrister,—and yet fearing him with so intense a fear!

Mr. Chaffanbrass smiled at his victim, and for a moment was quite soft with him,—as a cat is soft with a mouse. The reporters could hardly hear his first question,—'I believe you are an Under-Secretary of State?' Lord Fawn acknowledged the fact. Now it was the case that in the palmy days of our hero's former career he had filled the very office which Lord Fawn now occupied, and that Lord Fawn had at the time filled a similar position in another department. These facts Mr. Chaffanbrass extracted from his witness,—not without an appearance of unwillingness, which was produced, however, altogether by the natural antagonism of the victim to his persecutor; for Mr. Chaffanbrass, even when asking the simplest questions, in the simplest words, even when abstaining from that sarcasm of tone under which witnesses were wont to feel that they were being flayed alive, could so look at a man as to create an antagonism which no witness could conceal. In asking a man his name, and age, and calling, he could produce an impression that the man was unwilling to tell anything, and that, therefore, the jury were entitled to regard his evidence with suspicion. 'Then,' continued Mr. Chaffanbrass, 'you must have met him frequently in the intercourse of your business?'

'I suppose I did,—sometimes.'

'Sometimes? You belonged to the same party?'

'We didn't sit in the same House.'

'I know that, my lord. I know very well what House you sat in. But I suppose you would condescend to be acquainted with even a commoner who held the very office which you hold now. You belonged to the same club with him.'

'I don't go much to the clubs,' said Lord Fawn.

'But the quarrel of which we have heard so much took place at a club in your presence?' Lord Fawn assented. 'In fact you cannot but have been intimately and accurately acquainted with the personal appearance of the gentleman who is now on his trial. Is that so?'

'I never was intimate with him.'

Mr. Chaffanbrass looked up at the jury and shook his head sadly. 'I am not presuming, Lord Fawn, that you so far derogated as to be intimate with this gentleman,—as to whom, however, I shall be able to show by and by that he was the chosen friend of the very man under whose mastership you now serve. I ask whether his appearance is not familiar to you?' Lord Fawn at last said that it was. 'Do you know his height? What should you say was his height?' Lord Fawn altogether refused to give an opinion on such a subject, but acknowledged that he should not be surprised if he were told that Mr. Finn was over six feet high. 'In fact you consider him a tall man, my lord? There he is, you can look at him. Is he a tall man?' Lord Fawn did look, but wouldn't give an answer. 'I'll undertake to say, my lord, that there isn't a person in the Court at this moment, except yourself, who wouldn't be ready to express an opinion on his oath that Mr. Finn is a tall man. Mr. Chief Constable, just let the prisoner step out from the dock for a moment. He won't run away. I must have his lordship's opinion as to Mr. Finn's height.' Poor Phineas, when this was said, clutched hold of the front of the dock, as though determined that nothing but main force should make him exhibit himself to the Court in the manner proposed.

But the need for exhibition passed away. 'I know that he is a very tall man,' said Lord Fawn.

'You know that he is a very tall man. We all know it. There can be no doubt about it. He is, as you say, a very tall man,—with whose personal appearance you have long been familiar? I ask again, my lord, whether you have not been long familiar with his personal appearance?' After some further agonising delay Lord Fawn at last acknowledged that it had been so. 'Now we shall get on like a house on fire,' said Mr. Chaffanbrass.

But still the house did not burn very quickly. A string of questions was then asked as to the attitude of the man who had been seen coming out of the mews wearing a grey great coat,—as to his attitude, and as to his general likeness to Phineas Finn. In answer to these Lord Fawn would only say that he had not observed the man's attitude, and had certainly not thought of the prisoner when he saw the man. 'My lord,' said Mr. Chaffanbrass, very solemnly, 'look at your late friend and colleague, and remember that his life depends probably on the accuracy of your memory. The man you saw—murdered Mr. Bonteen. With all my experience in such matters,—which is great; and with all my skill,—which is something, I cannot stand against that fact. It is for me to show that that man and my client were not one and the same person, and I must do so by means of your evidence,—by sifting what you say to-day, and by comparing it with what you have already said on other occasions. I understand you now to say that there is nothing in your remembrance of the man you saw, independently of the colour of the coat, to guide you to an opinion whether that man was or was not one and the same with the prisoner?'

In all the crowd then assembled there was no man more thoroughly under the influence of conscience as to his conduct than was Lord Fawn in reference to the evidence which he was called upon to give. Not only would the idea of endangering the life of a human being have been horrible to him, but the sanctity of an oath was imperative to him. He was essentially a truth-speaking man, if only he knew how to speak the truth. He would have sacrificed much to establish the innocence of Phineas Finn,—not for the love of Phineas, but for the love of innocence;—but not even to do that would he have lied. But he was a bad witness, and by his slowness, and by a certain unsustained pomposity which was natural to him, had already taught the jury to think that he was anxious to convict the prisoner. Two men in the Court, and two only, thoroughly understood his condition. Mr. Chaffanbrass saw it all, and intended without the slightest scruple to take advantage of it. And the Chief Justice saw it all, and was already resolving how he could set the witness right with the jury.

'I didn't think of Mr. Finn at the time,' said Lord Fawn in answer to the last question.

'So I understand. The man didn't strike you as being tall.'

'I don't think that he did.'

'But yet in the evidence you gave before the magistrate in Bow Street I think you expressed a very strong opinion that the man you saw running out of the mews was Mr. Finn?' Lord Fawn was again silent. 'I am asking your lordship a question to which I must request an answer. Here is the *Times* report of the examination, with which you can refresh your memory, and you are of course aware that it was mainly on your evidence as here reported that my client stands there in jeopardy of his life.'

'I am not aware of anything of the kind,' said the witness.

'Very well. We will drop that then. But such was your evidence, whether important or not important. Of course your lordship can take what time you please for recollection.'

Lord Fawn tried very hard to recollect, but would not look at the newspaper which had been handed to him. 'I cannot remember what words I used. It seems to me that I thought it must have been Mr. Finn because I had been told that Mr. Finn could have been there by running round.'

'Surely, my lord, that would not have sufficed to induce you to give such evidence as is there reported?'

'And the colour of the coat,' said Lord Fawn.

'In fact you went by the colour of the coat, and that only?'

'Then there had been the quarrel.'

'My lord, is not that begging the question? Mr. Bonteen quarrelled with Mr. Finn. Mr. Bonteen was murdered by a man,—as we all believe,—whom you saw at a certain spot. Therefore you identified the man whom you saw as Mr. Finn. Was that so?'

'I didn't identify him.'

'At any rate you do not do so now? Putting aside the grey coat there is nothing to make you now think that that man and Mr. Finn were one and the same? Come, my lord, on behalf of that man's life, which is in great jeopardy,—is in great jeopardy because of the evidence given by you before the magistrate,—do not be ashamed to speak the truth openly, though it be at variance with what you may have said before with ill-advised haste.'

'My lord, is it proper that I should be treated in this way?' said the witness, appealing to the Bench.

'Mr. Chaffanbrass,' said the judge, again looking at the barrister over his spectacles, 'I think you are stretching the privilege of your position too far.'

'I shall have to stretch it further yet, my lord. His lordship in his evidence before the magistrate gave on his oath a decided opinion that the man he saw was Mr. Finn;—and on that evidence Mr. Finn was committed for murder. Let him say openly, now, to the jury,—when Mr. Finn is on his trial for his life before the Court, and for all his hopes in life before the country,—whether he thinks as then he thought, and on what grounds he thinks so.'

'I think so because of the quarrel, and because of the grey coat.'

'For no other reasons?'

'No;—for no other reasons.'

'Your only ground for suggesting identity is the grey coat?'

'And the quarrel,' said Lord Fawn.

'My lord, in giving evidence as to identity, I fear that you do not understand the meaning of the word.' Lord Fawn looked up at the judge, but the judge on this occasion said nothing. 'At any rate we have it from you at present that there was nothing in the appearance of the man you saw like to that of Mr. Finn except the colour of the coat.'

'I don't think there was,' said Lord Fawn, slowly.

Then there occurred a scene in the Court which no doubt

was gratifying to the spectators, and may in part have repaid them for the weariness of the whole proceeding. Mr. Chaffanbrass, while Lord Fawn was still in the witness-box, requested permission for a certain man to stand forward, and put on the coat which was lying on the table before him,—this coat being in truth the identical garment which Mr. Meager had brought home with him on the morning of the murder. This man was Mr. Wickerby's clerk, Mr. Scruby, and he put on the coat,—which seemed to fit him well. Mr. Chaffanbrass then asked permission to examine Mr. Scruby, explaining that much time might be saved, and declaring that he had but one question to ask him. After some difficulty this permission was given him, and Mr. Scruby was asked his height. Mr. Scruby was five feet eight inches, and had been accurately measured on the previous day with reference to the question. Then the examination of Lord Fawn was resumed, and Mr. Chaffanbrass referred to that very irregular interview to which he had so improperly enticed the witness in Mr. Wickerby's chambers. For a long time Sir Gregory Grogram declared that he would not permit any allusion to what had taken place at a most improper conference,—a conference which he could not stigmatize in sufficiently strong language. But Mr. Chaffanbrass, smiling blandly,—smiling very blandly for him,—suggested that the impropriety of the conference, let it have been ever so abominable, did not prevent the fact of the conference, and that he was manifestly within his right in alluding to it. 'Suppose, my lord, that Lord Fawn had confessed in Mr. Wickerby's chambers that he had murdered Mr. Finn himself, and had since repented of that confession, would Mr. Camperdown and Mr. Wickerby, who were present, and would I, be now debarred from stating that confession in evidence, because, in deference to some fanciful rules of etiquette, Lord Fawn should not have been there?' Mr. Chaffanbrass at last prevailed, and the evidence was resumed.

'You saw Mr. Scruby wear that coat in Mr. Wickerby's chambers.' Lord Fawn said that he could not identify the coat.

'We'll take care to have it identified. We shall get a great deal out of that coat yet. You saw that man wear a coat like that.'

'Yes; I did.'

'And you see him now.'

'Yes, I do.'

'Does he remind you of the figure of the man you saw come out of the mews?' Lord Fawn paused. 'We can't make him move about here as we did in Mr. Wickerby's room; but remembering that as you must do, does he look like the man?'

'I don't remember what the man looked like.'

'Did you not tell us in Mr. Wickerby's room that Mr. Scruby with the grey coat on was like the figure of the man?'

Questions of this nature were prolonged for near half an hour, during which Sir Gregory made more than one attempt to defend his witness from the weapons of their joint enemies; but Lord Fawn at last admitted that he had acknowledged the resemblance, and did, in some faint ambiguous fashion, acknowledge it in his present evidence.

'My lord,' said Mr. Chaffanbrass as he allowed Lord Fawn to go down, 'you have no doubt taken a note of Mr. Scruby's height.' Whereupon the judge nodded his head.

CHAPTER LXIII

Mr. Chaffanbrass for the Defence

THE case for the prosecution was completed on the Saturday evening, Mrs. Bunce having been examined as the last witness on that side. She was only called upon to say that her lodger had been in the habit of letting himself in and out of her house at all hours with a latch-key;—but she insisted on saying more, and told the judge and the jury and the barristers that if they thought that Mr. Finn had murdered nybody they didn't know anything about the world in general. Whereupon Mr. Chaffanbrass said that he would like to ask her a question or two, and with consummate flattery extracted from her her opinion of her lodger. She had known him for years, and thought that, of all the gentlemen that ever were born, he was the least likely to do such a bloody-minded action. Mr. Chaffanbrass was, perhaps, right in thinking that her evidence might be as serviceable as that of the lords and countesses.

During the Sunday the trial was, as a matter of course, the talk of the town. Poor Lord Fawn shut himself up, and was seen by no one;—but his conduct and evidence were discussed everywhere. At the clubs it was thought that he had escaped as well as could be expected; but he himself felt that he had been disgraced for ever. There was a very common opinion that Mr. Chaffanbrass had admitted too much when he had declared that the man whom Lord Fawn had seen was doubtless the murderer. To the minds of men generally it seemed to be less evident that the man so seen should have done the deed, than that Phineas Finn should have been that man. Was it probable that there should be two men going about in grey coats, in exactly the same vicinity, and at exactly the same hour of the night? And then the evidence which Lord Fawn had given before the magistrates was to the world at large at any rate as convincing as that given in the Court. The jury would, of course, be instructed to regard only the latter; whereas the general public would naturally be guided by the two combined. At the club it was certainly believed that the case was going against the prisoner.

'You have read it all, of course,' said the Duchess of Omnium to her husband, as she sat with the *Observer* in her hand on that Sunday morning. The Sunday papers were full of the report, and were enjoying a very extended circulation.

'I wish you would not think so much about it,' said the Duke.

'That's very easily said, but how is one to help thinking about it? Of course I am thinking about it. You know all about the coat. It belonged to the man where Mealyus was lodging.'

'I will not talk about the coat, Glencora. If Mr. Finn did commit the murder it is right that he should be convicted.'

'But if he didn't?'

'It would be doubly right that he should be acquitted. But the jury will have means of arriving at a conclusion without prejudice, which you and I cannot have; and therefore we should be prepared to take their verdict as correct.'

'If they find him guilty, their verdict will be damnable and false,' said the Duchess. Whereupon the Duke turned away in anger, and resolved that he would say nothing more about the trial,—which resolution, however, he was compelled to break before the trial was over.

'What do you think about it, Mr. Erle?' asked the other Duke.

'I don't know what to think;—I only hope.'

'That he may be acquitted?'

'Of course.'

'Whether guilty or innocent?'

'Well;—yes. But if he is acquitted I shall believe him to have been innocent. Your Grace thinks——?'

'I am as unwilling to think as you are, Mr. Erle.' It was thus that people spoke of it. With the exception of some very few, all those who had known Phineas were anxious for an acquittal, though they could not bring themselves to believe that an innocent man had been put in peril of his life.

On the Monday morning the trial was recommenced, and the whole day was taken up by the address which Mr. Chaffanbrass made to the jury. He began by telling them the history of the coat which lay before them, promising to prove by evidence all the details which he stated. It was not his intention, he said, to accuse any one of the murder. It was his business to defend the prisoner, not to accuse others. But, as he should prove to them, two persons had been arrested as soon as the murder had been discovered,—two persons totally unknown to each other, and who were never for a moment supposed to have acted together,—and the suspicion of the police had in the first instance pointed, not to his client, but to the other man. That other man had also quarrelled with Mr. Bonteen, and that other man was now in custody on a charge of bigamy chiefly through the instrumentality of Mr. Bonteen, who had been the friend of the victim of the supposed bigamist. With the accusation of bigamy they would have nothing to do, but he must ask them to take cognisance of that quarrel as well as of the quarrel at the club. He then named that formerly popular preacher, the Rev. Mr. Emilius, and explained that he would prove that this man, who had incurred the suspicion of the police in the first instance, had during the night of the murder been so circumstanced as to have been able to use the coat produced. He would prove also that Mr. Emilius was of precisely the same height as the man whom they had seen wearing the coat. God forbid that he should bring an accusation of murder against a man on such slight testimony. But if the evidence, as grounded on the coat, was slight against Emilius, how could it prevail at all against his client? The two coats were as different as chalk from cheese, the one being what would be called a gentleman's fashionable walking coat, and the other the wrap-rascal* of such a fellow as was Mr. Meager. And yet Lord Fawn, who attempted to identify the prisoner only by his coat, could give them no opinion as to which was the coat he had seen! But Lord Fawn, who had found himself to be debarred by his conscience from repeating the opinion he had given before the magistrate as to the identity of Phineas Finn with the man he had seen, did tell them that the figure of that man was similar to the figure of him who had worn the coat on Saturday in presence of them all. This man in the street had therefore been like Mr. Emilius, and could not in the least have resembled the prisoner. Mr. Chaffanbrass would not tell the jury that this point bore strongly against Mr. Emilius, but he took upon himself to assert that it was quite sufficient to snap asunder the thin thread of circumstantial evidence by which his client was connected with the murder. A great deal more was said about Lord Fawn, which was not complimentary to that nobleman. 'His lordship is an honest, slow man, who has doubtless meant to tell you the truth, but who does not understand the meaning of what he himself says. When he swore before the magistrate that he thought he could identify my client with the man in the street, he really meant that he thought that there must be identity, because he believed from other reasons that Mr. Finn was the man in the street. Mr. Bonteen had been murdered;—according to Lord Fawn's thinking had probably been murdered by Mr. Finn. And it was also probable to him that Mr. Bonteen had been murdered by the man in the street. He came thus to the conclusion that the prisoner was the man in the street. In fact, as far as the process of identifying is concerned, his lordship's evidence is altogether in favour of the prisoner. The figure seen by him we must suppose was the figure of a short man, and not of one tall and commanding in his presence, as is that of the prisoner.'

There were many other points on which Mr. Chaffanbrass insisted at great length;—but, chiefly, perhaps, on the improbability, he might say impossibility, that the plot for a murder so contrived should have entered into a man's head, have been completed and executed, all within a few minutes. 'But under no hypothesis compatible with the allegations of the prosecution can it be conceived that the murder should have been contemplated by my client before the quarrel at the club. No, gentlemen;—the murderer had been at his work for days. He had examined the spot and measured the distances. He had dogged the steps of his victim on previous nights. In the shade of some dark doorway he had watched him from his club, and had hurried by his secret path to the spot which he had appointed for the deed. Can any man doubt that the murder has thus been committed, let who will have been the murderer? But, if so, then my client could not have done the deed.' Much had been made of the words spoken at the club door. Was it probable,—was it possible,—that a man intending to commit a murder should declare how easily he could do it, and display the weapon he intended to use? The evidence given as to that part of the night's work was, he contended, altogether in the prisoner's favour. Then he spoke of the life-preserver, and gave a rather long account of the manner in which

which Phineas Finn had once taken two garotters prisoner in the street. All this lasted till the great men on the bench trooped out to lunch. And then Mr. Chaffanbrass, who had been speaking for nearly four hours, retired to a small room and there drank a pint of port wine. While he was doing so, Mr. Serjeant Birdbott spoke a word to him, but he only shook his head and snarled. He was telling himself at the moment how quick may be the resolves of the eager mind,—for he was convinced that the idea of attacking Mr. Bonteen had occurred to Phineas Finn after he had displayed the life-preserver at the club door; and he was telling himself also how impossible it is for a dull conscientious man to give accurate evidence as to what he had himself seen,—for he was convinced that Lord Fawn had seen Phineas Finn in the street. But to no human being had he expressed this opinion; nor would he express it,—unless his client should be hung.

After lunch he occupied nearly three hours in giving to the jury, and of course to the whole assembled Court, the details of about two dozen cases, in which apparently strong circumstantial evidence had been wrong in its tendency. In some of the cases quoted, the persons tried had been acquitted; in some, convicted and afterwards pardoned; in one pardoned after many years of punishment;—and in one the poor victim had been hung. On this he insisted with a pathetic eloquence which certainly would not have been expected from his appearance, and spoke with tears in his eyes,—real unaffected tears,—of the misery of those wretched jurymen who, in the performance of their duty, had been led into so frightful an error. Through the whole of this long recital he seemed to feel no fatigue, and when he had done with his list of judicial mistakes about five o'clock in the afternoon, went on to make what he called the very few remarks necessary as to the evidence which on the next day he proposed to produce as to the prisoner's character. He ventured to think that evidence as to the character of such a nature,—so strong, so convincing, so complete, and so free from all objection, had never yet been given in a criminal court. At six o'clock he completed his speech, and it was computed that the old man had been on his legs very nearly seven hours. It was said of him afterwards that he was taken home speechless by one of his daughters and immediately put to bed, that he roused himself about eight and ate his dinner and drank a bottle of port in his bedroom, that he then slept,—refusing to stir even when he was waked, till half-past nine in the morning, and that then he scrambled into his clothes, breakfasted, and got down to the Court in half an hour. At ten o'clock he was in his place, and nobody knew that he was any the worse for the previous day's exertion.

This was on a Tuesday, the fifth day of the trial, and upon the whole perhaps the most interesting. A long array of distinguished persons,—of women as well as men,—was brought up to give to the jury their opinion as to the character of Mr. Finn. Mr. Low was the first, who having been his tutor when he was studying at the bar, knew him longer than any other Londoner. Then came his countryman Laurence Fitzgibbon, and Barrington Erle, and others of his own party who had been intimate with him. And men, too, from the opposite side of the House were brought up, Sir Orlando Drought among the number, all of whom said that they had known the prisoner well, and from their knowledge would have considered it impossible that he should have become a murderer. The two last called were Lord Cantrip and Mr. Monk, one of whom was, and the other had been, a Cabinet Minister. But before them came Lady Cantrip,—and Lady Chiltern, whom we once knew as Violet Effingham, whom this very prisoner had in early days fondly hoped to make his wife, who was still young and beautiful, and who had never before entered a public Court.

There had of course been much question as to the witnesses to be selected. The Duchess of Omnium had been anxious to be one, but the Duke had forbidden it, telling his wife that she really did not know the man, and that she was carried away by a foolish enthusiasm. Lady Cantrip when asked had at once consented. She had known Phineas Finn, when he had served under her husband, and had liked him much. Then what other woman's tongue should be brought to speak of the man's softness and tender bearing! It was out of the question that Lady Laura Kennedy should appear. She did not even propose it when her brother with unnecessary sternness told her it could not be so. Then his wife looked at him. 'You shall go,' said Lord Chiltern, 'if you feel equal to it. It seems to be nonsense, but they say that it is important.'

'I will go,' said Violet, with her eyes full of tears. Afterwards when her sister-in-law besought her to be generous in her testimony, she only smiled as she assented. Could generosity go beyond hers?

Lord Chiltern preceded his wife. 'I have,' he said, 'known Mr. Finn well, and have loved him dearly. I have eaten with him and drank with him, have ridden with him, have lived with him, and have quarrelled with him; and I know him as I do my own right hand.' Then he stretched forth his arm with the palm extended.

'Irrespectively of the evidence in this case you would not have thought him to be a man likely to commit such a crime?' asked Serjeant Birdbott.

'I am quite sure from my knowledge of the man that he could not commit a murder,' said Lord Chiltern; 'and I don't care what the evidence is.'

Then came his wife, and it certainly was a pretty sight to see as her husband led her up to the box and stood close beside her as she gave her evidence. There were many there who knew much of the history of her life,—who knew that passage in it of her early love,—for the tale had of course been told when it was whispered about that Lady Chiltern was to be examined as a witness. Every ear was at first strained to hear her words;—but they were audible in every corner of the Court without any effort. It need hardly be said that she was treated with the greatest deference on every side. She answered the questions very quietly, but apparently without nervousness. 'Yes; she had known Mr. Finn long, and intimately, and had very greatly valued his friendship.

She did so still,—as much as ever. Yes; she had known him for some years, and in circumstances which she thought justified her in saying that she understood his character. She regarded him as a man who was brave and tenderhearted, soft in feeling and manly in disposition. To her it was quite incredible that he should have committed a crime such as this. She knew him to be a man prone to forgive offences, and of a sweet nature.' And it was pretty too to watch the unwonted gentleness of old Chaffanbrass as he asked the questions, and carefully abstained from putting any one that could pain her. Sir Gregory said that he had heard her evidence with great pleasure, but that he had no question to ask her himself. Then she stepped down, again took her husband's arm, and left the Court amidst a hum of almost affectionate greeting.

And what must he have thought as he stood there within the dock, looking at her and listening to her? There had been months in his life when he had almost trusted that he would succeed in winning that fair, highly-born, and wealthy woman for his wife; and though he had failed, and now knew that he had never really touched her heart, that she had always loved the man whom,—though she had rejected him time after time because of the dangers of his ways,—she had at last married, yet it must have been pleasant to him, even in his peril, to hear from her own lips how well she had esteemed him. She left the Court with her veil down, and he could not catch her eye; but Lord Chiltern nodded to him in his old pleasant familiar way, as though to bid him take courage, and to tell him that all things would even yet be well with him.

The evidence given by Lady Cantrip and her husband and by Mr. Monk was equally favourable. She had always regarded him as a perfect gentleman. Lord Cantrip had found him to be devoted to the service of the country,—modest, intelligent, and high-spirited. Perhaps the few words which fell from Mr. Monk were as strong as any that were spoken. 'He is a man whom I have delighted to call my friend, and I have been happy to think that his services have been at the disposal of his country.'

Sir Gregory Grogram replied. It seemed to him that the evidence was as he had left it. It would be for the jury to decide, under such directions as his lordship might be pleased to give them, how far that evidence brought the guilt home to the prisoner. He would use no rhetoric in pushing the case against the prisoner; but he must submit to them that his learned friend had not shown that acquaintance with human nature which the gentleman undoubtedly possessed in arguing that there had lacked time for the conception and execution of the crime. Then, at considerable length, he strove to show that Mr. Chaffanbrass had been unjustly severe upon Lord Fawn.

It was late in the afternoon when Sir Gregory had finished his speech, and the judge's charge was reserved for a sixth day.

CHAPTER LXIV

Confusion in the Court

On the following morning it was observed that before the judges took their seats Mr. Chaffanbrass entered the Court with a manner much more brisk than was expected from him now that his own work was done. As a matter of course he would be there to hear the charge, but, almost equally as a matter of course, he would be languid, silent, cross, and unenergetic. They who knew him were sure, when they saw his bearing on this morning, that he intended to do something more before the charge was given. The judges entered the Court nearly half an hour later than usual, and it was observed with surprise that they were followed by the Duke of Omnium. Mr. Chaffanbrass was on his feet before the Chief Justice had taken his seat, but the judge was the first to speak. It was observed that he held a scrap of paper in his hand, and that the barrister held a similar scrap. Then every man in the Court knew that some message had come suddenly by the wires. 'I am informed, Mr. Chaffanbrass, that you wish to address the Court before I begin my charge.'

'Yes, my lud; and I am afraid, my lud, that I shall have to ask your ludship to delay your charge for some days, and to subject the jury to the very great inconvenience of prolonged incarceration for another week;—either to do that or to call upon the jury to acquit the prisoner. I venture to assert, on my own peril, that no jury can convict the prisoner after hearing me read that which I hold in my hand.' Then Mr. Chaffanbrass paused, as though expecting that the judge would speak;—but the judge said not a word, but sat looking at the old barrister over his spectacles.

Every eye was turned upon Phineas Finn, who up to this moment had heard nothing of these new tidings,—who did not in the least know on what was grounded the singularly confident,—almost insolently confident assertion which Mr. Chaffanbrass had made in his favour. On him the effect was altogether distressing. He had borne the trying week with singular fortitude, having stood there in the place of shame hour after hour, and day after day, expecting his doom. It had been to him as a lifetime of torture. He had become almost numb from the weariness of his position and the agonising strain upon his mind. The gaoler had offered him a seat from day to day, but he had always refused it, preferring to lean upon the rail and gaze upon the Court. He had almost ceased to hope for anything except the end of it. He had lost count of the days, and had begun to feel that the trial was an eternity of torture in itself. At nights he could not sleep, but during the Sunday, after Mass, he had slept all day. Then it had begun again, and when the Tuesday came he hardly knew how long it had been since that vacant Sunday. And now he heard the advocate declare, without knowing on what ground the declaration was grounded, that the trial must be postponed, or that the jury must be instructed to acquit him.

'This telegram has reached us only this morning,' continued Mr. Chaffanbrass. ' "Mealyus had a house door-key made in Prague. We have the mould in our possession, and will bring the man who made the key to England." Now, my lud, the case in the hands of the police, as against this man Mealyus, or Emilius, as he has chosen to call himself, broke down altogether on the presumption that he could not have let himself in and out of the house in which he had put himself to

bed on the night of the murder. We now propose to prove that he had prepared himself with the means of doing so, and had done so after a fashion which is conclusive as to his having required the key for some guilty purpose. We assert that your ludship cannot allow the case to go to the jury without taking cognisance of this telegram; and we go further, and say that those twelve men, as twelve human beings with hearts in their bosoms and ordinary intelligence at their command, cannot ignore the message, even should your ludship insist upon their doing so with all the energy at your disposal.'

Then there was a scene in Court, and it appeared that no

less than four messages had been received from Prague, all to the same effect. One had been addressed by Madame Goesler to her friend the Duchess,—and that message had caused the Duke's appearance on the scene. He had brought his telegram direct to the Old Bailey, and the Chief Justice now held it in his hand. The lawyer's clerk who had accompanied Madame Goesler had telegraphed to the Governor of the gaol, to Mr. Wickerby, and to the Attorney-General. Sir Gregory, rising with the telegram in his hand, stated that he had received the same information. 'I do not see,' said he, 'that it at all alters the evidence as against the prisoner.'

'Let your evidence go to the jury, then,' said Mr. Chaffanbrass, 'with such observations as his lordship may choose to make on the telegram. I shall be contented. You have already got your other man in prison on a charge of bigamy.'

'I could not take notice of the message in charging the jury, Mr. Chaffanbrass,' said the judge. 'It has come, as far as we know, from the energy of a warm friend,—from that hearty friendship with which it seemed yesterday that this gentleman, the prisoner at the bar, has inspired so many men and women of high character. But it proves nothing. It is an assertion. And where should we all be, Mr. Chaffanbrass, if it should appear hereafter that the assertion is fictitious,—prepared purposely to aid the escape of a criminal?'

'I defy you to ignore it, my lord.'

'I can only suggest, Mr. Chaffanbrass,' continued the judge, 'that you should obtain the consent of the gentlemen on the other side to a postponement of my charge.'

Then spoke out the foreman of the jury. Was it proposed that they should be locked up till somebody should come from Prague, and that then the trial should be recommenced? The system, said the foreman, under which Middlesex juries were chosen for service in the City was known to be most horribly cruel;—but cruelty to jurymen such as this had never even been heard of. Then a most irregular word was spoken. One of the jurymen declared that he was quite willing to believe the telegram. 'Every one believes it,' said Mr. Chaffanbrass. Then the Chief Justice scolded the juryman, and Sir Gregory Grogram scolded Mr. Chaffanbrass. It seemed as though all the rules of the Court were to be set at defiance. 'Will my learned friend say that he doesn't believe it?' asked Mr. Chaffanbrass. 'I neither believe nor disbelieve it; but it cannot affect the evidence,' said Sir Gregory. 'Then send the case to the jury,' said Mr. Chaffanbrass. It seemed that everybody was talking, and Mr. Wickerby, the attorney, tried to explain it all to the prisoner over the bar of the dock, not in the lowest possible voice. The Chief Justice became angry, and the guardian of the silence of the Court bestirred himself energetically. 'My lud,' said Mr. Chaffanbrass, 'I maintain that it is proper that the prisoner should be informed of the purport of these telegrams. Mercy demands it, and justice as well.' Phineas Finn, however, did not understand, as he had known nothing about the latch-key of the house in Northumberland Street.

Something, however, must be done. The Chief Justice was of opinion that, although the preparation of a latch-key in Prague could not really affect the evidence against the prisoner,—although the facts against the prisoner would not be altered, let the manufacture of that special key be ever so clearly proved,—nevertheless the jury were entitled to have before them the facts now tendered in evidence before they could be called upon to give a verdict, and that therefore they should submit themselves, in the service of their country, to the very serious additional inconvenience which they would be called upon to endure. Sundry of the jury altogether disagreed with this, and became loud in their anger. They had already been locked up for a week. 'And we are quite prepared to give a verdict,' said one. The judge again scolded him very severely; and as the Attorney-General did at last assent, and as the unfortunate jurymen had no power in the matter, so it was at last arranged. The trial should be postponed till time should be given for Madame Goesler and the blacksmith to reach London from Prague.

If the matter was interesting to the public before, it became doubly interesting now. It was of course known to everybody that Madame Goesler had undertaken a journey to Bohemia,—and, as many supposed, a roving tour through all the wilder parts of unknown Europe, Poland, Hungary, and the Principalities for instance,—with the object of looking for evidence to save the life of Phineas Finn; and grandly romantic tales were told of her wit, her wealth, and her beauty. The story was published of the Duke of Omnium's will, only not exactly the true story. The late Duke had left her everything at his disposal, and, it was hinted that they had been privately married just before the Duke's death. Of course Madame Goesler became very popular, and the blacksmith from Prague who had made the key was expected with an enthusiasm which almost led to preparation for a public reception.

And yet, let the blacksmith from Prague be ever so minute in his evidence as to the key, let it be made as clear as running water that Mealyus had caused to be constructed for him in Prague a key that would open the door of the house in Northumberland Street, the facts as proved at the trial would not be at all changed. The lawyers were much at variance with their opinions on the matter, some thinking that the judge had been altogether wrong in delaying his charge. According to them he should not have allowed Mr. Chaffanbrass to have read the telegram in Court. The charge should have been given, and the sentence of the Court should have been pronounced if a verdict of guilty were given. The Home Secretary should then have granted a respite till the coming of the blacksmith, and have extended this respite to a pardon, if advised that the circumstances of the latch-key rendered doubtful the propriety of the verdict. Others, however, maintained that in this way a grievous penalty would be inflicted on a man who, by general consent, was now held to be innocent. Not only would he, by such an arrangement of circumstances, have been left for some prolonged period under the agony of a condemnation, but, by the necessity of the case, he would lose his seat for Tankerville. It would be imperative upon the House to declare vacant by its own action a seat held by a man condemned to death for murder, and no pardon from the Queen or from the Home Secretary would absolve the House from that duty. The House, as a House of Parliament, could only recognise the verdict of the jury as to the man's guilt. The Queen, of course, might pardon whom she pleased, but no pardon from the Queen would remove the guilt implied by the sentence. Many went much further than this, and were prepared to prove that were he once condemned he could not afterwards sit in the House, even if re-elected.

Now there was unquestionably an intense desire,—since the arrival of these telegrams,—that Phineas Finn should retain his seat. It may be a question whether he would not have been the most popular man in the House could he have sat there on the day after the telegrams arrived. The Attorney-General had declared,—and many others had declared with him,—that this information about the latch-key did not in the least affect the evidence as given against Mr. Finn. Could it have been possible to convict the other man, merely because he had surreptitiously caused a door-key of the house in which he lived to be made for him? And how would this new information have been received had Lord Fawn sworn unreservedly that the man he had seen running out of the mews had been Phineas Finn? It was acknowledged that the latch-key could not be accepted as sufficient evidence against Mealyus. But nevertheless the information conveyed by the telegrams altogether changed the opinion of the public as to the guilt or innocence of Phineas Finn. His life now might have been insured, as against the gallows, at a very low rate. It was felt that no jury could convict him, and he was much more pitied in being subjected to a prolonged incarceration than even those twelve unfortunate men who had felt sure that the Wednesday would have been the last day of their unmerited martyrdom.

Phineas in his prison was materially circumstanced precisely as he had been before the trial. He was supplied with a profusion of luxuries, could they have comforted him; and was allowed to receive visitors. But he would see no one but his sisters,—except that he had one interview with Mr. Low. Even Mr. Low found it difficult to make him comprehend the exact condition of the affair, and could not induce him to be comforted when he did understand it. What had he to do,—how could his innocence or his guilt be concerned,—with the manufacture of a paltry key by such a one as Mealyus? How would it have been with him and with his name for ever if this fact had not been discovered? 'I was to be hung or saved from hanging according to the chances of such a thing as this! I do not care for my life in a country where such injustice can be done.' His friend endeavoured to assure him that even had nothing been heard of the key the jury would have acquitted him. But Phineas would not believe him. It had seemed to him as he had listened to the whole proceeding that the Court had been against him. The Attorney and Solicitor-General had appeared to him resolved upon hanging him,—men who had been, at any rate, his intimate acquaintances, with whom he had sat on the same bench, who ought to have known him. And the judge had taken the part of Lord Fawn, who had seemed to Phineas to be bent on swearing away his life. He had borne himself very gallantly during that week, having in all his intercourse with his attorney, spoken without a quaver in his voice, and without a flaw in the perspicuity of his intelligence. But now, when Mr. Low came to him, explaining to him that it was impossible that a verdict should be found against him, he was quite broken down. 'There is nothing left of me,' he said at the end of the interview. 'I feel that I had better take to my bed and die. Even when I think of all that friends have done for me, it fails to cheer me. In this matter I should not have had to depend on friends. Had not she gone for me to that place every one would have believed me to be a murderer.'

And yet in his solitude he thought very much of the marvellous love shown to him by his friends. Words had been spoken which had been very sweet to him in all his misery,—

words such as neither men nor women can say to each other in the ordinary intercourse of life, much as they may wish that their purport should be understood. Lord Chiltern, Lord Cantrip, and Mr. Monk had alluded to him as a man specially singled out by them for their friendship. Lady Cantrip, than whom no woman in London was more discreet, had been equally enthusiastic. Then how gracious, how tender, how inexpressibly sweet had been the words of her who had been Violet Effingham! And now the news had reached him of Madame Goesler's journey to the continent. 'It was a wonderful thing for her to do,' Mr. Low had said. Yes, indeed! Remembering all that had passed between them he acknowledged to himself that it was very wonderful. Were it not that his back was now broken, that he was prostrate and must remain so, a man utterly crushed by what he had endured, it might have been possible that she should do more for him even than she yet had done.

CHAPTER LXV

'I hate her!'

LADY LAURA KENNEDY had been allowed to take no active part in the manifestations of friendship which at this time were made on behalf of Phineas Finn. She had, indeed, gone to him in his prison, and made daily efforts to administer to his comfort; but she could not go up into the Court and speak for him. And now this other woman, whom she hated, would have the glory of his deliverance! She already began to see a fate before her, which would make even her past misery as nothing to that which was to come. She was a widow,—not yet two months a widow; and though she did not and could not mourn the death of a husband as do other widows,—though she could not sorrow in her heart for a man whom she had never loved, and from whom she had been separated during half her married life,—yet the fact of her widowhood and the circumstances of her weeds were heavy on her. That she loved this man, Phineas Finn, with a passionate devotion of which the other woman could know nothing she was quite sure. Love him! Had she not been true to him and to his interests from the very first day in which he had come among them in London, with almost more than a woman's truth? She knew and recalled to her memory over and over again her own one great sin,—the fault of her life. When she was, as regarded her own means, a poor woman, she had refused to be this poor man's wife, and had given her hand to a rich suitor. But she had done this with a conviction that she could so best serve the interests of the man in regard to whom she had promised herself that her feeling should henceforth be one of simple and purest friendship. She had made a great effort to carry out that intention, but the effort had been futile. She had striven to do her duty to a husband whom she disliked,—but even in that she had failed. At one time she had been persistent in her intercourse with Phineas Finn, and at another had resolved that she would not see him. She had been madly angry with him when he came to her with the story of his love for another woman, and had madly shown her anger; but yet she had striven to get for him the wife he wanted, though in doing so she would have abandoned one of the dearest purposes of her life. She had moved heaven and earth for him,—her heaven and earth,—when there was danger that he would lose his seat in Parliament. She had encountered the jealousy of her husband with scorn,—and had then deserted him because he was jealous. And all this she did with a consciousness of her own virtue which was almost as sublime as it was ill-founded. She had been wrong. She confessed so much to herself with bitter tears. She had marred the happiness of three persons by the mistake she had made in early life. But it had not yet occurred to her that she had sinned. To her thinking the jealousy of her husband had been preposterous and abominable, because she had known,—and had therefore felt that he should have known,—that she would never disgrace him by that which the world calls falsehood in a wife. She had married him without loving him, but it seemed to her that he was in fault for that. They had become wretched, but she had never pitied his wretchedness. She had left him, and thought herself to be ill-used because he had ventured to reclaim his wife. Through it all she had been true in her regard to the one man she had ever loved, and,—though she admitted her own folly and knew her own shipwreck,—yet she had always drawn some woman's consolation from the conviction of her own constancy. He had vanished from her sight for a while with a young wife,—never from her mind,—and then he had returned a widower. Through silence, absence, and distance she had been true to him. On his return to his old ways she had at once welcomed him and strove to aid him. Everything that was hers should be his,—if only he would open his hands to take it. And she would tell it him all, —let him know every corner of her heart. She was a married woman, and could not be his wife. She was a woman of virtue, and would not be his mistress. But she would be to him a friend so tender that no wife, no mistress should ever have been fonder! She did tell him everything as they stood together on the ramparts of the old Saxon castle. Then he had kissed her, and pressed her to his heart,—not because he loved her, but because he was generous. She had partly understood it all,—but yet had not understood it thoroughly. He did not assure her of his love,—but then she was a wife, and would have admitted no love that was sinful. When she returned to Dresden that night she stood gazing at herself in the glass and saw that there was nothing there to attract the love of such a man as Phineas Finn,—of one who was himself glorious with manly beauty; but yet for her sadness there was some cure, some possibility of consolation in the fact that she was a wife. Why speak of love at all when marriage was so far out of the question? But now she was a widow and as free as he was,—a widow endowed with ample wealth; and she was the woman to whom he had sworn his love when they had stood together, both young, by the falls of the Linter! How often might they stand there again if only his constancy would equal hers?

She had seen him once since Fate had made her a widow; but then she had been but a few days a widow, and his life had at that moment been in strange jeopardy. There had certainly been no time then for other love than that which the circumstances and the sorrow of the hour demanded from their mutual friendship. From that day, from the first moment in which she had heard of his arrest, every thought, every effort of her mind had been devoted to his affairs. So great was his peril and so strange, that it almost wiped out from her mind the remembrance of her own condition. Should they hang him,—undoubtedly she would die. Such a termination to all her aspirations for him whom she had selected as her god upon earth would utterly crush her. She had borne much, but she could never bear that. Should he escape, but escape ingloriously;—ah, then he should know what the devotion of a woman could do for a man! But if he should leave his prison with flying colours, and come forth a hero to the world, how would it be with her then? She could foresee and understand of what nature would be the ovation with which he would be greeted. She had already heard what the Duchess was doing and saying. She knew how eager on his behalf were Lord and Lady Cantrip. She discussed the matter daily with her sister-in-law, and knew what her brother thought. If the acquittal were perfect, there would certainly be an ovation,—in which, was it not certain to her, that she would be forgotten? And she heard much, too, of Madame Goesler. And now there came the news. Madame Goesler had gone to Prague, to Cracow,—and where not?—spending her wealth, employing her wits, bearing fatigue, openly before the world on this man's behalf; and had done so successfully. She had found this evidence of the key, and now because the tracings of a key had been discovered by a woman, people were ready to believe that he was innocent, as to whose innocence she, Laura Kennedy, would have been willing to stake her own life from the beginning of the affair!

Why had it not been her lot to go to Prague? Would not she have drunk up Esil,* or swallowed a crocodile against any she-Laertes that would have thought to rival and to parallel her great love? Would not she have piled up new Ossas* had the opportunity been given her? Womanlike she had gone to him in her trouble,—had burst through his prison doors, had thrown herself on his breast, and had wept at his feet. But of what avail had been that? This strange female, this Moabitish* woman, had gone to Prague, and had found a key,—and everybody said that the thing was done! How she hated the strange woman, and remembered all the evil things that had been said of the intruder! She told herself over and over again that had it been any one else than this half-foreigner, this German Jewess, this intriguing unfeminine upstart, she could have borne it. Did not all the world know that the woman for the last two years had been the mistress of that old doting Duke who was now dead? Had one ever heard who was her father or who was her mother? Had it not always been declared of her that she was a pushing, dangerous, scheming creature? And then she was old enough to be his mother, though by some Medean* tricks known to such women, she was able to postpone,—not the ravages of age,—but the manifestation of them to the eyes of the world. In all of which charges poor Lady Laura wronged her rival foully;—in that matter of age especially, for, as it happened, Madame Goesler was by some months the younger of the two. But Lady Laura was a blonde, and trouble had told upon her outwardly, as it is wont to do upon those who are fair-skinned, and, at the same time, high-hearted. But Madame Goesler was a brunette, —swarthy, Lady Laura would have called her,—with bright eyes and glossy hair and thin cheeks, and now being somewhat over thirty she was at her best. Lady Laura hated her as a fair woman who has lost her beauty can hate the dark woman who keeps it.

'What made her think of the key?' said Lady Chiltern.

'I don't believe she did think of it. It was an accident.'

'Then why did she go?'

'Oh, Violet, do not talk to me about that woman any more, or I shall be mad.'

'She has done him good service.'

'Very well;—so be it. Let him have the service. I know they would have acquitted him if she had never stirred from London. Oswald says so. But no matter. Let her have her triumph. Only do not talk to me about her. You know what I have thought about her ever since she first came up in London. Nothing ever surprised me so much as that you should take her by the hand.'

'I do not know that I took her specially by the hand.'

'You had her down at Harrington.'

'Yes; I did. And I do like her. And I know nothing against her. I think you are prejudiced against her, Laura.'

'Very well. Of course you think and can say what you please. I hate her, and that is sufficient.' Then, after a pause, she added, 'Of course he will marry her. I know that well enough. It is nothing to me whom he marries—only,—only,—only, after all that has passed it seems hard upon me that his wife should be the only woman in London that I could not visit.'

'Dear Laura, you should control your thoughts about this young man.'

'Of course I should;—but I don't. You mean that I am disgracing myself.'

'No.'

'Yes, you do. Oswald is more candid, and tells me so openly. And yet what have I done? The world has been hard upon me, and I have suffered. Do I desire anything except that he shall be happy and respectable? Do I hope for anything? I will go back and linger out my life at Dresden, where my disgrace can hurt no one.' Her sister-in-law with all imaginable tenderness said what she could to console the miserable woman;—but there was no consolation possible. They both knew that Phineas Finn would never renew the offer which he had once made.

CHAPTER LXVI

The foreign bludgeon

IN the meantime Madame Goesler, having accomplished the journey from Prague in considerably less than a week, reached London with the blacksmith, the attorney's clerk, and the model of the key. The trial had been adjourned on Wednesday, the 24th of June, and it had been suggested that the jury should be again put into their box on that day week. All manner of inconvenience was to be endured by various members of the legal profession, and sundry irregularities were of necessity sanctioned on this great occasion. The sitting of the Court should have been concluded, and everybody concerned should have been somewhere else, but the matter was sufficient to justify almost any departure from routine. A member of the House of Commons was in custody, and it had already been suggested that some action should be taken by the House as to his speedy deliverance. Unless a jury could find him guilty, let him be at once restored to his duties and his privileges. The case was involved in difficulties, but in the meantime the jury, who had been taken down by train every day to have a walk in the country in the company of two sheriff's officers, and who had been allowed to dine at Greenwich one day and at Richmond on another in the hope that whitebait with lamb and salad might in some degree console them for their loss of liberty, were informed that they would be once again put into their box on Wednesday. But Madame Goesler reached London on the Sunday morning, and on the Monday the whole affair respecting the key was unravelled in the presence of the Attorney-General, and with the personal assistance of our old friend, Major Mackintosh. Without a doubt the man Mealyus had caused to be made for him in Prague a key which would open the door of the house in Northumberland Street. A key was made in London from the model now brought which did open the door. The Attorney-General seemed to think that it would be his duty to ask the judge to call upon the jury to acquit Phineas Finn, and that then the matter must rest for ever, unless further evidence could be obtained against Yosef Mealyus. It would not be possible to hang a man for a murder simply because he had fabricated a key,—even though he might possibly have obtained the use of a grey coat for a few hours. There was no tittle of evidence to show that he had ever had the great coat on his shoulders, or that he had been out of the house on that night. Lord Fawn, to his infinite disgust, was taken to the prison in which Mealyus was detained, and was confronted with the man, but he could say nothing. Mealyus, at his own suggestion, put on the coat, and stalked about the room in it. But Lord Fawn would not say a word. The person whom he now saw might have been the man in the street, or Mr. Finn might have been the man, or any other man might have been the man. Lord Fawn was very dignified, very reserved, and very unhappy. To his thinking he was the great martyr of this trial. Phineas Finn was becoming a hero. Against the twelve jurymen the finger of scorn would never be pointed. But his sufferings must endure for his life—might probably embitter his life to the very end. Looking into his own future from his present point of view he did not see how he could ever again appear before the eye of the public. And yet with what persistency of conscience had he struggled to be true and honest! On the present occasion he would say nothing. He had seen a man in a grey coat, and for the future would confine himself to that. 'You did not see me, my lord,' said Mr. Emilius with touching simplicity.

So the matter stood on the Monday afternoon, and the jury had already been told that they might be released on the following Tuesday,—might at any rate hear the judge's charge on that day,—when another discovery was made more wonderful than that of the key. And this was made without any journey to Prague, and might, no doubt, have been made on any day since the murder had been committed. And it was a discovery for not having made which the police force generally was subjected to heavy censure. A beautiful little boy was seen playing in one of those gardens through which the passage runs with a short loaded bludgeon in his hand. He came into the house with the weapon, the maid who was with him having asked the little lord no question on the subject. But luckily it attracted attention, and his little lordship took two gardeners and a coachman and all the nurses to the very spot at which he found it. Before an hour was over he was standing at his father's knee, detailing the fact with great open eyes to two policemen, having by this time become immensely proud of his adventure. This occurred late on the Monday afternoon, when the noble family were at dinner, and the noble family was considerably disturbed, and at the same time very much interested, by the occurrence. But on the Tuesday morning there was the additional fact established that a bludgeon loaded with lead had been found among the thick grass and undergrowth of shrubs in a spot to which it might easily have been thrown by any one attempting to pitch it over the wall. The news flew about the town like wildfire, and it was now considered certain that the real murderer would be discovered.

But the renewal of the trial was again postponed till the Wednesday, as it was necessary that an entire day should be devoted to the bludgeon. The instrument was submitted to the eyes and hands of persons experienced in such matters, and it was declared on all sides that the thing was not of English manufacture. It was about a foot long, with a leathern thong to the handle, with something of a spring in the shaft, and with the oval loaded knot at the end cased with leathern thongs very minutely and skilfully cut. They who understood modern work in leather gave it as their opinion that the weapon had been made in Paris. It was considered that Mealyus had brought it with him, and concealed it in preparation for this occasion. If the police could succeed in tracing the bludgeon into his hands, or in proving that he had purchased any such instrument, then,—so it was thought,—there would be evidence to justify a police magistrate in sending Mr. Emilius to occupy the place so lately and so long held by poor Phineas Finn. But till that had been done, there could be nothing to connect the preacher with the murder. All who had heard the circumstances of the case were convinced that Mr. Bonteen had been murdered by the weapon lately discovered, and not by that which Phineas had carried in his pocket,—but no one could adduce proof that it was so. This second bludgeon would no doubt help to remove the difficulty in regard to Phineas, but would not give atonement to the shade of Mr. Bonteen.

Mealyus was confronted with the weapon in the presence of Major Mackintosh, and was told its story;—how it was found in the nobleman's garden by the little boy. At the first moment, with instant readiness, he took the thing in his hand, and looked at it with feigned curiosity. He must have studied his conduct so as to have it ready for such an occasion, thinking that it might some day occur. But with all his presence of mind he could not keep the tell-tale blood from mounting.

'You don't know anything about it, Mr. Mealyus?' said one of the policemen present, looking closely into his face. 'Of course you need not criminate yourself.'

'What should I know about it? No;—I know nothing about the stick. I never had such a stick, or, as I believe, saw one before.' He did it very well, but he could not keep the blood from rising to his cheeks. The policemen were sure that he was the murderer,—but what could they do?

'You saved his life, certainly,' said the Duchess to her friend on the Sunday afternoon. That had been before the bludgeon was found.

'I do not believe that they could have touched a hair of his head,' said Madame Goesler.

'Would they not? Everybody felt sure that he would be hung. Would it not have been awful? I do not see how you are to help becoming man and wife now, for all the world are talking about you.' Madame Goesler smiled, and said that she was quite indifferent to the world's talk. On the Tuesday after the bludgeon was found, the two ladies met again. 'Now it was known that it was the clergyman,' said the Duchess.

'I never doubted it.'

'He must have been a brave man for a foreigner,—to have attacked Mr. Bonteen all alone in the street, when any one might have seen him. I don't feel to hate him so very much after all. As for that little wife of his, she has got no more than she deserved.'

'Mr. Finn will surely be acquitted now.'

'Of course he'll be acquitted. Nobody doubts about it. That is all settled, and it is a shame that he should be kept in prison even over to-day. I should think they'll make him a peer, and give him a pension,—or at the very least appoint him secretary to something. I do wish Plantagenet hadn't been in such a hurry about that nasty Board of Trade, and then he might have gone there. He couldn't very well be Privy Seal, unless they do make him a peer. You wouldn't mind,—would you, my dear?'

'I think you'll find that they will console Mr. Finn with something less gorgeous than that. You have succeeded in seeing him, of course?'

'Plantagenet wouldn't let me, but I know who did.'

'Some lady?'

'Oh, yes,—a lady. Half the men about the clubs went to him, I believe.'

'Who was she?'

'You won't be ill-natured?'

'I'll endeavour at any rate to keep my temper, Duchess.'

'It was Lady Laura.'

'I supposed so.'

'They say she is frantic about him, my dear.'

'I never believe those things. Women do not get frantic about men in these days. They have been very old friends, and have known each other for many years. Her brother, Lord Chiltern, was his particular friend. I do not wonder that she should have seen him.'

'Of course you know that she is a widow.'

'Oh, yes;—Mr. Kennedy had died long before I left England.'

'And she is very rich. She has got all Loughlinter for her

life, and her own fortune back again. I will bet you anything you like that she offers to share it with him.'

'It may be so,' said Madame Goesler, while the slightest blush in the world suffused her cheek.

'And I'll make you another bet, and give you any odds.'

'What is that?'

'That he refuses her. It is quite a common thing nowadays for ladies to make the offer, and for gentlemen to refuse. Indeed, it was felt to be so inconvenient while it was thought that gentlemen had not the alternative, that some men became afraid of going into society. It is better understood now.'

'Such things have been done, I do not doubt,' said Madame Goesler, who had contrived to avert her face without making the motion apparent to her friend.

'When this is all over we'll get him down to Matching, and manage better than that. I should think they'll hardly go on with the Session, as nobody has done anything since the arrest. While Mr. Finn has been in prison legislation has come to a standstill altogether. Even Plantagenet doesn't work above twelve hours a day, and I'm told that poor Lord Fawn hasn't been near his office for the last fortnight. When the excitement is over they'll never be able to get back to their business before the grouse. There'll be a few dinners of course, just as a compliment to the great man,—but London will break up after that, I should think. You won't come in for so much of the glory as you would have done if they hadn't found the stick. Little Lord Frederick must have his share, you know.'

'It's the most singular case I ever knew,' said Sir Simon Slope that night to one of his friends. 'We certainly should have hanged him but for the two accidents, and yet neither of them brings us a bit nearer to hanging any one else.'

'What a pity!'

'It shows the danger of circumstantial evidence,—and yet without it one never could get at any murder. I'm very glad, you know, that the key and the stick did turn up. I never thought much about the coat.'

CHAPTER LXVII

The Verdict

On the Wednesday morning Phineas Finn was again brought into the Court, and again placed in the dock. There was a general feeling that he should not again have been so disgraced; but he was still a prisoner under a charge of murder, and it was explained to him that the circumstances of the case and the stringency of the law did not admit of his being seated elsewhere during his trial. He treated the apology with courteous scorn. He should not have chosen, he said, to have made any change till after the trial was over, even had any change been permitted. When he was brought up the steps into the dock after the judges had taken their seats there was almost a shout of applause. The crier was very angry, and gave it to be understood that everybody would be arrested unless everybody was silent; but the Chief Justice said not a word, nor did those great men the Attorney and Solicitor-General express any displeasure. The bench was again crowded with Members of Parliament from both Houses, and on this occasion Mr. Gresham himself had accompanied Lord Cantrip. The two Dukes were there, and men no bigger than Laurence Fitzgibbon were forced to subject themselves to the benevolence of the Under-Sheriff.

Phineas himself was pale and haggard. It was observed that he leaned forward on the rail of the dock all the day, not standing upright as he had done before; and they who watched him closely said that he never once raised his eyes on this day to meet those of the men opposite to him on the bench, although heretofore throughout the trial he had stood with his face raised so as to look directly at those who were there seated. On this occasion he kept his eyes fixed upon the speaker. But the whole bearing of the man, his gestures, his gait, and his countenance were changed. During the first long week of his trial, his uprightness, the manly beauty of his countenance, and the general courage and tranquillity of his deportment had been conspicuous. Whatever had been his fatigue, he had managed not to show the outward signs of weariness. Whatever had been his fears, no mark of fear had disfigured his countenance. He had never once condescended to the exhibition of any outward show of effrontery. Through six weary days he had stood there, supported by a manhood sufficient for the terrible emergency. But now it seemed that at any rate the outward grace of his demeanour had deserted him. But it was known that he had been ill during the last few days, and it had been whispered through the Court that he had not slept at nights. Since the adjournment of the Court there had been bulletins as to his health, and everybody knew that the confinement was beginning to tell upon him.

On the present occasion the proceedings of the day were opened by the Attorney-General, who began by apologising to the jury. Apologies to the jury had been very frequent during the trial, and each apology had called forth fresh grumbling. On this occasion the foreman expressed a hope that the Legislature would consider the condition of things which made it possible that twelve gentlemen all concerned extensively in business should be confined for fourteen days because a mistake had been made in the evidence as to a murder. Then the Chief Justice, bowing down his head and looking at them over the rim of his spectacles with an expression of wisdom that almost convinced them, told them that he was aware of no mistake in the evidence. It might become their duty, on the evidence which they had heard and the further evidence which they would hear, to acquit the prisoner at the bar; but not on that account would there have been any mistake or erroneous procedure in the Court, other than such error on the part of the prosecution in regard to the alleged guilt of the prisoner as it was the general and special duty of jurors to remedy. Then he endeavoured to reconcile them to their sacrifice by describing the importance and glorious British nature of their position. 'My lord,' said one of the jurors, 'if you was a salesman, and hadn't got no partner, only a very young 'un, you'd know what it was to be kept out of your business for a fortnight.' Then that salesman wagged his head, and put his handkerchief up to his eyes, and there was pity also for him in the Court.

After that the Attorney-General went on. His learned friend on the other side,—and he nodded to Mr. Chaffanbrass,—had got some further evidence to submit to them on behalf of the prisoner who was still on his trial before them. He now addressed them with the view of explaining to them that if that evidence should be such as he believed, it would become his duty on behalf of the Crown to join with his learned friend in requesting the Court to direct the jury to acquit the prisoner. Not the less on that account would it be the duty of the jury to form their own opinion as to the credibility of the fresh evidence which would be brought before them.

'There won't be much doubt about the credibility,' said Mr. Chaffanbrass, rising in his place. 'I am not a bit afraid about the credibility, gentlemen; and I don't think that you need be afraid either. You must understand, gentlemen, that I am now going on calling evidence for the defence. My last witness was the Right Honourable M. Monk, who spoke as to character. My next will be a Bohemian blacksmith named Praska,—Peter Praska,—who naturally can't speak a word of English, and unfortunately can't speak a word of German either. But we have got an interpreter, and I daresay we shall find out without much delay what Peter Praska has to tell us.' Then Peter Praska was handed up to the rostrum for the witnesses, and the man learned in Czech and also in English was placed close to him, and sworn to give a true interpretation.

Mealyus the unfortunate one was also in Court, brought in between two policemen, and the Bohemian blacksmith swore that he had made a certain key on the instructions of the man he now saw. The reader need not be further troubled with all the details of the evidence about the key. It was clearly proved that in a village near to Prague a key had been made such as would open Mr. Meager's door in Northumberland Street, and it was also proved that it was made from a mould supplied by Mealyus. This was done by the joint evidence of Mr. Meager and of the blacksmith. 'And if I lose my key,' said the reverend gentleman, 'why should I not have another made? Did I ever deny it? This, I think, is very strange.' But Mr. Emilius was very quickly walked back out of the Court between the two policemen, as his presence would not be required in regard to the further evidence regarding the bludgeon.

Mr. Chaffanbrass, having finished his business with the key, at once began with the bludgeon. The bludgeon was produced, and was handed up to the bench, and inspected by the Chief Justice. The instrument excited great interest. Men rose on tiptoe to look at it even from a distance, and the Prime Minister was envied because for a moment it was placed in his hands. As the large-eyed little boy who had found it was not yet six years old, there was a difficulty in perfecting the thread of the evidence. It was not held to be proper to administer an oath to an infant. But in a roundabout way it was proved that the identical bludgeon had been picked up in the garden. There was an elaborate surveyor's plan produced of the passage, the garden, and the wall,—with the steps on which it was supposed that the blow had been struck; and the spot was indicated on which the child had said that he had found the weapon. Then certain workers in leather were questioned, who agreed in asserting that no such instrument as that handed to them had ever been made in England. After that, two scientific chemists told the jury that they had minutely examined the knob of the instrument with reference to the discovery of human blood,—but in vain. They were, however, of opinion that the man might very readily have been killed by the instrument without any effusion of blood at the moment of the blows. This seemed to the jury to be the less necessary, as three or four surgeons who had examined the murdered man's head had already told them that in all probability there had been no such effusion. When the judges went out to lunch at two o'clock the jury were trembling as to their fate for another night.

The fresh evidence, however, had been completed, and on the return of the Court Mr. Chaffanbrass said that he should only speak a very few words. For a few words he must ask indulgence, though he knew them to be irregular. But it was the speciality of this trial that everything in it was irregular, and he did not think that his learned friend the Attorney-General would dispute the privilege. The Attorney-General said nothing, and Mr. Chaffanbrass went on with his little speech,—with which he took up the greatest part of an hour. It was thought to have been unnecessary, as nearly all that he said was said again—and was sure to have been so said,—by the judge. It was not his business,—the business of him, Mr. Chaffanbrass,—to accuse another man of the murder of Mr. Bonteen. It was not for him to tell the jury whether there was or was not evidence on which any other man should be sent to trial. But it was his bounden duty in defence of his client to explain to them that a collection of facts tending to criminate another man,—which when taken together made a fair probability that another man had committed the crime,—rendered it quite out of the question that they should declare his client to be guilty. He did not believe that there was a single person in the Court who was not now convinced of the innocence of his client;—but it was not permitted to him to trust himself solely to that belief. It was his duty to show them that, of necessity, they must acquit his client. When Mr. Chaffanbrass sat down, the Attorney-General waived any right he might have of further reply.

It was half-past three when the judge began his charge. He would, he said, do his best to enable the jury to complete their tedious duty, so as to return to their families on that night. Indeed he would certainly finish his charge before he rose from the seat, let the hour be what it might; and though time must be occupied by him in going through the evidence and explaining the circumstances of this very singular trial, it might not be improbable that the jury would be able to find their verdict without any great delay among themselves. 'There won't be any delay at all, my lord,' said the suffering and very irrational salesman. The poor man was again rebuked, mildly, and the Chief Justice continued his charge.

As it occupied four hours in the delivery, of which by far the greater part was taken up in recapitulating and sifting evidence with which the careful reader, if such there be, has already been made too intimately acquainted, the account of it here shall be very short. The nature of circumstantial evidence was explained, and the truth of much that had been said in regard to such evidence by Mr. Chaffanbrass admitted; —but, nevertheless, it would be impossible,—so said his lordship,—to administer justice if guilt could never be held to have been proved by circumstantial evidence alone. In this case it might not improbably seem to them that the gentleman who had so long stood before them as a prisoner at the bar had been the victim of a most singularly untoward chain of circumstances, from which he would have to be liberated, should he be at last liberated, by another chain of circumstances as singular; but it was his duty to inform them now, after they had heard what he might call the double evidence, that he could not have given it to them as his opinion that the charge had been brought home against the prisoner, even had those circumstances of the Bohemian key and of the foreign bludgeon never been brought to light. He did not mean to say that the evidence had not justified the trial. He thought that the trial had been fully justified. Nevertheless, had nothing arisen to point to the possibility of guilt in another man, he should not the less have found himself bound in duty to explain to them that the thread of the evidence against Mr. Finn had been incomplete,—or, he would rather say, the weight of it had been, to his judgment, insufficient. He was the more intent on saying so much, as he was desirous of making it understood that, even had the bludgeon still remained buried beneath the leaves, had the manufacturer of that key never been discovered, the great evil would not, he thought, have fallen upon them of punishing the innocent instead of the guilty,— that most awful evil of taking innocent blood in their just attempt to punish murder by death. As far as he knew, to the best of his belief, that calamity had never fallen upon the country in his time. The administration of the law was so careful of life that the opposite evil was fortunately more common. He said so much because he would not wish that this case should be quoted hereafter as showing the possible danger of circumstantial evidence. It had been a case in which the evidence given as to character alone had been sufficient to make him feel that the circumstances which seemed to affect the prisoner injuriously could not be taken as establishing his guilt. But now other and imposing circumstances had been brought to light, and he was sure that the jury would have no difficulty with their verdict. A most frightful murder had no doubt been committed in the dead of the night. A gentleman coming home from his club had been killed,—probably by the hand of one who had himself moved in the company of gentlemen. A plot had been made,—had probably been thought of for days and weeks before,—and had been executed with extreme audacity, in order that an enemy might be removed. There could, he thought, be but little doubt that Mr. Bonteen had been killed by the instrument found in the garden, and if so, he certainly had not been killed by the prisoner, who could not be supposed to have carried two bludgeons in his pocket, and whose quarrel with the murdered man had been so recent as to have admitted of no preparation. They had heard the story of Mr. Meager's grey coat, and of the construction of the duplicate key for Mr. Meager's house-door. It was not for him to tell them on the present occasion whether these stories, and the evidence by which they had been supported, tended to affix guilt elsewhere. It was beyond his province to advert to such probability or possibility; but undoubtedly the circumstances might be taken by them as an assistance, if assistance were needed, in coming to a conclusion on the charge against the prisoner. 'Gentlemen,' he said at last, 'I think you will find no difficulty in acquitting the prisoner of the murder laid to his charge,' whereupon the jurymen put their heads together; and the foreman, without half a minute's delay, declared that they were unanimous, and that they found the prisoner Not Guilty. 'And we are of opinion,' said the foreman, 'that Mr. Finn should not have been put upon his trial on such evidence as has been brought before us.'

The necessity of liberating poor Phineas from the horrors of his position was too urgent to allow of much attention being given at the moment to this protest. 'Mr. Finn,' said the judge, addressing the poor broken wretch, 'you have been acquitted of the odious and abominable charge brought against you, with the concurrence, I am sure, not only of those who have heard this trial, but of all your countrymen and countrywomen. I need not say that you will leave that dock with no stain on your character. It has, I hope, been some consolation to you in your misfortune to hear the terms in which you have been spoken of by such friends as they who came here to give their testimony on your behalf. It is, and it has been, a great sorrow to me to see such a one as you subjected to so unmerited an ignominy; but a man educated in the laws of his country, as you have been, and understanding its constitution fundamentally, as you do, will probably have acknowledged that, great as has been the misfortune to you personally, nothing more than a proper attempt has been made to execute justice. I trust that you may speedily find yourself able to resume your place among the legislators of the country.' Thus Phineas Finn was acquitted, and the judges, collecting up their robes, trooped off from the bench, following the long line of their assessors who had remained even to that hour to hear the last word of the trial. Mr. Chaffanbrass collected his papers, with the assistance of Mr. Wickerby,—totally disregardful of his junior counsel, and the Attorney and Solicitor-General congratulated each other on the successful termination of a very disagreeable piece of business.

And Phineas was discharged. According to the ordinary meaning of the words he was now to go about his business as he pleased, the law having no further need of his person. We can understand how in common cases the prisoner discharged on his acquittal,—who probably in nine cases out of ten is conscious of his own guilt,—may feel the sweetness of his freedom and enjoy his immunity from danger with a light heart. He is received probably by his wife or young woman,—or perhaps, having no wife or young woman to receive him, betakes himself to his usual haunts. The interest which has been felt in his career is over, and he is no longer the hero of an hour;—but he is a free man, and may drink his gin-and-water where he pleases. Perhaps a small admiring crowd may welcome him as he passes out into the street, but he has become nobody before he reaches the corner. But it could not be so with this discharged prisoner,—either as regarded himself and his own feelings, or as regarded his friends. When the moment came he had hardly as yet thought about the immediate future,—had not considered how he would live, or where, during the next few months. The sensations of the moment had been so full, sometimes of agony and at others of anticipated triumph, that he had not attempted as yet to make for himself any schemes. The Duchess of Omnium had suggested that he would be received back into society with an elaborate course of fashionable dinners; but that view of his return to the world had certainly not occurred to him. When he was led down from the dock he hardly knew whither he was being taken, and when he found himself in a small room attached to the Court, clasped on one arm by Mr. Low and on the other by Lord Chiltern, he did not know what they would propose to him,—nor had he considered what answer he would make to any proposition. 'At last you are safe,' said Mr. Low.

'But think what he has suffered,' said Lord Chiltern.

Phineas looked round to see if there was any other friend present. Certainly among all his friends he had thought most of her who had travelled half across Europe for evidence to save him. He had seen Madame Goesler last on the evening preceding the night of the murder, and had not even heard from her since. But he had been told what she had done for him, and now he had almost fancied that he would have found her waiting for him. He smiled first at the one man and then at the other, and made an effort to carry himself with his ordinary tranquillity. 'It will be all right now, I dare say,' he said. 'I wonder whether I could have a glass of water.'

He sat down while the water was brought to him, and his two friends stood over him, hardly knowing how to do more than support him by their presence.

Then Lord Cantrip made his way into the room. He had sat on the bench to the last, whereas the other two had gone down to receive the prisoner when acquitted;—and with him came Sir Harry Coldfoot, the Home Secretary. 'My friend,' said the former, 'the bitter day has passed over you, and I hope that the bitterness will soon pass away also.' Phineas again attempted to smile as he held the hand of the man with whom he had formerly been associated in office.

'I should not intrude, Mr. Finn,' said Sir Harry, 'did I not feel myself bound in a special manner to express my regret at the great trouble to which you have been subjected.' Phineas rose, and bowed stiffly. He had conceived that every one connected with the administration of the law had believed him to be guilty, and none in his present mood could be dear to him but they who from the beginning trusted in his innocence. 'I am requested by Mr. Gresham,' continued Sir Harry, 'to express to you his entire sympathy, and his joy that all this is at last over.' Phineas tried to make some little speech, but utterly failed. Then Sir Harry left them, and he burst out into tears.

'Who can be surprised?' said Lord Cantrip. 'The marvel is that he should have been able to bear it so long.'

'It would have crushed me utterly, long since,' said the other lord. Then there was a question asked as to what he would do, and Mr. Low proposed that he should be allowed to take Phineas to his own house for a few days. His wife, he said, had known their friend so long and so intimately that she might perhaps be able to make herself more serviceable than any other lady, and at their house Phineas could receive his sisters just as he would at his own. His sisters had been lodging near the prison almost ever since the committal, and it had been thought well to remove them to Mr. Low's house in order that they might meet their brother there.

'I think I'll go to my—own room—in Marlborough Street.' These were the first intelligible words he had uttered since he had been led out of the dock, and to that resolution he adhered. Lord Cantrip offered the retirements of a country house belonging to himself within an hour's journey of London, and Lord Chiltern declared that Harrington Hall, which Phineas knew, was altogether at his service,—but Phineas decided in favour of Mrs. Bunce, and to Great Marlborough Street he was taken by Mr. Low.

'I'll come to you to-morrow,—with my wife,'—said Lord Chiltern, as he was going.

'Not to-morrow, Chiltern. But tell your wife how deeply I value her friendship.' Lord Cantrip also offered to come, but was asked to wait awhile. 'I am afraid I am hardly fit for visitors yet. All the strength seems to have been knocked out of me this last week.'

Mr. Low accompanied him to his lodgings, and then handed him over to Mrs. Bunce, promising that his two sisters should come to him early on the following morning. On that evening he would prefer to be quite alone. He would not allow the barrister even to go upstairs with him; and when he had entered his room, almost rudely begged his weeping landlady to leave him.

'Oh, Mr. Phineas, let me do something for you,' said the poor woman. 'You have not had a bit of anything all day. Let me get you just a cup of tea and a chop.'

In truth he had dined when the judges went out to their lunch,—dined as he had been wont to dine since the trial had been commenced,—and wanted nothing. She might bring him tea, he said, if she would leave him for an hour. And then at last he was alone. He stood up in the middle of the room, stretching forth his hands, and putting one first to his breast and then to his brow, feeling himself as though doubting his own identity. Could it be that the last week had been real,—that everything had not been a dream? Had he in truth been suspected of a murder and tried for his life? And then he thought of him who had been murdered, of Mr. Bonteen, his enemy. Was he really gone,—the man who the other day was to have been Chancellor of the Exchequer,—the scornful, arrogant, loud, boastful man? He had hardly thought of Mr. Bonteen before, during these weeks of his own incarceration. He had heard all the details of the murder with a fulness that had been at last complete. The man who had oppressed him, and whom he had at times almost envied, was indeed gone, and the world for awhile had believed that he, Phineas Finn, had been the man's murderer!

And now what should be his own future life? One thing seemed certain to him. He could never again go into the House of Commons, and sit there, an ordinary man of business, with other ordinary men. He had been so hacked and hewed about, so exposed to the gaze of the vulgar, so mauled by the public, that he could never more be anything but the wretched being who had been tried for the murder of his enemy. The pith had been taken out of him, and he was no longer a man fit for use. He could never more enjoy that freedom from self-consciousness, that inner tranquillity of spirit, which are essential to public utility. Then he remembered certain lines which had long been familiar to him, and he repeated them aloud, with some conceit that they were apposite to him:—

> The true gods sigh for the cost and pain,—
> For the reed that grows never more again
>   As a reed with the reeds in the river.*

He sat drinking his tea, still thinking of himself,—knowing how infinitely better it would be for him that he should indulge in no such thought, till an idea struck him, and he got up, and, drawing back the blinds from the open window, looked out into the night. It was the last day of June, and the weather was very sultry; but the night was dark, and it was now near midnight. On a sudden he took his hat, and feeling with a smile for the latchkey which he always carried in his pocket,—thinking of the latchkey which had been made at Prague for the lock of a house in Northumberland Street, New Road, he went down to the front door. 'You'll be back soon, Mr. Finn, won't you now?' said Mrs. Bunce, who had heard his step, and had remained up, thinking it better this, the first night of his return, not to rest till he had gone to his bed.

'Why should I be back soon?' he said, turning upon her. But then he remembered that she had been one of those who were true to him, and he took her hand and was gracious to her. 'I will be back soon, Mrs. Bunce, and you need fear nothing. But recollect how little I have had of liberty lately. I have not even had a walk for six weeks. You cannot wonder that I should wish to roam about a little.' Nevertheless she would have preferred that he should not have gone out all alone on that night.

He had taken off the black morning coat which he had worn during the trial, and had put on that very grey garment by which it had been sought to identify him with the murderer. So clad he crossed Regent Street into Hanover Square, and from thence went a short way down Bond Street, and by Bruton Street into Berkeley Square. He took exactly the reverse of the route by which he had returned home from the club on the night of the murder. Every now and then he trembled as he passed some figure which might be that of a man who would recognise him. But he walked fast, and went on till he came to the spot at which the steps descend from the street into the passage,—the very spot at which the murder had been committed. He looked down it with an awful dread, and stood there as though he were fascinated, thinking of all the details which he had heard throughout the trial. Then he looked around him, and listened whether there were any step approaching through the passage. Hearing none and seeing no one he at last descended, and for the first time in his life passed through that way into Bolton Row. Here it was that the wretch of whom he had now heard so much had waited for his enemy,—the wretch for whom during the last six weeks he had been mistaken. Heavens!—that men who had known him should have believed him to have done such a deed as that! He remembered well having shown the life-preserver to Erle and Fitzgibbon at the door of the club; and it had been thought that after having so shown it he had used it for the purpose to which in his joke he had alluded! Were men so blind, so ignorant of nature, so little capable of discerning the truth as this? Then he went on till he came to the end of Clarges Street, and looked up the mews opposite to it,—the mews from which the man had been seen to hurry. The place was altogether unknown to him. He had never thought whither it had led when passing it on his way up from Piccadilly to the club. But now he entered the mews so as to test the evidence that had been given, and found that it brought him by a turn close up to the spot at which he had been described as having been last seen by Erle and Fitzgibbon. When there he went on, and crossed the street, and looking back saw the club was lighted up. Then it struck him for the first time that it was the night of the week on which the members were wont to assemble. Should he pluck up courage, and walk in among them? He had not lost his right of entry there because he had been accused of murder. He was the same now as heretofore,—if he could only fancy himself to be the same. Why not go in, and have done with all this? He would be the wonder of the club for twenty minutes, and then it would all be over. He stood close under the shade of a heavy building as he thought of this, but he found that he could not do it. He had known from the beginning that he could not do it. How callous, how hard, how heartless, must he have been, had such a course been possible to him! He again repeated the lines to himself—

> The reed that grows never more again
> As a reed with the reeds in the river.

He felt sure that never again would he enter that room, in which no doubt all those assembled were now talking about him.

As he returned home he tried to make out for himself some plan for his future life,—but, interspersed with any idea that he could weave were the figures of two women, Lady Laura

Kennedy and Madame Max Goesler. The former could be nothing to him but a friend; and though no other friend would love him as she loved him, yet she could not influence his life. She was very wealthy, but her wealth could be nothing to him. She would heap it all upon him if he would take it. He understood and knew that. Taking no pride to himself that it was so, feeling no conceit in her love, he was conscious of her devotion to him. He was poor, broken in spirit, and almost without a future;—and yet could her devotion avail him nothing!

But how might it be with that other woman? Were she, after all that had passed between them, to consent to be his wife,—and it might be that she would consent,—how would the world be with him then? He would be known as Madame Goesler's husband, and have to sit at the bottom of her table, —and be talked of as the man who had been tried for the murder of Mr. Bonteen. Look at it in which way he might, he thought that no life could any longer be possible to him in London.

CHAPTER LXVIII

Phineas after the Trial

TEN days passed by, and Phineas Finn had not been out of his lodgings till after daylight, and then he only prowled about in the manner described in the last chapter. His sisters had returned to Ireland, and he saw no one, even in his own room, but two or three of his most intimate friends. Among those Mr. Low and Lord Chiltern were the most frequently with him, but Fitzgibbon, Barrington Erle, and Mr. Monk had also been admitted. People had called by the hundred, till Mrs. Bunce was becoming almost tired of her lodger's popularity; but they came only to inquire,—because it had been reported that Mr. Finn was not well after his imprisonment. The Duchess of Omnium had written to him various notes, asking when he would come to her, and what she could do for him. Would he dine, would he spend a quiet evening, would he go to Matching? Finally, would he become her guest and the Duke's next September for the partridge shooting? They would have a few friends with them, and Madame Goesler would be one of the number. Having had this by him for a week, he had not as yet answered the invitation. He had received two or three notes from Lady Laura, who had frankly explained to him that if he were really ill she would of course go to him, but that as matters stood she could not do so without displeasing her brother. He had answered each note by an assurance that his first visit should be made in Portman Square. To Madame Goesler he had written a letter of thanks,—a letter which had in truth cost him some pains. 'I know,' he said, 'for how much I have to thank you, but I do not know in what words to do it. I ought to be with you telling you in person of my gratitude; but I must own to you that for the present what has occurred has so unmanned me that I am unfit for the interview. I should only weep in your presence like a school-girl, and you would despise me.' It was a long letter, containing many references to the circumstances of the trial, and to his own condition of mind throughout its period. Her answer to him, which was very short, was as follows:—

'Park Lane, Sunday—

'MY DEAR MR. FINN,

'I can well understand that for a while you should be too agitated by what has passed to see your friends. Remember, however, that you owe it to them as well as to yourself not to sink into seclusion. Send me a line when you think that you can come to me that I may be at home. My journey to Prague was nothing. You forget that I am constantly going to Vienna on business connected with my own property there. Prague lies but a few hours out of the route.

'Most sincerely yours,

'M. M. G.'

His friends who did see him urged him constantly to bestir himself, and Mr. Monk pressed him very much to come down to the House. 'Walk in with me to-night, and take your seat as though nothing had happened,' said Mr. Monk.

'But so much has happened.'

'Nothing has happened to alter your outward position as a man. No doubt many will flock round you to congratulate you, and your first half-hour will be disagreeable; but then the thing will have been done. You owe it to your constituents to do so.' Then Phineas for the first time expressed an opinion that he would resign his seat,—that he would take the Chiltern Hundreds, and retire altogether from public life.

'Pray do nothing of the kind,' said Mr. Monk.

'I do not think you quite understand,' said Phineas, 'how such an ordeal as this works upon a man, how it may change a man, and knock out of him what little strength there ever was there. I feel that I am broken, past any patching up or mending. Of course it ought not to be so. A man should be made of better stuff;—but one is only what one is.'

'We'll put off the discussion for another week,' said Mr. Monk.

'There came a letter to me when I was in prison from one of the leading men in Tankerville, saying that I ought to resign. I know they all thought that I was guilty. I do not care to sit for a place where I was so judged,—even if I was fit any longer for a seat in Parliament.' He had never felt convinced that Mr. Monk had himself believed with confidence his innocence, and he spoke with soreness, and almost with anger.

'A letter from one individual should never be allowed to create interference between a member and his constituents. It should simply be answered to that effect, and then ignored. As to the belief of the townspeople in your innocence,—what is to guide you? I believed you innocent with all my heart.'

'Did you?'

'But there was always sufficient possibility of your guilt to prevent a rational man from committing himself to the expression of an absolute conviction.' The young member's brow became black as he heard this. 'I can see that I offend you by saying so,—but if you will think of it, I must be right. You were on your trial; and I as your friend was bound to await the result,—with much confidence, because I knew you; but with no conviction, because both you and I are human and fallible. If the electors at Tankerville, or any great proportion of them, express a belief that you are unfit to represent them because of what has occurred, I shall be the last to recommend you to keep your seat;—but I shall be surprised indeed if they should do so. If there were a general election to-morrow, I should regard your seat as one of the safest in England.'

Both Mr. Low and Lord Chiltern were equally urgent with him to return to his usual mode of life,—using different arguments for their purpose. Lord Chiltern told him plainly that he was weak and womanly,—or rather that he would be were he to continue to dread the faces of his fellow-creatures. The Master of the Brake hounds himself was a man less gifted than Phineas Finn, and therefore hardly capable of understanding the exaggerated feelings of the man who had recently been tried for his life. Lord Chiltern was affectionate, tender-hearted, and true;—but there were no vacillating fibres in his composition. The balance which regulated his conduct was firmly set, and went well. The clock never stopped, and wanted but little looking after. But the works were somewhat rough, and the seconds were not scored. He had, however, been quite true to Phineas during the dark time, and might now say what he pleased. 'I am womanly,' said Phineas. 'I begin to feel it. But I can't alter my nature.'

'I never was so much surprised in my life,' said Lord Chiltern. 'When I used to look at you in the dock, by heaven I envied you your pluck and strength.'

'I was burning up the stock of coals, Chiltern.'

'You'll come all right after a few weeks. You've been knocked out of time;—that's the truth of it.'

Mr. Low treated his patient with more indulgence; but he also was surprised, and hardly understood the nature of the derangement of the mechanism in the instrument which he was desirous of repairing. 'I should go abroad for a few months if I were you,' said Mr. Low.

'I should stick at the first inn I got to,' said Phineas. 'I think I am better here. By and bye I shall travel, I dare say,—all over the world, as far as my money will last. But for the present I am only fit to sit still.'

Mrs. Low had seen him more than once, and had been very kind to him; but she also failed to understand. 'I always thought that he was such a manly fellow,' she said to her husband.

'If you mean personal courage, there is no doubt that he possesses it,—as completely now, probably, as ever.'

'Oh yes;—he could go over to Flanders and let that lord shoot at him; and he could ride brutes of horses, and not care about breaking his neck. That's not what I mean. I thought that he could face the world with dignity;—but now it seems that he breaks down.'

'He has been very roughly used, my dear.'

'So he has,—and tenderly used too. Nobody has had better friends. I thought he would have been more manly.'

The property of manliness in a man is a great possession, but perhaps there is none that is less understood,—which is more generally accorded where it does not exist, or more frequently disallowed where it prevails. There are not many who ever make up their minds as to what constitutes manliness, or even inquire within themselves upon the subject. The woman's error, occasioned by her natural desire for a master, leads her to look for a certain outward magnificence of demeanour, a pretended indifference to stings and little torments, a would-be superiority to the bread-and-butter side of life, an unreal assumption of personal grandeur. But a robe of State such as this,—however well the garment may be worn with practice,—can never be the raiment natural to a man; and men, dressing themselves in women's eyes, have consented to walk about in buckram. A composure of the eye, which has been studied, a reticence as to the little things of life, a certain slowness of speech unless the occasion call for passion, an indifference to small surroundings, these,—joined, of course, with personal bravery,—are supposed to constitute manliness. That personal bravery is required in the composition of manliness must be conceded, though, of all the ingredients needed, it is the lowest in value. But the first requirement of all must be described by a negative. Manliness is not compatible with affectation. Women's virtues, all feminine attributes, may be marred by affectation, but the virtues and the vice may co-exist. An affected man, too, may be honest, may be generous, may be pious;—but surely he cannot be manly. The self-conscious assumption of any outward manner, the striving to add,—even though it be but a tenth of a cubit to the height,—is fatal, and will at once banish the all but divine attribute. Before the man can be manly, the gifts which make him so must be there, collected by him slowly, unconsciously, as are his bones, his flesh, and his blood. They cannot be put on like a garment for the nonce,—as may a little learning. A man cannot become faithful to his friends, unsuspicious before the world, gentle with women, loving with children, considerate to his inferiors, kindly with servants, tender-hearted with all,—and at the same time be frank, of open speech, with springing eager energies,—simply because he desires it. These things, which are the attributes of manliness, must come of training on a nature not ignoble. But they are the very opposites, the antipodes, the direct antagonism, of that staring, posed, bewhiskered and bewigged deportment, that *nil admirari*,* self-remembering assumption of manliness, that endeavour of twopence halfpenny to look as high as threepence, which, when you prod it through, has in it nothing deeper than deportment. We see the two things daily, side by side, close to each other. Let a man put his hat down, and you shall say whether he has deposited it with affectation or true nature. The natural man will probably be manly. The affected man cannot be so.

Mrs. Low was wrong when she accused our hero of being unmanly. Had his imagination been less alert in looking into the minds of men, and in picturing to himself the thoughts of others in reference to the crime with which he had been charged, he would not now have shrunk from contact with his fellow-creatures as he did. But he could not pretend to be other than he was. During the period of his danger, when men had thought that he would be hung,—and when he himself had believed that it would be so,—he had borne himself bravely without any conscious effort. When he had confronted the whole Court with that steady courage which had excited Lord Chiltern's admiration, and had looked the Bench in the face as though he at least had no cause to quail, he had known nothing of what he was doing. His features had answered the helm from his heart, but had not been played upon by his intellect. And it was so with him now. The reaction had overcome him, and he could not bring himself to pretend that it was not so. The tears would come to his eyes, and he would shiver and shake like one struck by palsy.

Mr. Monk came to him often, and was all but forgiven for the apparent defection in his faith. 'I have made up my mind to one thing,' Phineas said to him at the end of the ten days.

'And what is the one thing?'

'I will give up my seat.'

'I do not see a shadow of a reason for it.'

'Nevertheless I will do it. Indeed, I have already written to Mr. Ratler for the Hundreds. There may be and probably are men down at Tankerville who still think that I am guilty. There is an offensiveness in murder which degrades a man even by the accusation. I suppose it wouldn't do for you to move for the new writ.'

'Ratler will do it, as a matter of course. No doubt there will be expressions of great regret, and my belief is that they will return you again.'

'If so, they'll have to do it without my presence.'

Mr. Ratler did move for a new writ for the borough of Tankerville, and within a fortnight of his restoration to liberty Phineas Finn was no longer a Member of Parliament. It cannot be alleged that there was any reason for what he did, and yet the doing of it for the time rather increased than diminished his popularity. Both Mr. Gresham and Mr. Daubeny expressed their regret in the House, and Mr. Monk said a few words respecting his friend, which were very touching. He ended by expressing a hope that they soon might see him there again, and an opinion that he was a man peculiarly fitted by the tone of his mind, and the nature of his intellect, for the duties of Parliament.

Then at last, when all this had been settled, he went to Lord Brentford's house in Portman Square. He had promised that that should be the first house he would visit, and he was as good as his word. One evening he crept out, and walked slowly along Oxford Street, and knocked timidly at the door. As he did so he longed to be told that Lady Laura was not at home. But Lady Laura was at home,—as a matter of course. In those days she never went into society, and had not passed an evening away from her father's house since Mr. Kennedy's death. He was shown up into the drawing-room in which she sat, and there he found her—alone. 'Oh, Phineas, I am so glad you have come.'

'I have done as I said, you see.'

'I could not go to you when they told me that you were ill. You will have understood all that?'

'Yes; I understand.'

'People are so hard, and cold, and stiff, and cruel, that one can never do what one feels, oneself, to be right. So you have given up your seat.'

'Yes,—I am no longer a Member of Parliament.'

'Barrington says that they will certainly re-elect you.'

'We shall see. You may be sure at any rate of this,—that I shall never ask them to do so. Things seem to be so different now from what they did. I don't care for the seat. It all seems to be a bore and a trouble. What does it matter who sits in Parliament? The fight goes on just the same. The same falsehoods are acted. The same mock truths are spoken. The same wrong reasons are given. The same personal motives are at work.'

'And yet, of all believers in Parliament, you used to be the most faithful.'

'One has time to think of things, Lady Laura, when one lies in Newgate. It seems to me to be an eternity of time since they locked me up. And as for that trial, which they tell me lasted a week, I look back at it till the beginning is so distant that I can hardly remember it. But I have resolved that I will never talk of it again. Lady Chiltern is out probably.'

'Yes;—she and Oswald are dining with the Baldocks.'

'She is well?'

'Yes;—and most anxious to see you. Will you go to their place in September?'

He had almost made up his mind that if he went anywhere in September he would go to Matching Priory, accepting the offer of the Duchess of Omnium; but he did not dare to say so to Lady Laura, because she would have known that Madame Goesler also would be there. And he had not as yet accepted the invitation, and was still in doubt whether he would not escape by himself instead of attempting to return into the grooves of society. 'I think not;—I am hardly as yet sufficiently master of myself to know what I shall do.'

'They will be much disappointed.'

'And you?—what will you do?'

'I shall not go there. I am told that I ought to visit Loughlinter, and I suppose I shall. Oswald has promised to go down with me before the end of the month, but he will not remain above a day or two.'

'And your father?'

'We shall leave him at Saulsby. I cannot look it all in the face yet. It is not possible that I should remain all alone in that great house. The people all around would hate and despise me. I think Violet will come down with me, but of course she cannot remain there. Oswald must go to Harrington because of the hunting. It has become the business of his life. And she must go with him.'

'You will return to Saulsby.'

'I cannot say. They seem to think that I should live at Loughlinter;—but I cannot live there alone.'

He soon took leave of her, and did so with no warmer expressions of regard on either side than have here been given. Then he crept back to his lodgings, and she sat weeping alone in her father's house. When he had come to her during her husband's lifetime at Dresden, or even when she had visited him at his prison, it had been better than this.

CHAPTER LXIX
The Duke's first cousin

OUR pages have lately been taken up almost exclusively with the troubles of Phineas Finn, and indeed have so far not unfairly represented the feelings and interest of people generally at the time. Not to have talked of Phineas Finn from the middle of May to the middle of July in that year would have exhibited great ignorance or a cynical disposition. But other things went on also. Moons waxed and waned; children were born; marriages were contracted; and the hopes and fears of the little world around did not come to an end because Phineas Finn was not to be hung. Among others who had interests of their own there was poor Adelaide Palliser, whom we last saw under the affliction of Mr. Spooner's love,—but who before that had encountered the much deeper affliction of a quarrel with her own lover. She had desired him to free her,—and he had gone. Indeed, as to his going at that moment there had been no alternative, as he considered himself to have been turned out of Lord Chiltern's house. The red-headed lord, in the fierceness of his defence of Miss Palliser, had told the lover that under such and such circumstances he could not be allowed to remain at Harrington Hall. Lord Chiltern had said something about 'his roof.' Now, when a host questions the propriety of a guest remaining under his roof, the guest is obliged to go. Gerard Maule had gone; and, having offended his sweetheart by a most impolite allusion to Boulogne, had been forced to go as a rejected lover. From that day to this he had done nothing,—not because he was contented with the lot assigned to him, for every morning, as he lay on his bed, which he usually did till twelve, he swore to himself that nothing should separate him from Adelaide Palliser,—but simply because to do nothing was customary

with him. 'What is a man to do?' he not unnaturally asked his friend Captain Boodle at the club. 'Let her out on the grass for a couple of months,' said Captain Boodle, 'and she'll come up as clean as a whistle. When they get these humours there's nothing like giving them a run.' Captain Boodle undoubtedly had the reputation of being very great in council on such matters; but it must not be supposed that Gerard Maule was contented to take his advice implicitly. He was unhappy, ill at ease, half conscious that he ought to do something, full of regrets,—but very idle.

In the meantime Miss Palliser, who had the finer nature

of the two, suffered grievously. The Spooner affair was but a small addition to her misfortune. She could get rid of Mr. Spooner,—of any number of Mr. Spooners; but how should she get back to her the man she loved? When young ladies quarrel with their lovers it is always presumed, especially in books, that they do not wish to get them back. It is to be understood that the loss to them is as nothing. Miss Smith begs that Mr. Jones may be assured that he is not to consider her at all. If he is pleased to separate, she will be at any rate quite as well pleased,—probably a great deal better. No doubt she had loved him with all her heart, but that will make no difference to her, if he wishes,—to be off. Upon the whole Miss Smith thinks that she would prefer such an arrangement, in spite of her heart. Adelaide Palliser had said something of the kind. As Gerard Maule had regarded her as a 'trouble', and had lamented that prospect of 'Boulogne' which marriage had presented to his eyes, she had dismissed him with a few easily spoken words. She had assured him that no such troubles need weigh upon him. No doubt they had been engaged;—but, as far as she was concerned, the remembrance of that need not embarrass him. And so she and Lord Chiltern between them had sent him away. But how was she to get him back again?

When she came to think it over, she acknowledged to herself that it would be all the world to her to have him back. To have him at all had been all the world to her. There had been nothing peculiarly heroic about him, nor had she ever regarded him as a hero. She had known his faults and weaknesses, and was probably aware that he was inferior to herself in character and intellect. But, nevertheless, she had loved him. To her he had been, though not heroic, sufficiently a man to win her heart. He was a gentleman, pleasant-mannered, pleasant to look at, pleasant to talk to, not educated in the high sense of the word, but never making himself ridiculous by ignorance. He was the very antipodes of a Spooner, and he was,—or rather had been,—her lover. She did not wish to change. She did not recognise the possibility of changing. Though she had told him that he might go if he pleased, to her his going would be the loss of everything. What would life be without a lover,—without the prospect of marriage? And there could be no other lover. There could be no further prospect should he take her at her word.

Of all this Lord Chiltern understood nothing, but Lady Chiltern understood it all. To his thinking the young man had behaved so badly that it was incumbent on them all to send him away and so have done with him. If the young man wanted to quarrel with any one, there was he to be quarrelled with. The thing was a trouble, and the sooner they got to the end of it the better. But Lady Chiltern understood more than that. She could not prevent the quarrel as it came,—or was coming; but she knew that 'the quarrel of lovers is the renewal of love.'*At any rate, the woman always desires that it may be so, and endeavours to reconcile the parted ones. 'You'll see him in London,' Lady Chiltern had said to her friend.

'I do not want to see him,' said Adelaide proudly.

'But he'll want to see you, and then,—after a time,—you'll want to see him. I don't believe in quarrels, you know.'

'It is better that we should part, Lady Chiltern, if marrying will cause him—dismay. I begin to feel that we are too poor to be married.'

'A great deal poorer people than you are married every day. Of course people can't be equally rich. You'll do very well if you'll only be patient, and not refuse to speak to him when he comes to you.' This was said at Harrington after Lady Chiltern had returned from her first journey up to London. That visit had been very short, and Miss Palliser had been left alone at the hall. We already know how Mr. Spooner took advantage of her solitude. After that, Miss Palliser was to accompany the Chilterns to London, and she was there with them when Phineas Finn was acquitted. By that time she had brought herself to acknowledge to her friend Lady Chiltern that it would perhaps be desirable that Mr. Maule should return. If he did not do so, and that at once, there must come an end to her life in England. She must go away to Italy,—

altogether beyond the reach of Gerard Maule. In such case all the world would have collapsed for her, and she would become the martyr of a shipwreck. And yet the more that she confessed to herself that she loved the man so well that she could not part with him, the more angry she was with him for having told her that, when married, they must live at Boulogne.

The house in Portman Square had been practically given up by Lord Brentford to his son; but nevertheless the old Earl and Lady Laura had returned to it when they reached England from Dresden. It was, however, large, and now the two families,—if the Earl and his daughter can be called a family, —were lodging there together. The Earl troubled them but little, living mostly in his own rooms, and Lady Laura never went out with them. But there was something in the presence of the old man and the widow which prevented the house from being gay as it might have been. There were no parties in Portman Square. Now and then a few old friends dined there; but at the present moment Gerard Maule could not be admitted as an old friend. When Adelaide had been a fortnight in London she had not as yet seen Gerard Maule or heard a word from him. She had been to balls and concerts, to dinner parties and the play; but no one had as yet brought them together. She did know that he was in town. She was able to obtain so much information of him as that. But he never came to Portman Square, and had evidently concluded that the quarrel—was to be a quarrel.

Among other balls in London that July there had been one at the Duchess of Omnium's. This had been given after the acquittal of Phineas Finn, though fixed before that great era. 'Nothing on earth should have made me have it while he was in prison,' the Duchess had said. But Phineas was acquitted, and cakes and ale again became permissible. The ball had been given, and had been very grand. Phineas had been asked, but of course had not gone. Madame Goesler, who was a great heroine since her successful return from Prague, had shown herself there for a few minutes. Lady Chiltern had gone, and of course taken Adelaide. 'We are first cousins,' the Duke said to Miss Palliser,—for the Duke did steal a moment from his work in which to walk through his wife's drawing-room. Adelaide smiled and nodded, and looked pleased as she gave her hand to her great relative. 'I hope we shall see more of each other than we have done,' said the Duke. 'We have all been sadly divided, haven't we?' Then he said a word to his wife, expressing his opinion that Adelaide Palliser was a nice girl, and asking her to be civil to so near a relative.

The Duchess had heard all about Gerard Maule and the engagement. She always did hear all about everything. And on this evening she asked a question or two from Lady Chiltern. 'Do you know,' she said, 'I have an appointment to-morrow with your husband?'

'I did not know;—but I won't interfere to prevent it, now you are generous enough to tell me.'

'I wish you would, because I don't know what to say to him. He is to come about that horrid wood, where the foxes won't get themselves born and bred as foxes ought to do. How can I help it? I'd send down a whole Lying-in Hospital for the foxes if I thought that that would do any good.'

'Lord Chiltern thinks it's the shooting.'

'But where is a person to shoot if he mayn't shoot in his own woods? Not that the Duke cares about the shooting for himself. He could not hit a pheasant sitting on a haystack, and wouldn't know one if he saw it. And he'd rather that there wasn't such a thing as a pheasant in the world. He cares for nothing but farthings. But what is a man to do? Or, rather, what is a woman to do?—for he tells me that I must settle it.'

'Lord Chiltern says that Mr. Fothergill has the foxes destroyed. I suppose Mr. Fothergill may do as he pleases if the Duke gives him permission.'

'I hate Mr. Fothergill, if that'll do any good,' said the Duchess; 'and we wish we could get rid of him altogether. But that, you know, is impossible. When one has an old man on one's shoulders one never can get rid of him. He is my incubus; and then you see Trumpeton Wood is such a long way from us at Matching that I can't say I want the shooting for myself. And I never go to Gatherum if I can help it. Suppose we made out that the Duke wanted to let the shooting?'

'Lord Chiltern would take it at once.'

'But the Duke wouldn't really let it, you know. I'll lay awake at night and think about it. And now tell me about Adelaide Palliser. Is she to be married?'

'I hope so,—sooner or later.'

'There's a quarrel or something;—isn't there? She's the Duke's first cousin, and we should be so sorry that things shouldn't go pleasantly with her. And she's a very good-looking girl, too. Would she like to come down to Matching?'

'She has some idea of going back to Italy.'

'And leaving her lover behind her! Oh, dear, that will be very bad. She'd much better come to Matching, and then I'd ask the man to come too. Mr. Maud, isn't he?'

'Gerard Maule.'

'Ah, yes; Maule. If it's the kind of thing that ought to be, I'd manage it in a week. If you get a young man down into a country house, and there has been anything at all between them, I don't see how he is to escape. Isn't there some trouble about money?'

'They wouldn't be very rich, Duchess.'

'What a blessing for them! But then, perhaps, they'd be very poor.'

'They would be rather poor.'

'Which is not a blessing. Isn't there some proverb*about going safely in the middle? I'm sure it's true about money,—only perhaps you ought to be put a little beyond the middle. I don't know why Plantagenet shouldn't do something for her.'

As to this conversation Lady Chiltern said very little to Adelaide, but she did mention the proposed visit to Matching.

'The Duchess said nothing to me,' replied Adelaide, proudly.

'No; I don't suppose she had time. And then she is so very odd; sometimes taking no notice of one, and at others so very loving.'

'I hate that.'

'But with her it is neither impudence nor affectation. She says exactly what she thinks at the time, and she is always as good as her word. There are worse women than the Duchess.'

'I am sure I wouldn't like going to Matching,' said Adelaide.

Lady Chiltern was right in saying that the Duchess of Omnium was always as good as her word. On the next day, after that interview with Lord Chiltern about Mr. Fothergill and the foxes,—as to which no present further allusion need be made here,—she went to work and did learn a good deal about Gerard Maule and Miss Palliser. Something she learned from Lord Chiltern,—without any consciousness on his lordship's part, something from Madame Goesler, and something from the Baldock people. Before she went to bed on the second night she knew all about the quarrel, and all about the money. 'Plantagenet,' she said the next morning, 'what are you going to do about the Duke's legacy to Marie Goesler?'

'I can do nothing. She must take the things, of course.'

'She won't.'

'Then the jewels must remain packed up. I suppose they'll be sold at last for the legacy duty, and, when that's paid, the balance will belong to her.'

'But what about the money?'

'Of course it belongs to her.'

'Couldn't you give it to that girl who was here last night?'

'Give it to a girl!'

'Yes;—to your cousin. She's as poor as Job, and can't get married because she hasn't got any money. It's quite true; and I must say that if the Duke had looked after his own relations instead of leaving money to people who don't want it and won't have it, it would have been much better. Why shouldn't Adelaide Palliser have it?'

'How on earth should I give Adelaide Palliser what doesn't belong to me? If you choose to make her a present, you can, but such a sum as that would, I should say, be out of the question.'

The Duchess had achieved quite as much as she had anticipated. She knew her husband well, and was aware that she couldn't carry her point at once. To her mind it was 'all nonsense' his saying that the money was not his. If Madame Goesler wouldn't take it, it must be his; and nobody could make a woman take money if she did not choose. Adelaide Palliser was the Duke's first cousin, and it was intolerable that the Duke's first cousin should be unable to marry because she would have nothing to live upon. It became, at least, intolerable as soon as the Duchess had taken it into her head to like the first cousin. No doubt there were other first cousins as badly off, or perhaps worse, as to whom the Duchess would care nothing whether they were rich or poor,—married or single; but then they were first cousins who had not had the advantage of interesting the Duchess.

'My dear,' said the Duchess to her friend, Madame Goesler, 'you know all about those Maules?'

'What makes you ask?'

'But you do?'

'I know something about one of them,' said Madame Goesler. Now, as it happened, Mr. Maule, senior, had on that very day asked Madame Goesler to share her lot with his, and the request had been—almost indignantly, refused. The general theory that the wooing of widows should be quick had, perhaps, misled Mr. Maule. Perhaps he did not think that the wooing had been quick. He had visited Park Lane with the object of making his little proposition once before, and had then been stopped in his course by the consternation occasioned by the arrest of Phineas Finn. He had waited till Phineas had been acquitted, and had then resolved to try his luck. He had heard of the lady's journey to Prague, and was acquainted of course with those rumours which too freely connected the name of our hero with that of the lady. But rumours are often false, and a lady may go to Prague on a gentleman's behalf without intending to marry him. All the women in London were at present more or less in love with the man who had been accused of murder, and the fantasy of

Madame Goesler might be only as the fantasy of others. And then, rumour also said that Phineas Finn intended to marry Lady Laura Kennedy. At any rate a man cannot have his head broken for asking a lady to marry him,—unless he is very awkward in the doing of it. So Mr. Maule made his little proposition.

'Mr. Maule,' said Madame, smiling, 'is not this rather sudden?' Mr. Maule admitted that it was sudden, but still persisted. 'I think, if you please, Mr. Maule, we will say no more about it,' said the lady, with that wicked smile still on her face. Mr. Maule declared that silence on the subject had become impossible to him. 'Then, Mr. Maule, I shall have to leave you to speak to the chairs and tables,' said Madame Goesler. No doubt she was used to the thing, and knew how to conduct herself well. He also had been refused before by ladies of wealth, but had never been treated with so little consideration. She had risen from her chair as though about to leave the room, but was slow in her movement, showing him that she thought it was well for him to leave it instead of her. Muttering some words, half of apology and half of self-assertion, he did leave the room; and now she told the Duchess that she knew something of one of the Maules.

'That is, the father?'

'Yes,—the father.'

'He is one of your tribe, I know. We met him at your house just before the murder. I don't much admire your taste, my dear, because he's a hundred and fifty years old;—and what there is of him comes chiefly from the tailor.'

'He's as good as any other old man.'

'I dare say,—and I hope Mr. Finn will like his society. But he has got a son.'

'So he tells me.'

'Who is a charming young man.'

'He never told me that, Duchess.'

'I dare say not. Men of that sort are always jealous of their sons. But he has. Now I am going to tell you something and ask you to do something.'

'What was it the French Minister said. If it is simply difficult it is done. If it is impossible, it shall be done.'*

'The easiest thing in the world. You saw Plantagenet's first cousin the other night,—Adelaide Palliser. She is engaged to marry young Mr. Maule, and they neither of them have a shilling in the world. I want you to give them five-and-twenty thousand pounds.'

'Wouldn't that be peculiar?'

'Not in the least.'

'At any rate it would be inconvenient.'

'No it wouldn't, my dear. It would be the most convenient thing in the world. Of course I don't mean out of your pocket. There's the Duke's legacy.'

'It isn't mine, and never will be.'

'But Plantagenet says it never can be anybody else's. If I can get him to agree, will you? Of course there will be ever so many papers to be signed; and the biggest of all robbers, the Chancellor of the Exchequer, will put his fingers into the pudding and pull out a plum, and the lawyers will take more plums. But that will be nothing to us. The pudding will be very nice for them let ever so many plums be taken. The lawyers and people will do it all, and then it will be her fortune,—just as though her uncle had left it to her. As it is now, the money will never be of any use to anybody.' Madame Goesler said that if the Duke consented she also would consent. It was immaterial to her who had the money. If by signing any receipt she could facilitate the return of the money to any one of the Duke's family, she would willingly sign it. But Miss Palliser must be made to understand that the money did not come to her as a present from Madame Goesler.

'But it will be a present from Madame Goesler,' said the Duke.

'Plantagenet, if you go and upset everything by saying that, I shall think it most ill-natured. Bother about true! Somebody must have the money. There's nothing illegal about it.' And the Duchess had her own way. Lawyers were consulted, and documents were prepared, and the whole thing was arranged. Only Adelaide Palliser knew nothing about it, nor did Gerard Maule; and the quarrels of lovers had not yet become the renewal of love. Then the Duchess wrote the two following notes:—

'My dear Adelaide,

'We shall hope to see you at Matching on the 15th of August. The Duke, as head of the family, expects implicit obedience. You'll meet fifteen young gentlemen from the Treasury and the Board of Trade, but they won't incommode you, as they are kept at work all day. We hope Mr. Finn will be with us, and there isn't a lady in England who wouldn't give her eyes to meet him. We shall stay ever so many weeks at Matching, so that you can do as you please as to the time of leaving us.

'Yours affectionately,

'G. O.

'Tell Lord Chiltern that I have my hopes of making Trumpeton Wood too hot for Mr. Fothergill,—but I have to act with the greatest caution. In the meantime I am sending down dozens of young foxes, all labelled Trumpeton Wood, so that he shall know them.'

The other was a card rather than a note. The Duke and Duchess of Omnium presented their compliments to Mr. Gerard Maule, and requested the honour of his company to dinner on,—a certain day named. When Gerard Maule received this card at his club he was rather surprised, as he had never made the acquaintance either of the Duke or the Duchess. But the Duke was the first cousin of Adelaide Palliser, and of course he accepted the invitation.

CHAPTER LXX

'I will not go to Loughlinter'

THE end of July came, and it was settled that Lady Laura Kennedy should go to Loughlinter. She had been a widow now for nearly three months, and it was thought right that she should go down and see the house, and the lands, and the dependents whom her husband had left in her charge. It was now three years since she had seen Loughlinter, and when last she had left it, she had made up her mind that she would never place her foot upon the place again. Her wretchedness had all come upon her there. It was there that she had first been subjected to the unendurable tedium of Sabbath Day observances. It was there she had been instructed in the unpalatable duties that had been expected from her. It was there that she had been punished with the doctor from Callender whenever she attempted escape under the plea of a headache. And it was there, standing by the waterfall, the noise of which could be heard from the front-door, that Phineas Finn had told her of his love. When she accepted the hand of Robert Kennedy she had known that she had not loved him; but from the moment in which Phineas had spoken to her, she knew well that her heart had gone one way, whereas her hand was to go another. From that moment her whole life had quickly become a blank. She had had no period of married happiness,—not a month, not an hour. From the moment in which the thing had been done she had found that the man to whom she had bound herself was odious to her, and that the life before her was distasteful to her. Things which before had seemed worthy to her, and full at any rate of interest, became at once dull and vapid. Her husband was in Parliament, as also had been her father, and many of her friends,—and, by weight of his own character and her influence, was himself placed high in office; but in his house politics lost all the flavour which they had possessed for her in Portman Square. She had thought that she could at any rate do her duty as the mistress of a great household, and as the benevolent lady of a great estate; but household duties under the tutelage of Mr. Kennedy had been impossible to her, and that part of a Scotch Lady Bountiful which she had intended to play seemed to be denied to her. The whole structure had fallen to the ground, and nothing had been left to her.

But she would not sin. Though she could not bring herself to love her husband, she would at any rate be strong enough to get rid of that other love. Having so resolved, she became as weak as water. She at one time determined to be the guiding genius of the man she loved,—a sort of devoted elder sister, intending him to be the intimate friend of her husband; then she had told him not to come to her house, and had been weak enough to let him know why it was that she could not bear his presence. She had failed altogether to keep her secret, and her life during the struggle had become so intolerable to her that she had found herself compelled to desert her husband. He had shown her that he, too, had discovered the truth, and then she had become indignant, and had left him. Every place that she had inhabited with him had become disagreeable to her. The house in London had been so odious, that she had asked her intimate friends to come to her in that occupied by her father. But, of all spots upon earth, Loughlinter had been the most distasteful to her. It was there that the sermons had been the longest, the lessons in accounts the most obstinate, the lectures the most persevering, the dullness the most heavy. It was there that her ears had learned the sound of the wheels of Dr. Macnuthrie's gig. It was there that her spirit had been nearly broken. It was there that, with spirit not broken, she had determined to face all that the world might say of her, and fly from a tyranny which was insupportable. And now the place was her own, and she was told that she must go there as its owner;—go there and be potential, and beneficent, and grandly bland with persons, all of whom knew what had been the relations between her and her husband.

And though she had been indignant with her husband when at last she had left him,—throwing it in his teeth as an unmanly offence that he had accused her of the truth; though she had felt him to be a tyrant and herself to be a thrall; though the sermons, and the lessons, and the doctor had each, severally, seemed to her to be horrible cruelties; yet she had known through it all that the fault had been hers, and not his. He only did that which she should have expected when she married him;—but she had done none of that which he was entitled to expect from her. The real fault, the deceit, the fraud,—the sin had been with her,—and she knew it. Her life had been destroyed,—but not by him. His life had also been destroyed, and she had done it. Now he was gone, and she knew that his people,—the old mother who was still left alone, his cousins, and the tenants who were now to be her tenants, all said that had she done her duty by him he would still have been alive. And they must hate her the worse, because she had never sinned after such a fashion as to liberate him from his bond to her. With a husband's perfect faith in his wife, he had, immediately after his marriage, given to her for her life the lordship over his people, should he be without a child and should she survive him. In his hottest anger he had not altered that. His constant demand had been that she should come back to him, and be his real wife. And while making that demand,—with a persistency which had driven him mad,—he had died; and now the place was hers, and they told her that she must go and live there!

It is a very sad thing for any human being to have to say to himself,—with an earnest belief in his own assertion,—that all the joy of this world is over for him; and is the sadder because such conviction is apt to exclude the hope of other joy. This woman had said so to herself very often during the last two years, and had certainly been sincere. What was there in store for her? She was banished from the society of all those she liked. She bore a name that was hateful to her. She loved a man whom she could never see. She was troubled about money. Nothing in life had any taste for her. All the joys of the world were over,—and had been lost by her own fault. Then Phineas Finn had come to her at Dresden, and now her husband was dead!

Could it be that she was entitled to hope that the sun might rise again for her once more and another day be reopened for her with a gorgeous morning? She was now rich and still young,—or young enough. She was two and thirty, and had known many women,—women still honoured with the name of girls,—who had commenced the world successfully at that age. And this man had loved her once. He had told her so, and had afterwards kissed her when informed of her own engagement. How well she remembered it all. He, too, had gone through vicissitudes in life, had married and retired out of the world, had returned to it, and had gone through fire and water. But now everybody was saying good things of him, and all he wanted was the splendour which wealth would give him. Why should he not take it at her hands, and why should not the world begin again for both of them?

But though she would dream that it might be so, she was quite sure that there was no such life in store for her. The nature of the man was too well known to her. Fickle he might be;—or rather capable of change than fickle; but he was incapable of pretending to love when he did not love. She felt that in all the moments in which he had been most tender with her. When she had endeavoured to explain to him the state of her feelings at Königstein,—meaning to be true in what she said, but not having been even then true throughout, —she had acknowledged to herself that at every word he spoke she was wounded by his coldness. Had he then professed a passion for her she would have rebuked him, and told him that he must go from her,—but it would have warmed the blood in all her veins, and brought back to her a sense of youthful life. It had been the same when she visited him in the prison;—the same again when he came to her after his acquittal. She had been frank enough to him, but he would not even pretend that he loved her. His gratitude, his friendship, his services, were all hers. In every respect he had behaved well to her. All his troubles had come upon him because he would not desert her cause,—but he would never again say he loved her.

She gazed at herself in the glass, putting aside for the moment the hideous widow's cap which she now wore, and told herself that it was natural that it should be so. Though she was young in years her features were hard and worn with care. She had never thought herself to be a beauty, though she had been conscious of a certain aristocratic grace of manner which might stand in the place of beauty. As she examined herself she found that that was not all gone;—but she now lacked that roundness of youth which had been hers when first she knew Phineas Finn. She sat opposite the mirror, and pored over her own features with an almost skilful scrutiny, and told herself at last aloud that she had become an old woman. He was in the prime of life; but for her was left nothing but its dregs.

She was to go to Loughlinter with her brother and her brother's wife, leaving her father at Saulsby on the way. The Chilterns were to remain with her for one week, and no more. His presence was demanded in the Brake country, and it was with difficulty that he had been induced to give her so much of his time. But what was she to do when they should leave her? How could she live alone in that great house, thinking, as she ever must think, of all that had happened to her there? It seemed to her that everybody near to her was cruel in demanding from her such a sacrifice of her comfort. Her father had shuddered when she had proposed to him to accompany her to Loughlinter; but her father was one of those who insisted on the propriety of her going there. Then, in spite of that lesson which she had taught herself while sitting opposite to the glass, she allowed her fancy to revel in the idea of having him with her as she wandered over the braes. She saw him a day or two before her journey, when she told him her plans as she might tell them to any friend. Lady Chiltern and her father had been present, and there had been no special sign in her outward manner of the mingled tenderness and soreness of her heart within. No allusion had been made to any visit from him to the North. She would not have dared to suggest it in the presence of her brother, and was almost as much cowed by her brother's wife. But when she was alone, on the eve of her departure, she wrote to him as follows:—

'Sunday, 1st *August*——

'DEAR FRIEND,

'I thought that perhaps you might have come in this afternoon, and I have not left the house all day. I was so wretched that I could not go to church in the morning;—and when the afternoon came, I preferred the chance of seeing you to going out with Violet. We two were alone all the evening, and I did not give you up till nearly ten. I dare say you were right not to come. I should only have bored you with my complaints, and have grumbled to you of evils which you cannot cure.

'We start at nine to-morrow, and get to Saulsby in the afternoon. Such a family party as we shall be! I did fancy that Oswald would escape it; but, like everybody else, he has changed,—and has become domestic and dutiful. Not but that he is as tyrannous as ever; but his tyranny is now that of the responsible father of a family. Papa cannot understand him at all, and is dreadfully afraid of him. We stay two nights at Saulsby, and then go on to Scotland, leaving papa at home.

Of course it is very good in Violet and Oswald to come with me,—if, as they say, it be necessary for me to go at all. As to living there by myself, it seems to me to be impossible. You know the place well, and can you imagine me there all alone, surrounded by Scotch men and women, who, of course, must hate and despise me, afraid of every face that I see, and reminded even by the chairs and tables of all that is past? I have told papa that I know I shall be back at Saulsby before the middle of the month. He frets, and says nothing; but he tells Violet, and then she lectures me in that wise way of hers which enables her to say such hard things with so much seeming tenderness. She asks me why I do not take a companion with me, as I am so much afraid of solitude. Where on earth should I find a companion who would not be worse than solitude? I do feel now that I have mistaken life in having so little used myself to the small resources of feminine companionship. I love Violet dearly, and I used to be always happy in her society. But even with her now I feel but a half sympathy. That girl that she has with her is more to her than I am, because after the first half-hour I grow tired about her babies. I have never known any other woman with whom I cared to be alone. How then shall I content myself with a companion, hired by the quarter, perhaps from some advertisement in a newspaper?

'No companionship of any kind seems possible to me,—and yet never was a human being more weary of herself. I sometimes wonder whether I could go again and sit in that cage in the House of Commons*to hear you and other men speak,—as I used to do. I do not believe that any eloquence in the world would make it endurable to me. I hardly care who is in or out, and do not understand the things which my cousin Barrington tells me,—so long does it seem since I was in the midst of them all. Not but that I am intensely anxious that you should be back. They tell me that you will certainly be re-elected this week, and that all the House will receive you with open arms. I should have liked, had it been possible, to be once more in the cage to see that. But I am such a coward that I did not even dare to propose to stay for it. Violet would have told me that such manifestation of interest was unfit for my condition as a widow. But in truth, Phineas, there is nothing else now that does interest me. If, looking on from a distance, I can see you succeed, I shall try once more to care for the questions of the day. When you have succeeded, as I know you will, it will be some consolation to me to think that I also helped a little.

'I suppose I must not ask you to come to Loughlinter? But you will know best. If you will do so I shall care nothing for what any one may say. Oswald hardly mentions your name in my hearing, and of course I know of what he is thinking. When I am with him I am afraid of him, because it would add infinitely to my grief were I driven to quarrel with him; but I am my own mistress as much as he is his own master, and I will not regulate my conduct by his wishes. If you please to come you will be welcome as the flowers in May. Ah, how weak are such words in giving any idea of the joy with which I should see you!

'God bless you, Phineas.

'Your most affectionate friend,

'LAURA KENNEDY.

'Write to me at Loughlinter. I shall long to hear that you have taken your seat immediately on your re-election. Pray do not lose a day. I am sure that all your friends will advise you as I do.'

Throughout her whole letter she was struggling to tell him once again of her love, and yet to do it in some way of which she need not be ashamed. It was not till she had come to the last words that she could force her pen to speak of her affection, and then the words did not come freely as she would have had them. She knew that he would not come to Loughlinter. She felt that were he to do so he could come only as a suitor for her hand, and that such a suit, in these early days of her widowhood, carried on in her late husband's house, would be held to be disgraceful. As regarded herself, she would have faced all that for the sake of the thing to be attained. But she knew that he would not come. He had become wise by experience, and would perceive the result of such coming,—and would avoid it. His answer to her letter reached Loughlinter before she did:—

'Great Marlborough Street,

'Monday night.

'DEAR LADY LAURA,—

'I should have called in the Square last night, only that I feel that Lady Chiltern must be weary of the woes of so doleful a person as myself. I dined and spent the evening with the Lows, and was quite aware that I disgraced myself with them by being perpetually lachrymose. As a rule I do not think that I am more given than other people to talk of myself, but I am conscious of a certain incapability of getting rid of myself what has grown upon me since those weary weeks in Newgate and those frightful days in the dock; and this makes me unfit for society. Should I again have a seat in the House I shall be afraid to get up upon my legs, lest I should find myself talking of the time in which I stood before the judge with a halter round my neck.

'I sympathise with you perfectly in what you say about Loughlinter. It may be right that you should go there and show yourself,—so that those who knew the Kennedys in Scotland should not say that you had not dared to visit the place, but I do not think it possible that you should live there as yet. And why should you do so? I cannot conceive that your presence there should do good, unless you took delight in the place.

'I will not go to Loughlinter myself, although I know how warm would be my welcome.' When he had got so far with his letter he found the difficulty of going on with it to be almost insuperable. How could he give her any reasons for his not making the journey to Scotland? 'People would say that you and I should not be alone together after all the evil that has been spoken of us;—and would be specially eager in saying so were I now to visit you, so lately made a widow, and to sojourn with you in the house that did belong to your husband. Only think how eloquent would be the indignation of The People's Banner were it known that I was at Loughlinter.' Could he have spoken the truth openly, such were the reasons that he would have given; but it was impossible that such truths should be written by him in a letter to herself. And then it was almost equally difficult for him to tell her of a visit which he had resolved to make. But the letter must be completed, and at last the words were written. 'I could be of no real service to you there, as will be your brother and your brother's wife, even though their stay with you is to be so short. Were I you I would go out among the people as much as possible, even though they should not receive you cordially at first. Though we hear so much of clanship in the Highlands, I think the Highlanders are prone to cling to any one who has territorial authority among them. They thought a great deal of Mr. Kennedy, but they had never heard his name fifty years ago. I suppose you will return to Saulsby soon, and then, perhaps, I may be able to see you.

'In the meantime I am going to Matching.' This difficulty was worse even than the other. 'Both the Duke and Duchess have asked me, and I know that I am bound to make an effort to face my fellow-creatures again. The horror I feel at being stared at, as the man that was not—hung as a murderer, is stronger than I can describe; and I am well aware that I shall be talked to and made a wonder of on that ground. I am told that I am to be re-elected triumphantly at Tankerville without a penny of cost or the trouble of asking for a vote, simply because I didn't knock poor Mr. Bonteen on the head. This to me is abominable, but I cannot help myself, unless I resolve to go away and hide myself. That I know cannot be right, and therefore I had better go through it and have done with it. Though I am to be stared at, I shall not be stared at very long. Some other monster will come up and take my place, and I shall be the only person who will not forget it all. Therefore I have accepted the Duke's invitation, and shall go to Matching some time in the end of August. All the world is to be there.

'This re-election,—and I believe I shall be re-elected to-morrow,—would be altogether distasteful to me were it not that I feel that I should not allow myself to be cut to pieces by what has occurred. I shall hate to go back to the House, and have somehow learned to dislike and distrust all those things that used to be so fine and lively to me. I don't think that I believe any more in the party;—or rather in the men who lead it. I used to have a faith that now seems to me to be marvellous. Even twelve months ago, when I was beginning to think of standing for Tankerville, I believed that on our side the men were patriotic angels, and that Daubeny and his friends were all fiends or idiots,—mostly idiots, but with a strong dash of fiendism to control them. It has all come now to one common level of poor human interests. I doubt whether patriotism can stand the wear and tear and temptation of the front benches in the House of Commons. Men are flying at each other's throats, thrusting and parrying, making false accusations and defences equally false, lying and slandering,—sometimes picking and stealing,—till they themselves become unaware of the magnificence of their own position, and forget that they are expected to be great. Little tricks of sword-play engage all their skill. And the consequence is that there is no reverence now for any man in the House,—none of that feeling which we used to entertain for Mr. Mildmay.

'Of course I write—and feel—as a discontented man; and what I say to you I would not say to any other human being. I did long most anxiously for office, having made up my mind a second time to look to it as a profession. But I meant to earn my bread honestly, and give it up,—as I did before, when I could not keep it with a clear conscience. I knew that I was hustled out of the object of my poor ambition by that unfortunate man who has been hurried to his fate. In such a position I ought to distrust, and do, partly, distrust my own feelings. And I am aware that I have been soured by prison indignities. But still the conviction remains with me that parliamentary interests are not those battles of gods and giants which I used to regard them. Our Gyas*with the hundred hands is but a Three-fingered Jack,* and I sometimes think that we share our great Jove with the Strand Theatre.* Nevertheless I shall go back,—and if they will make me a joint lord to-morrow I shall be in heaven!

'I do not know why I should write all this to you except that there is no one else to whom I can say it. There is no one else who would give a moment of time to such lamentations. My friends will expect me to talk to them of my experiences in the dock rather than politics, and will want to know what rations I had in Newgate. I went to call on the Governor only yesterday, and visited the old room. "I never could really bring myself to think that you did it, Mr. Finn," he said. I looked at him and smiled, but I should have liked to fly at his throat. Why did he not know that the charge was a monstrous absurdity? Talking of that, not even you were truer to me than your brother. One expects it from a woman; —both the truth and the discernment.

'I have written to you a cruelly long letter; but when one's mind is full such relief is sometimes better than talking. Pray answer it before long, and let me know what you intend to do.

'Yours most affectionately,

'PHINEAS FINN.'

She did read the letter through,—read it probably more than once; but there was only one sentence in it that had for her any enduring interest. 'I will not go to Loughlinter myself.' Though she had known that he would not come her heart sank within her, as though now, at this moment, the really fatal wound had at last been inflicted. But, in truth, there was another sentence as a complement to the first, which rivetted the dagger in her bosom. 'In the meantime I am going to Matching.' Throughout his letter the name of that woman was not mentioned, but of course she would be there. The thing had all been arranged in order that they two might be brought together. She told herself that she had always hated that intriguing woman, Lady Glencora. She read the remainder of the letter and understood it; but she read it all in connection with the beauty, and the wealth, and the art,—and the cunning of Madame Max Goesler.

CHAPTER LXXI

Phineas Finn is re-elected

THE manner in which Phineas Finn was returned a second time for the borough of Tankerville was memorable among the annals of English elections. When the news reached the town that their member was to be tried for murder no doubt every elector believed that he was guilty. It is the natural assumption when the police and magistrates and lawyers, who have been at work upon the matter carefully, have come to that conclusion, and nothing but private knowledge or personal affection will stand against such evidence. At Tankerville there was nothing of either, and our hero's guilt was taken as a certainty. There was an interest felt in the whole matter which was full of excitement, and not altogether without delight to the Tankervillians. Of course the borough, as a borough, would never again hold up its head. There had never been known such an occurrence in the whole history of this country as the hanging of a member of the House of Commons. And this Member of Parliament was to be hung for murdering another member, which, no doubt, added much to the importance of the transaction. A large party in the borough declared that it was a judgment. Tankerville had degraded itself among boroughs by sending a Roman Catholic to Parliament, and had done so at the very moment in which the Church of England was being brought into danger. This was what had come upon the borough by not sticking to honest Mr. Browborough! There was a moment,—just before the trial was begun,—in which a large proportion of the electors was desirous of proceeding to work at once, and of sending Mr. Browborough back to his own place. It was thought that Phineas Finn should be made to resign. And very wise men in Tankerville were much surprised when they were told that a member of Parliament cannot resign his seat,—that when once returned he is supposed to be, as long as that Parliament shall endure, the absolute slave of his constituency and his country, and that he can escape from his servitude only by accepting some office under the Crown. Now it was held to be impossible that a man charged with murder should be appointed even to the stewardship of the Chiltern Hundreds. The House, no doubt, could expel a member, and would, as a matter of course, expel the member for Tankerville,—but the House could hardly proceed to expulsion before the member's guilt could have been absolutely established. So it came to pass that there was no escape for the borough from any part of the disgrace to which it had subjected itself by its unworthy choice, and some Tankervillians of sensitive minds were of opinion that no Tankervillian ever again ought to take part in politics.

Then, quite suddenly, there came into the borough the

tidings that Phineas Finn was an innocent man. This happened on the morning on which the three telegrams from Prague reached London. The news conveyed by the telegrams was at Tankerville almost as soon as in the Court at the Old Bailey, and was believed as readily. The name of the lady who had travelled all the way to Bohemia on behalf of their handsome young member was on the tongue of every woman in Tankerville, and a most delightful romance was composed. Some few Protestant spirits regretted the now assured escape of their Roman Catholic enemy, and would not even yet allow themselves to doubt that the whole murder had been arranged by Divine Providence to bring down the scarlet woman. It seemed to them to be so fitting a thing that Providence should interfere directly to punish a town in which the sins of the scarlet woman were not held to be abominable! But the multitude were soon convinced that their member was innocent; and as it was certain that he had been in great peril,—as it was known that he was still in durance, and as it was necessary that the trial should proceed, and that he should still stand at least for another day in the dock,—he became more than ever a hero. Then came the further delay, and at last the triumphant conclusion of the trial. When acquitted, Phineas Finn was still member for Tankerville and might have walked into the House on that very night. Instead of doing so he had at once asked for the accustomed means of escape from his servitude, and the seat for Tankerville was vacant. The most loving friends of Mr. Browborough perceived at once that there was not a chance for him. The borough was all but unanimous in resolving that it would return no one as its member but the man who had been unjustly accused of murder.

Mr. Ruddles was at once despatched to London with two other political spirits,—so that there might be a real deputation,—and waited upon Phineas two days after his release from prison. Ruddles was very anxious to carry his member back with him, assuring Phineas of an entry into the borough so triumphant that nothing like to it had ever been known at Tankerville. But to all this Phineas was quite deaf. At first he declined even to be put in nomination. 'You can't escape from it, Mr. Finn, you can't indeed,' said Ruddles. 'You don't at all understand the enthusiasm of the borough; does he, Mr. Gadmire?'

'I never knew anything like it in my life before,' said Gadmire.

'I believe Mr. Finn would poll two-thirds of the Church party to-morrow,' said Mr. Troddles, a leading dissenter in Tankerville, who on this occasion was the third member of the deputation.

'I needn't sit for the borough unless I please, I suppose,' pleaded Phineas.

'Well, no;—at least I don't know,' said Ruddles. 'It would be throwing us over a good deal, and I'm sure you are not the gentleman to do that. And then, Mr. Finn, don't you see that though you have been knocked about a little lately——'

'By George, he has,—most cruel,' said Troddles.

'You'll miss the House if you give it up; you will, after a bit, Mr. Finn. You've got to come round again, Mr. Finn, —if I may be so bold as to say so, and you shouldn't put yourself out of the way of coming round comfortably.'

Phineas knew that there was wisdom in the words of Mr. Ruddles, and consented. Though at this moment he was low in heart, disgusted with the world, and sick of humanity,—though every joint in his body was still sore from the rack on which he had been stretched, yet he knew that it would not be so with him always. As others recovered so would he, and it might be that he would live to 'miss the House,' should he now refuse the offer made to him. He accepted the offer, but he did so with a positive assurance that no consideration should at present take him to Tankerville.

'We ain't going to charge you, not one penny,' said Mr. Gadmire, with enthusiasm.

'I feel all that I owe to the borough,' said Phineas, 'and to the warm friends there who have espoused my cause; but I am not in a condition at present, either of mind or body, to put myself forward anywhere in public. I have suffered a great deal.'

'Most cruel!' said Troddles.

'And am quite willing to confess that I am therefore unfit in my present position to serve the borough.'

'We can't admit that,' said Gadmire, raising his left hand.

'We mean to have you,' said Troddles.

'There isn't a doubt about your re-election, Mr. Finn,' said Ruddles.

'I am very grateful, but I cannot be there. I must trust to one of you gentlemen to explain to the electors that in my present condition I am unable to visit the borough.'

Messrs. Ruddles, Gadmire, and Troddles returned to

Tankerville,—disappointed no doubt at not bringing with them him whose company would have made their feet glorious on the pavement of their native town,—but still with a comparative sense of their own importance in having seen the great sufferer whose woes forbade that he should be beheld by common eyes. They never even expressed an idea that he ought to have come, alluding even to their past convictions as to the futility of hoping for such a blessing; but spoke of him as a personage made almost sacred by the sufferings which he had been made to endure. As to the election, that would be a matter of course. He was proposed by Mr. Ruddles himself, and was absolutely seconded by the rector of Tankerville,—the staunchest Tory in the place, who on this occasion made a speech in which he declared that as an Englishman, loving justice, he could not allow any political or even any religious consideration to bias his conduct on this occasion. Mr. Finn had thrown up his seat under the pressure of a false accusation, and it was, the rector thought, for the honour of the borough that the seat should be restored to him. So Phineas Finn was re-elected for Tankerville without opposition and without expense; and for six weeks after the ceremony parcels were showered upon him by the ladies of the borough who sent him worked slippers, scarlet hunting waistcoats, pocket handkerchiefs, with 'P.F.' beautifully embroidered, and chains made of their own hair.

In this conjunction of affairs the editor of The People's Banner found it somewhat difficult to trim his sails. It was a rule of life with Mr. Quintus Slide to persecute an enemy. An enemy might at any time become a friend, but while an enemy was an enemy he should be trodden on and persecuted. Mr. Slide had striven more than once to make a friend of Phineas Finn; but Phineas Finn had been conceited and stiff-necked. Phineas had been to Mr. Slide an enemy of enemies, and by all his ideas of manliness, by all the rules of his life, by every principle which guided him, he was bound to persecute Phineas to the last. During the trial and the few weeks before the trial he had written various short articles with the view of declaring how improper it would be should a newspaper express any opinion of the guilt or innocence of a suspected person while under trial; and he gave two or three severe blows to contemporaries for having sinned in the matter; but in all these articles he had contrived to insinuate that the member for Tankerville, would, as a matter of course, be dealt with by the hands of justice. He had been very careful to recapitulate all circumstances which had induced Finn to hate the murdered man, and had more than once related the story of the firing of the pistol at Macpherson's Hotel. Then came the telegram from Prague, and for a day or two Mr. Slide was stricken dumb. The acquittal followed, and Quintus Slide had found himself compelled to join in the general satisfaction evinced at the escape of an innocent man. Then came the re-election for Tankerville, and Mr. Slide felt that there was opportunity for another reaction. More than enough had been done for Phineas Finn in allowing him to elude the gallows. There could certainly be no need for crowning him with a political chaplet because he had not murdered Mr. Bonteen. Among a few other remarks which Mr. Slide threw together, the following appeared in the columns of The People's Banner:—

'We must confess that we hardly understand the principle on which Mr. Finn has been re-elected for Tankerville with so much enthusiasm,—free of expense,—and without that usual compliment to the constituency which is implied by the personal appearance of the candidate. We have more than once expressed our belief that he was wrongly accused in the matter of Mr. Bonteen's murder. Indeed our readers will do us the justice to remember that, during the trial and before the trial, we were always anxious to allay the very strong feeling against Mr. Finn with which the public mind was then imbued, not only by the facts of the murder, but also by the previous conduct of that gentleman. But we cannot understand why the late member should be thought by the electors of Tankerville to be especially worthy of their confidence because he did not murder Mr. Bonteen. He himself, instigated, we hope, by a proper feeling, retired from Parliament as soon as he was acquitted. His career during the last twelve months has not enhanced his credit, and cannot, we should think, have increased his comfort. We ventured to suggest after that affair in Judd Street, as to which the police were so benignly inefficient, that it would not be for the welfare of the nation that a gentleman should be employed in the public service whose public life had been marked by the misfortune which had attended Mr. Finn. Great efforts were made by various ladies of the old Whig party to obtain official employment for him, but they were made in vain. Mr. Gresham was too wise, and our advice,—we will not say was followed,—but was found to agree with the decision of the Prime Minister. Mr. Finn was left out in the cold in spite of his great friends, —and then came the murder of Mr. Bonteen.

'Can it be that Mr. Finn's fitness for Parliamentary duties has been increased by Mr. Bonteen's unfortunate death, or by the fact that Mr. Bonteen was murdered by other hands than his own? We think not. The wretched husband, who, in the madness of jealousy, fired a pistol at this young man's head, has since died in his madness. Does that incident in the drama give Mr. Finn any special claim to consideration? We think not;—and we think also that the electors of Tankerville would have done better had they allowed Mr. Finn to return to that obscurity which he seems to have desired. The electors of Tankerville, however, are responsible only to their borough, and may do as they please with the seat in Parliament which is at their disposal. We may, however, protest against the employment of an unfit person in the service of his country,—simply because he has not committed a murder. We say so much now because rumours of an arrangement have reached our ears, which, should it come to pass,—would force upon us the extremely disagreeable duty of referring very forcibly to past circumstances, which may otherwise, perhaps, be allowed to be forgotten.'

CHAPTER LXXII

The End of the Story of Mr. Emilius and Lady Eustace

THE interest in the murder by no means came to an end when Phineas Finn was acquitted. The new facts which served so thoroughly to prove him innocent tended with almost equal weight to prove another man guilty. And the other man was already in custody on a charge which had subjected him to the peculiar ill-will of the British public. He a foreigner and a Jew, by name Yosef Mealyus,—as every one was now very careful to call him,—had come to England, had got himself to be ordained as a clergyman, had called himself Emilius, and had married a rich wife with a title, although he had a former wife still living in his own country. Had he called himself Jones it would have been better for him, but there was something in the name of Emilius which added a peculiar sting to his iniquities. It was now known that the bigamy could be certainly proved, and that his last victim,—our old friend, poor little Lizzie Eustace,—would be rescued from his clutches. She would once more be a free woman, and as she had been strong enough to defend her future income from his grasp, she was perhaps as fortunate as she deserved to be. She was still young and pretty, and there might come another lover more desirable than Yosef Mealyus. That the man would have to undergo the punishment of bigamy in its severest form, there was no doubt;—but would law, and justice, and the prevailing desire for revenge, be able to get at him in such a way that he might be hung? There certainly did exist a strong desire to prove Mr. Emilius to have been a murderer, so that there might come a fitting termination to his career in Great Britain.

The police seemed to think that they could make but little either of the coat or of the key, unless other evidence, that would be almost sufficient in itself, should be found. Lord Fawn was informed that his testimony would probably be required at another trial,—which intimation affected him so grievously that his friends for a week or two thought that he would altogether sink under his miseries. But he would say nothing which would seem to criminate Mealyus. A man hurrying along with a grey coat was all that he could swear to now,—professing himself to be altogether ignorant whether the man, as seen by him, had been tall or short. And then the manufacture of the key,—though it was that which made every one feel sure that Mealyus was the murderer,—did not, in truth, afford the slightest evidence against him. Even had it been proved that he had certainly used the false key and left Mrs. Meager's house on the night in question, that would not have sufficed at all to prove that therefore he had committed a murder in Berkeley Street. No doubt Mr. Bonteen had been his enemy,—and Mr. Bonteen had been murdered by an enemy. But so great had been the man's luck that no real evidence seemed to touch him. Nobody doubted; —but then but few had doubted before as to the guilt of Phineas Finn.

There was one other fact by which the truth might, it was hoped, still be reached. Mr. Bonteen had, of course, been killed by the weapon which had been found in the garden. As to that a general certainty prevailed. Mrs. Meager and Miss Meager, and the maid-of-all-work belonging to the Meagers, and even Lady Eustace, were examined as to this bludgeon. Had anything of the kind ever been seen in the possession of the clergyman? The clergyman had been so sly that nothing of the kind had been seen. Of the drawers and cupboards which he used, Mrs. Meager had always possessed duplicate keys, and Miss Meager frankly acknowledged that she had a general and fairly accurate acquaintance with the contents of these receptacles; but there had always been a big trunk with an impenetrable lock,—a lock which required that even if you had the key you should be acquainted with a certain combination of letters before you could open it,—and of that trunk no one had seen the inside. As a matter of course, the weapon, when brought to London, had been kept altogether hidden in the trunk. Nothing could be easier. But a man cannot be hung because he has had a secret hiding place in which a murderous weapon may have been stowed away.

But might it not be possible to trace the weapon? Mealyus, on his return from Prague, had certainly come through Paris. So much was learned,—and it was also learned as a certainty that the article was of French,—and probably of Parisian manufacture. If it could be proved that the man had bought this weapon, or even such a weapon, in Paris then,—so said all the police authorities,—it might be worth while to make an attempt to hang him. Men very skilful in unravelling such mysteries were sent to Paris, and the police of that capital entered upon the search with most praiseworthy zeal. But the number of life-preservers which had been sold altogether baffled them. It seemed that nothing was so common as that gentlemen should walk about with bludgeons in their pockets covered with leathern thongs. A young woman and an old man who thought that they could recollect something of a special sale were brought over,—and saw the splendour of London under very favourable circumstances;—but when confronted with Mr. Emilius, neither could venture to identify him. A large sum of money was expended,—no doubt justified by the high position which poor Mr. Bonteen had filled in the counsels of the nation; but it was expended in vain. Mr. Bonteen had been murdered in the streets at the West End of London. The murderer was known to everybody. He had been seen a minute or two before the murder. The motive which had induced the crime was apparent. The weapon with which it had been perpetrated had been found. The murderer's disguise had been discovered. The cunning with which he had endeavoured to prove that he was in bed at home had been unravelled, and the criminal purpose of his cunning made altogether manifest. Every man's eye could see the whole thing from the moment in which the murderer crept out of Mrs. Meager's house with Mr. Meager's coat upon his shoulders and the life-preserver in his pocket, till he was seen by Lord Fawn hurrying out of the mews to his prey. The blows from the bludgeon could be counted. The very moment in which they had been struck had been ascertained. His very act in hurling the weapon over the wall was all but seen. And yet nothing could be done. 'It is a very dangerous thing hanging a man on circumstantial evidence,' said Sir Gregory Grogram, who, a couple of months since, had felt almost sure that his honourable friend Phineas Finn would have to be hung on circumstantial evidence. The police and magistrates and lawyers all agreed that it would be useless, and indeed wrong, to send the case before a jury. But there had been quite sufficient evidence against Phineas Finn!

In the meantime the trial for bigamy proceeded in order that poor little Lizzie Eustace might be freed from the incubus which afflicted her. Before the end of July she was made once more a free woman, and the Rev. Joseph Emilius,—under which name it was thought proper that he should be tried,—was convicted and sentenced to penal servitude for five years. A very touching appeal was made for him to the jury by a learned serjeant, who declared that his client was to lose his wife and to be punished with extreme severity as a bigamist, because it was found to be impossible to bring home against him a charge of murder. There was, perhaps, some truth in what the learned serjeant said, but the truth had no effect upon the jury. Mr. Emilius was found guilty as quickly as Phineas Finn had been acquitted, and was, perhaps, treated with a severity which the single crime would hardly have elicited. But all this happened in the middle of the efforts which were being made to trace the purchase of the bludgeon, and when men hoped two or five or twenty-five years of threatened incarceration might be all the same to Mr. Emilius. Could they have succeeded in discovering where he had bought the weapon, his years of penal servitude would have afflicted him but little. They did not succeed; and though it cannot be said that any mystery was attached to the Bonteen murder, it has remained one of those crimes which are unavenged by the flagging law. And so the Rev. Mr. Emilius will pass away from our story.

There must be one or two words further respecting poor little Lizzie Eustace. She still had her income almost untouched, having been herself unable to squander it during her late married life, and having succeeded in saving it from the clutches of her pseudo husband. And she had her title, of which no one could rob her, and her castle down in Ayrshire, —which, however, as a place of residence she had learned to hate most thoroughly. Nor had she done anything which of itself must necessarily have put her out of the pale of society. As a married woman she had had no lovers; and, when a widow, very little fault in that line had been brought home against her. But the world at large seemed to be sick of her. Mrs. Bonteen had been her best friend, and, while it was still thought that Phineas Finn had committed the murder, with Mrs. Bonteen she had remained. But it was impossible that the arrangement should be continued when it became known,

—for it was known,—that Mr. Bonteen had been murdered by the man who was still Lizzie's reputed husband. Not that Lizzie perceived this,—though she was averse to the idea of her husband having been a murderer. But Mrs. Bonteen perceived it, and told her friend that she must—go. It was most unwillingly that the wretched widow changed her faith as to the murderer; but at last she found herself bound to believe as the world believed; and then she hinted to the wife of Mr. Emilius that she had better find another home.

'I don't believe it a bit,' said Lizzie.

'It is not a subject I can discuss,' said the widow.

'And I don't see that it makes any difference. He isn't my husband. You have said that yourself very often, Mrs. Bonteen.'

'It is better that we shouldn't be together, Lady Eustace.'

'Oh, I can go, of course, Mrs. Bonteen. There needn't be the slightest trouble about that. I had thought perhaps it might be convenient; but of course you know best.'

She went forth into lodgings in Half Moon Street, close to the scene of the murder, and was once more alone in the world. She had a child indeed, the son of her first husband, as to whom it behoved many to be anxious, who stood high in rank and high in repute; but such had been Lizzie's manner of life that neither her own relations nor those of her husband could put up with her, or endure her contact. And yet she was conscious of no special sins, and regarded herself as one who with a tender heart of her own, and a too-confiding spirit, had been much injured by the cruelty of those with whom she had been thrown. Now she was alone, weeping in solitude, pitying herself with deepest compassion; but it never occurred to her that there was anything in her conduct that she need alter. She would still continue to play her game as before, would still scheme, would still lie; and might still, at last, land herself in that Elysium of life of which she had been always dreaming. Poor Lizzie Eustace! Was it nature or education which had made it impossible to her to tell the truth, when a lie came to her hand? Lizzie, the liar! Poor Lizzie!

CHAPTER LXXIII

Phineas Finn returns to his Duties

THE election at Tankerville took place during the last week in July; and as Parliament was doomed to sit that year as late as the 10th of August, there was ample time for Phineas to present himself and take the oaths before the Session was finished. He had calculated that this could hardly be so when the matter of re-election was first proposed to him, and had hoped that his reappearance might be deferred till the following year. But there he was, once more member for Tankerville, while yet there was nearly a fortnight's work to be done, pressed by his friends, and told by one or two of those whom he most trusted, that he would neglect his duty and show himself to be a coward, if he abstained from taking his place. 'Coward is a hard word,' he said to Mr. Low, who had used it.

'So men think when this or that other man is accused of running away in battle or the like. Nobody will charge you with cowardice of that kind. But there is moral cowardice as well as physical.'

'As when a man lies. I am telling no lie.'

'But you are afraid to meet the eyes of your fellow-creatures.'

'Yes, I am. You may call me a coward if you like. What matters the name, if the charge be true? I have been so treated that I am afraid to meet the eyes of my fellow-creatures. I am like a man who has had his knees broken, or his arms cut off. Of course I cannot be the same afterwards as I was before.' Mr. Low said a great deal more to him on the subject, and all that Mr. Low said was true; but he was somewhat rough, and did not succeed. Barrington Erle and Lord Cantrip also tried their eloquence upon him; but it was Mr. Monk who at last drew from him a promise that he would go down to the House and be sworn in early on a certain Tuesday afternoon. 'I am quite sure of this,' Mr. Monk had said, 'that the sooner you do it the less will be the annoyance. Indeed there will be no trouble in the doing of it. The trouble is all in the anticipation, and is therefore only increased and prolonged by delay.' 'Of course it is your duty to go at once,' Mr. Monk had said again, when his friend argued that he had never undertaken to sit before the expiration of Parliament. 'You did consent to be put in nomination, and you owe your immediate services just as does any other member.'

'If a man's grandmother dies he is held to be exempted.'

'But your grandmother has not died, and your sorrow is not of the kind that requires or is supposed to require retirement.' He gave way at last, and on the Tuesday afternoon Mr. Monk called for him at Mrs. Bunce's house, and went down with him to Westminster. They reached their destination somewhat too soon, and walked the length of Westminster Hall two or three times while Phineas tried to justify himself. 'I don't think,' said he, 'that Low quite understands my position when he calls me a coward.'

'I am sure, Phineas, he did not mean to do that.'

'Do not suppose that I am angry with him. I owe him a great deal too much for that. He is one of the few friends I have who are entitled to say to me just what they please. But I think he mistakes the matter. When a man becomes crooked from age it is no good telling him to be straight. He'd be straight if he could. A man can't eat his dinner with a diseased liver as he could when he was well.'

'But he may follow advice as to getting his liver in order again.'

'And so am I following advice. But Low seems to think the disease shouldn't be there. The disease is there, and I can't banish it by simply saying that it is not there. If they had hung me outright it would be almost as reasonable to come and tell me afterwards to shake myself and be again alive. I don't think that Low realises what it is to stand in the dock for a week together, with the eyes of all men fixed on you, and a conviction at your heart that every one there believes you to have been guilty of an abominable crime of which you know yourself to have been innocent. For weeks I lived under the belief that I was to be made away by the hangman, and to leave behind me a name that would make every one who has known me shudder.'

'God in His mercy has delivered you from that.'

'He has;—and I am thankful. But my back is not strong enough to bear the weight without bending under it. Did you see Ratler going in? There is a man I dread. He is intimate enough with me to congratulate me, but not friend enough to abstain, and he will be sure to say something about his murdered colleague. Very well;—I'll follow you. Go up rather quick, and I'll come close after you.' Whereupon Mr. Monk entered between the two lamp-posts in the hall, and, hurrying along the passages, soon found himself at the door of the House. Phineas, with an effort at composure, and a smile that was almost ghastly at the door-keeper, who greeted him with some muttered word of recognition, held on his way close behind his friend, and walked up the House hardly conscious that the benches on each side were empty. There were not a dozen members present, and the Speaker had not as yet taken the chair. Mr. Monk stood by him while he took the oath, and in two minutes he was on a back seat below the gangway, with his friend by him, while the members, in slowly increasing numbers, took their seats. Then there were prayers, and as yet not a single man had spoken to him. As soon as the doors were again open gentlemen streamed in, and some few whom Phineas knew well came and sat near him. One or two shook hands with him, but no one said a word to him of the trial. No one at least did so in this early stage of the day's proceedings; and after half an hour he almost ceased to be afraid.

Then came up an irregular debate on the great Church question of the day, as to which there had been no cessation of the badgering with which Mr. Gresham had been attacked since he came into office. He had thrown out Mr. Daubeny by opposing that gentleman's stupendous measure for disestablishing the Church of England altogether, although,—as was almost daily asserted by Mr. Daubeny and his friends,—he was himself in favour of such total disestablishment. Over and over again Mr. Gresham had acknowledged that he was in favour of disestablishment, protesting that he had opposed Mr. Daubeny's Bill without any reference to its merits,—solely on the ground that such a measure should not be accepted from such a quarter. He had been stout enough, and, as his enemies had said, insolent enough, in making these assurances. But still he was accused of keeping his own hand dark, and of omitting to say what bill he would himself propose to bring in respecting the Church in the next Session. It was essentially necessary,—so said Mr. Daubeny and his friends,—that the country should know and discuss the proposed measure during the vacation. There was, of course, a good deal of retaliation. Mr. Daubeny had not given the country, or even his own party, much time to discuss his Church Bill. Mr. Gresham assured Mr. Daubeny that he would not feel himself equal to producing a measure that should change the religious position of every individual in the country, and annihilate the traditions and systems of centuries, altogether complete out of his own unaided brain; and he went on to say that were he to do so, he did not think that he should find himself supported in such an effort by the friends with whom he usually worked. On this occasion he declared that the magnitude of the subject and the immense importance of the interests concerned forbade him to anticipate the passing of any measure of general Church reform in the next Session. He was undoubtedly in favour of Church reform, but was by no means sure that the question was one which required immediate settlement. Of this he was sure,—that nothing in the way of legislative indiscretion could be so injurious to the country, as any attempt at a hasty and ill-considered measure on this most momentous of all questions.

The debate was irregular, as it originated with a question asked by one of Mr. Daubeny's supporters,—but it was allowed to proceed for a while. In answer to Mr. Gresham, Mr. Daubeny himself spoke, accusing Mr. Gresham of almost every known Parliamentary vice in having talked of a measure coming, like Minerva,* from his, Mr. Daubeny's, own brain. The plain and simple words by which such an accusation might naturally be refuted would be unparliamentary; but it would not be unparliamentary to say that it was reckless, unfounded, absurd, monstrous, and incredible. Then there were various very spirited references to Church matters, which concern us chiefly because Mr. Daubeny congratulated the House upon seeing a Roman Catholic gentleman with whom they were all well acquainted, and whose presence in the House was desired by each side alike, again take his seat for an English borough. And he hoped that he might at the same time take the liberty of congratulating that gentleman on the courage and manly dignity with which he had endured the unexampled hardships of the cruel position in which he had been placed by an untoward combination of circumstances. It was thought that Mr. Daubeny did the thing very well, and that he was right in doing it;—but during the doing of it poor Phineas winced in agony. Of course every member was looking at him, and every stranger in the galleries. He did not know at the moment whether it behoved him to rise and make some gesture to the House, or to say a word, or to keep his seat and make no sign. There was a general hum of approval, and the Prime Minister turned round and bowed graciously to the newly-sworn member. As he said afterwards, it was just this which he had feared. But there must surely have been something of consolation in the general respect with which he was treated. At the moment he behaved with natural instinctive dignity, though himself doubting the propriety of his own conduct. He said not a word, and made no sign, but sat with his eyes fixed upon the member from whom the compliment had come. Mr. Daubeny went on with his tirade, and was called violently to order. The Speaker declared that the whole debate had been irregular, but had been allowed by him in deference to what seemed to be the general will of the House. Then the two leaders of the two parties composed themselves, throwing off their indignation while they covered themselves well up with their hats,—and, in accordance with the order of the day, an honourable member rose to propose a pet measure of his own for preventing the adulteration of beer by the publicans. He had made a calculation that the annual average mortality of England would be reduced one and a half per cent., or in other words that every English subject born would live seven months longer if the action of the Legislature could provide that the publicans should sell the beer as it came from the brewers. Immediately there was such a rush of members to the door that not a word said by the philanthropic would-be purifier of the national beverage could be heard. The quarrels of rival Ministers were dear to the House, and as long as they could be continued the benches were crowded by gentlemen enthralled by the interest of the occasion. But to sink from that to private legislation about beer was to fall into a bathos which gentlemen could not endure; and so the House was emptied, and at about half-past seven there was a count-out. That gentleman whose statistics had been procured with so much care, and who had been at work for the last twelve months on his effort to prolong the lives of his fellow-countrymen, was almost broken-hearted. But he knew the world too well to complain. He would try again next year, if by dint of energetic perseverance he could procure a day.

Mr. Monk and Phineas Finn, behaving no better than the others, slipped out in the crowd. It had indeed been arranged that they should leave the House early, so that they might dine together at Mr. Monk's house. Though Phineas had been released from his prison now for nearly a month, he had not as yet once dined out of his own rooms. He had not been inside a club, and hardly ventured during the day into the streets about Pall Mall and Piccadilly. He had been frequently to Portman Square, but had not even seen Madame Goesler. Now he was to dine out for the first time; but there was to be no guest but himself.

'It wasn't so bad after all,' said Mr. Monk, when they were seated together.

'At any rate it has been done.'

'Yes;—and there will be no doing of it over again. I don't like Mr. Daubeny, as you know; but he is happy at that kind of thing.'

'I hate men who are what you call happy, but who are never in earnest,' said Phineas.

'He was earnest enough, I thought.'

'I don't mean about myself, Mr. Monk. I suppose he thought that it was suitable to the occasion that he should say something, and he said it neatly. But I hate men who can make capital out of occasions, who can be neat and appropriate at the spur of the moment,—having, however, probably had the benefit of some forethought,—but whose words never savour of truth. If I had happened to have been hung at this time,—as was so probable,—Mr. Daubeny would have devoted one of his half hours to the composition of a dozen tragic words which also would have been neat and appropriate. I can hear him say them now, warning young members around him to abstain from embittered words against each other, and I feel sure that the funereal grace of such an occasion would have become him even better than the generosity of his congratulations.'

'It is rather grim matter for joking, Phineas.'

'Grim enough; but the grimness and the jokes are always running through my mind together. I used to spend hours in thinking what my dear friends would say about it when they found that I had been hung in mistake;—how Sir Gregory Grogram would like it, and whether men would think about it as they went home from The Universe at night. I had various questions to ask and answer for myself,—whether they would pull up my poor body, for instance, from what unhallowed ground is used for gallows corpses, and give it decent burial, placing "M.P. for Tankerville" after my name on some more or less explicit tablet.'

'Mr. Daubeny's speech was, perhaps, preferable on the whole.'

'Perhaps it was;—though I used to feel assured that the

explicit tablet would be as clear to my eyes in purgatory as Mr. Daubeny's words have been to my ears this afternoon. I never for a moment doubted that the truth would be known before long,—but did doubt so very much whether it would be known in time. I'll go home now, Mr. Monk, and endeavour to get the matter off my mind. I will resolve, at any rate, that nothing shall make me talk about it any more.'

CHAPTER LXXIV

At Matching

For about a week in the August heat of a hot summer, Phineas attended Parliament with fair average punctuality, and then prepared for his journey down to Matching Priory. During that week he spoke no word to any one as to his past tribulation, and answered all allusions to it simply by a smile. He had determined to live exactly as though there had been no such episode in his life as that trial at the Old Bailey, and in most respects he did so. During this week he dined at the club, and called at Madame Goesler's house in Park Lane,—not, however, finding the lady at home. Once, and once only, did he break down. On the Wednesday evening he met Barrington Erle, and was asked by him to go to The Universe. At the moment he became very pale, but he at once said that he would go. Had Erle carried him off in a cab the adventure might have been successful; but as they walked, and as they went together through Clarges Street and Bolton Row and Curzon Street, and as the scenes which had been so frequently and so graphically described in Court appeared before him one after another, his heart gave way, and he couldn't do it. 'I know I'm a fool, Barrington; but if you don't mind I'll go home. Don't mind me, but just go on.' Then he turned and walked home, passing through the passage in which the murder had been committed.

'I brought him as far as the next street,' Barrington Erle

said to one of their friends at the club, 'but I couldn't get him in. I doubt if he'll ever be here again.'

It was past six o'clock in the evening when he reached Matching Priory. The Duchess had especially assured him that a brougham* should be waiting for him at the nearest station, and on arriving there he found that he had the brougham to himself. He had thought a great deal about it, and had endeavoured to make his calculations. He knew that Madame Goesler would be at Matching, and it would be necessary that he should say something of his thankfulness at their first meeting. But how should he meet her,—and in what way should he greet her when they met? Would any arrangement be made, or would all be left to chance? Should he go at once to his own chamber,—so as to show himself first when dressed for dinner, or should he allow himself to be taken into any of the morning rooms in which the other guests would be congregated? He had certainly not sufficiently considered the character of the Duchess when he imagined that she would allow these things to arrange themselves. She was one of those women whose minds were always engaged on such matters, and who are able to see how things will go. It must not be asserted of her that her delicacy was untainted, or her taste perfect; but she was clever,—discreet in the midst of indiscretions,—thoughtful, and good-natured. She had considered it all, arranged it all, and given her orders with accuracy. When Phineas entered the hall,—the brougham with the luggage having been taken round to some back door,—he was at once ushered by a silent man in black into the little sitting-room on the ground floor in which the old Duke used to take delight. Here he found two ladies,—but only two ladies,—waiting to receive him. The Duchess came forward to welcome him, while Madame Goesler remained in the background, with composed face,—as though she by no means expected his arrival and he had chanced to come upon them as she was standing by the window. He was thinking of her much more than of her companion, though he knew also how much he owed to the kindness of the Duchess. But what she had done for him had come from caprice, whereas the other had been instigated and guided by affection. He understood all that, and must have shown his feeling on his countenance. 'Yes, there she is,' said the Duchess, laughing. She had already told him that he was welcome to Matching, and had spoken some short word of congratulation at his safe deliverance from his troubles. 'If ever one friend was grateful to another, you should be grateful to her, Mr. Finn.' He did not speak, but walking across the room to the window by which Marie Goesler stood, took her right hand in his, and passing his left arm round her waist, kissed her first on one cheek and then on the other. The blood flew to her face and suffused her forehead, but she did not speak, or resist him or make any effort to escape from his embrace. As for him, he had no thought of it at all. He had made no plan. No idea of kissing her when they should meet had occurred to him till the moment came. 'Excellently well done,' said the Duchess, still laughing with silent pleasant laughter. 'And now tell us how you are, after all your troubles.'

He remained with them for half an hour, till the ladies went to dress, when he was handed over to some groom of the chambers to show him his room. 'The Duke ought to be here to welcome you, of course,' said the Duchess; 'but you know official matters too well to expect a President of the Board of Trade to do his domestic duties. We dine at eight; five minutes before that time he will begin adding up his last row of figures for the day. You never added up rows of figures, I think. You only managed colonies.' So they parted till dinner, and Phineas remembered how very little had been spoken by Madame Goesler, and how few of the words which he had spoken had been addressed to her. She had sat silent, smiling, radiant, very beautiful as he had thought, but contented to listen to her friend the Duchess. She, the Duchess, had asked questions of all sorts, and made many statements; and he had found that with those two women he could speak without discomfort, almost with pleasure, on subjects which he could not bear to have touched by men. 'Of course you knew all along who killed the poor man,' the Duchess had said. 'We did;—did we not, Marie?—just as well as if we had seen it. She was quite sure that he had got out of the house and back into it, and that he must have had a key. So she started off to Prague to find the key; and she found it. And we were quite sure too about the coat;—weren't we. That poor blundering Lord Fawn couldn't explain himself, but we knew that the coat he saw was quite different from any coat you would wear in such weather. We discussed it all over so often;—every point of it. Poor Lord Fawn! They say it has made quite an old man of him. And as for those policemen who didn't find the life-preserver; I only think that something ought to be done to them.'

'I hope that nothing will ever be done to anybody, Duchess.'

'Not to the Reverend Mr. Emilius;—poor dear Lady Eustace's Mr. Emilius? I do think that you ought to desire that an end should be put to his enterprising career! I'm sure I do.' This was said while the attempt was still being made to trace the purchase of the bludgeon in Paris. 'We've got Sir Gregory Grogram here on purpose to meet you, and you must fraternise with him immediately, to show that you bear no grudge.'

'He only did his duty.'

'Exactly;—though I think he was an addle-pated old ass not to see the thing more clearly. As you'll be coming into the Government before long, we thought that things had better be made straight between you and Sir Gregory. I wonder how it was that nobody but women did see it clearly? Look at that delightful woman, Mrs. Bunce. You must bring Mrs. Bunce to me some day,—or take me to her.'

'Lord Chiltern saw it clearly enough,' said Phineas.

'My dear Mr. Finn, Lord Chiltern is the best fellow in the world, but he has only one idea. He was quite sure of your innocence because you ride to hounds. If it had been found possible to accuse poor Mr. Fothergill, he would have been as certain that Mr. Fothergill committed the murder, because Mr. Fothergill thinks more of his shooting. However, Lord

Chiltern is to be here in a day or two, and I mean to go absolutely down on my knees to him,—and all for your sake. If foxes can be had, he shall have foxes. We must go and dress now, Mr. Finn, and I'll ring for somebody to show you your room.'

Phineas, as soon as he was alone, thought, not of what the Duchess had said, but of the manner in which he had greeted his friend, Madame Goesler. As he remembered what he had done, he also blushed. Had she been angry with him, and intended to show her anger by her silence? And why had he done it? What had he meant? He was quite sure that he would not have given those kisses had he and Madame Goesler been alone in the room together. The Duchess had applauded him, —but yet he thought that he regretted it. There had been matters between him and Marie Goesler of which he was quite sure that the Duchess knew nothing.

When he went downstairs he found a crowd in the drawing-room, from among whom the Duke came forward to welcome him. 'I am particularly happy to see you at Matching,' said the Duke. 'I wish we had shooting to offer you, but we are too far south for the grouse. That was a bitter passage of arms the other day, wasn't it? I am fond of bitterness in debate myself, but I do regret the roughness of the House of Commons. I must confess that I do.' The Duke did not say a word about the trial, and the Duke's guests followed their host's example.

The house was full of people, most of whom had before been known to Phineas, and many of whom had been asked specially to meet him. Lord and Lady Cantrip were there, and Mr. Monk, and Sir Gregory his accuser, and the Home Secretary, Sir Harry Coldfoot, with his wife. Sir Harry had at one time been very keen about hanging our hero, and was now of course hot with reactionary zeal. To all those who had been in any way concerned in the prosecution, the accidents by which Phineas had been enabled to escape had been almost as fortunate as to Phineas himself. Sir Gregory himself quite felt that had he prosecuted an innocent and very popular young

Member of Parliament to the death, he could never afterwards have hoped to wear his ermine in comfort. Barrington Erle was there, of course, intending, however, to return to the duties of his office on the following day,—and our old friend Laurence Fitzgibbon with a newly-married wife, a lady possessing a reputed fifty thousand pounds, by which it was hoped that the member for Mayo might be placed steadily upon his legs for ever. And Adelaide Palliser was there also, —the Duke's first cousin,—on whose behalf the Duchess was anxious to be more than ordinarily good-natured. Mr. Maule, Adelaide's rejected lover, had dined on one occasion with the Duke and Duchess in London. There had been nothing remarkable at the dinner, and he had not at all understood why he had been asked. But when he took his leave the Duchess had told him that she would hope to see him at Matching. 'We expect a friend of yours to be with us,' the Duchess had said. He had afterwards received a written invitation and had accepted it; but he was not to reach Matching till the day after that on which Phineas arrived. Adelaide had been told of his coming only on this morning, and had been much flurried by the news.

'But we have quarrelled,' she said. 'Then the best thing you can do is to make it up again, my dear,' said the Duchess. Miss Palliser was undoubtedly of that opinion herself, but she hardly believed that so terrible an evil as a quarrel with her lover could be composed by so rough a remedy as this. The Duchess, who had become used to all the disturbing excitements of life, and who didn't pay so much respect as some do to the niceties of a young lady's feelings, thought that it would be only necessary to bring the young people together again. If she could do that, and provide them with an income, of course they would marry. On the present occasion Phineas was told off to take Miss Palliser down to dinner. 'You saw the Chilterns before they left town, I know,' she said.

'Oh, yes. I am constantly in Portman Square.'

'Of course. Lady Laura has gone down to Scotland;—has she not;—and all alone?'

'She is alone now, I believe.'

'How dreadful! I do not know any one that I pity so much as I do her. I was in the house with her some time, and she gave me the idea of being the most unhappy woman I had ever met with. Don't you think that she is very unhappy?'

'She has had very much to make her so,' said Phineas. 'She was obliged to leave her husband because of the gloom of his insanity;—and now she is a widow.'

'I don't suppose she ever really—cared for him; did she?' The question was no sooner asked than the poor girl remembered the whole story which she had heard some time back,—the rumour of the husband's jealousy and of the wife's love, and she became as red as fire, and unable to help herself. She could think of no word to say, and confessed her confusion by her sudden silence.

Phineas saw it all, and did his best for her. 'I am sure she cared for him,' he said, 'though I do not think it was a well-assorted marriage. They had different ideas about religion, I fancy. So you saw the hunting in the Brake country to the end? How is our old friend, Mr. Spooner?'

'Don't talk of him, Mr. Finn.'

'I rather like Mr. Spooner;—and as for hunting the country, I don't think Chiltern could get on without him. What a capital fellow your cousin the Duke is.'

'I hardly know him.'

'He is such a gentleman;—and, at the same time, the most abstract and the most concrete man that I know.'

'Abstract and concrete!'

'You are bound to use adjectives of that sort now, Miss Palliser, if you mean to be anybody in conversation.'

'But how is my cousin concrete? He is always abstracted when I speak to him, I know.'

'No Englishman whom I have met is so broadly and intuitively and unceremoniously imbued with the simplicity of the character of a gentleman. He could no more lie than he could eat grass.'

'Is that abstract or concrete?'

'That's abstract. And I know no one who is so capable of throwing himself into one matter for the sake of accomplishing that one thing at a time. That's concrete.' And so the red colour faded away from poor Adelaide's face, and the unpleasantness was removed.

'What do you think of Laurence's wife?' Erle said to him late in the evening.

'I have only just seen her. The money is there, I suppose.'

'The money is there, I believe; but then it will have to remain there. He can't touch it. There's about £2,000 a-year, which will have to go back to her family unless they have children.'

'I suppose she's—forty?'

'Well; yes, or perhaps forty-five. You were locked up at the time, poor fellow,—and had other things to think of; but all the interest we had for anything beyond you through May and June was devoted to Laurence and his prospects. It was off and on, and on and off, and he was in a most wretched condition. At last she wouldn't consent unless she was to be asked here.'

'And who managed it?'

'Laurence came and told it all to the Duchess, and she gave him the invitation at once.'

'Who told you?'

'Not the Duchess,—nor yet Laurence. So it may be untrue, you know;—but I believe it. He did ask me whether he'd have to stand another election at his marriage. He has been going in and out of office so often, and always going back to the Co. Mayo at the expense of half a year's salary, that his mind had got confused, and he didn't quite know what did and what did not vacate his seat. We must all come to it sooner or later, I suppose, but the question is whether we could do better than an annuity of £2,000 a year on the life of the lady. Office isn't very permanent, but one has not to attend the House above six months a year, while you can't get away from a wife much above a week at a time. It has crippled him in appearance very much, I think.'

'A man always looks changed when he's married.'

'I hope, Mr. Finn, that you owe me no grudge,' said Sir Gregory, the Attorney-General.

'Not in the least; why should I?'

'It was a very painful duty that I had to perform,—the most painful that ever befel me. I had no alternative but to do it, of course, and to do it in the hope of reaching the truth. But a counsel for the prosecution must always appear to the accused and his friends like a hound running down his game, and anxious for blood. The habitual and almost necessary acrimony of the defence creates acrimony in the attack. If you were accustomed as I am to criminal courts you would observe this constantly. A gentleman gets up and declares in perfect faith that he is simply anxious to lay before the jury such evidence as has been placed in his hands. And he opens his case in that spirit. Then his witnesses are cross-examined with the affected incredulity and assumed indignation which the defending counsel is almost bound to use on behalf of his client, and he finds himself gradually imbued with pugnacity. He becomes strenuous, energetic, and perhaps eager for what must after all be regarded as success, and at last he fights for a verdict rather than for the truth.'

'The judge, I suppose, ought to put all that right?'

'So he does;—and it comes right. Our criminal practice does not sin on the side of severity. But a barrister employed on the prosecution should keep himself free from that personal desire for a verdict which must animate those engaged on the defence.'

'Then I suppose you wanted to—hang me, Sir Gregory.'

'Certainly not. I wanted the truth. But you in your position must have regarded me as a bloodhound.'

'I did not. As far as I can analyse my own feelings, I entertained anger only against those who, though they knew me well, thought that I was guilty.'

'You will allow me, at any rate, to shake hands with you,' said Sir Gregory, 'and to assure you that I should have lived a broken-hearted man if the truth had been known too late.

As it is I tremble and shake in my shoes as I walk about and think of what might have been done.' Then Phineas gave his hand to Sir Gregory, and from that time forth was inclined to think well of Sir Gregory.

Throughout the whole evening he was unable to speak to Madame Goesler, but to the other people around him he found himself talking quite at his ease, as though nothing peculiar had happened to him. Almost everybody, except the Duke, made some slight allusion to his adventure, and he, in spite of his resolution to the contrary, found himself driven to talk of it. It had seemed quite natural that Sir Gregory,—who had in truth been eager for his condemnation, thinking him to have been guilty,—should come to him and make peace with him by telling him of the nature of the work that had been imposed upon him;—and when Sir Harry Coldfoot assured him that never in his life had his mind been relieved of so heavy a weight as when he received the information about the key,—that also was natural. A few days ago he had thought that these allusions would kill him. The prospect of them had kept him a prisoner in his lodgings; but now he smiled and chatted, and was quiet and at ease.

'Good-night, Mr. Finn,' the Duchess said to him, 'I know the people have been boring you.'

'Not in the least.'

'I saw Sir Gregory at it, and I can guess what Sir Gregory was talking about.'

'I like Sir Gregory, Duchess.'

'That shows a very Christian disposition on your part. And then there was Sir Harry. I understood it all, but I could not hinder it. But it had to be done, hadn't it?—And now there will be an end of it.'

'Everybody has treated me very well,' said Phineas, almost in tears. 'Some people have been so kind to me that I cannot understand why it should have been so.'

'Because some people are your very excellent good friends. We,—that is, Marie and I, you know,—thought it would be the best thing for you to come down and get through it all here. We could see that you weren't driven too hard. By the bye, you have hardly seen her,—have you?'

'Hardly, since I was upstairs with your Grace.'

'My Grace will manage better for you to-morrow. I didn't like to tell you to take her out to dinner, because it would have looked a little particular after her very remarkable journey to Prague. If you ain't grateful you must be a wretch.'

'But I am grateful.'

'Well; we shall see. Good-night. You'll find a lot of men going to smoke somewhere, I don't doubt.'

CHAPTER LXXV

The Trumpeton Feud is Settled

IN these fine early autumn days spent at Matching, the great Trumpeton Wood question was at last settled. During the summer considerable acerbity had been added to the matter by certain articles which had appeared in certain sporting papers, in which the new Duke of Omnium was accused of neglecting his duty to the county in which a portion of his property lay. The question was argued at considerable length. Is a landed proprietor bound, or is he not, to keep foxes for the amusement of his neighbours? To ordinary thinkers, to unprejudiced outsiders,—to Americans, let us say, or Frenchmen,—there does not seem to be room even for an argument. By what law of God or man can a man be bound to maintain a parcel of injurious vermin on his property, in the pursuit of which he finds no sport himself, and which are highly detrimental to another sport in which he takes, perhaps, the keenest interest? Trumpeton Wood was the Duke's own,—to do just as he pleased with it. Why should foxes be demanded from him then any more than a bear to be baited, or a badger to be drawn, in, let us say, his London dining-room? But a good deal had been said which, though not perhaps capable of convincing the unprejudiced American or Frenchman, had been regarded as cogent arguments to country-bred Englishmen. The Brake Hunt had been established for a great many years, and was the central attraction of a district well known for its hunting propensities. The preservation of foxes might be an open question in such counties as Norfolk and Suffolk, but could not be so in the Brake country. Many things are, no doubt, permissible under the law, which, if done, would show the doer of them to be the enemy of his species,—and this destruction of foxes in a hunting country may be named as one of them. The Duke might have his foxes destroyed if he pleased, but he could hardly do so and remain a popular magnate in England. If he chose to put himself in opposition to the desires and very instincts of the people among whom his property was situated, he must live as a 'man forbid.'* That was the general argument, and then there was the argument special to this particular case. As it happened, Trumpeton Wood was, and always had been, the great nursery of foxes for that side of the Brake country. Gorse coverts make, no doubt, the charm of hunting, but gorse coverts will not hold foxes unless the woodlands be preserved. The fox is a travelling animal. Knowing well that 'home-staying youths have ever homely wits,'* he goes out and sees the world. He is either born in the woodlands, or wanders thither in his early youth. If all foxes so wandering be doomed to death, if poison, and wires, and traps, and hostile keepers await them there instead of the tender welcome of the loving fox-preserver, the gorse coverts will soon be empty, and the whole country will be afflicted with a wild dismay. All which Lord Chiltern understood well when he became so loud in his complaint against the Duke.

But our dear old friend, only the other day a duke, Planty Pall as he was lately called, devoted to work and to Parliament, an unselfish, friendly, wise man, who by no means wanted other men to cut their coats according to his pattern, was the last man in England to put himself forward as the enemy of an established delight. He did not hunt himself,—but neither did he shoot, or fish, or play cards. He recreated himself with Blue Books, and speculations on Adam Smith* had been his distraction;—but he knew that he was himself peculiar, and he respected the habits of others. It had fallen out in this wise. As the old Duke had become very old, the old Duke's agent had gradually acquired more than an agent's proper influence in the property; and as the Duke's heir would not shoot himself, or pay attention to the shooting, and as the Duke would not let the shooting of his wood, Mr. Fothergill, the steward, had gradually become omnipotent. Now Mr. Fothergill was not a hunting man,—but the mischief did not at all lie there. Lord Chiltern would not communicate with Mr. Fothergill. Lord Chiltern would write to the Duke, and Mr. Fothergill became an established enemy. Hinc illæ iræ.* From this source sprung all those powerfully argued articles in *The Field*, *Bell's Life*, and *Land and Water*;* —for on this matter all the sporting papers were of one mind.

There is something doubtless absurd in the intensity of the worship paid to the fox by hunting communities. The animal becomes sacred, and his preservation is a religion. His irregular destruction is a profanity, and words spoken to his injury are blasphemous. Not long since a gentleman shot a fox running across a woodland ride in a hunting country. He had mistaken it for a hare, and had done the deed in the presence of keepers, owner, and friends. His feelings were so acute and his remorse so great that, in their pity, they had resolved to spare him; and then, on the spot, entered into a solemn compact that no one should be told. Encouraged by the forbearing tenderness, the unfortunate one ventured to return to the house of his friend, the owner of the wood, hoping that, in spite of the sacrilege committed, he might be able to face a world that would be ignorant of his crime. As the vulpicide, on the afternoon of the day of the deed, went along the corridor to his room, one maid-servant whispered to another, and the poor victim of an imperfect sight heard the words—'That's he as shot the fox!' The gentleman did not appear at dinner, nor was he ever again seen in those parts.

Mr. Fothergill had become angry. Lord Chiltern, as we know, had been very angry. And even the Duke was angry.

The Duke was angry because Lord Chiltern had been violent;—and Lord Chiltern had been violent because Mr. Fothergill's conduct had been, to his thinking, not only sacrilegious, but one continued course of wilful sacrilege. It may be said of Lord Chiltern that in his eagerness as a master of hounds he had almost abandoned his love of riding. To kill a certain number of foxes in the year, after the legitimate fashion, had become to him the one great study of life;—and he did it with an energy equal to that which the Duke devoted to decimal coinage. His huntsman was always well mounted, with two horses; but Lord Chiltern would give up his own to the man and take charge of a weary animal as a common groom when he found that he might thus further the object of the day's sport. He worked as men work only at pleasure. He never missed a day, even when cub-hunting required that he should leave his bed at 3 A.M. He was constant at his kennel. He was always thinking about it. He devoted his life to the Brake Hounds. And it was too much for him that such a one as Mr. Fothergill should be allowed to wire foxes in Trumpeton Wood! The Duke's property, indeed! Surely all that was understood in England by this time. Now he had consented to come to Matching, bringing his wife with him, in order that the matter might be settled. There had been a threat that he would give up the country, in which case it was declared that it would be impossible to carry on the Brake Hunt in a manner satisfactory to masters, subscribers, owners of coverts, or farmers, unless a different order of things should be made to prevail in regard to Trumpeton Wood.

The Duke, however, had declined to interfere personally. He had told his wife that he should be delighted to welcome Lord and Lady Chiltern,—as he would any other friends of hers. The guests, indeed, at the Duke's house were never his guests, but always hers. But he could not allow himself to be brought into an argument with Lord Chiltern as to the management of his own property. The Duchess was made to understand that she must prevent any such awkwardness. And she did prevent it. 'And now, Lord Chiltern,' she said, 'how about the foxes?' She had taken care there should be a council of war around her. Lady Chiltern and Madame Goesler were present, and also Phineas Finn.

'Well;—how about them?' said the lord, showing by the fiery eagerness of his eye, and the increased redness of his face, that though the matter had been introduced somewhat jocosely, there could not really be any joke about it.

'Why couldn't you keep it all out of the newspapers?'

'I don't write the newspapers, Duchess. I can't help the newspapers. When two hundred men ride through Trumpeton Wood, and see one fox found, and that fox with only three pads, of course the newspapers will say that the foxes are trapped.'

'We may have traps if we like it, Lord Chiltern.'

'Certainly;—only say so, and we shall know where we are.' He looked very angry, and poor Lady Chiltern was covered with dismay. 'The Duke can destroy the hunt if he pleases, no doubt,' said the lord.

'But we don't like traps, Lord Chiltern;—nor yet poison, nor anything that is wicked. I'd go and nurse the foxes myself if I knew how, wouldn't I, Marie?'

'They have robbed the Duchess of her sleep for the last six months,' said Madame Goesler.

'And if they go on being not properly brought up and educated, they'll make an old woman of me. As for the Duke, he can't be comfortable in his arithmetic for thinking of them. But what can one do?'

'Change your keepers,' said Lord Chiltern energetically.

'It is easy to say,—change your keepers. How am I to set about it? To whom can I apply to appoint others? Don't you know what vested interests mean, Lord Chiltern?'

'Then nobody can manage his own property as he pleases?'

'Nobody can,—unless he does the work himself. If I were to go and live in Trumpeton Wood I could do it; but you see I have to live here. I vote that we have an officer of State, to go in and out with the Government,—with a seat in the Cabinet or not according as things go, and that we call him Foxmaster-General. It would be just the thing for Mr. Finn.'

'There would be a salary, of course,' said Phineas.

'Then I suppose that nothing can be done,' said Lord Chiltern.

'My dear Lord Chiltern, everything has been done. Vested interests have been attended to. Keepers shall prefer foxes to pheasants, wires shall be unheard of, and Trumpeton Wood shall once again be the glory of the Brake Hunt. It won't cost the Duke above a thousand or two a year.'

'I should be very sorry indeed to put the Duke to any unnecessary expense,' said Lord Chiltern solemnly,—still fearing that the Duchess was only playing with him. It made him angry that he could not imbue other people with his idea of the seriousness of the amusement of a whole county.

'Do not think of it. We have pensioned poor Mr. Fothergill, and he retires from the administration.'

'Then it'll be all right,' said Lord Chiltern.

'I am so glad,' said his wife.

'And so the great Mr. Fothergill falls from power, and goes down into obscurity,' said Madame Goesler.

'He was an impudent old man, and that's the truth,' said the Duchess;—'and he has always been my thorough detestation. But if you only knew what I have gone through to get rid of him,—and all on account of Trumpeton Wood,—you'd send me every brush taken in the Brake country during the next season.'

'Your Grace shall at any rate have one of them,' said Lord Chiltern.

On the next day Lord and Lady Chiltern went back to Harrington Hall. When the end of August comes, a Master of Hounds,—who is really a master,—is wanted at home. Nothing short of an embassy on behalf of the great coverts of his country would have kept this master away at present; and now, his diplomacy having succeeded, he hurried back to make the most of its results. Lady Chiltern, before she went, made a little speech to Phineas Finn.

'You'll come to us in the winter, Mr. Finn?'

'I should like.'

'You must. No one was truer to you than we were, you know. Indeed, regarding you as we do, how should we not have been true? It was impossible to me that my old friend should have been——'

'Oh, Lady Chiltern!'

'Of course you'll come. You owe it to us to come. And may I say this? If there be anybody to come with you, that will make it only so much the better. If it should be so, of course there will be letters written?' To this question, however, Phineas Finn made no answer.

CHAPTER LXXVI

Madame Goesler's Legacy

One morning, very shortly after her return to Harrington, Lady Chiltern was told that Mr. Spooner of Spoon Hall had called, and desired to see her. She suggested that the gentleman had probably asked for her husband,—who, at that moment, was enjoying his recovered supremacy in the centre of Trumpeton Wood; but she was assured that on this occasion Mr. Spooner's mission was to herself. She had no quarrel with Mr. Spooner, and she went to him at once. After the first greeting he rushed into the subject of the great triumph. 'So we've got rid of Mr. Fothergill, Lady Chiltern.'

'Yes; Mr. Fothergill will not, I believe, trouble us any more. He is an old man, it seems, and has retired from the Duke's service.'

'I can't tell you how glad I am, Lady Chiltern. We were afraid that Chiltern would have thrown it up, and then I don't know where we should have been. England would not have been England any longer, to my thinking, if we hadn't won the day. It'd have been just like a French revolution. Nobody would have known what was coming or where he was going.'

That Mr. Spooner should be enthusiastic on any hunting

question was a matter of course; but still it seemed to be odd that he should have driven himself over from Spoon Hall to pour his feelings into Lady Chiltern's ear. 'We shall go on very nicely now, I don't doubt,' said she; 'and I'm sure that Lord Chiltern will be glad to find that you are pleased.'

'I am very much pleased, I can tell you.' Then he paused, and the tone of his voice was changed altogether when he spoke again. 'But I didn't come over only about that, Lady Chiltern. Miss Palliser has not come back with you, Lady Chiltern?'

'We left Miss Palliser at Matching. You know she is the Duke's cousin.'

'I wish she wasn't, with all my heart.'

'Why should you want to rob her of her relations, Mr. Spooner?'

'Because—— because——. I don't want to say a word against her, Lady Chiltern. To me she is perfect as a star;—beautiful as a rose.' Mr. Spooner as he said this pointed first to the heavens and then to the earth. 'But perhaps she wouldn't have been so proud of her grandfather hadn't he been a Duke.'

'I don't think she is proud of that.'

'People do think of it, Lady Chiltern; and I don't say that they ought not. Of course it makes a difference, and when a man lives altogether in the country, as I do, it seems to signify so much more. But if you go back to old county families, Lady Chiltern, the Spooners have been here pretty nearly as long as the Pallisers,—if not longer. The Desponders, from whom we come, came over with William the Conqueror.'

'I have always heard that there isn't a more respectable family in the county.'

'That there isn't. There was a grant of land, which took their name, and became the Manor of Despond; there's where Spoon Hall is now. Sir Thomas Desponder was one of those who demanded the Charter, though his name wasn't always given because he wasn't a baron. Perhaps Miss Palliser does not know all that.'

'I doubt whether she cares about those things.'

'Women do care about them,—very much. Perhaps she has heard of the two spoons crossed, and doesn't know that that was a stupid vulgar practical joke. Our crest is a knight's head bowed, with the motto, "Desperandum." Soon after the Conquest one of the Desponders fell in love with the Queen, and never would give it up, though it wasn't any good. Her name was Matilda, and so he went as a Crusader and got killed. But wherever he went he had the knight's head bowed, and the motto on the shield.'

'What a romantic story, Mr. Spooner!'

'Isn't it? And it's quite true. That's the way we became Spooners. I never told her of it, but, somehow I wish I had now. It always seemed that she didn't think that I was anybody.'

'The truth is, Mr. Spooner, that she was always thinking that somebody else was everything. When a gentleman is told that a lady's affections have been pre-engaged, however much he may regret the circumstances, he cannot, I think, feel any hurt to his pride. If I understand the matter, Miss Palliser explained to you that she was engaged when first you spoke to her.'

'You are speaking of young Gerard Maule.'

'Of course I am speaking of Mr. Maule.'

'But she has quarrelled with him, Lady Chiltern.'

'Don't you know what such quarrels come to?'

'Well, no. That is to say, everybody tells me that it is really broken off, and that he has gone nobody knows where. At any rate he never shows himself. He doesn't mean it, Lady Chiltern.'

'I don't know what he means.'

'And he can't afford it, Lady Chiltern. I mean it, and I can afford it. Surely that might go for something.'

'I cannot say what Mr. Maule may mean to do, Mr. Spooner, but I think it only fair to tell you that he is at present staying at Matching, under the same roof with Miss Palliser.'

'Maule staying at the Duke's!' When Mr. Spooner heard this there came a sudden change over his face. His jaw fell, and his mouth was opened, and the redness of his cheeks flew up to his forehead.

'He was expected there yesterday, and I need hardly suggest to you what will be the end of the quarrel.'

'Going to the Duke's won't give him an income.'

'I know nothing about that, Mr. Spooner. But it really seems to me that you misinterpret the nature of the affections of such a girl as Miss Palliser. Do you think it likely that she should cease to love a man because he is not so rich as another?'

'People, when they are married, want a house to live in, Lady Chiltern. Now at Spoon Hall——'

'Believe me, that is in vain, Mr. Spooner.'

'You are quite sure of it?'

'Quite sure.'

'I'd have done anything for her,—anything! She might have had what settlements she pleased. I told Ned that he must go, if she made a point of it. I'd have gone abroad, or lived just anywhere. I'd come to that, that I didn't mind the hunting a bit.'

'I'm sorry for you,—I am indeed.'

'It cuts a fellow all to pieces so! And yet what is it all about? A slip of a girl that isn't anything so very much out of the way after all. Lady Chiltern, I shouldn't care if the horse kicked the trap all to pieces going back to Spoon Hall, and me with it.'

'You'll get over it, Mr. Spooner.'

'Get over it! I suppose I shall; but I shall never be as I was. I've been always thinking of the day when there must be a lady at Spoon Hall, and putting it off, you know. There'll never be a lady there now;—never. You don't think there's any chance at all?'

'I'm sure there is none.'

'I'd give half I've got in all the world,' said the wretched man, 'just to get it out of my head. I know what it will come to.' Though he paused, Lady Chiltern could ask no question respecting Mr. Spooner's future prospects. 'It'll be two bottles of champagne at dinner, and two bottles of claret afterwards, every day. I only hope she'll know that she did it. Good-bye, Lady Chiltern. I thought that perhaps you'd have helped me.'

'I cannot help you.'

'Good-bye.' So he went down to his trap, and drove himself violently home,—without, however, achieving the ruin which he desired. Let us hope that as time cures his wound that threat as to increased consumption of wine may fall to the ground unfulfilled.

In the meantime Gerard Maule had arrived at Matching Priory.

'We have quarrelled,' Adelaide had said when the Duchess told her that her lover was to come. 'Then you had better make it up again,' the Duchess had answered,—and there had been an end of it. Nothing more was done; no arrangement was made, and Adelaide was left to meet the man as best she might. The quarrel to her had been as the disruption of the heavens. She had declared to herself that she would bear it; but the misfortune to be borne was a broken world falling about her own ears. She had thought of a nunnery, of Ophelia among the water-lilies,* and of an early death-bed. Then she had pictured to herself the somewhat ascetic and very laborious life of an old maiden lady whose only recreation fifty years hence should consist in looking at the portrait of him who had once been her lover. And now she was told that he was coming to Matching as though nothing had been the matter! She tried to think whether it was not her duty to have her things at once packed, and ask for a carriage to take her to the railway station. But she was in the house of her nearest relative,—of him and also of her who were bound to see that things were right; and then there might be a more pleasureable existence than that which would have to depend on a photograph for its keenest delight. But how should she meet him? In what way should she address him? Should she ignore the quarrel, or recognize it, or take some milder course? She was half afraid of the Duchess, and could not ask for assistance. And the Duchess, though good-natured, seemed to her to be rough.

There was nobody at Matching to whom she could say a word; —so she lived on, and trembled, and doubted from hour to hour whether the world would not come to an end.

The Duchess was rough, but she was very good-natured. She had contrived that the two lovers should be brought into the same house, and did not doubt at all but what they would be able to adjust their own little differences when they met. Her experiences of the world had certainly made her more alive to the material prospects than to the delicate aroma of a love adventure. She had been greatly knocked about herself, and the material prospects had come uppermost. But all that had happened to her had tended to open her hand to other people, and had enabled her to be good-natured with delight, even when she knew that her friends imposed upon her. She didn't care much for Laurence Fitzgibbon; but when she was told that the lady with money would not consent to marry the aristocratic pauper except on condition that she should be received at Matching, the Duchess at once gave the invitation. And now, though she couldn't go into the 'fal-lallery,'—as she called it, to Madame Goesler,—of settling a meeting between two young people who had fallen out, she worked hard till she accomplished something perhaps more important to their future happiness. 'Plantagenet,' she said, 'there can be no objection to your cousin having that money.'

'My dear!'

'Oh come; you must remember about Adelaide, and that young man who is coming here to-day.'

'You told me that Adelaide is to be married. I don't know anything about the young man.'

'His name is Maule, and he is a gentleman, and all that. Some day when his father dies he'll have a small property somewhere.'

'I hope he has a profession.'

'No, he has not. I told you all that before.'

'If he has nothing at all, Glencora, why did he ask a young lady to marry him?'

'Oh, dear; what's the good of going into all that? He has

got something. They'll do immensely well, if you'll only listen. She is your first cousin.'

'Of course she is,' said Plantagenet, lifting up his hand to his hair.

'And you are bound to do something for her.'

'No; I am not bound. But I'm very willing,—if you wish it. Put the thing on a right footing.'

'I hate footings,—that is, right footings. We can manage this without taking money out of your pocket.'

'My dear Glencora, if I am to give my cousin money I shall do so by putting my hand into my own pocket in preference to that of any other person.'

'Madame Goesler says that she'll sign all the papers about the Duke's legacy,—the money, I mean,—if she may be allowed to make it over to the Duke's niece.'

'Of course Madame Goesler may do what she likes with her own. I cannot hinder her. But I would rather that you should not interfere. Twenty-five thousand pounds is a very serious sum of money.'

'You won't take it.'

'Certainly not.'

'Nor will Madame Goesler; and therefore there can be no reason why these young people should not have it. Of course Adelaide being the Duke's niece does make a difference. Why else should I care about it? She is nothing to me,—and as for him, I shouldn't know him again if I were to meet him in the street.'

And so the thing was settled. The Duke was powerless against the energy of his wife, and the lawyer was instructed that Madame Goesler would take the proper steps for putting herself into possession of the Duke's legacy,—as far as the money was concerned,—with the view of transferring it to the Duke's niece, Miss Adelaide Palliser. As for the diamonds, the difficulty could not be solved. Madame Goesler still refused to take them, and desired her lawyer to instruct her as to the form by which she could most thoroughly and conclusively renounce that legacy.

Gerard Maule had his ideas about the meeting which would of course take place at Matching. He would not, he thought, have been asked there had it not been intended that he should marry Adelaide. He did not care much for the grandeur of the Duke and Duchess, but he was conscious of certain profitable advantages which might accrue from such an acknowledgement of his position from the great relatives of his intended bride. It would be something to be married from the house of the Duchess, and to receive his wife from the Duke's hand. His father would probably be driven to acquiesce, and people who were almost omnipotent in the world would at any rate give him a start. He expected no money; nor did he possess that character, whether it be good or bad, which is given to such expectation. But there would be encouragement, and the thing would probably be done. As for the meeting,—he would take her in his arms if he found her alone, and beg her pardon for that cross word about Boulogne. He would assure her that Boulogne itself would be a heaven to him if she were with him,—and he thought that she would believe him. When he reached the house he was asked into a room in which a lot of people were playing billiards or crowded round a billiard-table. The Chilterns were gone, and he was at first ill at ease, finding no friend. Madame Goesler, who had met him at Harrington, came up to him, and told him that the Duchess would be there directly, and then Phineas, who had been playing at the moment of his entrance, shook hands with him, and said a word or two about the Chilterns. 'I was so delighted to hear of your acquittal,' said Maule.

'We never talk about that now,' said Phineas, going back to his stroke. Adelaide Palliser was not present, and the difficulty of the meeting had not yet been encountered. They all remained in the billiard-room till it was time for the ladies to dress, and Adelaide had not yet ventured to show herself. Somebody offered to take him to his room, and he was conducted upstairs, and told that they dined at eight,—but nothing had been arranged. Nobody had as yet mentioned her name to him. Surely it could not be that she had gone away when she heard that he was coming, and that she was really determined to make the quarrel perpetual? He had three quarters of an hour in which to get ready for dinner, and he felt himself to be uncomfortable and out of his element. He had been sent to his chamber prematurely, because nobody had known what to do with him; and he wished himself back in London. The Duchess, no doubt, had intended to be good-natured, but she had made a mistake. So he sat by his open window, and looked out on the ruins of the old Priory, which were close to the house, and wondered why he mightn't have been allowed to wander about the garden instead of being shut up there in a bedroom. But he felt that it would be unwise to attempt any escape now. He would meet the Duke or the Duchess, or perhaps Adelaide herself, in some of the passages,—and there would be an embarrassment. So he dawdled away the time, looking out of the window as he dressed, and descended to the drawing room at eight o'clock. He shook hands with the Duke, and was welcomed by the Duchess, and then glanced round the room. There she was, seated on a sofa between two other ladies,—of whom one was his friend, Madame Goesler. It was essentially necessary that he should notice her in some way, and he walked up to her, and offered her his hand. It was impossible that he should allude to what was past, and he merely muttered something as he stood over her. She had blushed up to her eyes, and was absolutely dumb. 'Mr. Maule, perhaps you'll take our cousin Adelaide out to dinner,' said the Duchess, a moment afterwards, whispering in his ear.

'Have you forgiven me?' he said to her, as they passed from one room to the other.

'I will,—if you care to be forgiven.' The Duchess had been quite right, and the quarrel was all over without any arrangement.

On the following morning he was allowed to walk about the grounds without any impediment, and to visit the ruins which had looked so charming to him from the window. Nor was he alone. Miss Palliser was now by no means anxious as she had been yesterday to keep out of the way, and was willingly persuaded to show him all the beauties of the place.

'I shouldn't have said what I did, I know,' pleaded Maule. 'Never mind it now, Gerard.'

'I mean about going to Boulogne.'

'It did sound so melancholy.'

'But I only meant that we should have to be very careful how we lived. I don't know quite whether I am so good at being careful about money as a fellow ought to be.'

'You must take a lesson from me, sir.'

'I have sent the horses to Tattersall's,'* he said in a tone that was almost funereal.

'What!—already?'

'I gave the order yesterday. They are to be sold,—I don't know when. They won't fetch anything. They never do. One always buys bad horses there for a lot of money, and sells good ones for nothing. Where the difference goes to I never could make out.'

'I suppose the man gets it who sells them.'

'No; he don't. The fellows get it who have their eyes open. My eyes never were open,—except as far as seeing you went.'

'Perhaps if you had opened them wider you wouldn't have to go to——'

'Don't, Adelaide. But, as I was saying about the horses, when they're sold of course the bills won't go on. And I suppose things will come right. I don't owe so very much.'

'I've got something to tell you,' she said.

'What about?'

'You're to see my cousin to-day at two o'clock.'

'The Duke?'

'Yes,—the Duke; and he has got a proposition. I don't know that you need sell your horses, as it seems to make you so very unhappy. You remember Madame Goesler?'

'Of course I do. She was at Harrington.'

'There's something about a legacy which I can't understand

at all. It is ever so much money, and it did belong to the old Duke. They say it is to be mine,—or yours rather, if we should ever be married. And then you know, Gerard, perhaps, after all, you needn't go to Boulogne.' So she took her revenge, and he had his as he pressed his arm round her waist and kissed her among the ruins of the old Priory.

Precisely at two to the moment he had his interview with the Duke, and very disagreeable it was to both of them. The Duke was bound to explain that the magnificent present which was being made to his cousin was a gift, not from him, but from Madame Goesler; and, though he was intent on making this as plain as possible, he did not like the task. 'The truth is, Mr. Maule, that Madame Goesler is unwilling, for reasons with which I need not trouble you, to take the legacy which was left to her by my uncle. I think her reasons to be insufficient, but it is a matter in which she must, of course, judge for herself. She has decided,—very much, I fear, at my wife's instigation, which I must own I regret,—to give the money to one of our family, and has been pleased to say that my cousin Adelaide shall be the recipient of her bounty. I have nothing to do with it. I cannot stop her generosity if I would, nor can I say that my cousin ought to refuse it. Adelaide will have the entire sum as her fortune, short of the legacy duty, which, as you are probably aware, will be ten per cent., as Madame Goesler was not related to my uncle. The money will, of course, be settled on my cousin and on her children. I believe that will be all I shall have to say, except that Lady Glencora,—the Duchess, I mean,—wishes that Adelaide should be married from our house. If this be so I shall, of course, hope to have the honour of giving my cousin away.' The Duke was by no means a pompous man, and probably there was no man in England of so high rank who thought so little of his rank. But he was stiff and somewhat ungainly, and the task which he was called upon to execute had been very disagreeable to him. He bowed when he had finished his speech, and Gerard Maule felt himself bound to go, almost without expressing his thanks.

'My dear Mr. Maule,' said Madame Goesler, 'you literally must not say a word to me about it. The money was not mine, and under no circumstances would or could be mine. I have given nothing, and could not have presumed to make such a present. The money, I take it, does undoubtedly belong to the present Duke, and, as he does not want it, it is very natural that it should go to his cousin. I trust that you may both live to enjoy it long, but I cannot allow any thanks to be given to me by either of you.'

After that he tried the Duchess, who was somewhat more gracious. 'The truth is, Mr. Maule, you are a very lucky man to find twenty thousand pounds and more going begging about the country in that way.'

'Indeed I am, Duchess.'

'And Adelaide is lucky, too, for I doubt whether either of you are given to any very penetrating economies. I am told that you like hunting.'

'I have sent my horses to Tattersall's.'

'There is enough now for a little hunting, I suppose, unless you have a dozen children. And now you and Adelaide must settle when it's to be. I hate things to be delayed. People go on quarrelling and fancying this and that, and thinking that the world is full of romance and poetry. When they get married they know better.'

'I hope the romance and poetry do not all vanish.'

'Romance and poetry are for the most part lies, Mr. Maule, and are very apt to bring people into difficulty. I have seen something of them in my time, and I much prefer downright honest figures. Two and two make four; idleness is the root of all evil; love your neighbour like yourself, and the rest of it. Pray remember that Adelaide is to be married from here, and that we shall be very happy that you should make every use you like of our house until then.'

We may so far anticipate in our story as to say that Adelaide Palliser and Gerard Maule were married from Matching Priory at Matching Church early in that October, and that as far as the coming winter was concerned, there certainly was no hunting for the gentleman. They went to Naples instead of Boulogne, and there remained till the warm weather came in the following spring. Nor was that peremptory sale at Tattersall's countermanded as regarded any of the horses. What prices were realised the present writer has never been able to ascertain.

CHAPTER LXXVII

Phineas Finn's Success

When Phineas Finn had been about a week at Matching, he received a letter, or rather a very short note, from the Prime Minister, asking him to go up to London; and on the same day the Duke of Omnium spoke to him on the subject of the letter. 'You are going up to see Mr. Gresham. Mr. Gresham has written to me, and I hope that we shall be able to congratulate ourselves in having your assistance next Session.' Phineas declared that he had no idea whatever of Mr. Gresham's object in summoning him up to London. 'I have his permission to inform you that he wishes you to accept office.' Phineas felt that he was becoming very red in the face, but he did not attempt to make any reply on the spur of the moment. 'Mr. Gresham thinks it well that so much should be said to you before you see him, in order that you may turn the matter over in your own mind. He would have written to you probably, making the offer at once, had it not been that there must be various changes, and that one man's place must depend on another. You will go, I suppose.'

'Yes; I shall go, certainly. I shall be in London this evening.'

'I will take care that a carriage is ready for you. I do not presume to advise, Mr. Finn, but I hope that there need be no doubt as to your joining us.' Phineas was somewhat confounded, and did not know the Duke well enough to give expression to his thoughts at the moment. 'Of course you will return to us, Mr. Finn.' Phineas said that he would return and trespass on the Duke's hospitality for yet a few days. He was quite resolved that something must be said to Madame Goesler before he left the roof under which she was living. In the course of the autumn she purposed, as she had told him, to go to Vienna, and to remain there almost up to Christmas. Whatever there might be to be said should be said at any rate before that.

He did speak a few words to her before his journey to London, but in those words there was no allusion made to the great subject which must be discussed between them. 'I am going up to London,' he said.

'So the Duchess tells me.'

'Mr. Gresham has sent for me,—meaning, I suppose, to offer me the place which he would not give me while that poor man was alive.'

'And you will accept it of course, Mr. Finn?'

'I am not at all so sure of that.'

'But you will. You must. You will hardly be so foolish as to let the peevish animosity of an ill-conditioned man prejudice your prospects even after his death.'

'It will not be any remembrance of Mr. Bonteen that will induce me to refuse.'

'It will be the same thing;—rancour against Mr. Gresham because he had allowed the other man's counsel to prevail with him. The action of no individual man should be to you of sufficient consequence to guide your conduct. If you accept office, you should not take it as a favour conferred by the Prime Minister; nor if you refuse it, should you do so from personal feelings in regard to him. If he selects you, he is presumed to do so because he finds that your services will be valuable to the country.'

'He does so because he thinks that I should be safe to vote for him.'

'That may be so, or not. You can't read his bosom quite distinctly;—but you may read your own. If you go into office you become the servant of the country,—not his servant, and should assume his motive in selecting you to be the same as your own in submitting to the selection. Your foot must be on the ladder before you can get to the top of it.'

'The ladder is so crooked.'

'Is it more crooked now than it was three years ago;—worse than it was six months ago, when you and all your friends looked upon it as certain that you would be employed? There is nothing, Mr. Finn, that a man should fear so much as some twist in his convictions arising from a personal accident to himself. When we heard that the Devil in his sickness wanted to be a monk, we never thought that he would become a saint in glory. When a man who has been rejected by a lady expresses a generally ill opinion of the sex, we are apt to ascribe his opinions to disappointment rather than to judgment. A man falls and breaks his leg at a fence, and cannot be induced to ride again,—not because he thinks the amusement to be dangerous, but because he cannot keep his mind from dwelling on the hardship that has befallen himself. In all such cases self-consciousness gets the better of the judgment.'

'You think it will be so with me?'

'I shall think so if you now refuse—because of the misfortune which befell you—that which I know you were most desirous of possessing before that accident. To tell you the truth, Mr. Finn, I wish Mr. Gresham had delayed his offer till the winter.'

'And why?'

'Because by that time you will have recovered your health. Your mind now is morbid, and out of tune.'

'There was something to make it so, Madame Goesler.'

'God knows there was; and the necessity which lay upon you of bearing a bold front during those long and terrible weeks of course consumed your strength. The wonder is that the fibres of your mind should have retained any of their elasticity after such an ordeal. But as you are so strong, it would be a pity that you should not be strong altogether. This thing that is now to be offered to you is what you have always desired.'

'A man may have always desired that which is worthless.'

'You tried it once, and did not find it worthless. You found yourself able to do good work when you were in office. If I remember right, you did not give it up then because it was irksome to you, or contemptible, or, as you say, worthless; but from difference of opinion on some political question. You can always do that again.'

'A man is not fit for office who is prone to do so.'

'Then do not you be prone. It means success or failure in the profession which you have chosen, and I shall greatly regret to see you damage your chance of success by yielding to scruples which have come upon you when you are hardly as yet yourself.'

She had spoken to him very plainly, and he had found it to be impossible to answer her, and yet she had hardly touched the motives by which he believed himself to be actuated. As he made his journey up to London he thought very much of her words. There had been nothing said between them about money. No allusion had been made to the salary of the office which would be offered to him, or to the terrible shortness of his own means of living. He knew well enough himself that he must take some final step in life, or very shortly return into absolute obscurity. This woman who had been so strongly advising him to take a certain course as to his future life, was very rich;—and he had fully decided that he would sooner or later ask her to be his wife. He knew well that all her friends regarded their marriage as certain. The Duchess had almost told him so in as many words. Lady Chiltern, who was much more to him than the Duchess, had assured him that if he should have a wife to bring with him to Harrington, the wife would be welcome. Of what other wife could Lady Chiltern have thought? Laurence Fitzgibbon, when congratulated on his own marriage, had returned counter congratulations. Mr. Low had said that it would of course come to pass. Even Mrs. Bunce had hinted at it, suggesting that she would lose her lodger and be a wretched woman. All the world had heard of the journey to Prague, and all the world expected the marriage. And he had come to love the woman with excessive affection, day by day, ever since the renewal of their intimacy at Broughton Spinnies. His mind was quite made up;—but he was by no means sure of her mind as the rest of the world might be. He knew of her, what nobody else in all the world knew,—except himself. In that former period of his life, on which he now sometimes looked back as though it had been passed in another world, this woman had offered her hand and fortune to him. She had done so in the enthusiasm of her love, knowing his ambition and knowing his poverty, and believing that her wealth was necessary to the success of his career in life. He had refused the offer,—and they had parted without a word. Now they had come together again, and she was certainly among the dearest of his friends. Had she not taken that wondrous journey to Prague in his behalf, and been the first among those who had striven,—and had striven at last successfully,—to save his neck from the halter? Dear to her! He knew well as he sat with his eyes closed in the railway carriage that he must be dear to her! But might it not well be that she had resolved that friendship should take the place of love? And was it not compatible with her nature,—with all human nature,—that in spite of her regard for him she should choose to be revenged for the evil which had befallen her, when she offered her hand in vain? She must know by this time that he intended to throw himself at her feet; and would hardly have advised him as she had done as to the necessity of following up that success which had hitherto been so essential to him, had she intended to give him all that she had once offered him before. It might well be that Lady Chiltern, and even the Duchess, should be mistaken. Marie Goesler was not a woman, he thought, to reveal the deeper purposes of her life to any such friend as the Duchess of Omnium.

Of his own feelings in regard to the offer which was about to be made to him he had hardly succeeded in making her understand anything. That a change had come upon himself was certain, but he did not at all believe that it had sprung from any weakness caused by his sufferings in regard to the murder. He rather believed that he had become stronger than weaker from all that he had endured. He had learned when he was younger,—some years back,—to regard the political service of his country as a profession in which a man possessed of certain gifts might earn his bread with more gratification to himself than in any other. The work would be hard, and the emolument only intermittent; but the service would in itself be pleasant; and the rewards of that service,—should he be so successful as to obtain reward,—would be dearer to him than anything which could accrue to him from other labours. To sit in the Cabinet for one Session would, he then thought, be more to him than to preside over the Court of Queen's Bench as long as did Lord Mansfield. But during the last few months a change had crept across his dream,—which he recognized but could hardly analyse. He had seen a man whom he despised promoted, and the place to which the man had been exalted had at once become contemptible in his eyes. And there had been quarrels and jangling, and the speaking of evil words between men who should have been quiet and dignified. No doubt Madame Goesler was right in attributing the revulsion in his hopes to Mr. Bonteen and Mr. Bonteen's enmity; but Phineas Finn himself did not know that it was so.

He arrived in town in the evening, and his appointment with Mr. Gresham was for the following morning. He breakfasted at his club, and there he received the following letter from Lady Laura Kennedy:—

'*Saulsby 28th August*, 18——

'MY DEAR PHINEAS,

'I have just received a letter from Barrington in which he tells me that Mr. Gresham is going to offer you your old place at the Colonies. He says that Lord Fawn has been so upset by this affair of Lady Eustace's husband, that he is obliged to resign and go abroad.'—This was the first intimation that Phineas had heard of the nature of the office to be offered to him.—'But Barrington goes on to say that he thinks you won't accept Mr. Gresham's offer, and he asks me to write to you. Can this possibly be true? Barrington writes most kindly,—with true friendship,—and is most anxious for you to join. But he thinks that you are angry with Mr. Gresham because he passed you over before, and that you will not forgive him for having yielded to Mr. Bonteen. I can hardly believe this possible. Surely you will not allow the shade of that unfortunate man to blight your prospects? And, after all, of what matter to you is the friendship or enmity of Mr. Gresham? You have to assert yourself, to make your own way, to use your own opportunities, and to fight your own battle without reference to the feelings of individuals. Men act together in office constantly, and with constancy, who are known to hate each other. When there are so many to get what is going, and so little to be given, of course there will be struggling and trampling. I have no doubt that Lord Cantrip has made a point of this with Mr. Gresham;—has in point of fact insisted upon it. If so, you are lucky to have such an ally as Lord Cantrip. He and Mr. Gresham are, as you know, sworn friends, and if you get on well with the one you certainly may with the other also. Pray do not refuse without asking for time to think about it;—and if so, pray come here, that you may consult my father.

'I spent two weary weeks at Loughlinter, and then could stand it no longer. I have come here, and here I shall remain for the autumn and winter. If I can sell my interest in the Loughlinter property I shall do so, as I am sure that neither the place nor the occupation is fit for me. Indeed I know not what place or what occupation will suit me! The dreariness of the life before me is hardly preferable to the disappointments I have already endured. There seems to be nothing left for me but to watch my father to the end. The world would say that such a duty in life is fit for a widowed childless daughter; but to you I cannot pretend to say that my bereavements or misfortunes reconcile me to such a fate. I cannot cease to remember my age, my ambition, and I will say, my love. I suppose that everything is over for me,—as though

I were an old woman, going down into the grave, but at my time of life I find it hard to believe that it must be so. And then the time of waiting may be so long! I suppose I could start a house in London, and get people around me by feeding and flattering them, and by little intrigues,—like that woman of whom you are so fond. It is money that is chiefly needed for that work, and of money I have enough now. And people would know at any rate who I am. But I could not flatter them, and I should wish the food to choke them if they did not please me. And you would not come, and if you did,—I may as well say it boldly,—others would not. An ill-natured sprite has been busy with me, which seems to deny me everything which is so freely granted to others.

'As for you, the world is at your feet. I dread two things for you,—that you should marry unworthily, and that you should injure your prospects in public life by an uncompromising stiffness. On the former subject I can say nothing to you. As to the latter, let me implore you to come down here before you decide upon anything. Of course you can at once accept Mr. Gresham's offer; and that is what you should do unless the office proposed to you be unworthy of you. No friend of yours will think that your old place at the Colonies should be rejected. But if your mind is still turned towards refusing, ask Mr. Gresham to give you three or four days for decision, and then come here. He cannot refuse you,—nor after all that is passed can you refuse me.

'Yours affectionately,

'L. K.'

When he had read this letter he at once acknowledged to himself that he could not refuse her request. He must go to Saulsby, and he must do so at once. He was about to see Mr. Gresham immediately,—within half an hour; and as he could not expect at the most above twenty-four hours to be allowed to him for consideration, he must go down to Saulsby on the same evening. As he walked to the Prime Minister's house he called at a telegraph office and sent down his message. 'I will be at Saulsby by the train arriving at 7 P.M. Send to meet me.' Then he went on, and in a few minutes found himself in the presence of the great man.

The great man received him with an excellent courtesy. It is the special business of Prime Ministers to be civil in detail, though roughness, and perhaps almost rudeness in the gross, becomes not unfrequently a necessity of their position. To a proposed incoming subordinate a Prime Minister is, of course, very civil, and to a retreating subordinate he is generally more so,—unless the retreat be made under unfavourable circumstances. And to give good things is always pleasant, unless there be a suspicion that the good thing will be thought to be not good enough. No such suspicion as that now crossed the mind of Mr. Gresham. He had been pressed very much by various colleagues to admit this young man into the Paradise of his government, and had been pressed very much also to exclude him; and this had been continued till he had come to dislike the name of the young man. He did believe that the young man had behaved badly to Mr. Robert Kennedy, and he knew that the young man on one occasion had taken to kicking in harness, and running a course of his own. He had decided against the young man,—very much no doubt at the instance of Mr. Bonteen,—and he believed that in so doing he closed the Gates of Paradise against a Peri* most anxious to enter it. He now stood with the key in his hand and the gate open,—and the seat to be allotted to the re-accepted one was that which he believed the Peri would most gratefully fill. He began by making a little speech about Mr. Bonteen. That was almost unavoidable. And he praised in glowing words the attitude which Phineas had maintained during the trial. He had been delighted with the re-election at Tankerville, and thought that the borough had done itself much honour. Then came forth his proposition. Lord Fawn had retired, absolutely broken down by repeated examinations respecting the man in the grey coat, and the office which Phineas had before held with so much advantage to the public, and comfort to his immediate chief, Lord Cantrip, was there for his acceptance.

Mr. Gresham went on to express an ardent hope that he might have the benefit of Mr. Finn's services. It was quite manifest from his manner that he did not in the least doubt the nature of the reply which he would receive.

Phineas had come primed with his answer,—so ready with it that it did not even seem to be the result of any hesitation at the moment. 'I hope, Mr. Gresham, that you will be able to give me a few hours to think of this.' Mr. Gresham's face fell, for, in truth, he wanted an immediate answer; and though he knew from experience that Secretaries of State, and First Lords, and Chancellors, do demand time, and will often drive very hard bargains before they will consent to get into harness, he considered that Under-Secretaries, Junior Lords, and the like, should skip about as they were bidden, and take the crumbs offered them without delay. If every underling wanted a few hours to think about it, how could any Government ever be got together? 'I am sorry to put you to inconvenience,' continued Phineas, seeing that the great man was but ill-satisfied, 'but I am so placed that I cannot avail myself of your flattering kindness without some little time for consideration.'

'I had hoped that the office was one which you would like.'

'So it is, Mr. Gresham.'

'And I was told that you are now free from any scruples,—political scruples, I mean,—which might make it difficult for you to support the Government.'

'Since the Government came to our way of thinking,—a year or two ago,—about Tenant Right, I mean,—I do not know that there is any subject on which I am likely to oppose it. Perhaps I had better tell you the truth, Mr. Gresham.'

'Oh, certainly,' said the Prime Minister, who knew very well that on such occasions nothing could be worse than the telling of disagreeable truths.

'When you came into office, after beating Mr. Daubeny on the Church question, no man in Parliament was more desirous of place than I was,—and I am sure that none of the disappointed ones felt their disappointment so keenly. It was aggravated by various circumstances,—by calumnies in newspapers, and by personal bickerings. I need not go into that wretched story of Mr. Bonteen, and the absurd accusation which grew out of those calumnies. These things have changed me very much. I have a feeling that I have been ill-used,—not by you, Mr. Gresham, specially, but by the party; and I look upon the whole question of office with altered eyes.'

'In filling up the places at his disposal, a Prime Minister, Mr. Finn, has a most unenviable task.'

'I can well believe it.'

'When circumstances, rather than any selection of his own, indicate the future occupant of any office, this abrogation of his patronage is the greatest blessing in the world to him.'

'I can believe that also.'

'I wish it were so with every office under the Crown. A Minister is rarely thanked, and would as much look for the peace of heaven in his office as for gratitude.'

'I am sorry that I should have made no exception to such thanklessness.'

'We shall neither of us get on by complaining;—shall we, Mr. Finn? You can let me have an answer perhaps by this time to-morrow.'

'If an answer by telegraph will be sufficient.'

'Quite sufficient. Yes or No. Nothing more will be wanted. You understand your own reasons, no doubt, fully; but if they were stated at length they would perhaps hardly enlighten me. Good-morning.' Then as Phineas was turning his back, the Prime Minister remembered that it behoved him as Prime Minister to repress his temper. 'I shall still hope, Mr. Finn, for a favourable answer.' Had it not been for that last word Phineas would have turned again, and at once rejected the proposition.

From Mr. Gresham's house he went by appointment to

Mr. Monk's, and told him of the interview. Mr. Monk's advice to him had been exactly the same as that given by Madame Goesler and Lady Laura. Phineas, indeed, understood perfectly that no friend could or would give him any other advice. 'He has his troubles, too,' said Mr. Monk, speaking of the Prime Minister.

'A man can hardly expect to hold such an office without trouble.'

'Labour of course there must be,—though I doubt whether it is so great as that of some other persons;—and responsibility. The amount of trouble depends on the spirit and nature of the man. Do you remember old Lord Brock? He was never troubled. He had a triple shield,—a thick skin, an equable temper, and perfect self-confidence. Mr. Mildmay was of a softer temper, and would have suffered had he not been protected by the idolatry of a large class of his followers. Mr. Gresham has no such protection. With a finer intellect than either, and a sense of patriotism quite as keen, he has a self-consciousness which makes him sore at every point. He knows the frailty of his temper, and yet cannot control it. And he does not understand men as did these others. Every word from an enemy is a wound to him. Every slight from a friend is a dagger in his side. But I can fancy that self-accusations make the cross on which he is really crucified. He is a man to whom I would extend all my mercy, were it in my power to be merciful.'

'You will hardly tell me that I should accept office under him by way of obliging him.'

'Were I you I should do so,—not to oblige him, but because I know him to be an honest man.'

'I care but little for honesty,' said Phineas, 'which is at the disposal of those who are dishonest. What am I to think of a Minister who could allow himself to be led by Mr. Bonteen?'

CHAPTER LXXVIII

The Last Visit to Saulsby

PHINEAS, as he journeyed down to Saulsby, knew that he had in truth made up his mind. He was going thither nominally that he might listen to the advice of almost his oldest political friend before he resolved on a matter of vital importance to himself; but in truth he was making the visit because he felt that he could not excuse himself from it without unkindness and ingratitude. She had implored him to come, and he was bound to go, and there were tidings to be told which he must tell. It was not only that he might give her his reasons for not becoming an Under-Secretary of State that he went to Saulsby. He felt himself bound to inform her that he intended to ask Marie Goesler to be his wife. He might omit to do so till he had asked the question,—and then say nothing of what he had done should his petition be refused; but it seemed to him that there would be cowardice in this. He was bound to treat Lady Laura as his friend in a special degree, as something more than his sister,—and he was bound above all things to make her understand in some plainest manner that she could be nothing more to him than such a friend. In his dealings with her he had endeavoured always to be honest,—gentle as well as honest; but now it was specially his duty to be honest to her. When he was young he had loved her, and had told her so,—and she had refused him. As a friend he had been true to her ever since, but that offer could never be repeated. And the other offer,—to the woman whom she was now accustomed to abuse,—must be made. Should Lady Laura choose to quarrel with him it must be so; but the quarrel should not be of his seeking.

He was quite sure that he would refuse Mr. Gresham's offer, although by doing so he would himself throw away the very thing which he had devoted his life to acquire. In a foolish, soft moment,—as he now confessed to himself,—he had endeavoured to obtain for his own position the sympathy of the Minister. He had spoken of the calumnies which had hurt him, and of his sufferings when he found himself excluded from place in consequence of the evil stories which had been told of him. Mr. Gresham had, in fact, declined to listen to him;—had said Yes or No was all that he required, and had gone on to explain that he would be unable to understand the reasons proposed to be given even were he to hear them. Phineas had felt himself to be repulsed, and would at once have shown his anger, had not the Prime Minister silenced him for the moment by a civilly-worded repetition of the offer made.

But the offer should certainly be declined. As he told himself that it must be so, he endeavoured to analyse the causes of this decision, but was hardly successful. He had thought that he could explain the reasons to the Minister, but found himself incapable of explaining them to himself. In regard to means of subsistence he was no better off now than when he began the world. He was, indeed, without incumbrance, but was also without any means of procuring an income. For the last twelve months he had been living on his little capital, and two years more of such life would bring him to the end of all that he had. There was, no doubt, one view of his prospects which was bright enough. If Marie Goesler accepted him, he need not, at any rate, look about for the means of earning a living. But he assured himself with perfect confidence that no hope in that direction would have any influence upon the answer he would give to Mr. Gresham. Had not Marie Goesler herself been most urgent with him in begging him to accept the offer; and was he not therefore justified in concluding that she at least had thought it necessary that he should earn his bread? Would her heart be softened towards him,—would any further softening be necessary,—by his obstinate refusal to comply with her advice? The two things had no reference to each other,—and should be regarded by him as perfectly distinct. He would refuse Mr. Gresham's offer,—not because he hoped that he might live in idleness on the wealth of the woman he loved,—but because the chicaneries and intrigues of office had become distasteful to him. 'I don't know which are the falser,' he said to himself, 'the mock courtesies or the mock indignations of statesmen.'

He found the Earl's carriage waiting for him at the station, and thought of many former days, as he was carried through the little town for which he had sat in Parliament, up to the house which he had once visited in the hope of wooing Violet Effingham. The women whom he had loved had all, at any rate, become his friends, and his thorough friendships were almost all with women. He and Lord Chiltern regarded each other with warm affection; but there was hardly ground for real sympathy between them. It was the same with Mr. Low and Barrington Erle. Were he to die there would be no gap in their lives;—were they to die there would be none in his. But with Violet Effingham,—as he still loved to call her to himself,—he thought it would be different. When the carriage stopped at the hall door he was thinking of her rather than of Lady Laura Kennedy.

He was shown at once to his bedroom,—the very room in which he had written the letter to Lord Chiltern which had brought about the duel at Blankenberg. He was told that he would find Lady Laura in the drawing-room waiting for dinner for him. The Earl had already dined.

'I am so glad you are come,' said Lady Laura, welcoming him. 'Papa is not very well and dined early, but I have waited for you, of course. Of course I have. You did not suppose I would let you sit down alone? I would not see you before you dressed because I knew that you must be tired and hungry, and that the sooner you got down the better. Has it not been hot?'

'And so dusty! I only left Matching yesterday, and seem to have been on the railway ever since.'

'Government officials have to take frequent journeys, Mr. Finn. How long will it be before you have to go down to Scotland twice in one week, and back as often to form a Ministry? Your next journey must be into the dining-room;—in making which will you give me your arm?'

She was, he thought, lighter in heart and pleasanter in manner than she had been since her return from Dresden. When she had made her little joke about his future ministerial duties the servant had been in the room, and he had not, therefore, stopped her by a serious answer. And now she was solicitous about his dinner,—anxious that he should enjoy the good things set before him, as is the manner of loving women, pressing him to take wine, and playing the good hostess in all things. He smiled, and ate, and drank, and was gracious under her petting; but he had a weight on his bosom, knowing, as he did, that he must say that before long which would turn all her playfulness either to anger or to grief. 'And who had you at Matching?' she asked.

'Just the usual set.'

'Minus the poor old Duke?'

'Yes; minus the old Duke certainly. The greatest change is in the name. Lady Glencora was so specially Lady Glencora that she ought to have been Lady Glencora to the end. Everybody calls her Duchess, but it does not sound half so nice.'

'And is he altered?'

'Not in the least. You can trace the lines of lingering regret upon his countenance when people be-Grace him; but that is all. There was always about him a simple dignity which made it impossible that any one should slap him on the back; and that of course remains. He is the same Planty Pall; but I doubt whether any man ever ventured to call him Planty Pall to his face since he left Eton.'

'The house was full, I suppose?'

'There were a great many there; among others Sir Gregory Grogram, who apologised to me for having tried to—put an end to my career.'

'Oh, Phineas!'

'And Sir Harry Coldfoot, who seemed to take some credit to himself for having allowed the jury to acquit me. And Chiltern and his wife were there for a day or two.'

'What could take Oswald there?'

'An embassy of State about the foxes. The Duke's property

runs into his country. She is one of the best women that ever lived.'

'Violet?'

'And one of the best wives.'

'She ought to be, for she is one of the happiest. What can she wish for that she has not got? Was your great friend there?'

He knew well what great friend she meant. 'Madame Max Goesler was there.'

'I suppose so. I can never quite forgive Lady Glencora for her intimacy with that woman.'

'Do not abuse her, Lady Laura.'

'I do not intend,—not to you at any rate. But I can better understand that she should receive the admiration of a gentleman than the affectionate friendship of a lady. That the old Duke should have been infatuated was intelligible.'

'She was very good to the old Duke.'

'But it was a kind of goodness which was hardly likely to recommend itself to his nephew's wife. Never mind; we won't talk about her now. Barrington was there?'

'For a day or two.'

'He seems to be wasting his life.'

'Subordinates in office generally do, I think.'

'Do not say that, Phineas.'

'Some few push through, and one can almost always foretell who the few will be. There are men who are destined always to occupy second-rate places, and who seem also to know their fate. I never heard Erle speak even of an ambition to sit in the Cabinet.'

'He likes to be useful.'

'All that part of the business which distresses me is pleasant to him. He is fond of arrangements, and delights in little party successes. Either to effect or to avoid a count-out is a job of work to his taste, and he loves to get the better of the Opposition by keeping it in the dark. A successful plot is as dear to him as to a writer of plays. And yet he is never bitter as is Ratler, or unscrupulous as was poor Mr. Bonteen, or full of wrath as is Lord Fawn. Nor is he idle like Fitzgibbon. Erle always earns his salary.'

'When I said he was wasting his life, I meant that he did not marry. But perhaps a man in his position had better remain unmarried.' Phineas tried to laugh, but hardly succeeded well. 'That, however, is a delicate subject, and we will not touch it now. If you won't drink any wine we might as well go into the other room.'

Nothing had as yet been said on either of the subjects which had brought him to Saulsby, but there had been words which made the introduction of them peculiarly unpleasant. His tidings, however, must be told. 'I shall not see Lord Brentford to-night?' he asked, when they were together in the drawing-room.

'If you wish it you can go up to him. He will not come down.'

'Oh, no. It is only because I must return to-morrow.'

'To-morrow, Phineas!'

'I must do so. I have pledged myself to see Mr. Monk,—and others also.'

'It is a short visit to make to us on my first return home! I hardly expected you at Loughlinter, but I thought that you might have remained a few nights under my father's roof.' He could only reassert his assurance that he was bound to be back in London, and explain as best he might that he had come to Saulsby for a single night, only because he would not refuse her request to him. 'I will not trouble you, Phineas, by complaints,' she said.

'I would give you no cause for complaint if I could avoid it.'

'And now tell me what has passed between you and Mr. Gresham,' she said as soon as the servant had given them coffee. They were sitting by a window which opened down to the ground, and led on to the terrace and to the lawns below. The night was soft, and the air was heavy with the scent of many flowers. It was now past nine, and the sun had set; but there was a bright harvest moon, and the light, though pale, was clear as that of day. 'Will you come and take a turn round the garden? We shall be better there than sitting here. I will get my hat; can I find yours for you?' So they both strolled out, down the terrace steps, and went forth, beyond the gardens, into the park, as though they had both intended from the first that it should be so. 'I know you have not accepted Mr. Gresham's offer, or you would have told me so.'

'I have not accepted.'

'Nor have you refused?'

'No; it is still open. I must send my answer by telegram to-morrow—Yes or No,—Mr. Gresham's time is too precious to admit of more.'

'Phineas, for Heaven's sake do not allow little feelings to injure you at such a time as this. It is of your own career, not of Mr. Gresham's manners, that you should think.'

'I have nothing to object to in Mr. Gresham. Yes or No will be quite sufficient.'

'It must be Yes.'

'It cannot be Yes, Lady Laura. That which I desired so ardently six months ago has now become so distasteful to me that I cannot accept it. There is an amount of hustling on the Treasury Bench which makes a seat there almost ignominious.'

'Do they hustle more than they did three years ago?'

'I think they do, or if not it is more conspicuous to my eyes. I do not say that it need be ignominious. To such a one as was Mr. Palliser it certainly is not so. But it becomes so when a man goes there to get his bread, and has to fight his way as though for bare life. When office first comes, unasked for, almost unexpected, full of the charms which distance lends, it is pleasant enough. The new-comer begins to feel that he too is entitled to rub his shoulders among those who rule the world of Great Britain. But when it has been expected, longed for as I longed for it, asked for by my friends and refused, when all the world comes to know that you are a suitor for that which should come without any suit,—then the pleasantness vanishes.'

'I thought it was to be your career.'

'And I hoped so.'

'What will you do, Phineas? You cannot live without any income.'

'I must try,' he said, laughing.

'You will not share with your friend, as a friend should.'

'No, Lady Laura. That cannot be done.'

'I do not see why it cannot. Then you might be independent.'

'Then I should indeed be dependent.'

'You are too proud to owe me anything.'

He wanted to tell her that he was too proud to owe such obligation as she had suggested to any man or any woman; but he hardly knew how to do so, intending as he did to inform her before they returned to the house of his intention to ask Madame Goesler to be his wife. He could discern the difference between enjoying his wife's fortune and taking gifts of money from one who was bound to him by no tie;—but to her in her present mood he could explain no such distinction. On a sudden he rushed at the matter in his mind. It had to be done, and must be done before he brought her back to the house. He was conscious that he had in no degree ill-used her. He had in nothing deceived her. He had kept back from her nothing which the truest friendship had called upon him to reveal to her. And yet he knew that her indignation would rise hot within her at his first word. 'Laura,' he said, forgetting in his confusion to remember her rank, 'I had better tell you at once that I have determined to ask Madame Goesler to be my wife.'

'Oh, then;—of course your income is certain.'

'If you choose to regard my conduct in that light I cannot help it. I do not think that I deserve such reproach.'

'Why not tell it all? You are engaged to her?'

'Not so. I have not asked her yet.'

'And why do you come to me with the story of your intentions,—to me of all persons in the world? I sometimes think that of all the hearts that ever dwelt within a man's bosom yours is the hardest.'

'For God's sake do not say that of me.'

'Do you remember when you came to me about Violet,—

to me,—to me? I could bear it then because she was good and earnest, and a woman that I could love even though she robbed me. And I strove for you even against my own heart, —against my own brother. I did; I did. But how am I to bear it now? What shall I do now? She is a woman I loathe.'

'Because you do not know her.'

'Not know her! And are your eyes so clear at seeing that you must know her better than others? She was the Duke's mistress.'

'That is untrue, Lady Laura.'

'But what difference does it make to me? I shall be sure that you will have bread to eat, and horses to ride, and a seat in Parliament without being forced to earn it by your labour. We shall meet no more, of course.'

'I do not think that you can mean that.'

'I will never receive that woman, nor will I cross the sill of her door. Why should I?'

'Should she become my wife,—that I would have thought might have been the reason why.'

'Surely, Phineas, no man ever understood a woman so ill as you do.'

'Because I would fain hope that I need not quarrel with my oldest friend?'

'Yes, sir; because you think you can do this without quarrelling. How should I speak to her of you; how listen to what she would tell me? Phineas, you have killed me at last.' Why could he not tell her that it was she who had done the wrong when she gave her hand to Robert Kennedy? But he could not tell her, and he was dumb. 'And so it's settled!'

'No; not settled.'

'Psha! I hate your mock modesty! It is settled. You have become far too cautious to risk fortune in such an adventure. Practice has taught you to be perfect. It was to tell me this that you came down here.'

'Partly so.'

'It would have been more generous of you, sir, to have remained away.'

'I did not mean to be ungenerous.'

Then she suddenly turned upon him, throwing her arms round his neck, and burying her face upon his bosom. They were at the moment in the centre of the park, on the grass beneath the trees, and the moon was bright over their heads. He held her to his breast while she sobbed, and then relaxed his hold as she raised herself to look into his face. After a moment she took his hat from his head with one hand, and with the other swept the hair back from his brow. 'Oh, Phineas,' she said, 'Oh, my darling! My idol that I have worshipped when I should have worshipped my God!'

After that they roamed for nearly an hour backwards and forwards beneath the trees, till at last she became calm and almost reasonable. She acknowledged that she had long expected such a marriage, looking forward to it as a great sorrow. She repeated over and over again her assertion that she could not 'know' Madame Goesler as the wife of Phineas, but abstained from further evil words respecting the lady. 'It is better that we should be apart,' she said at last. 'I feel that it is better. When we are both old, if I should live, we may meet again. I knew that it was coming, and we had better part.' And yet they remained out there, wandering about the park for a long portion of the summer night. She did not reproach him again, nor did she speak much of the future; but she alluded to all the incidents of their past life, showing him that nothing which he had done, no words which he had spoken, had been forgotten by her. 'Of course it has been my fault,' she said, as at last she parted with him in the drawing-room. 'When I was younger I did not understand how strong the heart can be. I should have known it, and I pay for my ignorance with the penalty of my whole life.' Then he left her, kissing her on both cheeks and on her brow, and went to his bedroom with the understanding that he would start for London on the following morning before she was up.

CHAPTER LXXIX
At Last—At Last

As he took his ticket Phineas sent his message to the Prime Minister, taking that personage literally at his word. The message was, No. When writing it in the office it seemed to him to be uncourteous, but he found it difficult to add any other words that should make it less so. He supplemented it with a letter on his arrival in London, in which he expressed his regret that certain circumstances of his life which had occurred during the last month or two made him unfit to undertake the duties of the very pleasant office to which Mr. Gresham had kindly offered to appoint him. That done, he remained in town but one night, and then set his face again towards Matching. When he reached that place it was already known that he had refused to accept Mr. Gresham's offer, and he was met at once with regrets and condolements. 'I am sorry that it must be so,' said the Duke,—who was sorry, for he liked the man, but who said not a word more upon the subject. 'You are still young, and will have further opportunities,' said Lord Cantrip, 'but I wish that you could have consented to come back to your old chair.' 'I hope that at any rate we shall not have you against us,' said Sir Harry Coldfoot. Among themselves they declared one to another that he had been so completely upset by his imprisonment and subsequent trial as to be unable to undertake the work proposed to him. 'It is not a very nice thing, you know, to be accused of murder,' said Sir Gregory, 'and to pass a month or two under the full conviction that you are going to be hung. He'll come right again some day. I only hope it may not be too late.'

'So you have decided for freedom?' said Madame Goesler to him that evening,—the evening of the day on which he had returned.

'Yes, indeed.'

'I have nothing to say against your decision now. No doubt your feelings have prompted you right.'

'Now that it is done, of course I am full of regrets,' said Phineas.

'That is simple human nature, I suppose.'

'Simple enough; and the worst of it is that I cannot quite explain even to myself why I have done it. Every friend I had in the world told me that I was wrong, and yet I could not help myself. The thing was offered to me, not because I was thought to be fit for it, but because I had become wonderful by being brought near to a violent death! I remember once, when I was a child, having a rocking-horse given to me because I had fallen from the top of the house to the bottom without breaking my neck. The rocking-horse was very well then, but I don't care now to have one bestowed upon me for any such reason.'

'Still, if the rocking-horse is in itself a good rocking-horse——'

'But it isn't.'

'I don't mean to say a word against your decision.'

'It isn't good. It is one of those toys which look to be so very desirable in the shop-windows, but which give no satisfaction when they are brought home. I'll tell you what occurred the other day. The circumstances happen to be known to me, though I cannot tell you my authority. My dear old friend Laurence Fitzgibbon, in the performance of his official duties, had to give an opinion on a matter affecting an expenditure of some thirty or forty thousand pounds of public money. I don't think that Laurence has generally a very strong bias this way or that on such questions, but in the case in question he took upon himself to be very decided. He wrote, or got some one to write, a report proving that the service of the country imperatively demanded that the money should be spent, and in doing so was strictly within his duty.'

'I am glad to hear that he can be so energetic.'

'The Chancellor of the Exchequer got hold of the matter, and told Fitzgibbon that the thing couldn't be done.'

'That was all right and constitutional, I suppose.'

'Quite right and constitutional. But something had to be said about it in the House, and Laurence, with all his usual fluency and beautiful Irish brogue, got up and explained that the money would be absolutely thrown away if expended on a purpose so futile as that proposed. I am assured that the great capacity which he has thus shown for official work and official life will cover a multitude of sins.'

'You would hardly have taken Mr. Fitzgibbon as your model statesman.'

'Certainly not;—and if the story affected him only it would hardly be worth telling. But the point of it lies in this;—that he disgusted no one by what he did. The Chancellor of the Exchequer thinks him a very convenient man to have about him, and Mr. Gresham feels the comfort of possessing tools so pliable.'

'Do you think that public life then is altogether a mistake, Mr. Finn?'

'For a poor man I think that it is, in this country. A man of fortune may be independent; and because he has the power of independence those who are higher than he will not expect him to be subservient. A man who takes to parliamentary office for a living may live by it, but he will have but a dog's life of it.'

'If I were you, Mr. Finn, I certainly would not choose a dog's life.'

He said not a word to her on that occasion about herself, having made up his mind that a certain period of the following day should be chosen for the purpose, and he had hardly yet arranged in his mind what words he would use on that occasion. It seemed to him that there would be so much to be said that he must settle beforehand some order of saying it. It was not as though he had merely to tell her of his love. There had been talk of love between them before, on which occasion he had been compelled to tell her that he could not accept that which she offered to him. It would be impossible, he knew, not to refer to that former conversation. And then he had to tell her that he, now coming to her as a suitor and knowing her to be a very rich woman, was himself all but penniless. He was sure, or almost sure, that she was as well aware of this fact as he was himself; but, nevertheless, it was necessary that he should tell her of it,—and if possible so tell her as to force her to believe him when he assured her that he asked her to be his wife, not because she was rich, but because he loved her. It was impossible that all this should be said as they sat side by side in the drawing-room with a crowd of people almost within hearing, and Madame Goesler had just been called upon to play, which she always did directly she was asked. He was invited to make up a rubber, but he could not bring himself to care for cards at the present moment. So he sat apart and listened to the music.

If all things went right with him to-morrow that music,—or the musician who made it,—would be his own for the rest of his life. Was he justified in expecting that she would give him so much? Of her great regard for him as a friend he had no doubt. She had shown it in various ways, and after a fashion that had made it known to all the world. But so had Lady Laura regarded him when he first told her of his love at Loughlinter. She had been his dearest friend, but she had declined to become his wife; and it had been partly so with Violet Effingham, whose friendship to him had been so sweet as to make him for a while almost think that there was more than friendship. Marie Goesler had certainly once loved him; —but so had he once loved Laura Standish. He had been wretched for a while because Lady Laura had refused him. His feelings now were altogether changed, and why should not the feelings of Madame Goesler have undergone a similar change? There was no doubt of her friendship; but then neither was there any doubt of his for Lady Laura. And in spite of her friendship would not revenge be dear to her,—revenge of that nature which a slighted woman must always desire? He had rejected her, and would it not be fair also that he should be rejected? 'I suppose you'll be in your own room before lunch to-morrow,' he said to her as they separated for the night. It had come to pass from the constancy of her visits to Matching in the old Duke's time, that a certain small morning-room had been devoted to her, and this was still supposed to be her property,—so that she was not driven to herd with the public or to remain in her bedroom during all the hours of the morning. 'Yes,' she said; 'I shall go out immediately after breakfast, but I shall soon be driven in by the heat, and then I shall be there till lunch. The Duchess always comes about half-past twelve, to complain generally of the guests.' She answered him quite at her ease, making arrangement for privacy if he should desire it, but doing so as though she thought that he wanted to talk to her about his trial, or about politics, or the place he had just refused. Surely she would hardly have answered him after such a fashion had she suspected that he intended to ask her to be his wife.

At a little before noon the next morning he knocked at her door, and was told to enter. 'I didn't go out after all,' she said. 'I hadn't courage to face the sun.'

'I saw that you were not in the garden.'

'If I could have found you I would have told you that I should be here all the morning. I might have sent you a message, only——only I didn't.'

'I have come——'

'I know why you have come.'

'I doubt that. I have come to tell you that I love you.'

'Oh Phineas;—at last, at last!' And in a moment she was in his arms.

It seemed to him that from that moment all the explanations, and all the statements, and most of the assurances were made by her and not by him. After this first embrace he found himself seated beside her, holding her hand. 'I do not know that I am right,' said he.

'Why not right?'

'Because you are rich and I have nothing.'

'If you ever remind me of that again I will strike you,' she said, raising up her little fist and bringing it down with gentle pressure on his shoulder. 'Between you and me there must be nothing more about that. It must be an even partnership. There must be ever so much about money, and you'll have to go into dreadful details, and make journeys to Vienna to see that the houses don't tumble down;—but there must be no question between you and me of whence it came.'

'You will not think that I have to come to you for that?'

'Have you ever known me to have a low opinion of myself? Is it probable that I shall account myself to be personally so mean and of so little value as to imagine that you cannot love me? I know you love me. But Phineas, I have not been sure till very lately that you would ever tell me so. As for me——! Oh, heavens! when I think of it.'

'Tell me that you love me now.'

'I think I have said so plainly enough. I have never ceased to love you since I first knew you well enough for love. And I'll tell you more,—though perhaps I shall say what you will think condemns me;—you are the only man I ever loved. My husband was very good to me,—and I was, I think, good to him. But he was many years my senior, and I cannot say I loved him,—as I do you.' Then she turned to him, and put her head on his shoulder. 'And I loved the old Duke, too, after a fashion. But it was a different thing from this. I will tell you something about him some day that I have never yet told to a human being.'

'Tell me now.'

'No; not till I am your wife. You must trust me. But I will tell you,' she said, 'lest you should be miserable. He asked me to be his wife.'

'The old Duke?'

'Yes, indeed, and I refused to be a—duchess. Lady Glencora knew it all, and, just at the time I was breaking my heart, —like a fool, for you! Yes, for you! But I got over it, and am not broken-hearted a bit. Oh, Phineas, I am so happy now.'

Exactly at the time she had mentioned on the previous evening, at half-past twelve, the door was opened, and the Duchess entered the room. 'Oh dear,' she exclaimed, 'perhaps I am in the way; perhaps I am interrupting secrets.'

'No, Duchess.'

'Shall I retire? I will at once if there be anything confidential going on.'

'It has gone on already, and been completed,' said Madame Goesler rising from her seat. 'It is only a trifle. Mr. Finn has asked me to be his wife.'

'Well?'

'I couldn't refuse Mr. Finn a little thing like that.'

'I should think not, after going all the way to Prague to find a latch-key! I congratulate you, Mr. Finn, with all my heart.'

'Thanks, Duchess.'

'And when is it to be?'

'We have not thought about that yet, Mr. Finn,—have we?' said Madame Goesler.

'Adelaide Palliser is going to be married from here some time in the autumn,' said the Duchess, 'and you two had better take advantage of the occasion.' This plan, however, was considered as being too rapid and rash. Marriage is a very serious affair, and many things would require arrangement. A lady with the wealth which belonged to Madame Goesler cannot bestow herself off-hand as may a curate's daughter, let her be ever so willing to give her money as well as herself. It was impossible that a day should be fixed quite at once; but the Duchess was allowed to understand that the affair might be mentioned. Before dinner on that day every one of the guests at Matching Priory knew that the man who had refused to be made Under-Secretary of State had been accepted by that possessor of fabulous wealth who was well known to the world as Madame Goesler of Park Lane. 'I am very glad that you did not take office under Mr. Gresham,' she said to him when they first met each other again in London. 'Of course when I was advising you I could not be sure that this would happen. Now you can bide your time, and if the opportunity offers you can go to work under better auspices.'

CHAPTER LXXX

Conclusion

THERE remains to us the very easy task of collecting together the ends of the thread of our narrative, and tying them into a simple knot, so that there may be no unravelling. Of Mr. Emilius it has been already said that his good fortune clung to him so far that it was found impossible to connect him with the tragedy of Bolton Row. But he was made to vanish for a certain number of years from the world, and dear little Lizzie Eustace was left a free woman. When last we heard of her she was at Naples, and there was then a rumour that she was about to join her fate to that of Lord George de Bruce Carruthers, with whom pecuniary matters had lately not been going comfortably. Let us hope that the match, should it be a match, may lead to the happiness and respectability of both of them.

As all the world knows, Lord and Lady Chiltern still live at Harrington Hall, and he has been considered to do very well with the Brake country. He still grumbles about Trumpeton Wood, and says that it will take a lifetime to repair the injuries done by Mr. Fothergill;—but then who ever knew a Master of Hounds who wasn't ill-treated by the owners of coverts?

Of Mr. Tom Spooner it can only be said that he is still a bachelor, living with his cousin Ned, and that none of the neighbours expect to see a lady at Spoon Hall. In one winter, after the period of his misfortune, he became slack about his hunting, and there were rumours that he was carrying out that terrible threat of his as to the crusade which he would go to find a cure for his love. But his cousin took him in hand somewhat sharply, made him travel abroad during the summer, and brought him out the next season, 'as fresh as paint,' as the members of the Brake Hunt declared. It was known to every sportsman in the country that poor Mr. Spooner had been in love; but the affair was allowed to be a mystery, and no one ever spoke to Spooner himself upon the subject. It is probable that he now reaps no slight amount of gratification from his memory of the romance.

The marriage between Gerard Maule and Adelaide Palliser was celebrated with great glory at Matching, and was mentioned in all the leading papers as an alliance in high life. When it became known to Mr. Maule, Senior, that this would be so, and that the lady would have a very considerable fortune from the old Duke, he reconciled himself to the marriage altogether, and at once gave way in that matter of Maule Abbey. Nothing he thought would be more suitable than that the young people should live at the old family place. So Maule Abbey was fitted up, and Mr. and Mrs. Maule have taken up their residence there. Under the influence of his wife he has promised to attend to his farming, and proposes to do no more than go out and see the hounds when they come into his neighbourhood. Let us hope that he may prosper. Should the farming come to a good end more will probably have been due to his wife's enterprise than to his own. The energetic father is, as all the world knows, now in pursuit of a widow with three thousand a year who has lately come out in Cavendish Square.

Of poor Lord Fawn no good account can be given. To his thinking, official life had none of those drawbacks with which the fantastic feelings of Phineas Finn had invested it. He could have been happy for ever at the India Board or at the Colonial Office;—but his life was made a burden to him by the affair of the Bonteen murder. He was charged with having nearly led to the fatal catastrophe of Phineas Finn's condemnation by his erroneous evidence, and he could not bear the accusation. Then came the further affair of Mr. Emilius, and his mind gave way;—and he disappeared. Let us hope that he may return some day with renewed health, and again be of service to his country.

Poetical justice reached Mr. Quintus Slide of The People's Banner The acquittal and following glories of Phineas Finn were gall and wormwood to him; and he continued his attack upon the member for Tankerville even after it was known that he had refused office, and was about to be married to Madame Goesler. In these attacks he made allusions to Lady Laura which brought Lord Chiltern down upon him, and there was an action for libel. The paper had to pay damages and costs, and the proprietors resolved that Mr. Quintus Slide was too energetic for their purposes. He is now earning his bread in some humble capacity on the staff of The Ballot Box, —which is supposed to be the most democratic daily newspaper published in London. Mr. Slide has, however, expressed his intention of seeking his fortune in New York.

Laurence Fitzgibbon certainly did himself a good turn by his obliging deference to the opinion of the Chancellor of the Exchequer. He has been in office ever since. It must be acknowledged of all our leading statesmen that gratitude for such services is their characteristic. It is said that he spends much of his eloquence in endeavouring to make his wife believe that the air of County Mayo is the sweetest in the world. Hitherto, since his marriage, this eloquence has been thrown away, for she has always been his companion through the Session in London.

It is rumoured that Barrington Erle is to be made Secretary for Ireland, but his friends doubt whether the office will suit him.

The marriage between Madame Goesler and our hero did not take place till October, and then they went abroad for the greater part of the winter, Phineas having received leave of absence officially from the Speaker and unofficially from his constituents. After all that he had gone through it was acknowledged that so much ease should be permitted to him. They went first to Vienna, and then back into Italy, and were unheard of by their English friends for nearly six months. In April they reappeared in London, and the house in Park Lane was opened with great *éclat*. Of Phineas every one says that of all living men he has been the most fortunate. The present writer will not think so unless he shall soon turn his hand to some useful task. Those who know him best say that he will of course go into office before long.

Of poor Lady Laura hardly a word need be said. She lives at Saulsby the life of a recluse, and the old Earl her father is still alive.

The Duke, as all the world knows, is on the very eve of success with the decimal coinage. But his hair is becoming grey, and his back is becoming bent; and men say that he will never live as long as his uncle. But then he will have done a great thing,—and his uncle did only little things. Of the Duchess no word need be said. Nothing will ever change the Duchess.

APPENDIX

TROLLOPE'S ELECTION ADDRESS AT BEVERLEY, 1868

This excerpt is taken from the account of Trollope's speech given in the *Beverley Weekly Recorder and General Advertiser* Saturday, 14 November 1868:

Mr. Gladstone, as you all well know, and all his followers in the House of Commons that lived the other day and which is about to live again, took in hand that great work the disendowment and disestablishment of that monstrous anomaly the Irish Church. There is no doubt that it will take place, but should there be among my hearers at the present moment any man who thinks that the Irish Church should not be disestablished and disendowed, that gentleman and I disagree, and he will of course vote against me. I do not think that because we wish to disendow and disestablish the Irish Church that we are the enemies of any Protestant Church. I am a member, and a very sincere member . . . of a Protestant Church; but I would never be a member of any Church which is mixed up with and looks upon the state as its support./Cheers/If a Church can't exist by the faith which is in the hearts of the people—if it can't exist except by endowment and establishment,—that church can never bring souls to the enjoyment of heaven./Cheers/ This church of which we are speaking now is a Protestant endowed church—the church of a trifling minority. No man wishes to rob that minority. No man wishes to take from Ireland any privilege any Protestant now has . . . You all know, probably, how this question of the disendowment and disestablishment of the Church first came up. Many accusations have been made against Mr. Gladstone on the score, that he brought it forward in the House of Commons, not from any good feeling, but because he wished to get rid of Mr. Disraeli. The truth of the matter was this; at the beginning of the session a motion was made in the House by an Irish member as to the grievances in Ireland, and the most interesting debate on the whole subject was that which took place on the motion, which was made by Mr. Maguire, an Irish member. No subject of greater interest could have occupied their attention. The Reform Bill had been already passed, and this measure—this question as to what would have to be done, came up in its natural position./Cheers/

APPENDIX

GLADSTONE'S REPLY ON THE SECOND READING OF THE IRISH CHURCH BILL

This speech, made on 23 March 1869, is taken from *Gladstone's Speeches*, edited by A. T. Bassett, 1910, pp. 399–400:

It is for the interest of us all that we should not keep this Establishment of religion in a prolonged agony. Nothing can come from that prolongation but an increase of pain, an increase of exasperation, and a diminution of that temper which now happily prevails—a temper which is disposed to mitigate the adjustment of this great question in details. There may also come from that prolongation the very evil which the right hon. gentleman opposite made it a charge against us that we were labouring to produce, but which we think likely to be rather the probable consequence of his line of argument—namely, the drawing into the Irish controversy the English question which we conceive to be wholly different. We think so, because, although in the two countries there may be and there are Establishments of religion, we never can admit that an Establishment which we think, in the main, good and efficient for its purposes, is to be regarded as being endangered by the course which we may adopt in reference to an Establishment which we look upon as being inefficient and bad. The day, therefore, it seems to me, is rapidly approaching when this controversy will come to an end, and I feel that I am not wrong in appealing to that silent witness to the justice of my anticipations which I am satisfied exists on both sides of the House. Not now are we opening this great question. Opened, perhaps, it was, when the Parliament which expired last year pronounced upon it that emphatic judgment which can never be recalled. Opened it was further, when in the months of autumn the discussions which were held in every quarter of the country turned mainly on the subject of the Irish Church. Prosecuted another stage it was, when the completed elections discovered to us a manifestation of the national verdict more emphatic than, with the rarest exceptions, has been witnessed during the whole of our Parliamentary history . . .

EXPLANATORY NOTES

DESPITE dealing with the occult activities of fox-hunting and party politics *Phineas Redux* contains relatively few obscurities. Learned allusions and quotations are frequently explained, though not identified, by their context. Similarly, links with the related novels, and with *Phineas Finn* in particular, are likely to cause little difficulty and do not ordinarily seem to justify notes. Although information and statements on important contemporary issues have been offered to the reader, no attempt has been made to construct exact historical, biographical or geographical parallels to the fiction.

Part Title. Redux: Latin adjective, meaning 'brought back'.

VOLUME I

CHAPTER I

Page 1. (1) *18—*: *Phineas Redux* reflects indirectly many of the important political actions and debates of the late 1860s. After the Reform Bill of 1867 under the Conservatives led by Disraeli, and government by a minority, Gladstone came to power in 1868 to head an administration which had Irish policy as a priority. The disestablishment of the Irish Church was of key significance in the General Election of 1868 and took up most of the session of 1869. (See Appendices.) Typically, Trollope refers to the specific events obliquely, and projects the major issues and implications of these events with a good deal of flexibility, mixing policy and party according to his own design. In this instance the Church question is given an English context in keeping with much of his fiction.

The change in political climate evident in the second half of this decade was noted by Walter Bagehot in the introduction to the second edition of *The English Constitution*:

In Lord Palmerston's time Sir George Grey said that the disestablishment of the Irish Church would be an 'act of revolution': it has now been disestablished by great majorities, with Sir George Grey himself assenting. A new world has arisen which is not as the old world; and we naturally ascribe the change to the Reform Act. But this is a complete mistake. If there had been no Reform Act at all there would, nevertheless, have been a great change in English politics. There has been a change of the sort which, above all, generates other changes—a change of generation. (*The Collected Works*, edited by Norman St. John-Stevas, vol. V, p. 166.)

(2) *the Ballot*: the question of the Ballot had been the subject of various committees and societies since 1832. Surprisingly, Liberals were committed to secret ballot only after 1870.

Page 2. *entitled by numerical strength*: in an essay entitled 'The Necessary Consequence of Government by a Minority' (1868) Bagehot wrote:

The present Government and the present House of Commons can hardly long coexist, because the Government not only has no predominant influence in the House, but is regarded by a warm majority with constant suspicion and incessant dislike. (*The Collected Works*, vol VII, p. 172.)

Page 3. *decimal coinage*: this was considered by various commissions and committees from 1841 onwards. Following an international conference in 1867 a commission was set up in 1868 to assess possible changes in coinage for the sake of uniformity.

Page 5. *a measure*: the measure was Tenant Right in Ireland. (See *Phineas Finn.*) The Irish Land Act of 1870 gave tenant right legal force wherever the custom was thought to prevail, but the measure was not a success. It did not satisfy demands for fixity of tenure and fair rents.

Page 7. *Tibur . . . Rome*: a reference to Horace *Epistles*. I. viii. 12: 'Romae Tibur amem ventosus, Tibure Romam': 'When at Rome I love Tivoli, when there I prefer Rome.'

Page 9. *bare bodkin*: *Hamlet*, III. i. 76: 'When he himself might his quietus make/With a bare bodkin'.

Page 12. *brass*: presumably 'brass' here refers to bold self-confidence rather than money.

CHAPTER II

Page 14. *former book*: *Phineas Finn.*

Page 15. *cub-hunting*: the hunting of young foxes at the beginning of the season, usually taking place early in the morning between the beginning of August and the end of September.

Page 18. *Job's comforter*: one who aggravates.

Page 23. (1) *penny post*: Rowland Hill's Post Office Reform in 1837 led to an act in 1839 establishing the penny postage and replacing variation in prices by a flat rate of one penny per half ounce.

(2) *Horace Walpole . . . Mr. Mann*: Horace Walpole (1717–97), famous for *The Castle of Otranto*, was also noted as a letter-writer. Horace Mann was one of the correspondents.

CHAPTER IV

Page 31. *Oil City*: the centre of the Pennsylvania oil region. Founded in 1860 it was partly destroyed by fire and flood in 1866.

Page 32. *Loughton and Loughshane were gone*: pocket boroughs like these were still present even after the 1832 and 1867 Reform Bills, even though their number had been reduced.

Page 35. *Shibboleth*: in modern usage refers to a party catchword. It was the word used as a test by Jepthah to distinguish the Gileadites from the Ephraimites (Judges 12:6).

Page 37. *every collier was now a voter*: the Reform Bill of 1867 added 938,000 voters to an electorate of 1,057,000. Working-class voters were in the majority in the towns, but most of the new county voters were from the middle class.

CHAPTER V

Page 38. *Irish Church as a State establishment*: after the activities of the Fenian movement in the mid-1860s Gladstone, contrary to his former position, was persuaded that disestablishment and disendowment of the Church of Ireland was necessary. Within two months of office he brought forward his bill. On 1 January 1871 the Church of Ireland became a voluntary body and disendowment amounted to £16 million.

In an article entitled 'The Irish Church' published by the *Fortnightly Review* in 1865 Trollope wrote:

We cannot ruthlessly cut down the half-dead tree of the grove, and tear asunder the roots, and plough and sow the soil, where the spot has been hallowed by ancient piety. The work of removal has, indeed, to be done; but it must be done tenderly, not ruthlessly. With loving hands must the old timber be dragged away, and the ground cleared for purposes of new utility.

The Irish Church Establishment, as it at present exists, is an institution thus dear to us, thus allowed to live among us a while longer, and thus absolutely and avowedly indefensible.

Page 40. *Nestor*: in Greek mythology the son of Neleus who lived to a great age. He appears in the *Iliad* as a respected, ineffective old man.

Page 42. *quidnuncs*: inquisitive persons or gossips.

Page 43. (1) *when the bran was bolted*: a condensed form of 'Fancy may bolt bran and think it flour'.

(2) *Catholic Emancipation . . . Corn Laws . . . household suffrage*: Catholic Emancipation took place in 1829, the repeal of the Corn Laws in 1846, and household suffrage as a result of the Second Reform Bill in 1867.

(3) *Israel*: used figuratively it refers to the chosen people of God—the elect.

Page 44. *Peel's bill for the Corn Laws*: Robert Peel repealed the Corn Laws in 1846 and fell from power soon afterwards.

Page 47. *thin edge*: an interesting variation on 'The thin end of the wedge is to be feared'.

Page 48. *Shaftesbury . . . Lord Russell and others suffered*: Anthony Ashley Cooper 1st Earl of Shaftesbury (1621–83) considered rebellion in 1682 with Russell, Monmouth, and others. Soon after he fled the country to Holland where he died. Russell was supposed to be connected with the 'Rye-house plot' and was executed in 1683 after being convicted on hearsay evidence.

CHAPTER VI

Page 51. (1) *existing law*: by the Parliamentary Elections Act of 1868 the laws relating to election petitions were amended, and special judges were nominated to handle them.

(2) *the Reform*: this club was formed in 1838 by the radicals to counter the Carlton Club.

Page 53. *junketing*: a picnic excursion.

Page 54. *gauds*: gaieties.

Page 57. *traces*: the straps which harness a horse.

Page 58. *flaming in the forehead of the morning sky*: a reference to Milton's *Lycidas*, 170–71:

> And tricks his beams, and with new spangled ore,
> Flames in the forehead of the morning sky.

CHAPTER VII

Page 60. *have the wood drawn*: to 'draw' in hunting language is either to take a single hound from the pack or, as in this instance, to search a covert for a fox.

Page 62. *drag*: usually a private vehicle like a stage-coach.

CHAPTER VIII

Page 69. (1) *Ichabod*: Biblical in origin (see I Samuel 4:19–22) and means literally 'no glory'.

(2) *Free Trade*: the term is used narrowly to describe the abolition of tax on imported corn under the Corn Laws.

Page 70. (1) *Von Moltke*: Count Helmut Von Moltke (1800–91) was a German military strategist who realized the importance of sea power and railways. He was prominent in the wars with Denmark in 1863–4, with Austria in 1866, and with France in 1870.

(2) *his Bœotia*: Bœotia was a supposedly backward agricultural region of ancient Greece.

Page 71. *Lord Liverpool's time*: Robert Banks Jenkinson, 2nd Earl of Liverpool, was Prime Minister from 1812 to 1827.

Page 72. (1) *the woman of Babylon*: in Revelation 17:3–5 Babylon is introduced in the following way:

' . . . I saw a woman sit upon a scarlet coloured beast, full of names of blasphemy, having seven heads and ten horns. And the woman was arrayed in purple and scarlet colour, and decked with gold and precious stones and pearls, having a golden cup in her hand full of abominations and filthiness of her fornication: And upon her forehead *was* a name written, MYSTERY, BABYLON THE GREAT, THE MOTHER OF HARLOTS AND ABOMINATIONS OF THE EARTH'.

(2) *'men, not measures'*: the popularity of the phrase is attributed to

George Canning, following Burke's example, in a speech of 1807 in the House of Commons: 'Away with the cant of "Measures not men"'.

Page 73. *status pupillaris*: the state of being a pupil or a minor.

Page 75. *Quod minime . . . ab urbe*: a quotation from Virgil's *Aeneid,* VI. 96: 'via prima salutis, quod minime reris, Graia pandetur ab urbe': 'The first step to safety will appear, where it is least expected, from a Grecian town.'

CHAPTER X

Page 84. *fly*: a light vehicle usually hired from a livery stable.

Page 86. (1) *tenuis ratio saporum*: Horace, *Satires,* II. iv. 36: 'the subtle theory of flavours'.

(2) *Caleb Balderstones*: the servant of the Ravenswoods in Scott's *The Bride of Lammermoor* (1819).

(3) *goblets of Gladstone*: the name given to the cheap French wines more freely available as a result of Gladstone's reduction of customs duty in 1860.

Page 88. *Belial*: a demon usually identified with the devil.

Page 91. *Apollyon*: 'the destroyer'—the angel of the abyss in Revelation 9:11.

CHAPTER XIII

Page 113. *Cagliostro*: Count Alessandro Cagliostro (1743–95) was a celebrated Italian necromancer and alchemist. Carlyle relates his adventures in his *Miscellanies*.

Page 114. (1) *casus belli*: the cause of a quarrel/the occasion of war.

(2) *Quarter Sessions*: formerly the justices of the peace for a county or part of a county. They were replaced by Crown Courts in 1972.

Page 118. (1) *personated*: to personate is to feign or counterfeit.

(2) Lady Laura Standish: Lady Laura's maiden name.

Page 119. *never more be officer of ours*: an adaptation of Othello's speech to Cassio in *Othello,* II. iii. 239: 'But never more be officer of mine'.

CHAPTER XIV

Page 120. *the Duke of Omnium*: he appears in both Palliser and Barchester novels.

CHAPTER XVI

Page 135. *rowels*: the spiked rotating wheel at the end of the spur.

Page 137. *Tankerville and hunting . . . were equally matters of course*: in 'Mr. Freeman On the Morality of Hunting', published by the *Fortnightly Review* in December 1869, Trollope wrote:

> The recreation or 'delectatio' experienced in the hunting field is very various in its nature; but a promiscuous intercourse with the mangled limbs of the quarry is not a part of it. Men are thrown together who would not otherwise meet, and converse on all subjects common to men. Politics are discussed, and agriculture, social habits, the affairs of the country, the preservation of foxes, the enmity of this enemy to the sport, and the devoted friendship of that friend. Perhaps of all the delights of the hunting field conversation is the most general . . . A community is formed in which equality prevails, and the man with small means and no rank holds his own against the lord or the millionaire as he can do nowhere else amidst the scenes of our life.

Page 138. *butts*: the embankment behind rifle targets.

Page 140. *a round robin*: a complaint with the names of the subscribers arranged in a circle to conceal the order of their signing.

CHAPTER XVII

Page 148. *the Duke did make an offer to Madame Goesler*: this offer of marriage is made at the end of II. Chapter LX of *Phineas Finn*.

Page 149. *fainéant*: a lazy person.

CHAPTER XVIII

Page 153. *chignon*: a coil of false hair worn at the back of the head.

Page 155. *Cavour . . . Garibaldi . . . Bismarck*: Cavour (1810–61) was Premier of Piedmont and a prominent figure in the unification of

Italy. So, too, was Garibaldi (1807–82) who conquered Sicily and Naples for the emergent Italy in 1860. Bismarck (1815–98), Prime Minister of Prussia, became the first chancellor of the Reich in 1871.

Page 159. (1) *stick covert*: an area of land covered with thorn bushes for fox-cover.

(2) *leash*: a sporting term for a set of three.

CHAPTER XIX

Page 164. *screw*: a crazed or unsound, worthless horse.

CHAPTER XX

Page 170. *1870*: the only explicit date in the novel: all others are in the form '18—'.

Page 171. *the present Divorce Court*: until 1857 divorce required a special act of parliament. Despite opposition from the clergy and Gladstone, the Act of 1857 created a special court for divorce petitions. By the Matrimonial Causes Act (1866) guilty husbands could be ordered to pay allowances to their wives. Proof of adultery was sufficient to secure the divorce of a wife, but a woman could not divorce her husband without proof not only of adultery, but also of cruelty and desertion.

Page 175. *retricked his beams*: as on p. 58, *Lycidas*, 170.

CHAPTER XXI

Page 183. (1) *the Tracts and Mr. Newman*: a reference to the Oxford Movement begun in 1833 by the Tractarians to insist on the Catholic character of the Church of England, and involving the theologian John Henry Newman.

(2) *Mr. Justice Shallow*: Robert Shallow is a country justice in Shakespeare's *Henry IV, Part 2*.

Page 188. *all the blood of all the Howards*: a quotation from Pope's *Essay on Man*, IV. 215–16:

> What can ennoble sots, or slaves, or cowards?
> Alas! not all the blood of all the *Howards*.

Page 192. *act the king and the prince*: presumably an oblique allusion to *Henry IV, Part 2*, IV. V.

CHAPTER XXII

Page 195. *'Point de zèle' of Talleyrand*: the phrase 'Pas de zèle', meaning 'don't overdo it', is attributed to the French foreign minister by Saint-Beuve in *Portrait de femme: Madame de Stael.*

Page 200. *Upper Ten*: short for the upper 10,000.

CHAPTER XXII

Page 203. *Judd Street*: this opens on to Euston Road almost opposite Midland Road near to St. Pancras and King's Cross stations.

Page Page 211. *Mr. Septimus Slope*: an interesting example of Trollope's cross-reference between novels. In this instance a character's mistaken reference to Quintus Slide presents us instead with the aspiring curate whose plans fail in *Barchester Towers*, and the first name of 'the Warden' Septimus Harding.

CHAPTER XXIV

Page 213. *strawberry-leaves*: decoration on the coronet of a duke, marquis or an earl.

Page 216. *certain memoirs*: probably those of Cardinal Retz 1613–79) which were first published in 1715.

CHAPTER XXVI

Page 230. *in former days*: in *The Small House At Allington.*

CHAPTER XXVII

Page 239. *Cerberus*: in Greek mythology the monstrous dog guarding Hades.

CHAPTER XXIX

Page 257. *'Nuptiarum expers'*: Ned quotes lines 10 and 11 of a passage from Horace's *Odes,* III. xi, which describes a young filly that, 'leaps playfully, and fears being touched, having no part in marriage'.

CHAPTER XXX

Page 264. *another great jewel robbery*: an allusion to *The Eustace Diamonds.*

CHAPTER XXXII

Page 279. *meed*: reward.

Page 283. *The Woolsack*: the seat of the Lord Chancellor in the House of Lords.

CHAPTER XXXIII

Page 290. *the Athenaeum*: Trollope was a member of the Athenaeum; the club was founded in 1824.

Page 297. (1) *Eli*: high priest and teacher of Samuel. The misappropriation referred to is related in I Samuel 2: 12–17.

(2) *Levites . . . Melchisedek*: the Levites were a priestly tribe, Melchisedek a priestly king.

(3) *Thomas à Becket*: Becket was murdered after Henry II's attempt to control the clergy.

Page 300. *Dead Sea apples*: this dry and bitter fruit grows in the arid Dead Sea region and is symbolic of the desolation of the cities of Sodom and Gomorrah.

CHAPTER XXXIV

Page 311. *The higher a monkey climbs*—: 'The higher the ape goes the more he shows his tail.'

CHAPTER XXXVII

Page 328. *Lord North's time*: Lord North, Frederick Earl of Guildford (1732–92), Prime Minister from 1770 to 1782.

Page 336. *Lord Melbourne*: William Lamb Melbourne (1779–1848), Prime Minister in 1834 and from 1835 to 1841.

CHAPTER XXXIX

Page 348. *Chiltern Hundreds*: an MP cannot resign his seat unless he is disqualified and so he applies for the stewardship of the Hundreds of Burnham, Desborough, and Stoke in Buckinghamshire, which is a place of honour under the Crown and therefore incompatible with holding a seat.

Page 354. *our Chatham*: William Pitt 1st Earl of Chatham (1708–78) resigned in 1761 over his proposal to attack Spain without notice of war.

CHAPTER XL

Page 356. *Juno . . . nectar*: Romulus, abandoned with his twin brother Remus, was the mythical founder of Rome. Eventually he was admitted to the company of the gods of which Juno was the queen.

Page 358. (1) *Charles James Fox*: (1749–1806) leader of the Whigs and a famous orator who held the position of Secretary of State under Rockingham.

(2) *Cavendish*: the Cavendish family, including the Earls and Dukes of Devonshire, was one of the great aristocratic families with a Whig interest.

Page 359. *But whither would'st thou, Muse? . . . mysteries*: Sir Theodore Martin's translation of Horace *Odes,* III. iii. 70. Martin contributed to *Fraser's* and *Tait's* magazines. His work included many translations of the classics and parodies of contemporary verse. He published a translation of Horace's *Odes* in 1860 and a more complete set of translations of Horace's work in 1870.

VOLUME II

CHAPTER XLI

Page 7. (1) *stuff*: woollen material.

(2) *Pandemonium*: the capital of Hell, from Milton's coinage in *Paradise Lost*.

Page 8. *paladin*: a knightly champion.

CHAPTER XLIII

Page 22. *lion*: a celebrity.

CHAPTER XLV

Page 40. *limited mail*: a mail train in which a limited number of passengers were carried.

Page 46. *Galicia*: a region of eastern Europe now falling between south-east Poland and the Soviet Union.

CHAPTER XLVI

Page 52. (1) *'In vino veritas'*: 'In wine is truth'.

(2) *blackballs*: to blackball is to exclude a person from a club, originally by placing black balls in a ballot-box.

Page 57. (1) *life-preserver*: a stick loaded with lead used for self-preservation.

(2) *garotting*: there were a number of garotting incidents in the early 1860s. Phineas saves Robert Kennedy from garotters in I Chapter XXX of *Phineas Finn*.

CHAPTER XLIX

Page 80. *Gibus hat*: a collapsible or 'crush' opera hat.

CHAPTER L

Page 89. *that woman's necklace*: in *The Eustace Diamonds*.

CHAPTER LIII

Page 112. *the Carabineers*: a carbine is a rifle used by cavalry.

Page 114. (1) '*Labor omnia vincit*': 'work conquers all'.

(2) *put through his facings*: a phrase of military origin, meaning to turn or face in another direction.

(3) *phaeton*: a light open carriage.

Page 117. *Diana . . . Bacchus*: the beautiful hunter Orion was killed by Diana. The riotous god Bacchus married Ariadne after she had been deserted by Theseus.

CHAPTER LV

Page 136. '*No one at an instant . . . becomes most base*': the quotation is from Juvenal, *Satire* ii. 83: 'Nemo repente fuit turpissimus.'

CHAPTER LVIII

Page 163. *the gods would at last vouchsafe an ending*: the last phrase of the sentence is taken from the *Aeneid*, I. 199: 'Dabit deus his quoque finem.'

CHAPTER LX

Page 176. (1) *Caveat emptor*: 'let the buyer beware'.

(2) *Caveat lex*: 'let the law beware'.

Page 178. *Palmer . . . Cook*: William Palmer (1824–56) was a medical man who poisoned his wife, his brother, and his friend John Cook, all for monetary gain. His trial attracted great interest, and he was hanged after being convicted entirely on circumstantial evidence.

CHAPTER LXI

Page 194. (1) *Amy Robsart*: in Scott's *Kenilworth* (1821) she falls through a trap door after being wrongly suspected of infidelity.

(2) *Eugene Aram*: in Bulwer Lytton's novel *Eugene Aram* (1832) Aram consents under the pressure of poverty to the murder of

Clarke by his accomplice Houseman, after which he suffers remorse.

(3) *Dirk Hatteraick . . . in The Antiquary*: Dirk Hatteraick murders Glossin in Scott's *Guy Mannering* (1815).

CHAPTER LXII

Page 195. *resent upon*: Trollope uses the verb here in a way which suggests its rare meaning of 'revenge'.

CHAPTER LXIII

Page 207. *wrap-rascal*: a loose overcoat or great-coat.

CHAPTER LXV

Page 224. *drunk up Esil*: the analogy in the first part of the paragraph is with *Hamlet*, v. i, which also contains references to Esil, and Ossa. It is the scene of confrontation between Hamlet and Laertes at Ophelia's grave. Esil is vinegar.

Page 225. (1) *Ossas*: Ossa, a mountain in Thessaly, is famous in mythology because of the attempt by the giants Otus and Ephialtes to reach heaven by piling the mountains Olympus and Pelion on top of Ossa.

(2) *Moabitish*: in this context the word seems to refer to strangeness as well as Jewishness. This could be a result of the fact that the Moabite Stone, possessing a fragmentary inscription in Moabite—a Hebrew dialect—was discovered in 1868.

(3) *Medean*: Medea was the enchantress who helped Jason and the Argonauts in their quest for the Golden Fleece.

CHAPTER LXVII

Page 244. *'The true gods . . . in the river'*: these are the last lines of Elizabeth Barrett Browning's poem 'A Musical Instrument' which was published in the *Cornhill Magazine* in 1860.

EXPLANATORY NOTES

CHAPTER LXVIII

Page 252. *nil admirari*: 'marvelling at nothing'.

CHAPTER LXIX

Page 259. *'the quarrel of lovers is the renewal of love.'*: Terence, *Andria*, 555: 'Amantium irae amoris integratio est.'

Page 262. *some proverb*: Ovid, *Metamorphoses*, II. 137: 'Medio tutissimus ibis': 'The middle way is safest.'

Page 266. *the French Minister said . . . shall be done'*: presumably the words of Charles de Calonne (1734–1802): 'si c'est possible, c'est fait; impossible? cela fera': 'if a thing is possible, consider it done; impossible? that will be done'.

CHAPTER LXX

Page 274. *cage in the House of Commons*: the 'cage' is the familiar name for the Ladies Gallery.

Page 278. (1) *Gyas*: Gyes in Greek mythology one of the offspring of Uranus and Ge, like the Cyclopes. Each of these monstrous offspring had only one eye and some, like Gyes, had a hundred arms and hands. Gyas, used incorrectly here, figures in the boat race in Book V of the *Aeneid*.

(2) *Three-fingered Jack*: unidentified. It could be Trollope's own coinage or a contemporary stage act rather than a character in a play. There is, of course, a Crook-fingered Jack in Gay's *The Beggar's Opera*.

(3) *Strand Theatre*: noted for its popular burlesques and extravaganzas.

CHAPTER LXXIII

Page 297. *Minerva*: the goddess who sprang fully grown from the head of her father Zeus after he had swallowed Metis when she was pregnant.

EXPLANATORY NOTES

CHAPTER LXXIV

Page 301. *brougham*: a closed carriage drawn by one horse.

CHAPTER LXXV

Page 311. (1) *'man forbid'*: *Macbeth*, I. iii. 21: 'He shall live a man forbid.'

(2) 'home-staying youths have ever homely wits': *The Two Gentlemen of Verona*, I. i. 2: 'Home-keeping youth have ever homely wits'.

(3) *Adam Smith*: (1723–90) the economist and philosopher who wrote *The Wealth of Nations*.

Page 312. (1) *Hinc illæ iræ*: 'hence those rages'.

(2) *The Field, Bell's Life, and Land and Water*: contemporary sporting magazines. *Bell's Life in London and Sporting Chronicle* ran from 1822, *The Field* from 1853, and *Land and Water* from 1866.

CHAPTER LXXVI

Page 320. *Ophelia among the water-lilies*: Gertrude's description of Ophelia's death in *Hamlet*, IV. vii does not include water-lilies.

Page 325. *Tattersall's*: a London horse auction and sporting centre founded in 1776.

CHAPTER LXXVII

Page 336. *Peri*: a beautiful fairy-like creature in Persian folklore.

WHO'S WHO
IN
PHINEAS REDUX

[Characters whose names are in capital letters appear also in the other novels indicated]

C.Y.F.H. = Can You Forgive Her; *D.C.* = The Duke's Children; *E.D.* = The Eustace Diamonds; *Ph. F.* = Phineas Finn; *P.M.* = The Prime Minister.

Advance, George, of the City, i. 214.
Asdrabal, Jacob, an Old Bailey barrister, i. 318.
Attenbury, or Atterbury, of Florence; his wife, elder half-sister of Adelaide Palliser, i. 20, 154.

BALDOCK, Lord: Gustavus Boreham; his wife, *née* Mouser, i. 127.
See also *Ph. F.*
BALDOCK, dowager Lady; her d. Hon. Augusta Boreham, later Sister Veronica John, i. 24.
See also *Ph. F.*
Ballance, of the City, i. 214, *see* Discount.
Barchester, Bishop of (perhaps identical with Madam Goesler's friend the Bishop of Dorchester, *Ph. F.* ii. 33), ii. 150.
Birdbott, Mr. Serjeant, junior counsel for Phineas, ii. 152.
Blinks, ——, of the Carabineers, ii. 112.
Bobwig, Sir Samuel, a judge, i. 276.
BOFFIN, ——, Conservative President of the Board of Trade, i. 77, ii. 88.
See also *P.M.*
BONEBREAKER, Lord Chiltern's horse, i. 18.
See also *Ph. F.*
BONTEEN, J——, Liberal M.P.; various minor offices, i. 284, 311; President of the Board of Trade, i. 362; murdered, ii. 58; of St. James's Place, ii. 58; his widow, ii. 106.
See also *Ph. F.*, *E.D.*
BOODLE, Captain, ii. 257.
See also *The Claverings*, *E.D.*
Boultby, Mr. Baron, ii. 31.
Bouncer, ——, an author, ii. 53, 189.
BRENTFORD, Earl of: of Saulsby and Portman Square, i. 282.
See also *Ph. F.*, *D.C.*

BROCK, Lord, former Liberal Prime Minister; 'a thick skin, an equable temper, and perfect self-confidence' (Monk), ii. 339.
See also *C.Y.F.H.*, *P.M.*, *Framley Parsonage.*

Browborough, Conservative M.P. for Tankerville, Durham, i. 8; loses his seat on petition, i. 118.

Brudi, Count, i. 155.

BUNCE, JACOB, Phineas's landlord in Great Marlborough St., i. 51; Mrs. B., i. 51.
See also *Ph. F.*

Burnaby, Mrs., i. 163.

CAMPERDOWN, SAMUEL, attorney to the Eustace family, ii. 170.
See also *E.D.*

CANTRIP, Earl of, i. 45; Liberal Colonial Secretary, i. 284; K.G., ii. 151; Lady C.
See also *Ph. F.*, *P.M.*, *D.C.*

CARRUTHERS, Lord GEORGE DE BRUCE, rumoured about to marry Lady Eustace, ii. 357.
See also *E.D.*

CHAFFANBRASS, ——, Q.C., Phineas's counsel, ii. 152.
See also *The Three Clerks*, *Orley Farm.*

Chief Justice, the Lord, ii. 153.

CHILTERN, Lord: Oswald Standish, o.s. of the Earl of Brentford; of Harrington Hall; Master of the Brake Hounds; his wife Violet (*née* Effingham), i. 15.
See also *Ph. F.*, *E.D.*, *American Senator*, *P.M.*, *D.C.*

CLADDAGH, Lord; *see* Fitzgibbon.
See also *Ph. F.*

COLDFOOT, Sir HARRY, Liberal Home Secretary, i. 284, ii. 138.
See also *Ph. F.*, *E.D.*

COLEPEPPER, Captain, i. 98.
See also *Ph. F.*

Coulson, a lawyer's clerk, ii. 43, 45.

Cox, the Brake huntsman, i. 63.

Dandolo, Chiltern's horse, i. 125.

DAUBENY, ——, Conservative Prime Minister, i. 1; M.P. for East Barsetshire, i. 41; 'this audacious Cagliostro among statesmen', i. 113.
See also *Ph. F.*, *P.M.*

DE TERRIER, Lord: presumably intended in the description of Daubeny's former leader 'who, though thoroughly trusted, was very idle', i. 45.
See also *Ph. F.*, *Framley Parsonage.*

Dick, Col. the Hon. Mowbray, Conservative M.P. for West Bustard, i. 74.
Dickers, Charlie, a hunting man, i. 164.
Dido, a hound that never lied, i. 142.
Discount, Duke of, m. —— Ballance of Lombard Street, i. 214.
Doggett, of the Brake Hunt, i. 122.
DROUGHT, Sir ORLANDO, Conservative Secretary of State, i. 315.
See also *The Way we Live now, P.M., D.C.*
DRUMMOND, Lord, Conservative Secretary at War, i. 77, ii. 88.
See also *American Senator* (Foreign Secretary), *P.M., D.C.*

EMILIUS, Rev. JOSEPH, a Bohemian Jew; m. Lady Eustace, q.v., ii. 38; of the London Chapel, ii. 41; originally Yosef Mealyus, ii. 41; his alleged wife, of Cracow, ii. 46, 169; of Lowndes Square, later of Northumberland St., ii. 59; of 3 Jellybag St., Edgware Road, ii. 166; convicted of bigamy but not of murder, ii. 291.
See also *E.D.*
ERLE, BARRINGTON, Liberal M.P., i. 6; cousin of Lord Chiltern, i. 17.
See also *Ph. F., E.D., P.M., D.C.*
EUSTACE, Lady, i. 229; Elizabeth ('Lizzy'), widow of Sir Florian E. and reputed wife of Emilius, q.v.; of Portray Castle, Ayrshire, ii. 40, 291; her son, heir to the Eustace estate, ii. 170; *see* Carruthers.
See also *E.D., P.M.*

Farren, Dick, Spooner's groom, ii. 115.
FAWN, Viscount, i. 129, 280.
See also *Ph. F., E.D., P.M.*
FINN, PHINEAS, m. (1) Mary Flood Jones (see *Phineas Finn*); her death, i. 6; inherits 'about four thousand pounds' from his father, 'routine duties in Dublin,' i. 7 (but in *Ph. F.* his office was in Cork); rejected at Tankerville, i. 38; of Brooks's (see *Ph. F.* ii. 345) and Reform Club, i. 51; 'no touch of vanity' in his character (contrast *Autobiog.* quoted in *Ph. F.*, ii. 358), i. 96; returned for Tankerville on petition, i. 118; committed for trial on charge of murdering Bonteen, ii. 83; 'his two sisters' (here see *Ph. F.*), ii. 132 ('two of his sisters', ii. 179); acquitted, ii. 240; resigns his seat, ii. 253; and is re-elected, ii. 279; refuses office, ii. 350; m. (2) Marie Goesler, ii. 359.
See also *Ph. F., E.D., P.M., D.C.*
Fitzaskerly, Lady Gertrude, i. 139.
FITZGIBBON, Hon. LAURENCE, e.s. of Lord Claddagh, Liberal M.P., formerly Under-Secretary for the Colonies, i. 245, 273; his s. ASPASIA, i. 273; married, ii. 305.
See also *Ph. F., E.D., P.M.*
Forster, Lord Brentford's lawyer, i. 281.

FOTHERGILL, ——, the Duke of Omnium's man of business, i. 121; pensioned, ii. 315; see Trumpeton Wood.
See also *Doctor Thorne*, *Framley Parsonage*, *Small House at Allington*, *Last Chronicle of Barset*, *Ph. F.*, *P.M.*
Frederick ——, Lord, ii. 228, 232.

Gadmire, of Tankerville, ii. 282.
GARNETT, jeweller, i. 231. See also *E.D.*
GOESLER, MARIE, 'Madame Max', of Park Lane, i. 133; described, i. 125. 'This Moabitish woman ... This half-foreigner, This German Jewess' (Lady Laura), ii. 225; m. (2) Phineas Finn.
See also *Ph. F.* (and s.v. Finn).
Golightly, junior counsel for Phineas, ii. 153.
GOWRAN, ANDREW, Lady Eustace's servant, ii. 40.
See also *E.D.*
GRESHAM, ——, Liberal Prime Minister, i. 1.
See also *Ph. F.*, *E.D.*, *P.M.*
GROGRAM, Sir GREGORY, Liberal Attorney-General, i. 172, ii. 28; M.P. for Clovelly, ii. 28;
(Cf. 'Gogram' in *C.Y.F.H.*). See also *P.M.*
Gunthorpe, Lord, i. 19.

Harry, of the Brake Hunt, i. 123.
HARTLETOP, Dowager Marchioness of, i. 223.
See also *Framley Parsonage*, *Small House at Allington*, *Ph. F.*

Kennedy, George, cousin of Robert K., i. 244.
KENNEDY, Lady LAURA, *née* Standish, d. of the Earl of Brentford, w. of Robert K., i. 8.
See also *Ph. F.*, *P.M.*
KENNEDY, ROBERT, Liberal M.P. for Dunross-shire; of Loughlinter, Perthshire, i. 13; 282; his widowed mother, Sarah, i. 83, ii. 104; his death and will, ii. 104.
See also *Ph. F.*

LEGGE WILSON, ——, Liberal Chancellor of the Exchequer, i. 284, 362.
See also, *Ph. F.*, *E.D.*, *P.M.* ('Wilson'), *The Way We Live Now*.
LOW, ——, Q.C., Conservative M.P. for North Broughton, i. 54; of Baker St. (but see *Ph. F.*, i. 48), i. 54; Mrs. L., i. 54 (Georgiana, i. 211).
See also *Ph. F.*

Macaw, Lord, M.P., i. 8.
MACKINTOSH, Major, head of the London constabulary, i. 247.
See also *E.D.*

Macpherson, Mr. and Mrs., hotel-keepers in Judd Street, i. 204.
McSilk, Sir Alexander, a judge, i. 276.
Maltanops, Marquis of, shareholder in a Burton brewery, i. 214.
Mary, a maid at Harrington Hall, ii. 117.
Mary Jane, Mrs. Bunce's maid, i. 53.
MAULE, GERARD, i. 26; e.s. of Maurice M., i. 154; his relations, i. 154; £800 a year, i. 28; m. Adelaide Palliser, ii. 327.
See also *D.C.*
Maule, Maurice, of M. Abbey, Herefordshire, i. 154, 182; about 55, i. 184.
Maxwell, Polly, ii. 112.
Meager, Mrs., of Northumberland St., Marylebone Road, Emilius's landlady; her husband, and daughter Amelia, ii. 141, 144.
Merthyr, Lord, mine-owner in partnership with the Earl of Tydvil, i. 214.
MILDMAY, WILLIAM, the Prime Minister succeeding De Terrier.
See also *Ph. F.*, *D.C.*
Molescroft, Liberal organizer, i. 11.
MONK, JOSHUA, Radical M.P., i. 23, 78.
See also *Ph. F.*, *P.M.*, *D.C.*
MOUNT THISTLE, Lord, i. 283.
See also *E.D.*, and 'Morecombe' in *Ph. F.*

OMNIUM, Duke of: George Palliser, K.G.; 'now almost fallen into second childhood', i. 120; of Gatherum Castle, Barsetshire, i. 150; Lord Lieutenant of Barsetshire, i. 214; his death and will, i. 227.
See also *Doctor Thorne*, *Framley Parsonage*, *Small House at Allington*, *C.Y.F.H.*, *Ph. F.*, *E.D.*

PALLISER, ADELAIDE, first cousin of Plantagenet P., i. 20, 122; £4,000, i. 28; described, i. 153; £20,000, ii. 327; m. Gerard Maule, q.v.
See also *D.C.*
PALLISER, PLANTAGENET, Liberal Chancellor of the Exchequer, i. 3; nephew and heir of the Duke of Omnium, i. 121; of Matching Priory, Yorkshire, i. 122; his wife, Lady GLENCORA, i. 126; succeeds his uncle as Duke of Omnium, i. 227; of Carlton Gardens, i. 273 but 'Carlton Terrace', ii. 45, 74, and so in *P.M.* and *D.C.*; Lord Privy Seal, ii. 23; Lord Lieutenant of Barsetshire, ii. 151; President of the Board of Trade, ii. 156.
See also *Small House at Allington*, *C.Y.F.H.*, *Ph. F.*, *E.D.*, *P.M.*, *D.C.*
PALLISER, PLANTAGENET, e.s. of Plantagenet, Duke of Omnium, styled Earl of Silverbridge, i. 222.
See also *C.Y.F.H.*, *Ph. F.*, *P.M.*, *D.C.*

Pickering, ——, a Vice Chancellor, i. 211.
PIE, Sir OMICRON, 'that very distinguished but now aged physician', i. 213.
See also *Barchester Towers, The Bertrams, Doctor Thorne, Small House at Allington, P.M.*
Platters, the, of Platter House, i. 168.
PLINLIMMON, Lord, i. 360.
See also *Ph. F.*
Praska, Peter, a Bohemian blacksmith, ii. 235.
Pratt, a lawyer's clerk, ii. 43, 45.
PRINCE of Wales, the, i. 150, ii. 54.
See also *C.Y.F.H., Ph. F.*
Putt, Confucius, R.A., ii. 56, 88.

Quin, an opponent of Reform, i. 44.

RATLER, or RATTLER, Liberal Whip, i. 5.
See also *Ph. F., P.M., D.C.*
ROBY, Thomas, Conservative Patronage Secretary, i. 137.
See also *Ph. F., P.M., D.C.*
Ruddles, Liberal agent at Tankerville, i. 31.

ST. BUNGAY, Duke of : —— Fitzhoward, i. 115; of Longroyston, i. 116.
See also *C.Y.F.H., Ph. F., E.D., P.M., D.C.*
Scruby, Wickerby's clerk, ii. 172–5. Perhaps not the Scruby of *C.Y.F.H.*
Seymour, Parkinson, a clubman, i. 214.
SLIDE, QUINTUS, editor of *The People's Banner*, i. 52; of Quartpot Alley, Fleet Street, and Camden Town, i. 234–5.
See also *Ph. F., P.M.*
SLOPE, Sir SIMON, Liberal Solicitor-General, ii. 152.
See also *E.D.*
Smith, of Gartlow, 'a gentleman who did not hunt', i. 61.
Snow, junior counsel for Phineas, ii. 153.
SPOONER, THOMAS, of Spoon Hall, i. 131; his cousin Edward, i. 256.
See also *D.C.*

Tankerville, Durham, Phineas's seat, i. 31.
Thrift, Lord, i. 284.
Troddles, of Tankerville, ii. 282.
TRUMPETON WOOD, in the Duke of Omnium's property, i. 60, 120, 222, 274, ii. 49, 261, 310.
See also *P.M.* and 'Trumpington Wood' in *D.C.*